AF706134

INTERVIEWS WITH RUDOLPH A. MARCUS ON ELECTRON TRANSFER REACTIONS

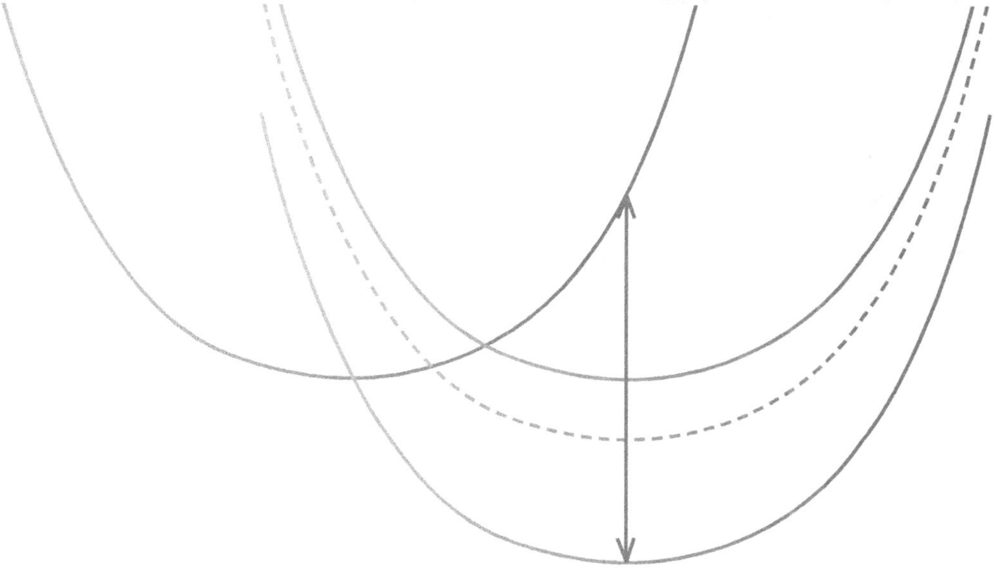

INTERVIEWS WITH RUDOLPH A. MARCUS ON ELECTRON TRANSFER REACTIONS

Francesco Di Giacomo

Sapienza University of Rome

World Scientific

NEW JERSEY • LONDON • SINGAPORE • BEIJING • SHANGHAI • HONG KONG • TAIPEI • CHENNAI • TOKYO

Published by

World Scientific Publishing Co. Pte. Ltd.

5 Toh Tuck Link, Singapore 596224

USA office: 27 Warren Street, Suite 401-402, Hackensack, NJ 07601

UK office: 57 Shelton Street, Covent Garden, London WC2H 9HE

British Library Cataloguing-in-Publication Data
A catalogue record for this book is available from the British Library.

INTERVIEWS WITH RUDOLPH A. MARCUS ON ELECTRON TRANSFER REACTIONS

Copyright © 2020 by World Scientific Publishing Co. Pte. Ltd.

All rights reserved. This book, or parts thereof, may not be reproduced in any form or by any means, electronic or mechanical, including photocopying, recording or any information storage and retrieval system now known or to be invented, without written permission from the publisher.

For photocopying of material in this volume, please pay a copying fee through the Copyright Clearance Center, Inc., 222 Rosewood Drive, Danvers, MA 01923, USA. In this case permission to photocopy is not required from the publisher.

ISBN 978-981-121-756-2 (hardcover)
ISBN 978-981-121-824-8 (paperback)
ISBN 978-981-121-757-9 (ebook for institutions)
ISBN 978-981-121-758-6 (ebook for individuals)

For any available supplementary material, please visit
https://www.worldscientific.com/worldscibooks/10.1142/11753#t=suppl

Typeset by Stallion Press
Email: enquiries@stallionpress.com

Preface

This volume presents interviews with Professor Marcus on the theory of electron transfer reactions. The interviews refer to Marcus' articles published from 1987 to the present. The Marcus theory of electron transfer reactions and his articles from 1956 to 1986 have been object of the previous book "Introduction to Marcus Theory of Electron Transfer Reactions." I believe that Professor Marcus' comments, his discussing, explaining, correcting, and adding to his work and that of his coworkers, are an interesting supplement to the articles.

Abbreviations

Marcus' papers are reported in the form **MN** where **M** stands for Marcus and **N** is the order number in the official list of Marcus' publications. The Figures from Marcus' papers appear with their original numbers and legends. The numbers of Figures from the author are followed by an asterisk, like, say, Fig. 1*. They have no legends, their meaning being explained in the text. Other abbreviations

A acceptor orbital
BEBO bond energy–bond order
CR charge recombination
CS charge separation
CSH charge shift
CT charge transfer
CV conduction band
D donor
DAE dissociative attachment energy
DCET diffusion-controlled ET
ET electron transfer
FC Franck–Condon factor
FWHM full width at half maximum
HOMO highest occupied molecular orbital
KIE for kinetic isotope effect
LE locally excited state
LUMO lowest occupied molecular orbital
LZ Landau–Zener
PES potential energy surface
QD quantum dot
QM quantum mechanics
ST singlet triplet
STM scanning tunneling microscope
TS transition state
VAE for vertical attachment energy
VB valence band

Acknowledgements

I am very much indebted to Doctor Marshall Newton of Brookhaven National Laboratories for his friendly review and advice and to Professor Robert J. Cave of Harvey Mudd College for his detailed critical review. The line drawings of my figures have been beautifully drawn by Architect Andrea Cataldi.

Contents

Preface v

Abbreviations vii

Acknowledgements ix

List of Marcus' Papers Considered in the Chapters of the Book xvii

Chapter 1 Solvent Dynamical and Symmetrized Potential Aspects of ET Rates; Superexchange versus an Intermediate $BChl^-$ Mechanism in Reaction Centers of Photosyntetic Bacteria; Dynamical Effects in ET Reactions. II. Numerical Solution; Recent Developments in ET Reaction; Some Recent Developments in ET: Charge Separation, Long Distances, Solvent Dynamics, and Free Energy Aspects; Nonexponential Time Behavior of ET in an Inhomogeneous Polar Medium; An Internal Consistency Test and Its Implications for the Initial Steps in Bacterial Photosynthesis; Early Steps in Bacterial Photosynthesis. Comparison of Three Mechanisms 1

Chapter 2 Early Steps in Bacterial Photosynthesis, Charge Transfer Absorption, Fluorescence Spectra and the Inverted Region, Reorganization Free Energy for ETs at Liquid–Liquid and Dielectric Semiconductor–Liquid Interfaces, Photosynthetic ET, ET in Dynamically Disordered Polar Medium, Dynamics of ET for a Nonsuperexchange Coherent Mechanism 43

Chapter 3 Matrix Elements for Long-range ET, ET across Liquid–Liquid Interfaces, Charge Transfer Spectra in Frozen Media, ET in Proteins: Calculation of Electronic Coupling, ET across Liquid–Liquid Interfaces 2. Schrödinger Equation for Strongly Interacting ET Systems 99

Chapter 4 ET Matrix Elements of Bridged Systems: A Molecular Fragment Approach; Theory of ET, Comparison with Experiments; Tight-binding for Semi-infinite solids; ET in Proteins: An Artificial Intelligence Approach to Electronic Coupling; Surface Properties of Solids Using a Semi-infinite Approach and the Tight-binding Approximation; Theoretical Model of Scanning Tunneling Microscopy 147

Chapter 5 ET in Proteins: Electronic Coupling in Myoglobin, Quantum Corrections for ET Rates. Polarizable vs. Nonpolarizable Description of Solvent, ET in Proteins: Electronic Coupling in Modified Cytochrome *c* and Myoglobin Derivatives, Free Energy of Nonequilibrium Polarization Systems. 4: A Formalism Based on the Nonequilibrium Dielectric Displacement, ET in Ferrocytochromes 193

Chapter 6 Tunneling Matrix Elements in Ru-modified Blue Copper Proteins: Pruning the Protein in Search of ET Pathways; Gaussian Field Model of Dielectric Solvation Dynamics; Symmetry or Asymmetry of k_{ET} and i_{STM} vs. Potential Curves; A Sequential Formula for Electronic Coupling in Long-Range Bridge-Assisted ET: Formulation of the Theory and Application to Alkanethiol Monolayers; Time-dependent Stokes Shift and Its Calculation from Solvent Dielectric Dispersion Data 245

Chapter 7 ET Model for the Electric Field Effect on Quantum Yield of Charge Separation in Bacterial Photosynthetic Reaction Centers; Theory of Rates of S_N2 Reactions and Relation to Those of Outer Sphere Bond Rupture Electron Transfers; Source of Image Contrast in STM Images of Functionalized Alkanes on Graphite: A Systematic Functional Group Approach 267

Chapter 8	Dynamic Stokes Shift in Solution: Effect of Finite Pump Pulse Duration; Time-dependent Fluorescence Spectra of Large Molecules in Polar Solvents; Remarks on Dissociative Anion Potential Energy Curves for Organic Electron Transfers; Ion Pairing and Electron Transfer	335
Chapter 9	Linear Response Theory of ET Reactions as an Alternative to the Molecular Harmonic Oscillator Model; Nonadiabatic Electron Transfer at Metal Surfaces; On the Theory of Electron Transfer Reactions at Semiconductors Electrode/Liquid Interfaces; On the Theory of Electron Transfer Reactions at Semiconductors Electrode/Liquid Interfaces. II. A Free Electron Model; Temperature Dependence of the Electronic Factor in the Nonadiabatic Electron Transfer at Metal and Semiconductor Electrodes	413
Chapter 10	Theoretical Investigation of the Directional Electron Transfer in 4-Aminonaphtalimide Compounds; Variable-range Hopping Electron Transfer through Disordered Bridge States: Application to DNA; A Model for Charge Transfer Inverse Photoemission; Mechanisms of Fluorescence Blinking in Semiconductor Nanocrystals Quantum Dots; Diffusion-controlled Electron Transfer Processes and Power-law Statistics of Fluorescence Intermittency of Nanoparticles	469

Chapter 11 Single Particle versus Ensemble Average: From Power-law Intermittency of a Single Quantum Dot to Quasistretched Exponential Fluorescence Decay of an Ensemble; Explanation of Quantum Dot Blinking without the Long-lived Trap Hypothesis; Photoinduced Spectral Diffusion and Diffusion-controlled Electron Transfer Reactions in Fluorescence Intermittency of Quantum Dots; Chain Dynamics and Power-law Distance Fluctuations of Single-molecule Systems; Determination of Energetics and Kinetics from Single-particle Intermittency and Ensemble-averaged Fluorescence Intensity Decay of Quantum Dots 531

Chapter 12 Summarizing Lecture: Factors Influencing Enzymatic H-transfers, Analysis of Nuclear Tunneling Isotope Effects and Thermodynamic versus Specific Effects; Enzymatic Catalysis and Transfers in Solution. I. Theory and Computations: A Unified View; H and Other Transfers in Enzymes and in Solution: Theory and Computations: A Unified View. 2. Applications to Experiment and Computations 625

Chapter 13 Evidence for a Diffusion-controlled Mechanism for Fluorescence Blinking of Colloidal Quantum Dots, Universal Emission Intermittency in Quantum Dots, Nanorods, and Nanowires, ET Past and Future, Beyond the Historical Perspective on Hydrogen and Electron Transfers, Interaction between Experiments, Analytical Theories, and Computation, Interaction of Theory and Experiment: Examples from Single Molecule Studies of Nanoparticles, Theory of a Single Dye Molecule Blinking with a Diffusion-based Power Law Distribution, Extension of the Diffusion Controlled Electron Transfer Theory for Intermittent Fluorescence of Quantum Dots: Inclusion of Biexcitons and the Difference of "On" and "Off" Time Distributions 701

List of Marcus' Papers Considered in the Chapters of the Book

The papers are numbered as in the official list of Marcus' publications.

217. R. A. Marcus. "Solvent Dynamical and Symmetrized Potential Aspects of Electron Transfer Rates." In *Understanding Molecular Properties*, A. E. Hansen, J. Avery, and J. P. Dahl, eds. (Reidel, Boston, 1987) p. 229. **M217**, Ch. 1
223. R. A. Marcus. "Superexchange versus an Intermediate BChlø Mechanism in Reaction Centers of Photosynthetic Bacteria." *Chem. Phys. Lett.* **133**, 471, (1987) **M223**, Ch. 1
224. W. Nadler and R. A. Marcus. "Dynamical Effects in Electron Transfer Reactions. II. Numerical Solution." *J. Chem. Phys.* **86**, 3906, (1987) **M224**, Ch. 1
225. R. A. Marcus. "Recent Developments in Electron Transfer Reactions." *Nouveau J. Chim.* **11**, 79, (1987) **M225**, Ch. 1
229. R. A. Marcus. "Some Recent Developments in Electron Transfer: Charge Separation, Long Distances, Solvent Dynamics, and Free Energy Aspects." In *Supramolecular Chemistry*, NATO ASI Ser., Ser. C, **214**, 45, (1987) **M229**, Ch. 1
231. W. Nadler and R. A. Marcus. "Non-exponential Time Behavior of Electron Transfer in an Inhomogeneous Polar Medium." *Chem. Phys. Lett.* **144**, 24, (1988) **M231**, Ch. 1
237. R. A. Marcus. "An Internal Consistency Test and Its Implications for the Initial Steps in Bacterial Photosynthesis." *Chem. Phys. Lett.* **146**, 13, (1988) **M237**, Ch. 1

238. R. A. Marcus. "Early Steps in Bacterial Photosynthesis. Comparison of Three Mechanisms." In *The Photosynthetic Bacterial Reaction Center—Structure and Dynamics*, NATO ASI Ser., Ser. A: Life Sciences **149**, 389, (1988) **M238**, Ch. 1
244. R. A. Marcus. "Mechanisms of the Early Steps in Bacterial Photosynthesis and Their Implications for Experiment." *Israel J. Chem.* **28**, 205, (1988) **M244**, Ch. 2
245. R. A. Marcus. "Relation between Charge Transfer Absorption and Fluorescence Spectra and the Inverted Region." *J. Phys. Chem.* **93**, 3078, (1989) **M245**, Ch. 2
249. R. A. Marcus. "Reorganization Free Energy for Electron Transfers at Liquid–Liquid and Dielectric Semiconductor-Liquid Interfaces." *J. Phys. Chem.* **94**, 1050, (1990) **M249**, Ch. 2
250. R. A. Marcus. "Theory and Experiment in Photosynthetic Electron Transfer." In *Current Research in Photosynthesis: Proceedings of VIIIth International Congress on Photosynthesis*, M. Baltscheffsky, ed. (Kluwer, Dordrecht, 1990) p. 1. **M250**, Ch. 2
251. W. Nadler and R. A. Marcus. "Electron Transfer in a Dynamically Disordered Polar Medium." *Israel J. Chem.* **30**, 69, (1990) **M251**, Ch. 2
252. R. Almeida and R. A. Marcus. "Dynamics of Electron Transfer for a Nonsuperexchange Coherent Mechanism." *J. Phys. Chem.* **94**, 2973, (1990) **M252**, Ch. 2
253. R. Almeida and R. A. Marcus. "Dynamics of Electron Transfer for a Nonsuperexchange Coherent Mechanism. 2. Numerical Calculations." *J. Phys. Chem.* **94**, 2978, (1990) **M253**, Ch. 2
254. P. Siddarth and R. A. Marcus. "Comparison of Experimental and Theoretical Electronic Matrix Elements for Long-range Electron Transfer." *J. Phys. Chem.* **94**, 2985, (1990) **M254**, Ch. 3
255. R. A. Marcus. "Theory of Electron-transfer Rates." *J. Phys. Chem.* **94**, 4152, 7742, (1990) **M255**, Ch. 3. a. Erratum: **99**, 5742, (1995)

256. R. A. Marcus. "Theory of Charge-transfer Spectra in Frozen Media." *J. Phys. Chem.* **94**, 4963, (1990) **M256**, Ch. 3
259. P. Siddarth and R. A. Marcus. "Electron Transfer Reactions in Proteins: A Calculation of Electronic Coupling." *J. Phys. Chem.* **94**, 8430, (1990) **M259**, Ch. 3
261. R. A. Marcus. "Theory of Electron-transfer Rates Across Liquid–Liquid Interfaces. 2. Relationships and Application." *J. Phys. Chem.* **95**, 2010, (1991) **M261**, Ch. 3 a. Erratum: **99**, 5742, (1995)
266. R. A. Marcus. "Schrödinger Equation for Strongly Interacting Electron-transfer Systems." *J. Phys. Chem.* **96**, 1753, (1992) **M266**, Ch. 3
269. P. Siddarth and R. A. Marcus. "Calculation of Electron-transfer Matrix Elements of Bridged Systems Using a Molecular Fragment Approach." *J. Phys. Chem.* **96**, 3213, (1992) **M269**, Ch. 4
270. R. A. Marcus and P. Siddarth. "Theory of Electron Transfer Reactions and Comparison with Experiments." In *Photoprocesses in Transition Metal Complexes, Biosystems and Other Molecules: Experiment and Theory*, NATO ASI Ser. C, **376**, 49, (1992) **M270**, Ch. 4
272. R. A. Marcus. "Tight-binding Approximation for Semi-infinite Solids. Application of a Transform Method and of Delta Function Normalization." *J. Chem. Phys.* **98**, 5604, (1993) **M272**, Ch. 4
273. P. Siddarth and R. A. Marcus. "Electron-transfer Reactions in Proteins: An Artificial Intelligence Approach to Electronic Coupling." *J. Phys. Chem.* **97**, 2400, (1993) **M273**, Ch. 4
274. H. Ou-Yang, B. K. llebring, and R. A. Marcus. "Surface Properties of Solids Using a Semi-infinite Approach and the Tight-binding Approximation." *J. Chem. Phys.* **98**, 7405, (1993) **M274**, Ch. 4
275. H. Ou-Yang, B. Källebring, and R. A. Marcus. "A Theoretical Model of Scanning Tunneling Microscopy: Application to the

Graphite (0001) and Au(111) Surfaces." *J. Chem. Phys.* **98**, 7565, (1993) **M275**, Ch. 4

279. P. Siddarth and R. A. Marcus. "Electron-transfer Reactions in Proteins: Electronic Coupling in Myoglobin." *J. Phys. Chem.* **97**, 6111, (1993) **M279**, Ch. 5
283. X. Song and R. A. Marcus. "Quantum Correction for Electron Transfer Rates. Comparison of Polarizable versus Nonpolarizable Descriptions of Solvent." *J. Chem. Phys.* 99, 7768, (1993) **M283**, Ch. 5
286. P. Siddarth and R. A. Marcus. "Correlation between Theory and Experiment in Electron-transfer Reactions in Proteins: Electronic Couplings in Modified Cytochrome *c* and Myoglobin Derivatives." *J. Phys. Chem.* **97**, 13078, (1993) **M286**, Ch. 5
289. R. A. Marcus. "Free Energy of Nonequilibrium Polarization Systems 4. A Formalism Based on the Nonequilibrium Dielectric Displacement." *J. Phys. Chem.* **98**, 7170, (1994) **M289**, Ch. 5
293. A. A. Stuchebrukhov and R. A. Marcus. "Theoretical Study of Electron Transfer in Ferrocytochromes." *J. Phys. Chem.* **99**, 7581, (1995) **M293**, Ch. 5
296. J. N. Gehlen, I. Daizadeh, A. A. Stuchebrukhov, and R. A. Marcus. "Tunneling Matrix Element in Ru-modified Blue Copper Proteins: Pruning the Protein in Search of Electron Transfer Pathways." *Inorg. Chim. Acta* **243**, 271, (1996) **M296**, Ch. 6
299. X. Song, D. Chandler, and R. A. Marcus. "Gaussian Field Model of Dielectric Solvation Dynamics." *J. Phys. Chem.* **100**, 11954, (1996) **M299**, Ch. 6
300. R. A. Marcus. "Symmetry or Asymmetry of k_{ET} and i_{STM} vs. Potential Curves." *J. Chem. Soc. Faraday Trans.* **92**, 3905, (1996) **M300**, Ch. 6
301. C.-P. Hsu and R. A. Marcus. "A Sequential Formula for Electronic Coupling in Long Range Bridge-assisted Electron Transfer: Formulation of Theory and Application to Alkanethiol Monolayers." *J. Chem. Phys.* **106**, 584, (1997) **M301**, Ch. 6

303. C.-P. Hsu, X. Song, and R. A. Marcus. "Time-dependent Stokes Shift and Its Calculation from Solvent Dielectric Dispersion Data." *J. Phys. Chem. B* **101**, 2546, (1997) **M303**, Ch. 6
304. S. Tanaka and R. A. Marcus. "Electron Transfer Model for the Electric Field Effect on Quantum Yield of Charge Separation in Bacterial Photosynthetic Reaction Centers." *J. Phys. Chem. B* **101**, 5031, (1997) **M304**, Ch. 7
305. R. A. Marcus. "Theory of Rates of S_N2 Reactions and Relation to Those of Outer Sphere Bond Rupture Electron Transfers." *J. Phys. Chem. A* **101**, 4072, (1997) **M305**, Ch. 7
308. C. L. Claypool, F. Faglioni, W. A. Goddard III, H. B. Gray, N. S. Lewis, and R. A. Marcus. "Source of Image Contrast in STM Images of Functionalized Alkanes on Graphite: A Systematic Functional Group Approach. *J. Phys. Chem. B* **101**, 5978, (1997) **M308**, Ch. 7
309. Y. Georgievskii, C.-P. Hsu, and R. A. Marcus. "Dynamic Stokes Shift in Solution: Effect of Finite Pump Pulse Duration." *J. Chem. Phys.* **108**, 7356, (1998) **M309**, Ch. 8
310. C.-P. Hsu, Y. Georgievskii, and R. A. Marcus. "Time-Dependent Fluorescence Spectra of Large Molecules in Polar Solvents." *J. Phys. Chem. A* **102**, 2658, (1998) **M310**, Ch. 8
311. R. A. Marcus. "Remarks on Dissociative Anion Potential Energy Curves For Organic Electron Transfers." *Acta Chem. Scand.* **52**, 858, (1998) **M311**, Ch. 8
312. R. A. Marcus. "Ion Pairing and Electron Transfer." *J. Phys. Chem. B* **102**, 10071, (1998) **M312**, Ch. 8
314. R. A. Marcus. "Electron Transfer Past and Future." *Adv. Chem. Phys.* **106**, 1, (1999) **M314**, Ch. 8
315. Y. Georgievskii, C.-P. Hsu, and R. A. Marcus. "Linear Response in Theory of Electron Transfer Reactions as an Alternative to the Molecular Harmonic Oscillator Model." *J. Chem. Phys.* **110**, 5307, (1999) **M315**, Ch. 9

317. Shachi Gosavi and R. A. Marcus. "Nonadiabatic Electron Transfer at Metal Surfaces." *J. Phys. Chem. B* **104**, 2067, (2000) **M317**, Ch. 9
318. Y. Q. Gao, Y. Georgievskii, and R. A. Marcus. "On the Theory of Electron Transfer Reactions at Semiconductor Electrode/Liquid Interfaces." *J. Chem. Phys.* **112**, 3358, (2000) **M318**, Ch. 9
322. Y. Q. Gao and R. A. Marcus. "On the Theory of Electron Transfer at Semiconductor Electrode/Liquid Interfaces. II. A Free Electron Model." *J. Chem. Phys.* **113**, 6351, (2000) **M322**, Ch. 9
323. S. Gosavi, Y. Q. Gao, and R. A. Marcus. "Temperature Dependence of the Electronic Factor in Nonadiabatic Electron Transfer at Metal and Semiconductor Electrodes." *J. Electroanal. Chem.* **500**, 71, (2001) **M323**, Ch. 9
331. Y. Q. Gao and R. A. Marcus. "Theoretical Investigation of the Directional Electron Transfer in 4-Aminonaphthalimide Compounds." *J. Phys. Chem. A* **106**, 1956, (2002) **M331**, Ch. 10
337. T. Renger and R. A. Marcus "Variable-Range Hopping Electron Transfer through Disordered Bridge States: Application to DNA." *J. Phys. Chem. A* **107**, 8404, (2003) **M337**, Ch. 10
339. S. Gosavi and R. A. Marcus. "A Model for Charge Transfer Inverse Photoemission." *Electrochim. Acta*, **49**, 3, (2003) **M339**, Ch. 10
341. J. Tang and R. A. Marcus. "Mechanisms of Fluorescence Blinking in Semiconductor Nanocrystal Quantum Dots." *J. Chem. Phys.* **123**, 054704, (2005). **M341**, Ch. 10
342. J. Tang and R. A. Marcus. "Diffusion-controlled Electron Transfer Processes and Power-law Statistics of Fluorescence Intermittency of Nanoparticles." *Phys. Rev. Lett.* **95**, 107401, (2005) **M342**, Ch. 10
344. J. Tang and R. A. Marcus. "Single Particle vs. Ensemble Average: From Power-law Intermittency of a Single Quantum Dot to Quasistretched Exponential Fluorescence Decay of an Ensemble." *J. Chem. Phys.*, **123**, 204511, (2005) **M344**, Ch. 11

345. P. A. Frantsuzov and R. A. Marcus. "Explanation of Quantum Dot Blinking without the Long-lived Trap Hypothesis." *Phys. Rev. B* **72**, 155321, (2005) **M345**, Ch. 11
350. J. Tang and R. A. Marcus. "Photoinduced Spectral Diffusion and Diffusion-controlled Electron Transfer Reactions in Fluorescence Intermittency of Quantum Dots." *J. Chin. Chem. Soc.* **53**, 1, (2006) **M350**, Ch. 11
351. J. Tang and R. A. Marcus. "Chain Dynamics and Power-law Fluctuations of Single-molecule Systems." *Phys. Rev. E* **73**, 022102, (2006) **M351**, Ch. 11
352. J. Tang and R. A. Marcus. "Determination of Energetics and Kinetics from Single-particle Intermittency and Ensemble-averaged Fluorescence Intensity Decay of Quantum Dots." **M352**, Ch. 11
353. R. A. Marcus. "Summarizing Lecture: Quantum Catalysis in Enzymes—Beyond the Transition State Theory Paradigm." *Phil. Trans. Roy. Soc. B* **361**, 1445, (2006) **M353**, Ch. 12
354. R. A. Marcus. "Enzymatic Catalysis and Transfers in Solution. 1. Theory and Computations, a Unified View." *J. Chem. Phys.* **125**, 194504, (2006) **M354**, Ch. 12
357. R. A. Marcus. "H and Other Transfers in Enzymes and in Solution: Theory and Computations, a Unified View. II. Applications to Experiment and Computations." *J. Phys. Chem. B* **111**, 6643, (2007) **M357**, Ch. 12
358. M. Pelton, G. Smith, N. F. Scherer, and R. A. Marcus. "Evidence for a Diffusion-controlled Mechanism for Fluorescence Blinking of Colloidal Quantum Dots." *Proc. Nat. Acad. Sci. U.S.A.*, **104**, 14249, (2007) **M358**, Ch. 13
366. P. Frantsuzov, M. Kuno, B. Janko, and R. A. Marcus. "Universal Emission Intermittency in Quantum Dots, Nanorods, and Nanowires." *Nature Phys.* **4**, 521, (2008) **M366**, Ch. 13
367. R. A. Marcus. "Beyond the Historical Perspective on Hydrogen and Electron Transfers." In *Quantum Tunnelling in Enzyme-Catalysed Reactions*, R. K. Allemann and N. S. Scrutton, eds.

(Royal Society of Chemistry, Cambridge, UK, 2009) p. v. **M367**, Ch. 13

368. R. A. Marcus. "Interaction between Experiments, Analytical Theories, and Computation." *J. Phys. Chem. C* **113**, 14598, (2009) **M368**, Ch. 13
373. R. A. Marcus. "Interaction of Theory and Experiment: Examples from Single Molecule Studies of Nanoparticles." *Phil. Trans. R. Soc. A* **368**, 1109–1124, (2010) **M373**, Ch. 13
374. R. A. Marcus. "Spiers Memorial Lecture—Interplay of Theory and Computation in Chemistry—Examples from On-water Organic Catalysis, Enzyme Catalysis, and Single-molecule Fluctuations. *Faraday Discuss.* **145**, 9–14, (2010) **M374**, Ch. 13
379. Wei-Chen Chen and R. A. Marcus. "Theory of a Single Dye Molecule Blinking with a Diffusion-based Power Law Distribution." *J. Phys. Chem. C* **116**, 15782–15789, (2012) **M379**, Ch. 13
389. Z. Zhu and R. A. Marcus. "Extension of the Diffusion Controlled Electron Transfer Theory for Intermittent Fluorescence of Quantum Dots: Inclusion of Biexcitons and the Difference of "On" and "Off" Time Distributions." *Phys. hem. Chem. Phys.*, 10.1039/C4CP01274G (2014) **M389**, Ch. 13

CHAPTER 1

Solvent Dynamical and Symmetrized Potential Aspects of ET Rates; Superexchange versus an Intermediate $BChl^-$ Mechanism in Reaction Centers of Photosyntetic Bacteria; Dynamical Effects in ET Reactions. II. Numerical Solution; Recent Developments in ET Reaction; Some Recent Developments in ET: Charge Separation, Long Distances, Solvent Dynamics, and Free Energy Aspects; Nonexponential Time Behavior of ET in an Inhomogeneous Polar Medium; An Internal Consistency Test and Its Implications for the Initial Steps in Bacterial Photosynthesis; Early Steps in Bacterial Photosynthesis. Comparison of Three Mechanisms

Interviews on M217, M223, M224, M225, M229, M231, M237, M238

M217 Solvent Dynamical and Symmetrized Potential Aspects of ET Rates

R. A. Marcus

NOTES

1. p. 230 top: The advent of picosecond techniques has permitted the study of intrinsically very fast intramolecular electron transfer reaction—so fast that the slow step in some cases is not the *activation process* itself but rather the dynamics of dielectric reorganization of the solvent."

M: Maybe that's not the correct way to phrase it... but let me try to say or try an interpretation of it... You have the two parabolas, to reach the intersection you need some solvent reorganization. Supposing that the rate of fluctuation to get there is pretty slow. Then there is a certain time associated with trying to get there... you have sort of diffusing back and forth of the solvent so many times and the rate then depends actually on the relaxation time. Because to follow the complete activation process you don't have to worry about solvent dynamics, in this case the rate would become proportional to $1/\tau_L$, and then $e^{\Delta G^{\ddagger}}$, would not depend on solvent polarization... It is not correct to say that the solvent polarization doesn't enter into the barrier... the solvent relaxation frequency, the relaxation time comes in.... to get to that intersection you can say that the rate is determined by the relaxation frequency. That statement could have been better phrased. Note that τ is not the reciprocal of a rate constant but rather a relaxation time, there is a danger of confusion with thinking of τ as being a lifetime.

2. p. 231 middle-bottom: "For the solutes studied in Refs. 4 and 5, some *twisting* of an amino group relative to the aromatic ring is expected to accompany the intramolecular charge transfer, and some (minor) equilibrium bond length changes are also expected. Thus, λ_i is non-zero."

Q: The differences in bond lengths in ET are taken care by λ_i. How does one take care of other conformational changes like the above twisting?

M: By some change in angle... then one could write $\frac{1}{2}\vartheta^2$ multiplied by some force constant.

3. p. 233 bottom: "We consider as an example the case where the reactant B in Eq. (2)

$$A_{ox} + B_{red} \xrightarrow{k_{12}} A_{red} + B_{ox} \tag{2}$$

is an aquo cation, such as Fe^{2+}, with a relatively large Δq_k^0, and where reactant A is some large ion ML_3^{3+}, such as $Ru(bpy)_3^{3+}$, with a small Δq_k^0.... In the case of reactant B there is only one normal mode which need be considered for the ET, i.e. the symmetric stretching mode."

Q: Is it so because $\Delta q_k^0 \approx 0$ for the others?

M: Some Δq_k^0's may be greater than zero, they may be increasing going from reactants to products, some may be decreasing and their sum may equal zero, they cancel with each other. Certainly, certain bonds increase their length, certain bonds decrease their length. In general, in the case of the $Ru(bpy)_3^{3+}$ it is the symmetric stretching vibration that is involved, so in effect there is only one normal coordinate that involves six bond coordinates. Δq_k^0 does not refer to the bonds but to the normal coordinate. For example, in the asymmetric stretching vibrations, maybe a bond is increasing and the other decreasing, so there are maybe changes in the single bonds, but Δq_k^0 is zero for the asymmetric stretching, because some bond length is increasing, the other is decreasing.

M223 Superexchange vs. an Intermediate $BChl^-$ Mechanism in Reaction Centers of Photosyntetic Bacteria

R. A. Marcus

NOTES

1. p. 471 2nd column top-middle: "To distinguish between the two mechanisms we make use of some magnetic data. The latter indicate a very weak exchange interaction between the $(BChl)_2^+$ and BPh^-, a result which is in marked contrast with the extremely rapid ET from I." In Eq. (1)

$$\underset{\text{I}}{(BChl)_2^*} \xrightarrow{e} \underset{\text{II}}{BChl} \xrightarrow{e} \underset{\text{III}}{Bph} \xrightarrow{e} \underset{\text{IV}}{Q}, \tag{1}$$

where $(BChl)_2^*$ is photoexcited.

M: The point is this: for the electron to go from the dimer to the pheophytin there are two possible mechanisms, one mechanism is to hop from $(BChl)_2^*$ to BChl and then hop to the pheophytin, that's one mechanism in which one considers an intermediate. On the other hand in the superexchange mechanism the intermediate doesn't exist for any significant length of time, the wave function overlaps the wave function of the monomer... there is superexchange. The concentration of the intermediate is fantastically small. You have to describe the process in terms of wave functions. Either you have *a true intermediate* or you have a *superexchange*. The key issue was: do you have a real intermediate or you have an electron exchange from one to the other by a superexchange mechanism? The magnetic interactions tell something about the direct electronic coupling of the BChl and the dimeric system.

Robert J. Cave's Note: "It now appears that that the initial excitation occurs for BChl, and a hole is transferred to the dimer, with an electron going to the BPh.

Q: Is the monomer BChl?

M: Yes, that's right. You describe both processes in terms of wave functions, but the point is that if we have just A going to B, one could formulate the rate constant for that, and if you use classical mechanics for the trajectories, and a PES, you can capture that approximately with classical mechanics, but to capture the electronic part of the superexchange you need wave functions, you either need wave functions or some equivalent formulations in QM, but I just want to emphasize that you couldn't just do on a single PES . . .

Q: p. 472 1st column middle: "We consider first the use of Eq. (3)

$$\begin{vmatrix} H_{11}-E & H_{12} & 0 \\ H_{12} & H_{22}-E & H_{23} \\ 0 & H_{23} & H_{33}-E \end{vmatrix} = 0 \tag{3}$$

to treat the superexchange behaviour of electronic configurations 1 and 3

1. $(\mathrm{BChl})_2^*$ BChl BPh
2. $(\mathrm{BChl})_2^+$ BChl^- BPh
3. $(\mathrm{BChl})_2^+$ BChl BPh^-

Using a partitioning technique the problem in Eq. (3) is reduced to a 2×2 secular determinant. We have

$$\begin{vmatrix} \overline{H}_{11}-E & \overline{H}_{13} \\ \overline{H}_{13} & \overline{H}_{33}-E \end{vmatrix} = 0 \tag{4}$$

where

$$\overline{H}_{ii} = H_{ii} + H_{i2}^2(H_{22}-E) \quad (i = 1, 3), \tag{5a}$$

$$\overline{H}_{13} = H_{12}H_{23}/(H_{22}-E). \tag{6a}$$

In the case of a superexchange mechanism for ET the latter can occur when the system crosses the intersection of the $\overline{H}_{11}$ and $\overline{H}_{33}$ surfaces, i.e. where $\overline{H}_{11} = \overline{H}_{33}$ in nuclear configuration space, as in fig. 1, and the relevant ET matrix element is $\overline{H}_{13}$. We can thereby approximate the E in Eqs. (5a) and (6a) by $\overline{H}_{33}(=\overline{H}_{11})$, thus yielding

$$\overline{H}_{ii} \approx H_{ii} + H_{i2}^2/(H_{22} - \overline{H}_{33}) \quad (i = 1, 3) \tag{5b}$$

and

$$\overline{H}_{13} \approx H_{12}H_{23}/(H_{22} - \overline{H}_{33})." \tag{6b}$$

Q: Can you give a physical explanation of the difference between, say, H_{11} and $\overline{H}_{11}$ and of why ET happens now at the intersection of the $\overline{H}_{11}$ and $\overline{H}_{33}$ surfaces?

M: This partitioning is used a tremendous amount in statistical mechanics.... When you have any number of coordinates if you want to focus on a couple...Because of that intermediate state the real wave function is not that of the reactant system, one that is localized on site 1 but one that has a small delocalization on site 2, and you consider the matrix element between this wave function and one in which there is an overlap between site 3 and site 2.

I am not sure if partitioning went back to Löwdin, it's certainly used in Feshbach resonances, where you partition the Hamiltonian, the most common application is called Feshbach resonances, and there you see...a Hamiltonian partitioned focusing on one part and on some other part, that's related to connecting those parts. Now, as far as the physical interpretation, the H_{11} would be related to the sort of isolated wave function that's on site one, the $\overline{H}_{11}$ would be sort of the modified wave function, where that original wave function is perturbed by having the presence of another site, so that $\overline{H}_{11}$ is just simply perturbation theory, more or less.

2. p. 472, 2nd column, caption to Fig. 1: "Superexchange mechanism: profile of $\overline{H}_{11}$ and $\overline{H}_{33}$ potential energy surfaces in many-dimensional nuclear configuration space for an activationless transfer (no energy barrier). *The H_{22} may have its minimum, when plotted along this coordinate, to the left or right of the intersection of $\overline{H}_{11}$ and $\overline{H}_{33}$*, and only a possible value of H_{22} at the intersection is indicated."

Q: Is there some reason for the position of the minimum of H_{22} to be to the left or to the right relative to the position of the intersection of $\overline{H}_{11}$ and $\overline{H}_{33}$ or it is so that the position of the minimum of H_{22} exactly at the position of the intersection of $\overline{H}_{11}$ and $\overline{H}_{33}$ is just mathematically very improbable?

M: Yes, the measure is zero.

3. p. 472 2nd column middle: "In the case of the calculation of the energies E of the singlet and triplet states of the $\mathrm{BChl}_2^+\mathrm{BPh}^-$ radical pair when their interaction is due to a superexchange, the E in Eqs. (5a) and (6a) is again approximated by $\overline{H}_{33}$, but now the relevant nuclear configurations are those appropriate to *this modified electronic configuration 3* rather than those where $\overline{H}_{11} = \overline{H}_{33}$, namely at the minimum of the $\overline{H}_{33}$ surface in Fig. 1. Thus, Eqs. (5b) and (6b) again apply, but with *different nuclear configurations involved.* To obtain a perturbation expression for the energy E of the singlet or triplet state of the separated radical pair using Eq. (4), the first E in the latter is approximated by $\overline{H}_{33}$, thereby yielding

$$E \approx \overline{H}_{33} + \overline{H}_{13}^2/(\overline{H}_{11} - \overline{H}_{33}) \tag{7}$$

(config. 3 perturbed by superexchange).

In contrast, the energy $E^\dagger$ of the transition state for a superexchange ET mechanism for I $\rightarrow$ III (I $=$ $(\mathrm{BChl})_2^*$, III $=$ BPh) is obtained by setting $\overline{H}_{11} = \overline{H}_{33}$ in Eq. (4), whence $E^\dagger = \overline{H} - |\overline{H}_{13}|$."

Q: How is the electronic configuration modified? And which are the different nuclear configurations here involved? Please explain this, in particular the role of the crossing of $\overline{H}_{11}$ and $\overline{H}_{33}$ vs. the minimum of $\overline{H}_{33}$.

M: Yes, because there you are observing that triplet... that ion pair state... it should be $\overline{H}_{33}$, the correct value for 3. There is a slight perturbation by interaction with the intermediate, that's $\overline{H}_{33}$. And so that's the true state. If that system 3 would be left to itself it would relax and fluctuate about the minimum of $\overline{H}_{33}$. Because that's in fact the potential.

H_{33} is the final state, the triplet state should be the H_{33}. But the actual ion pair state is the state which includes any sort of perturbations... so its energy function is really $\overline{H}_{33}$, in other words it's a state which has been modified by the presence of the other state.

4. p. 473 2nd column middle: "From the values of the various energy differences, one may thereby infer from Eq. (10)

$$E^S - E^T \approx \frac{(\overline{H}_{13})^2(\overline{H}^S_{11} - \overline{H}^T_{11})}{(\overline{H}^S_{11} - \overline{H}_{33})(\overline{H}^T_{11} - \overline{H}_{33})} \tag{10}$$

a value for $\overline{H}_{13}$. Related expressions are given in Refs. [10, 18], the former containing a wayward factor of 2."

Q: What do you mean by "wayward"?

M: Probably that it shouldn't be there. It means that I don't want to make a big deal with that.

5. p. 473 from note ‡: "We have $H_{ij}(i \neq j) \propto \exp(-\frac{1}{2}\beta d)$... The β changed from 1.12 to 1.15 A^{-1} when *the binding energy of the electron in a solute relative to the intervening environment* was changed from 2.0 to 2.2 eV...

Q: What do you mean by "binding energy of the electron in a solute relative to the intervening environment"? How is this magnitude defined?

M: If you think of the electron moving from one potential energy well to another potential energy well, then you have some barrier to the electron moving, due to the presence of the solvent or of a carbon chain.... Now if the exchange is large the barrier in this model is lowered, so you can change this sort of barrier by changing the height of the barrier or by changing β, β is a reflection of the ease of going from one well to the other, and so the barrier height is a reflection of the ease of going from one barrier to the other, β is related to the barrier height. It is really relative to the intervening material, the solvent being the intervening material. Energy differences of different intervening environments.

Q: What do you mean by "exchange mechanism, self-exchange"?

M: The exchange *mechanism* is superexchange, that's a mechanism where you transfer one particle from one site to a distance some distance away, maybe exchange wasn't the right terminology, but you are transferring it from here to way over there, you have your intermediate things, it is *a superexchange kind of barrier* because the electron doesn't reside really on the intermediate solvent molecules, *it only senses them*, and so it's called superexchange, so *a direct thing* would be *exchange*, but this is really where one thing were in contact with another by a bunch of atoms, by a bunch of molecules and, whatever, *the mechanism* for describing it is *superexchange, in that mechanism the electron never really resides on those intervening things, they are too high in energy for to reside on, but the particle senses their presence by the wave function.* The β in superexchange is supposing solvent molecules in between, the β is related to a matrix element H_{12} divided by the energy difference $H_{11} - H_{22}$, β is like $\frac{H_{12}}{H_{11}-H_{ss}}$, β is a perturbation.

Why should I call self-exchange? If you go from A to B, some distance away, one mechanism is simply the direct thing, as though there was vacuum in between, in the other you have a whole bunch of molecules in between, they can enhance by their presence the wave function at every point, and *the formalism* associated with all that is called *the superexchange mechanism.* I don't know why I said self-exchange, I think what I was probably meaning was *the direct process as though the other intermediates weren't present, as if there was vacuum in between the reactants, and a wave function of a reactant dies very readily there, but if there is material in between then the perturbation theory allows the tail of the wave function to be extended a bit, you know, and that's what gives rise to the* β, with the n particles entering almost with an exponential dependence and the barrier presents the $e^{-\beta}$ dependence of distance.

M224 Dynamical Effects in ET Reactions. II. Numerical Solution

W. Nadler and R. A. Marcus

NOTES

1. p. 3906 2nd column middle: "For the probability distribution $P(X,t)dX$, i.e., the fraction of molecules that have not transferred their electron at time $t$and which experience a solvent polarization coordinate in the interval $(X, X + dX)$, this equation has the form

$$\frac{\partial}{\partial t}P(X,t) = [\mathbf{L}(X) - k(X)]P(X,t). \qquad (2.1)$$

$\mathbf{L}(X)$ is a Fokker–Planck operator that determines the stochastic motion along the polarization coordinate and has the form

$$\mathbf{L}(X) = D\frac{\partial}{\partial X}\left\{\frac{\partial}{\partial X} + \beta\left[\frac{d}{dX}U(X)\right]\right\}, \qquad (2.2)\text{''}$$

Q: In your first paper with Sumi appear the equations

$$\frac{\partial P}{\partial t} = D\frac{\partial^2 P}{\partial X^2} + \frac{D}{k_BT}\frac{\partial}{\partial X}\left(P\frac{dV}{dX}\right) - k(X)P \qquad (4.2)$$

$$g(X) = e^{-\frac{1}{2}V(X)/k_BT} \Big/ \left[\int e^{-V(X)/k_BT}dX\right]^{1/2} \qquad (5.1)$$

$$\int g(X)^2 dX = 1 \qquad (5.2)$$

$$P(X;t) = g(X)p(X;t) \qquad (5.3)$$

$$\frac{\partial}{\partial t}p(X;t) = -[H + k(X)]p(X;t) \tag{5.4}$$

$$H = -D\frac{\partial^2}{\partial X^2} + \frac{D}{2k_BT}\left[\frac{1}{2k_BT}\left(\frac{dV}{dX}\right)^2 - \frac{d^2V}{dX^2}\right] \tag{5.5}$$

Q: Apart from the fact of using $P(X;t)$ in Eq. (4.2) and $p(X;t)$ in Eq. (5.4) the two equations look the same.

M: The equations are the same. If you go from Eq. (4.2) and insert Eq. (5.3) then when you do that and go through all the arguments...

2. p. 3907 1st column top: "In deriving the reactive term the assumption was made that equilibration in the vibrational coordinates is fast, the motion across the transition state is ballistic rather than diffusive"

Q: What do you mean by ballistic motion?

M: Ballistic means not diffusive. Local velocity times a probability gives a flux. It is a motion like a ball, not diffusive.

3. Q: There is now a series of equations in this paper that differ from the same equations in your first paper with Sumi. I shall write the list in which the first equation is that of Marcus-Nadler and the second that of Marcus–Sumi.

Marcus-Nadler

$$k(X) = \nu_q \exp[-\beta\Delta G(X)] \tag{2.6}$$

Marcus–Sumi

$$k(X) = \nu_q \exp[-\Delta G^*(X)/k_BT] \tag{4.1}$$

$$\Delta G(X) = \frac{1}{2}\left(\frac{\lambda_0}{\lambda_i}\right)(X - X_c)^2 \tag{2.7}$$

$$\Delta G^*(X) = \frac{1}{2}(X - X_c)^2(\lambda_0/\lambda_i) \tag{3.6}$$

On the same page where, in the first column, we read Eq. (2.7), we read, in the second column, Eq. (2.7')

$$\beta\Delta G(X) = \frac{1}{2}\left(\frac{\lambda_0}{\lambda_i}\right)(X - X_c)^2. \tag{2.7'}$$

The β is missing in Eq. (2.7).

Consider now the following two Marcus-Nadler equations

$$\Delta G^* = \frac{1}{2}\left(1 + \frac{\lambda_i}{\lambda_0}\right)^{-1} X_c^2 \tag{2.10a}$$

$$X_c = \sqrt{2\left(1 + \frac{\lambda_i}{\lambda_0}\right)\beta\Delta G^*}. \tag{2.10'}$$

In the second, a β appears that is missing in the first one, see Eqs. (2.7) and (2.7').

Let now write the Marcus–Sumi (3.6) equation for $X = 0$. We get $\Delta G^*(0) = \frac{1}{2}(X_c)^2(\lambda_0/\lambda_i)$. Now the question: how is the relation between $\Delta G^*(0)$ and ΔG^*?

From Eq. (2.10a) we see that $\Delta G^* = \frac{1}{2}\left(\frac{\lambda_0}{\lambda}\right) X_c^2$ and from Eq. (3.6) we see that $\Delta G^*(0) = \frac{1}{2}(X_c)^2(\lambda_0/\lambda_i)$. Notice the λ in the first equation vs. the λ_i in the second one. Do you have a physical explanation for that?

At first I thought that in the first two couple of equations, (2.6) and (4.1), (2.7) and (3.6) it was only a matter of a misprint, of a * missing in (2.6) and (2.7). But apparently it is not so because you consider the equation

$$k(X) = k_e\sqrt{1 + \frac{\lambda_0}{\lambda_i}}\exp\{-\beta[\Delta G(X) - \Delta G^*]\}. \tag{2.6'}$$

M: The * is just a matter of symbols that have been used. Just a notation. The one decided to put a star, the other decided not to. I think the β shouldn't be present in Eq. (2.10'), we would have wrong units. Even in Eq. (2.7') the β shouldn't be there. Which is the physical significance of writing $\Delta G^*(0)$? If you are looking at that slanting line, then you have $\Delta G^* = 0$ for $X = X_c$, right? So, this $\Delta G^*(0)$ means counting the free energy barrier starting from that point. As you go along the slanting line, you have the values of $\Delta G^*(X)$ starting from $\Delta G^*(X_c) = 0$.

4. Q: In Marcus–Sumi paper, we have the Eq. (3.6): p. 3907 1st column middle, Eq. (2.7)

$$\Delta G^*(X) = \frac{1}{2}(X - X_c)^2(\lambda_0/\lambda_i)$$

In this equation, in which ΔG^* is given as function of the polarization variable X, λ_0, and λ_i appear as a quotient, contrary to their usual appearing on the same footing in the sum $\lambda = \lambda_i + \lambda_0$, as in your classical $\Delta G^*(R) = w^r + \frac{\lambda}{4} + \frac{\Delta G^{0\prime}(R)}{2} + \frac{\Delta G^{0\prime}(R)^2}{4\lambda}$ in which ΔG^* is function of R. Can you give a physical explanation of the different algebraic relations of λ_0 and λ_i in the two cases?

M: The above equation is not symmetrical in the X and q coordinates, so it should not be symmetrical in the λ_0 and λ_i. One asymmetry induces the other asymmetry.

5. p. 3907 2nd column middle: "the survival probability $Q(t) = \int_{-\infty}^{+\infty} dXP(X,t)\ldots$the operator $[\mathbf{L}(X) - k(X)]$ is expected to have a discrete spectrum of eigenvalues $-\gamma_n\ldots$a spectral expansion of $Q(t)\ldots$

$$Q(t) = \sum_{n=1}^{\infty} a_n e^{-\gamma_n t}\text{''}$$

Q: Is the eigenvalue equation $[\mathbf{L}(X)-k(X)]P(X,t) = -\gamma_n P(X,t)$? And of which kind of physical magnitude are the $-\gamma_n$ eigenvalues? Are they particular values of polarization?

M: If you have an operator you have its eigenvalues under certain conditions and eigenfunctions. A spectral expansion is the solution of the eigenvalue equation in terms of eigenvalues and eigenfunctions. Expanding in the eigenvalues of that operator. The equation you wrote is correct. The eigenvalues are probably the relaxation times. Well, the eigenvalues γ_n are reciprocals of relaxation times as far as units are concerned, wouldn't they? Because k has the units of 1/seconds, I guess, yes, I believe that's correct, but the point is that is the formalism of Sumi or Nadler and they would know the answer better, but my impression is yes. The eigenvalues are reciprocals of relaxation times...I think that would be not for a P but for a P_n, in other words if you decompose P into components $P_n(X, t)$, then you would have an eigenvalue equation. I think you would have P_n there, the nth component of it, because the only place where n appears is in the γ_n, so you have that n appears someplace else too.

6. p. 3908 1st column bottom: "the parameter λ_i/λ_0 which regulates the width of the Gaussian reactive term $k(X)$ and, therefore, will be called *reaction window parameter*"

Q: Please check if my understanding is correct: from Eqs. (2.6) and (2.7) we get the Gaussian

$$k(X) = \nu_q \exp\left[-\beta\frac{1}{2}\left(\frac{\lambda_0}{\lambda_i}\right)(X - X_c)^2\right]$$

which measures how peaked is $k(X)$ around X_c.

M: If λ_i were zero, for example, X would be localized at a value of X, so the fact that you have some λ_i causes the X not be localized, there are various values of X at the intersection surface. It is a Gaussian, OK.

7. p. 3908 2nd column top: "***4. Wide reaction window limit,*** $\frac{\lambda_i}{\lambda_o} \to \infty$.

In this other extreme case of the reaction window parameter the contributions from the solvent to the electron transfer process vanish, λ_0 being zero. The reactive term is practically constant and equal to k_e over the relevant range of the potential"

Q: From the Eqs. (2.9) and (2.10a) p. 3907

$$k_e = \nu_q \left(1 + \frac{\lambda_0}{\lambda_i}\right)^{-1/2} \exp(-\beta \Delta G^*) \tag{2.9}$$

$$\Delta G^* = \frac{1}{2}\left(1 + \frac{\lambda_i}{\lambda_0}\right)^{-1} X_c^2 \tag{2.10a}$$

(1) How is this point in accord with $\Delta G^*(R) = w^r + \frac{\lambda}{4} + \frac{\Delta G^{0\prime}(R)}{2} + \frac{\Delta G^{0\prime}(R)^2}{4\lambda}$?

(2) How is it possible to have an ET transfer process in solution with no contribution from the solvent? Remembering Fig. 1 of your first paper with Sumi would this mean that the ET is in this case the one happening for $X = 0$, that is, the one happening along the ordinates q axis? Is in this case ET in solution similar to that in the gas phase or in a nonpolar solvent?

(3) Looking at Fig. 1 in the first Marcus–Sumi paper one sees that even when $X = 0$ it is possible to cross the C–C line of transition states but the products system can relax toward the equilibrium coordinates (q_0, X_0) only if the X may tend (may move toward) X_0. Is it so?

M: (1) When λ_i is very large the whole range of X contributes to the reaction coordinate, it causes the intersection line with many Xs. I have to think about k_e ... the starting point, the starting value of X, so you don't need any fluctuation but λ_X, localized about the starting X, where the reactants are.

$\lambda_i/\lambda_0 \to \infty$ means not that λ_0 vanishes but that it may be small. That k_e depends on λ_0. ΔG^* in Eq. (2.10a) depends on X_c and it is depending on λ_i and λ_0. *X_c is reported as* $X_c = \sqrt{2\left(1 + \frac{\lambda_i}{\lambda_0}\right)\beta \Delta G^*}$. You see that X_c is not independent of λ_i, so if you have X_c^2 *depending on λ_i/λ_0 one should consider it inside* Eq. (2.10a). Same for Eq. (2.9). ΔG^* depends on λ_i/λ_0, $\Delta G^*(X) = \frac{1}{2}(X - X_c)^2(\lambda_0/\lambda_i)$. You have to insert how ΔG^* depends specifically on λ_i/λ_0. So can you come to a conclusion. I don't remember the physical meaning of k_e. You can use Eq. (2.9) after you put in how ΔG^* depends on λ_0/λ_i.

(2) When you have an intersection of the surfaces the vibrations can take care of that. Things can go on that q axis without polarization, absolutely. In the gas phase as far as the reorganization part (*along* q) it is the same. Of course in the gas phase you may have long range forces, collisions...

(3) Along the q axis you have ET if you cross the transition states line at $X = 0$. Sometimes the system just bounces back, or it may go on and eventually relax, remember that this is a two coordinates description of a many coordinates problem, so, to relax internally, the relaxing coordinate is an internal coordinate. Remember, this is the reaction coordinate. For example, consider ET between two symmetrical ions: the combinations of the symmetric stretching vibrations *brings the system to reaction*, another coordinate brings to *relaxation.*

From your question (3): "even when $X = 0$ it is possible to cross the C–C line": That's right, at a very high energy, you notice, much higher than the energy at the saddle point. If you cross the TS line at $X = 0$, now the system would be going up vertically, at $X = 0$ would be going up vertically from the point 0. The q axis, that's the vertical axis, so if you only had a vibrational motion occurring when you go along the q axis, sort of the ordinate, you go all along there where $X = 0$, and you'd reach that intersection line at a very high energy, so that's one thing that would happen there. You see, if you go up then...supposing you go up, and maybe you don't reach that intersection line, then you'd just fall back down. Now supposing that you'd cross that intersection line, now when you fall down you fall toward the point on the up right hand corner. In other words, sometimes the system bounces back because you don't actually reach the TS, but there is also what can happen where supposing you do go across, then the question is to what extent will the system continue toward the valley in the upper right hand corner and to what extent it might just roll back down. The reaction coordinate normally is a coordinate for what would be leading from O to the saddle point, so that'd be the reaction coordinate, but supposing that the X were very slow motion, then the only way of getting across that line would be to go up and then eventually you'd have to relax down, and I suppose as you're going down it may be that you instead of... you might even fluctuate back up a little bit, go down to the products at the reactants part, but normally eventually relax and go down to the point O', in the upper right hand corner, but for reaction to complete eventually you have to reach the O' region, all the rest would be a very transient region.

"the relaxing coordinate is an internal coordinate." Yes, you relax thermally, these are two coordinates, in the presence of a bath of 10^{23} coordinates. These are two coordinates of the system, and there are maybe 10^{23} coordinates, which is the bath. Typically you are going to relax toward the lowest energy in these coordinates, the energy going into the bath which has many more ways of distributing that lost energy.

8. p. 3908 2nd column middle: "Reaction-*diffusion* models for ET processes coupled to solvent polarization *fluctuations*..."

M: When you have diffusion you have a diffusion constant, we have the diffusion along the q coordinate, it is relatively unimportant, you get fluctuations but we don't put into a diffusion constant, so we treat that differently, for you can't regard the motion along that C–C line as being a diffusion because one component of it is not a diffusion, it is a matter of how you treat the fluctuations, in one case you treat the fluctuations in a diffusion way, in the other case you treat the fluctuations as being internal equilibrium fluctuations.

Q: What do you mean by "internal equilibrium fluctuations"?

M: Fluctuations of the... I don't know why I called them internal... because almost everything is internal... Probably there I meant the internal coordinate q... I imagine, yes, the point is that we assume that to reach one value of q from another, you do it very quickly if you have the energy involved, you don't diffuse slowly from one to the other, so there is no diffusion constant associated with that, in other words it is called the fast coordinate, so effectively its diffusion constant is infinity, if it gets enough energy then it bounces up there, the process is not a slow diffusion process.

"Internal equilibrium fluctuations." That just means that when one speaks of fluctuations in the q coordinate without calling it diffusion at all, because there is no slow diffusion of that, it's just sort of, by one means or another, suddenly gets a certain amount of energy from the surroundings, sometimes it doesn't, but not in a diffusion-like fashion, you don't give a diffusion equation to calculate how its energy changes with time.

These are all important questions, sometimes you want sort of to get results out and you don't necessarily speak all the details out.

9. Following: "Though different approaches were employed in these investigations, the models derived had quite comparable features. They resulted in the description of the transfer process as a 'diffusion' of the solvent polarization (or an equivalent coordinate) along a one-dimensional free energy surface of the reactant, with the reaction occurring via the crossing over to the free energy surface of the product at the intersection point of the two free energy surfaces. In the *adiabatic* case the crossing over to the product free energy surface occurs with unit probability. In this case, neglecting the backreaction, the reaction process can be viewed as diffusion in a one-dimensional potential well with an *absorptive boundary* at the intersection point of the two free energy surfaces"

Q: (1) I realize that when the intersection passes through the minimum of the free energy surface, and so there is no reaction barrier, it is possible to go by polarization diffusion from reactant to product. But if there is a reaction barrier is it not a polarization fluctuation necessary in order to overcome the barrier?

(2) What do you mean by "absorptive boundary"?

M: (1) Yes, if the polarization is stopped you only have to use internal degrees of freedom and how you describe those? You describe by a time scale and imagine that you have *a fast coordinate so diffusing. You could consider diffusion with fast degrees of freedom but you would end up largely to have a local equilibrium so diffusion is not a slow step*, so no diffusion constant comes in by diffusing but diffusion is so rapid that the thermal fluctuations are important rather than the slow diffusion. If you have a fast coordinate you don't really speak about diffusion, you just speak about fluctuations. Of course in a sense you diffuse through space whether it is a fast coordinate or a slow coordinate but the diffusion is so fast that you just speak of its net results, the fluctuations, and not of the slowness of getting there. You just have: What is the probability that I got there? And that's all, and not the time scale of it.

(2) The absorption boundary is by the line of transition states. If there is a recrossing then describing it as an absorption boundary is not quite right, that's a subtle detail.

10. p. 3909 2nd column top-middle: "...the *adjoint* Fokker-Planck operator, which has the form

$$\mathbf{L}^{+}(X) = \tau_L^{-1}\left[\frac{\partial}{\partial X} - X\right]\frac{\partial}{\partial X}, \qquad \text{(2.19)''}$$

Q: The formula of for the Fokker–Planck operator is

$$\mathbf{L}(X) = D\frac{\partial}{\partial X}\left\{\frac{\partial}{\partial X} + \beta\left[\frac{d}{dX}U(X)\right]\right\} \qquad (2.2)$$

Considering that $\tau_L = (\beta D)^{-1}$ (2.5) and $U(X) = \frac{1}{2}X^2$ (2.3)

we have $\mathbf{L}^{+}(X) = \beta D\left[\frac{\partial}{\partial X} - X\right]\frac{\partial}{\partial X}$ and $\mathbf{L}(X) = D\frac{\partial}{\partial X}\left\{\frac{\partial}{\partial X} + \beta X\right\}$ how would you get $\mathbf{L}^{+}(X)$ from $\mathbf{L}(X)$?

M: There is a well-defined procedure for going from an operator to its adjoint. $\langle YL \mid X\rangle = \langle L^{+}Y \mid X\rangle$ is a definition of the adjoint operator and in this case you can get it by integration by parts to see what the adjoint operator is. You integrate by parts, by parts you get that minus sign.

11. p. 3920 1st column middle: "in the nondiffusing limit $\tau_L k_e \to \infty$ .... in this limiting case no solvent *dynamical* effects influence the ET process."

Q: Can you give examples of physical systems where this happens? Glasses? Solids? How can ET happen without a previous appropriate solvent fluctuation? Only through intramolecular coordinates fluctuation? And can there be a *static* solvent influence just through the dielectric constants?

M: Yes, glasses. Well, a solid, yes, a solid. In a glass there is no reorientation. Static effects appear only through ΔG^0. Through dielectric constants only if they affect ΔG^0.

12. p. 3923 2nd column from Note: "This reactive term, Eq. (2.6)

$$k(X) = \nu_q \exp[-\beta \Delta G(X)] \tag{2.6}$$

is, in effect, the transition state theory expression for the reaction rate at a certain value of the polarization coordinate X. The exponential factor in Eq. (2.6) is the Boltzmann probability that a coordinate q, which represents the totality of the intramolecular coordinates, has the transition state value if the system sees a specified value of the polarization coordinate X."

M: You remember that in this case at every value of X you can cross the slanting line by a fluctuation in the q coordinate of X. If you start at a particular value of X and if, at that value of X, you now fluctuate in q to reach the intersection line, then there is the $\Delta G(X)$. That is the free energy difference between the value at that X and the value of that X at the $q = 0$ line. The $\Delta G(X)$ is contained in the diffusion equation, to go from $X = 0$ to that value of X so that part of the free energy barrier is separately in the equation. When you specify X, then because of the slanting line of the transition states the value of q is automatically specified by the line there.

13. Following: "Hereby the assumption enters that during the reaction the distribution of states along a q *hypersurface* of the (X, q) space on the reactant side of the transition state line is a Boltzmann distribution."

M: In other words, draw a line parallel to the X axis at different values of q. That line at that value of q is a q hypersurface. At another value of q you have another hypersurface. A hypersurface is defined as a surface with one less dimension than the space. It is not just a segment, it goes to infinity, it is in the entire space, it is in the reactants region and in the products region, the q hypersurface occupies the whole space.

14. Following: "For $\lambda_i/\lambda_0 \rightarrow 0$ the transition state line gradually approaches the line $X = X_c$ which is perpendicular to the X axis."

M: I would delete the word perpendicular. I wouldn't even call it a line, I would say that it is $X = X_c$, that would a better way of describing it. The line reduces to a point, the point $X = X_c$.

REVIEW: M225 Recent Developments in ET Reaction

R. A. Marcus

NOTES

1. p. 79, 2nd column middle-bottom: "2) the rate constant k of a series of *similar* reactions (i.e., those possessing a similar reorganization term λ)"

Q: Is the earlier statement a *definition* of the concept of similar reactions in ET?

M: Yes, certainly.

2. p. 80, 1st column top: "... $k_{el}/Z_{el} \leq (k_{ex}/Z_{ex})^{1/2}$ ... (the equality sign occurs when no adsorbed species layer separates the reactant from the electrode)"

Q: Can this be taken as a *definition* of adsorption (or nonadsorption) to an electrode in ET?

M: Yes, I guess so because the ion touches the surface of the electrode just like reacting in solution. The theory isn't correct, it should be corrected by the work terms.

3. p. 79 2nd column bottom and p. 80 1st column middle: "3) there is a dependence of $\ln k$ on the dielectric properties of the solvent $(D_{op}^{-1} - D_s^{-1})$ with a known slope, in the absence of specific solvent–solute interactions... *the quantitative effect of a nonspecific solvent (3) depend on the assumption of a dielectric continuum* used for the solvent outside the innermost coordination shell of the reactants"

Q: Do you have a complex in mind when writing of the innermost coordination shell?

M: Yes, sure. If you have, say, an ion with coordination shell, outside of it the solvent molecules are weakly bound, so they are not part of the radius.

4. p. 80 2nd column middle: "In the treatment of electronic orbital orientation effects on ETs, there are two types of effects which can occur, "through bond" or "through space" (or "through the solvent medium"). The first of these becomes possible when the two reactants are linked by actual chemical bonds.

Q: To have ET between reactants linked by chemical bonds: is it the same as to have an *intramolecular* ET?

M: Yes.

5. p. 81 2nd column top: "... making the λ in equation (1)

$$\Delta G^* = \frac{1}{4}\lambda(1 + \Delta G^0/\lambda)^2 \tag{1}$$

small, and so creating the conditions for an 'inverted effect' for a given ΔG^0"

M: The inverted effect occurs when you have a $\Delta G^0 = -\lambda$. If λ is small it is easy for ΔG^0 to be more negative.

REVIEW M229 Some Recent Developments in ET: Charge Separation, Long distances, Solvent Dynamics, and Free Energy Aspects

R. A. Marcus

NOTES

1. p. 47 top-middle: "Since reformation of an excited singlet $BChl_2^*$ in the back reaction may be somewhat uphill, in terms of free energy, and since *re-reaction* (~3 ps) is much faster than fluorescence (nanoseconds) this second possible path for ET from BPh^- also does not effectively compete with ET from BPh^- to Q (~200 ps)."

Q: What do you mean by re-reaction?

M: **Typo**: It is a typographical error. Read: "since reaction."

2. p. 47 middle: "The environment of these pigments $BChl_2$, BChl, and BPh is largely hydrophobic, leading to only a small environmental λ_0."

Q: Which one is the relation between hydrophobicity and λ_0?

M: The key question is whether you have an environment polar or not polar. If you have water molecules, then you have a difference between D_{op} and D_s. If you have a hydrophobic environment, essentially nonpolar, there is very little difference between D_{op} and D_s.

3. p. 49 top-middle: "More generally, when $\lambda_i = \Delta G^* = 0$, the reciprocal of the electron transfer rate constant is given approximately by τ:

$$\tau \cong k_e^{-1} + \tau_L \qquad \text{(2a)''}$$

Q: How would you compare this formula to your former equation

$$\frac{1}{k_{\text{rate}}} = \frac{1}{k_{\text{act}}} + \frac{1}{k_{\text{diff}}}\ ?$$

Is it just the same expressed in terms of times instead of rate constants?

M: Yes, it is equivalent. Except that here one talks of an actual diffusion of particles toward each other and in the other case it is diffusion in solvent space, instead of reaction space, intramolecular. But in the same spirit. The second equation was referring to a bimolecular instead of unimolecular process.

4. p. 49 bottom: "In a purely activation-controlled, i.e. a k_e controlled electron transfer reaction, the survival probability of a reactant should decay exponentially with time, when the electron transfer is intramolecular or when it occurs bimolecularly but between reactants fixed in position."

Q: Back in a sense to the fourth note of the preceding **M225**. How and when you decide to treat the system as unimolecular or bimolecular? Is there some kind of continuity between the two possible systems? Is it a matter of convenience?

M: To some extent it is a matter of choice. You have a kind of equilibrium between complexes that are sort of near each other and complexes that are far apart. You have an equilibrium constant for that.

Now if you treat a reaction of that complex as unimolecular, then if you multiply by the equilibrium constant of that complex then you have a bimolecular rate constant. Approximately, the bulk rate constant times a unimolecular rate constant.

Q: How do you calculate the formula for the equilibrium constant?

M: The point is this: supposing you have an actual bimolecular reaction, now think of the two reactants coming together and forming a transient but not stable complex. And think of having an equilibrium constant for that. Now, it's a strange kind of equilibrium constant because it is a nonstable thing, but maybe you think of an equilibrium constant associated with being some small integral of distances of separation. Now you can make an equilibrium constant out of it. Now, imagine that once the two reactants are fixed in that position, you have a unimolecular constant, and the bimolecular constant is equal to the equilibrium constant times the unimolecular constant. In other words, this is an approximate way of doing things.... Then that complex can react and the overall rate then is the equilibrium constant to form the complex times the rate of reaction of the complex, a first-order process, but the overall process is bimolecular, so in this way you have expressed the bimolecular rate constant in terms of a kind of equilibrium constant times a first-order rate constant.

5. Following: "...Another source of unusual time behaviour can occur when there are *significant deviations from the Debye relaxation behaviour*"

M: The point is this, that if you have a system decaying by a single exponential, a relaxation time, that is one behavior, if you have a system having a more complex time relaxation, not single exponential, much more complicated than single exponential, that means that the whole treatment which is based upon a single exponential relaxation has to be modified.

6. p. 50 bottom: "What has been assumed in the theory for the solvent motion is not that the potential energy surface is quadratic but rather that the solvation free energy of the entire ensemble of solvent molecules is a quadratic function along the 'reaction coordinate.' This assumption, which is *the statistical mechanical counterpart of the dielectric polarization being proportional to an external field...*"

M: If that polarization is proportional to an external field, then the free energy which is related to the work done by the polarization times the change of field times the change of polarization integrated becomes quadratic. The key idea is that it is quadratic. The statistical mechanical level is probably an example that is quadratic when the individual underlying motions are not quadratic, maybe an example of the central limit there, it is a very subtle example. When the electric polarization is proportional to an external field, then the energy associated with it is related to the square of that dielectric polarization, so you get a quadratic expression. So the quadratic comes from polarization times a force, essentially, maybe it's integrated over a volume, polarization times a force, times a field really, and that gives you an energy, a free energy.

7. Following: "... is much milder than assuming harmonic potential energy surfaces."

Q: Why is it much milder?

M: The point is that the assumption can be valid even when the potentials are strongly anharmonic. The free energy is a quadratic function even though at the molecular level the potential energy may not be quadratic. Milder because for the free energy to be quadratic is not necessary to assume harmonic motion at the molecular level.

M231 Nonexponential Time Behavior of ET in an Inhomogeneous Polar Medium

W. Nadler and R. A. Marcus

NOTES

1. p. 24 1st column, top: "When intramolecular contributions to the ET are *negligible*..."

M: The λ_i would be of the *order of vibrational relaxation time*, pretty fast, and the λ_0 would be virtually of the *order of relaxation times of orientation polarization.* Some fast vibrations that make the dielectric constants at high frequencies differ from 2 by being about 4.5 is the bending vibration polarization. It's probably almost entirely the polarization of that bending vibration that's responsible for the difference between 2 and 4.5. I mean, I almost certainly said it's what you might call vibrational polarization, it's not the stretching vibrations, they are not in that frequency range, it is more like bending vibrations. Now, whether it is the bending vibration of an individual water molecule.... I think that's the bending vibration of an individual water molecule, rather than of water structure.

2. p. 24 2nd column bottom: "For ET processes with non-vanishing intramolecular contributions the time behaviour of the fraction $Q(t)$ of non-reacted donors can deviate strongly from a single-exponential form (**M213**, **M214**, **M224**). This behaviour is also expected for ET in a medium inhomogeneous with respect to the dielectric relaxation times, even in the absence of intramolecular contributions. In such a situation *the unambiguous definition of a (time independent) ET rate constant is no longer possible*"

M: If the conditions are such that the survival probability is not a single exponential, then, you know, you have something that cannot be defined by some rate constant. You have a more complicated time evolution. *The concept of rate constant is not something which is always applicable.* For example, remember some numerical work that Nadler did, he didn't have something that was constant then. Of course the concept of rate constant applies to most chemical reactions but there are those where there is slow solvent behavior where it doesn't.

3. Following: "Recently, McGuire and McLendon have reported results on intermolecular ET in solid glycerol which indicated a fractional power-law dependence of the ET rate on the dielectric relaxation time, $k_{ET} \propto \tau^{-0.6}$. The analysis of their experiment is somewhat indirect since they observed ET *between randomly distributed donors and acceptors*, i.e. *they saw a superposition of many different ET processes*"

Q: How does ET in a system like this differ from that in a liquid solvent? Aren't donors and acceptors randomly distributed even in this case?

M: Yes but *in the liquid solvent case* they somehow react when they diffuse close to each other. That's not possible in a solid solvent. That becomes a very slow step but when the diffusion to each other is fast then it doesn't matter that they are randomly distributed, the chance of being close together is given by a Markovian distribution, it isn't given if donors and acceptors are close or not. They're randomly distributed but if diffusion is slow, as you see from looking at the many different paths that could occur in that figure, that figure of the ellipses and so on, you get a whole distribution of reaction times, the rate constants, you cross at many different places, not just around the saddle point region, when you have a sluggish motion, where you cross is sort of kind of a balance between the sluggishness of getting to the right solvent coordinate, and that's slow vs. a slowness due to overcoming a high energy barrier, in any case you have *a distribution of rate constants* and *not a single rate constant*. Any time you have that you will get some kind of *fractional exponential behavior*...in other words not a simple first order, second order...

4. p. 25 2nd column top-middle: "...the Davidson-Cole behavior of the *dielectric spectrum*"

Q: Is the dielectric spectrum the set of all possible dielectric constants of a system?

M: Yes, that's when you plot the non-imaginary part vs. frequency.

5. p. 25 2nd column middle: "An algorithm...the non-exponential time behaviour of ET in an *inhomogeneous* medium...In this Letter we restrict our attention to the case with a vanishing intramolecular contribution, where the *local time behaviour*..."

Q: Do the different relaxation times τ's refer to different solvent layers around a donor–acceptor couple or are there different τ's in different local microscopic environments when the distances between donor and acceptor are the same in each local microscopic environment or are the τ's depending on the donor–acceptor distances?

M: Probably when one has the solvent nonhomogeneously distributed all over...

Q: In different local environments there may be different relaxation times....

M: Well, even in a single place you have an all spectrum of relaxation times, even in a single homogeneous solvent you can have a distribution of relaxation times, water for example, has a distribution of relaxation times, alcohol has a distribution of relaxation times not just one relaxation time, there need not to be any selective solvation around some ion for the solvent to have relaxation times.

Q: So it is not a matter of different layers of solvent around the couple.

M: Not necessarily, there can be a difference too, yes, there can be a difference in different layers but things become a little bit hairy because dielectric constant is really a bulk property and you try to apply at the molecular level, so, you know, it is not quite correct. Certainly if you have any selective solvation, in other words if the distribution of solvent is inhomogeneous around there, that's a definition of selective solvation, you can have inhomogeneity in the dielectric background.

Q: So all those three possibilities are possible.

M: Absolutely.

6. p. 26 2nd column top-middle: "... a distribution of dielectric relaxation times

$$g_{DC}(\tau) = \frac{\sin(\pi\beta)}{\pi\tau}\left(\frac{\tau}{\tau_0 - \tau}\right)^{\beta}, \quad \text{for } \tau \leq \tau_0\text{"}$$

Q: Is this the distribution of dielectric relaxation times relative to each microscopic environment?

M: Yes, except that you have to be careful because the situation is microscopic and you are using dielectric constant... it becomes a question of concept.

M237 An Internal Consistency Test and its Implications for the Initial Steps in Bacterial Photosynthesis

R. A. Marcus

NOTES

1. Q: Consider Fig. 1(b) for the "intermediate" mechanism. Here the reaction path represented by a point moving along the reaction coordinate and going from the minimum, say, of curve 1, crossing then to 2, going along 2 to the crossing with 3 and finally relaxing to the minimum of 3 is clear. Which one would be the path of the representative point in the case of the superexchange mechanism in Fig. 1(a)? Considering the minimum energy path in going from 1 to 3, which is the use of curve 2 in superexchange?

M: You follow the path of the representative point only if you do the process adiabatically. In the case where 1 and 3 intersect, just going from one to the other would not be superexchange. The path would be going up to the intersection of 1 to 3 and then continuing along the 3. Then if you make use of number 2 that would be at the intersection of 1 and 3, you use the data on 2 at that point to calculate the integrals in superexchange. If you talk of superexchange, the reaction is from 1 to 3, one would go along the path to the transition state where they intersect and you use the data there at that intersection to calculate the superexchange rate using the data on 2. You use the energy of 2 at that point and you insert it into the integrals . . . So, one doesn't go from 1 to 2, then from 2 to 3, not if the mechanism is superexchange. I think in my first paper on superexchange I was not careful in specifying what energy one should use.

2. p. 13 2nd column middle: "There seems to be a real bleaching of the BChl band when a higher energy photon is used for the excitation"

M: I'm just thinking of some higher frequencies and not talking about higher intensities. If the intermediate is there, then it would absorb light and bleach. The key question is if the intermediate absorbed at a higher frequency the you can bleach it if you had high intensity . . . It might have to do with the absorption spectrum of the intermediate . . .

3. p. 14 2nd column top: "Experimentally, the initial ET reaction in the bacterial photosynthetic reaction center is known to be activationless. More precisely, the reaction is even faster at low temperature than at room temperature"

M: You have different species of bacteria, and that's certainly true for one of them. The Franck–Condon factors overlap in this case presumably best when you are in the ground vibrational state.

4. p. 14 2nd column bottom in caption to Fig. 1 "The reaction coordinate in (a) may differ from that in (b) and fig. 2. In (a) the minimum for curve 2 is not necessarily a global minimum in the space of all reaction coordinates."

M: You have to look at the species, I'll call them 1, 2, and 3, there are three different species there. You have to look at how each of the species differs from the other, and if you compare 1 and 2 you have to look in the vibrational space and deduce a reaction coordinate from that, and, similarly, if you go *directly from 1 to 3, by superexchange, you would use a different reaction coordinate*, so for each of the possibilities 1 to 2, 2 to 3, or 1 to 3 you would have some combination of the vibrations of each of the species involved in 1, in 2, in 3 and then you make some sort of a difference in going from one vibration to another... the reaction coordinate would be different for each of the three possibilities, going directly to the intermediate, from the intermediate to the product, or going from the reactant to the intermediate by superexchange.

Q: So different combinations of vibrational states.

M: That's right.

5. p. 15 1st column top: "The effective superexchange matrix element $\overline{H}_{13}$ is given later by Eq. (13)

$$\overline{H}_{13} = H_{13} + H_{12}H_{23}/(H_{22} - E). \tag{13}$$

In our case the H_{13} there is negligible and, introducing an appropriate value for E into Eq. (13), Eq. (13) reduces to

$$\overline{H}_{13} = H_{12}H_{23}/(H_{22} - H_{11}). \tag{3}$$

Here $H_{22} - H_{11}$ is a vertical energy difference of states 1 and 2 at the activationless superexchange transition state."

Q: Some questions: (1) Does a negligible H_{13} mean that a direct passage from 1 to 3 is very improbable even if the path is activationless?

(2) If so the path from 1 to 3 would be closed and the system (its representative point) would first pass through the crossing between curves 1 and 2 and successively through that of 2 and 3? (3) If so is the crossing between curves 1 and 3 a purely theoretical transition state, I mean a transition state through which the system would not pass, a fictitious transition state without transit?

M: The intersection in the 1 and 3 is the transition state for the superexchange. Yes, it depends on the overlap integrals. In fact activationless means that if you going from 1 to 3 you go through the minimum of 1, essentially you are talking of the diabatic curves, and so the matrix elements don't affect that at all, the effective matrix element is the superexchange one. You use the superexchange off diagonal matrix element.

Q: On one side there are matrix elements that don't affect: Which ones are they? On the other hand there are the superexchange off diagonal matrix elements, apparently different from the former ones.

M: The superexchange off diagonal matrix element is the combination of several matrix elements. The overlap between the wave functions for 1 and 3 would be very small, that'd be the direct process, so that's not involved, and it makes use instead of 2, the process goes like superexchange. The direct reaction would be going directly from 1 to 3, without the benefit of making use of curve 2 and the intervening wave functions, that'd be the direct thing, that's negligible apparently, and so the system is making use of curve 2 but really just of the wave functions there and does it by superexchange, so would actually follow the path that going from 1 to the intersection with 3, then going down the 3, but the matrix element that's involved at that intersection of 1 and 3 is not the direct matrix element it's the one that invokes superexchange using the matrix element going from 1 to 2 and 2 to 3 divided by the energy difference between 2 and the intersection of 1 and 3. There it is in Eq. (13). And the idea is that the first matrix element H_{13} is negligible.

6. p. 15 2nd column middle: "The rate constant k_{12} is given by Eq. (1)

$$k_{12} = \frac{2\pi}{\hbar}\frac{|H_{12}|^2}{(4\pi\lambda_{12}RT)^{1/2}}\exp\left(-\frac{(\Delta G^0_{12}+\lambda_{12})^2}{4\lambda_{12}RT}\right) \qquad (1)$$

for small values of $|H_{12}|$ and by Eq. (9)

$$k_{12} = \nu\exp(-\Delta G^*_{12}/RT) \qquad (9)$$

for large values. Whenever, for the *given* $|H_{12}|$, the k_{12} calculated by the former exceeds the latter, Eq. (9) should be used.

M: When the k_{12} calculated from Eq. (1) for the nonadiabatic case is very large, that means that the coupling is very large and so the process to be used is the one for the adiabatic case. In other words, think of tunneling, plot the k_{12} rate against $|H_{12}|^2$, then initially you get a straight line, and then eventually when things get adiabatic you get a curve parallel to the $|H_{12}|^2$ axis, in other words *the* k_{12} *becomes independent on* $|H_{12}|$ *essentially.* Except that it is also true that if you increase $|H_{12}|$ you eventually lower the barrier, you get more splitting, but let's forget about that for the moment. So, the idea is that the formula, the right formula, is Eq. (1) with $|H_{12}|^2$, but then when k_{12} begins to flatten off, then clearly you are in the other regime, when k_{12} is much larger then you are in the adiabatic regime. *The diabatic regime is when you are in the region where the rates are proportional to the* $|H_{12}|^2$.

7. p. 16 1st column top-middle: "However, Plato has estimated the sum S of the overlap integrals between the various reactants and we consider the postulate that the ET matrix elements are proportional to these overlap integrals."

Q: Is it a postulate or an approximation or an approximation taken as postulate? Moreover: is this postulate always true in ET processes?

M: You see, the point is that the electron transfer matrix element is not just an overlap integral, it has some Hamiltonian sandwiched between the two wave functions. So it may be that that ET is proportional to the overlap integral but that's an assumption. Because in the ET matrix element you have the two wave functions involved in the overlap, but you have the Hamiltonian in between them.

Q: So this assumption is not always true.

M: No, I suppose not. You may not assume in general that if you got a Hamiltonian between two wave functions you will get a simple proportionality.

8. p. 16 2nd column middle: "...the singlet $BChl_2^+BPh^-$ ... via an *internal* ET...forms the ground state $BChl_2BPh$."

Q: Is "internal ET" the same as unimolecular ET or intramolecular ET? Does the decision of considering ET unimolecular or bimolecular depend on the distance between the two ions or on some other features of the reaction?

M: I think it is related to whether the ions are free to move large distances with respect to each other. If they are not and they are more or less fixed in position then it is more like a unimolecular process. Internal, unimolecular, or intramolecular are the same.

9. p. 17 1st column bottom: "(It may be added, however, that for a reaction to be activationless it is not essential that $-\Delta G^0 = \lambda$, but rather that these quantities be fairly close"

Q: How much close?

M: A zero point energy or something.

10. p. 17 2nd column middle: "If the situation is as depicted later in Fig. 2, there is no $BChl^-$ as such in the vicinity of the avoided intersection of curves 2 and 3 in Fig. 2."

Q: Why so? Curve 2 refers to the system $BChl_2^+ - BChl^- - BPh$ and curve 3 to the system $BChl_2^+ - BChl - BPh^-$.

M: It is not even temporarily stable, to be stable means to be on an adiabatic curve with a hump going the other side, a barrier, in order to be an intermediate you would have to have a minimum some place, but in that figure it is not existing on a minimum, it is only existing on the side of the slope, it reacts immediately, it just goes down the slope there. It is not existing as an intermediate, it is just a flight by night. I mean, if you go down after some reaction you don't call a point in between an intermediate if it resides there only for some time.

11. p. 18 1st column: "... $\overline{H}_{ii} = H_{ii} - H_{i2}^2/(H_{22} - E)$, $i = 1, 3$ (12).

Each of the H_{ii} terms in these equations varies with the nuclear configuration (the H_{ij} terms with $i \neq j$ vary only weakly so)."

Q: Why so? Maybe because the distance between dimer, monomer, and pheophytin don't change much during ET?

M: They are less sensitive than the others, that's right. Because the electronic wave functions aren't changing that much with distance.

12. p. 18 2nd column top-middle: "One finds in this way that for the singlet state of the radical pair at its equilibrium configuration the $\overline{H}_{13}^2/(\overline{H}_{11} - \overline{H}_{33})$ term in Eq. (14)

$$E \approx \overline{H}_{33} - \overline{H}_{13}^2/(\overline{H}_{11} - \overline{H}_{33}) \tag{14}$$

is about 0.03 cm^{-1}. In calculating the corresponding term for the triplet state we note that $\overline{H}_{11} - \overline{H}_{33}$ is of *opposite sign.*

Q: Why so?

M: It may be that you have a splitting, the perturbation... You have two diabatic curves and if you have a splitting there, then you get two new states, one which is raised and the other that is lowered, and that's maybe the situation here.

Q: The situation of lower and higher are exchanged in the two cases.

M: Yes, that could be.

13. p. 19 1st column middle: "One possible alternative mechanism for the initial step is depicted in fig. 2. In it the transition at the first intersection is rate controlling and nonadiabatic, while that at the second intersection is adiabatic."

Q: When there is a sequence of processes nonadiabatic → adiabatic which one is the rate controlling one?

M: If you have a nonadiabatic and then the adiabatic has a high activation energy, then the adiabatic would be the rate controlling one. It depends on the activation energy and on the details that occur.

14. p. 19 2nd column bottom: "There is yet another possible mechanism which should be considered: If, for reasons not yet clear, the dependence of the ET rate on separation distance were considerably smaller than that typically (**M204**) found, there would be a possibility of an ET from $^1BChl_2^*$ to BPh through the intervening material, with no role for the BChl. However, the small singlet-triplet splitting of the $BChl_2^+BPh^-$ state appears to eliminate this possibility also; Due to the presence *of the* $^1BChl_2^*$ *state* the singlet $BChl_2^+BPh^-$ state would be shifted by an amount.... The triplet $BChl_2^+BPh^-$ is shifted by the $^3BChl_2^*BChl$ state...

Q: Remember that the electronic configurations of the three states model you consider in the paper are

1. $BChl_2^* - BChl - BPh$
2. $BChl_2^+ - BChl^- - BPh$
3. $BChl_2^+ - BChl - BPh^-$

M: States numbers 2 and 3 could be singlets or triplets. I suppose that there would be no role for the BChl. Suppose there would be no role for BChl, then you forget about BChl and you put there only those molecules that interact and BChl is omitted from the configuration you consider. I don't know why I consider it for the triplet, it may be that the BChl that is missing there, the BChl, just doesn't interact with the triplet, perhaps. In the case of the singlet you may not overlook the interaction, but in the triplet state then none of the wave functions involving the triplet acts on that BChl, maybe that's the reason.

15. p. 20 1st column middle: "**8. Effect of applied electric fields on the initial steps**"

Q: Please explain the effect of applied electric fields on ET.

M: I wrote a paper with someone from Japan . . . on the effect of electric fields . . . If you have an electric field, then you change your potential, the energy of interaction of the negative and positive charges with the electric field. When you put that in, effectively, approximately, if you use the point charge approximation, if you take, say, the dipole associated with the material there and multiply by the electric field . . . consider the dot product, then you have an energy of interaction, and so influence the energy before an electron transfers in a way different from the way it affects the energy of the final state, after the ET. That comes in and we treated it in the paper.

16. p. 20 1st column middle bottom: "We shall suppose that the C_2 axis of each close packed reaction center is oriented parallel to and antiparallel to the field . . . Following, bottom: "The imposition of an adverse field of 0.15 eV/nm parallel to the C_2 axis makes an adverse contribution to ΔG^0_{12} of 0.06 eV and to ΔG^0_{13} of 0.22 eV . . .

Following 2nd column "If the ΔG^0_{13} for the unrelaxed $BChl_2^+BPh^-$ at room temperature were about −0.19 eV in the quinone-free system, its value in the presence of this applied field would now become +0.03 eV."

Q: How do you get these numbers? Please explain the relationship between free energy changes and applied electric field in this case.

M: The numbers for the ΔG of interaction... well, first you obtain all these numbers if you impose an electric field and take a dipole moment, the direction parallel to the field should give you the 0.06. It is dipole time field, a dot product.

M238 Early Steps in Bacterial Photosynthesis. Comparison of Three Mechanisms

R. A. Marcus

NOTES

1. p. 389 middle: "One puzzle which has arisen,... concerns the apparent discrepancy between the fast forward rate of ET to form a radical pair $BChl_2^+BPh^-$ and the small singlet-triplet splitting of that pair."

Q: Can you explain the relation between reaction rate and singlet-triplet splitting?

M: The reaction rate would have the matrix element involving that radical pair. Assuming an H_{12} related to H_{13}, H_{23}, and so on, if one looks up an expression for singlet triplet splitting, in the paper there are relations between ST splittings and the matrix elements. If you look at that expression, then if the ST splitting is small that probably implies that some of those matrix elements are small, if you look at the forward rate, the matrix elements can't be too small... That was the essence of the paper that Michel-Beyerle wrote with me, she and Haberkorn, this is one of the motivations that led to that paper actually, the contrast between some splittings... if you look at the forward rate matrix element the effect is really fast... In other words if we went on the basis of some splittings which reflected interaction of wave functions, maybe at different sites, because the singlet–triplet splitting is probably one electron on one site and one electron on another site, and the amount of interaction of those two systems is reflected in the singlet–triplet splitting, and it's very small, probably meaning that the direct 1 to 3 matrix element is very small, but the reaction was very fast and so I think that's why we invoked superexchange. In other words, the singlet–triplet splittings, if I remember rightly, involve really the wave functions 1 and 3 but somehow I think probably superexchange may be involved in the singlet–triplet splittings, but *essentially you have two electrons one is on 1 and the other is on 3, and the spins can be opposite or can they be parallel. If there are no interactions the energy difference of those two things would be zero, if there is no interaction between 1 and 3, wouldn't matter which way one spin was pointing relative to the other, the spin up and down and... in an isolated environment is degenerate in its energy, it's not so in an external magnetic field, it is then only because of the coupling that you get a splitting between singlet and triplet, and apparently that splitting is so small that if you think of it in terms of the direct process, not counting superexchange, you couldn't understand why the reaction itself is pretty fast going, an electron jumping from 1 to 3. So that's why we invoked superexchange, it was to understand a magnetic problem and a fast reaction problem.*

2. Consider Eq. (5) on p. 391 (see Fig. 1)

$$\Delta E_{12}^{\text{vert}} = \lambda_{12} + \Delta G_{12}^{0} \tag{5}$$

Q: (1) Is this equation always valid in ET theory?

(2) Moreover: the magnitude of a typical electronic $\Delta E_{12}^{\text{vert}}$ is of the order of the electron volt. Which one is the range of possible values of H_{12} in nonadiabatic and in adiabatic ET?

M: (1) The equation is always valid. It depends on having parabolas with the same curvature.

(2) In the case of the adiabatic it can be, just ballpark, of the order of kT, 0.025 eV, in the case of nonadiabatic could be tremendously small, if donor and acceptor are separated by quite large distance, if H_{12} is

less than 0.025 the process is certainly nonadiabatic. By tremendously small I mean maybe a factor of 10 smaller. In principle, if you have a donor and an acceptor separated by very long distance then if there is superexchange the matrix element decreases by a factor of 3 for every Å increase in distance.

Superexchange goes back to some magnetic work of physicists about some ions, in the 30 s. Some solid-state systems.

3. p. 395 middle: "We have also considered elsewhere the implications of a superexchange mechanism for a $BPh^{-} \rightarrow BChl$ charge transfer band."

Q: Is the symbol $BPh^{-} \rightarrow BChl$ correct for a charge transfer band? Normally one uses symbols like $DA \rightarrow D^{+}A^{-}$.

M: I don't know if there is any convention there, but it says that if you have an electron it will finally end up on the BChl.

4. p. 395 middle: "Electric fields effects... it was deduced that with an adverse field of about 150 mV/nm the energy of state 2 and that of the (unrelaxed) state 3 at their equilibrium configurations would approach each other, if the situation at no field were as depicted in Fig. 2. If there were no 'unrelaxed' state or if the situation were, instead, that given in Fig. 1a, the energy of state 2 would still be higher than that of state 3 and there would still be no BChl bleaching. Thus, with a sufficiently adverse field there would be the possibility of observing $BChl^{-}$ and a bleaching of BChl if the mechanism in Fig. 2 prevailed."

M: In other words, if you lower that BChl appreciably, then you might see it. It may be that the unrelaxed state is just immediately after the ET, and then it relaxes, so maybe that's what I was talking about.

Q: In the last line of the paper there is a reference #18 which is not in the References list (which ends up with #17).

M: I guess some reference got chopped off.

CHAPTER 2

Early Steps in Bacterial Photosynthesis, Charge Transfer Absorption, Fluorescence Spectra and the Inverted Region, Reorganization Free Energy for ETs at Liquid–Liquid and Dielectric Semiconductor–Liquid Interfaces, Photosynthetic ET, ET in Dynamically Disordered Polar Medium, Dynamics of ET for a Nonsuperexchange Coherent Mechanism

Interviews on M244, M245, M249, M250, M251, M252, M253

REVIEW M244 Mechanisms of the Early Steps in Bacterial Photosynthesis and Their Implications for Experiment

R. A. Marcus

NOTES

1. p. 207, 2nd column middle: "λ_{13} is roughly about the same magnitude as this $-\Delta G_{13}^0$, *since the rate constant* k_f *of the reaction of* ${}^1\mathrm{BChl}_2^*\mathrm{BPh}$ *has a negative temperature coefficient.* (Compare Eqs. (1) and (2)

$$k = \frac{2\pi}{\hbar}\frac{1}{(4\pi\lambda kT)^{1/2}}|H|^2 \exp[-(\Delta G^0 + \lambda)^2/4\lambda k_B T] \qquad (1)$$

$$k = \frac{2\pi}{\hbar}\frac{1}{(2\pi|\Delta G^0|\hbar\omega)}|H|^2(\lambda/|\Delta G^0|)^{|\Delta G^0/\hbar\omega|} \times \exp[-(\Delta G^0 + \lambda)/\hbar\omega] \quad (2)$$

and a plot for k_f vs. λ similar to Fig. 1."

M: At that time there were at least two photosynthetic systems that had been looked at, one began I think with Virdis, and the other with Spheroides, there were two and maybe there are more, and the results showed different temperature coefficients at low temperature, one had a negative coefficient and the other didn't, I think it had a positive coefficient, so that using this explanation about the effect of temperature, one would explain just one of them and not the other. The idea under the explanation is that when you look at the overlap of the vibrational wave functions in the first and last electronic structure, that of the reactants and that of the products, . . . I think that the second excited state somehow overlapped better than the first with the ground state, and so when you raise the temperature you get less overlap, I think that was it, I don't remember the details though, but there was some point about the overlap of the initial wave function with the final wave function and how, because of the effect of temperature on exciting the initial state, there was less electronic overlap, less vibrational overlap, at the slightly higher temperatures, and that gave rise to a negative temperature coefficient. I think I have written about that in one of those articles, I'm sure I must have written about that because I thought about that at that time, in one of the photosynthetic articles. Basically it was a question of the ease of . . . suppose you start off in the ground state of the reactant and you look at the overlap of it with the ground state of the product, then you have some excitation and a certain probability of being in the upper state, and that overlapped less with a corresponding state of the product, something like that, so that meant that *the rate constant was less even though the temperature was someone higher.* Experimentally that was observed for one photosystem . . . oh, Virdis is the bacterial photosynthetic system I was thinking of. Anyway, one would have to look at the experimental data there, which go back many years. And I think in one of the articles I surely commented if anyone had thought at the time on this question of the overlap of the wave functions and *the main idea is that if you get less overlap at a higher temperature because of the nature of the wave functions then the rate would decrease instead of increasing with temperature.*

The transfer ultimately is from 1 to 3, with the 2 just being a *transient* sort of intermediate that is not even a real intermediate . . . the system never spends any real time there, so what counts in that particular argument is that if the temperature coefficient is approximately zero, then one parabola of the three parabolas is approximately intersecting the bottom of the first parabola. The only way you're going to get an activationless reaction, is when the products' parabola, which in this case is 3, intersects the bottom of the reactants' parabola which in this case is 1.

2. p. 209, 1st column middle: "Mechanism IV is one in which there are two sequential steps, i.e.,

$$ {}^{1}\mathrm{BChl}_2^{*}\mathrm{BChl\,BPh} - k_1 \rightarrow \mathrm{BChl}_2\mathrm{BChl}^{+}\mathrm{BPh}^{-} - k_2 $$
$$ \rightarrow \mathrm{BChl}_2^{+}\mathrm{BChl\,BPh}^{-} \ldots \text{"} $$

Q: In going from reactants to products either the excitation "*" disappears.

M: The excitation disappears. You go from an electronic excited state to the ground state. The $BChl_2$ loses an electron. So first the $BChl_2$ becomes a +, either if you have superexchange and a single process or you have consecutive things and you have two processes. The excitation has to go, absolutely. *If, as the first step, you transfer an electron from the excited thing to the BChl, you know, then the* $BChl_2$ *would have a (+) after it, BChl would have a (−) and the BPh would have nothing*, that'd be the first step. And then the (−) can go to the BPh except that the (+) may still be on the $BChl_2$. The (+) remains there, the BChl may never see a (+).

Typo: There is a typographical error then. The above reactions should be written

$$^1BChl_2^* \, BChl \, BPh - k_1 \rightarrow BChl_2^+ BChl^- BPh - k_2$$
$$\rightarrow BChl_2^+ BChl \, BPh^-$$

3. p. 209, 1st column bottom: "To explain the smallness of ΔE_{ST}, mechanism I has been modified by introducing a fourth state: an *internal $BChl_2$* charge transfer state, both as singlet and triplet $BChl^+$and $BChl^-$ states."

Q: What is an *internal* charge transfer state?

M: An internal ET in $BChl_2$ maybe involves a charge transfer between one BChl of the dimer and the other one. I assume so.

4. p. 210 2nd column top-middle: "The study of electric-field-induced fluorescence anisotropy,[18c]"

Typo: Misprint: reference 18c is not in the list.

M: When you have a field and polar molecules, the polar molecules get oriented somehow and if you look at the fluorescence, the fluorescence of those ordered molecules is different in different directions, the molecules are no longer random. The fluorescence is no longer uniform over the 4π region.

5. p. 210 2nd column middle: "*5.6 Electric Field Effect on ΔE_{ST} and k_T*

For most of the mechanisms, a substantial effect of the electric field on small energy denominators, such as that in Eq. (3) for $\overline{H}_{13}^{\mathrm{rp,T}}$,

$$\overline{H}_{13}^{rp} = H_{12}H_{23}/(H_{22} - H_{33}) + H_{13} \tag{3}$$

is predicted to occur. Existing arguments can be used to estimate an electric effect on ΔE_{ST}. *The effect on k_T itself is estimated to be small (cf. Fig. 1).*"

Q: How is it possible to infer the smallness of the effect from Fig. 1?

M: Maybe it refers to a Fig. 1 in some other paper... In Fig. 1 you can see what effect λ has on k_T. If instead of plotting vs. λ you plot vs. ΔG^0 there would be probably something similar. And if you look at the range of values... for the electric field effect one may see that it is going to change ΔG^0 very much... One should think of ΔG^0. And you are near the $\lambda = -\Delta G^0$.

6. Following: "*5.7. Effect of Electric Field on BChl depletion*

In the presence of a suitable electric field in an oriented reaction center, *the level of* BPh^- *might be raised sufficiently* that an actual BChl depletion might be observed."

Q: To which kind of "level" do you refer?

M: The level is the level of the energy relative to the excited BChl_2. If you make the product's energy very high, then you may not even reach that product. It is the level of the third electronic configuration to which BPh^- belongs. If you raise the level high enough, the ET might not go to BPh, it might go only as far as BChl.

7. p. 211 1st column middle: "*5.6 Possible Charge Transfer Band for* $\mathit{BPh}^- \rightarrow \mathit{BChl}^-$

The implications of mechanism I lead to a predicted charge transfer absorption BChl BPh^- $\rightarrow$ $BChl^-$ BPh if the mechanism is applicable."

Q: Is the symbol $BPh^- \rightarrow BChl^-$ an abbreviation for BChl BPh^- $\rightarrow$ $BChl^-$ BPh?

M: A shorthand.

8. Following, middle-bottom: "In particular, when *the Q_y transitions* of BChl and BPh were excited by an optical pulse, a transfer of the excitation to $BChl_2$ occurred within the time resolution of the apparatus, about 100 fs. While this process is an excitation transfer, the principles of excitation transfers and electron transfer have much in common. Perhaps some high frequency motion is available, at least for the excitation transfer, or perhaps *considerations of coherency vs. incoherency arise.*"

Q: What is it a Q_y transition? Can you say something about this point of coherency vs. incoherency?

M: If the transfer goes consecutively, if you have an intermediate for a very short time that would be an incoherent transfer. If on the other hand it goes by superexchange that would be a coherent transfer.

9. p. 211 2nd column middle: "Mechanism III (nonadiabatic/ adiabatic, $BChl^-$).... Thus far, in a one-dimensional calculation, mechanism III is also consistent with the absence of BCh depletion"

M: The one-dimensional calculation is just a preliminary calculation. In other words there are several bonds involved anyway, just treating all of them as one collective coordinate . . . that's a simplification by far.

M245 Relation between Charge Transfer Absorption and Fluorescence Spectra and the Inverted Region

R. A. Marcus

NOTES

1. p. 3079 2nd column middle: "As can be seen from (3) and (4),

$$k_{\mathrm{rp}}^{\mathrm{ET}}/k_{\mathrm{rp}}^{\mathrm{ET,max}} = (\varepsilon_\nu/\nu)/(\varepsilon_\nu/\nu)_{\mathrm{max}} \tag{3}$$

$$\Delta G_{\mathrm{rp}}^{0} = \Delta G_{\mathrm{rp}}^{0\,\mathrm{max}} + h\nu_{\mathrm{a}}^{\mathrm{max}} - h\nu_{\mathrm{a}} \tag{4}$$

the decrease of ε_υ/ν with increasing ν on the high frequency side of the charge transfer absorption maximum corresponds to the decrease of $k_{\mathrm{rp}}^{\mathrm{ET}}$ with increasingly negative ΔG^0, i.e. to the inverted region for the r → p ET reaction"

Q: (1) Did you first think of the inverted free energy effect because inspired by the maximum in CT spectra or because of playing with parabolas, downshifting the products parabola with respect to the reactants parabola?

(2) Were you the first to discover this relation between CT maximum and ET inverted effect?

(3) For some time the existence of the inverted free energy effect was not sure for ET reactions. But didn't the existence of the maximum in CT spectra strongly suggest the existence of the inverted free energy effect for ET reactions? Why was this suggestion of an existence not enough to assure of its effective existence?

M: (1) The second, not the first, I didn't make the connection with the absorption until much later, in fact, if I remember, Jortner made the connection to absorption.

(2) No, I believe that Jortner had that, but I am not 100% sure.

(3) It should have been but I think they didn't realize until late. The paper published in '89 has been written after the key experiments of '84 or '85. I didn't realize until those papers, actually. Maybe Jortner realized but he didn't tell anything about it in terms of evidence for the inverted effect. Here we deal with CT spectra, it was an interpretation, now the CT spectra is a type of ET but it doesn't need to be a complete ET, it is just a change in dipole, in limiting case they may involve a charge moving from one part to the other, but in general those CT spectra are extremely weak. The existence of the maximum in CT spectra doesn't automatically assure of the existence of the inverted free energy effect because a lot of charge transfers are just partial charge transfers, they are just not an electron going from here to over there, most are more restricted than that.

Q: Basically they are charge density transfers.

M: That's right, that's right.

2. p. 3079 1st column bottom: "Associated with the absorbance ε_ν at the indicated absorption frequency $\nu_s(q^\ddagger)$ is the probability factor $\exp[-\Delta G^\ddagger_{\mathrm{rp}}(q^\ddagger)/k_B T]$ discussed earlier. This $\Delta G^\ddagger_{\mathrm{rp}}(q^\ddagger)$ equals the vertical distance B'A' in Figure 1b"

Q: Does this exponential Boltzmann factor take care of the decreasing population with energy of the states which can absorb photons of $\nu_s(q^\ddagger)$ frequency?

M: Yes, the decreasing absorption is due to a decreasing population. Yes, you get a Gaussian distribution of the spectrum. If it would not look like a Gaussian, then the answer is no. For example, if you have discrete levels, then if you look at spectra the distribution is not necessarily Gaussian. If you have discrete levels you have to talk of energy and not just ΔG^0. When you have the discrete levels, then the distribution usually is more, I think, like the Poisson distribution, not a Gaussian. And if you are talking of going from one level transitioning to another level of products, then you have to look at the individual levels and the ΔG^0 associated with each of those levels, you see, and not to the overall ΔG^0.

3. p. 3080 1st column bottom: "We adapt the results derived elsewhere (**M194**) for ET reactions. Namely, the general semiclassical treatment of ref. **M194** is used for the cited modes, together with the molecular (statistical mechanical) form of dielectric unsaturation approximation. A quantum mechanical treatment is retained for any high-frequency modes. To this end we first convert the starting expression in ref. **M194**.

$$k_{\alpha\beta} = \frac{2\pi V^2}{\hbar}|\langle\varphi_f \mid \varphi_i\rangle|^2\delta(E_f - E_i + \Delta E^0) \tag{1}$$

to one resembling (7)–(11). Equation 1 of ref **M194**, after being multiplied by a Boltzmann factor and integrated over a small range of energies δE, can be written (in the notation of ref **M194**) as

$$k_{\alpha\beta}(T) = \frac{2\pi V^2}{\hbar}\frac{1}{\delta E}\sum_{\beta}(\Psi_f \mid \Psi_i)^2 e^{-E_i/k_BT}/Q_r \tag{12}$$

where the sum is over the initial and final states i and f for which $0 \leq E_f - E_i + \Delta E^0 \leq \delta E$. (The i of ref **M194** denotes the present n for electronic state r, and f denotes the present m for state p.)

Q: How you go from Eq. (1) of ref **M194** to Eq. (12)?

M: States are in a certain range and suppose you have degeneracy, and roughly you can get at that by summing over the discrete states and dividing by that range. In other words, that delta function . . . one way of expressing it is to write it as a sum over states divided by the range. It is roughly a way of expressing a delta function. The sum is over states whose energy difference is related by δE. The sum over i's means that you have to sum over initial states, that's where the Boltzmann factor comes in. The term $e^{-E_i/k_BT}/Q_r$ is the probability of being in the i state. When you think of a delta function, the delta function picks out states that are in a certain region divided by the width of that region and then that ratio goes to some constant value as the width of the region goes to zero. This is just another way of rewriting a delta function. If you have a delta function you can write it as a sum of states over a certain region divided by that region. Looking at many possible different limiting situations, one is like a limiting Gaussian, another way is just a step function. I mean a function that goes up, is constant for a little while, then comes down, and if you pick up the states in that region and then divide by the width of that region, you have essentially a delta function. That's really what it is. You can write a delta function in a million ways, anything that has a sharp peak.

4. Following: "Comparison with the present (7) and (8) for ε_ν shows that, apart from notation, the $2\pi V^2/\hbar$ in (12) is replaced by $C_a h$ (since $\delta E = h\delta\nu$) and ΔE^0 is replaced by $\Delta E^0 - h\nu$. With this change we can immediately apply (25) of ref. **M194**

$$k(T) = \frac{2\pi V^2}{\hbar} \sum_{I,F} \frac{|\langle F \mid I\rangle|^2 e^{-E_I/kT}}{Q_I(4\pi\lambda_0 kT)^{1/2}} \times \exp\left[-\frac{\lambda_0}{4kT}\left(1 + \frac{\Delta G^0_{IF}}{\lambda_0}\right)^2\right] \tag{25}$$

yielding (13):

$$\varepsilon_\nu/\nu = C_a h \sum_{n,m} |\langle\varphi_n^r \mid \varphi_m^p\rangle|^2 (4\pi\lambda_0 k_B T)^{-1/2} Q_{\nu r}^{-1} \times \exp[-\{\varepsilon_n^r + (\varepsilon_m^p - \varepsilon_n^r + \Delta G_{rp}^{0\,s} + \lambda_0 - h\nu)^2\}/4\lambda_0 k_B T]$$

where m and n now refer only to the high frequency modes. The φ's and energies ε's denote any high frequency vibrational wave functions and their energy levels.... $\Delta G_{rp}^{0\,s}$ denotes the free energy change for the r → p transition, apart from the enthalpy and entropy change due to excitations in the quantized modes."

Q7: What does it mean the superscript s in $\Delta G_{rp}^{0\,s}$ in Eq. (13)? Maybe "solvation"?

M: My guess is that s is intended to denote the solvation contribution to ΔG^0.

5. Following: "When the vibration frequency of those quantized modes is high, they make a negligible contribution to ΔS_{rp}^0 and to the thermal energy change (the energy change in excess of the $0 \to 0$ value) and $\Delta G_{rp}^{0\,s}$ becomes ΔG_{rp}^0."

Q: Why is the contribution negligible?

M: If you have a low frequency vibration, say for frequency $h\nu$ much less than kT, then *thermally* you would have many states occupied because the energy tends to distribute among the degrees of freedom in order to maximize the entropy, that means that you have high entropy contrary to the first case of high frequency vibration where the entropy is small, that's all, it's basically number of states available.

6. Following: "One feature of (13) and (14) is the occurrence of the ΔG^0, rather than the ΔE^0 in related expressions *derived under more restrictive assumptions.*"

M: That goes back to 1959, and in the theories of Levich and Dogonadze. In their treatment they didn't consider any low frequency modes, they considered just high frequency modes, so they had no entropy effects, they didn't have any ΔG^0 in there. That is an error that they made, because they only considered high frequency modes, so they would not have $\Delta G^0 + \lambda$ but instead a $\Delta E^0 + \lambda$. I had a long discussion about it with Levich before being able to convince him.

7. p. 3081 1st column middle-bottom: "We first consider the solvent inside the partially saturated shell and the *high frequency vibrations* and later include the solvent outside the shell. The model is chosen to be the same as that employed in ref 5a (Kakitani & Mataga...)."

p. 3081 2nd column top: "For any given quantized vibrational states m and n of electronic states p and r, respectively, and for an absorption or fluorescence frequency ν the value of x is the solution of the equation

$$h\nu(x) = G_p^s(x) - G_t^s(x) + m\hbar\omega_p - m\hbar\omega_r + \Delta G_{rp}^0 \qquad (16)$$

for the case of a single *high frequency vibrational mode*"

Q: Why a high frequency mode and not a low or medium frequency mode?

M: Because those would be treated classically. The high frequency modes you treat discretely. You lump together the low frequency modes.

8. Following: "The absorbance ε_ν divided by ν is given by

$$\varepsilon_\nu/\nu = C_a \sum_{m,n} |\langle\phi_n^r \mid \phi_m^p\rangle|^2 e^{-G_r^s(x)/k_BT} \times p_n(T)(dx/d\nu) \Big/ \int_{-\infty}^{+\infty} e^{-G_r^s(x)/k_BT}\,dx \qquad (17)$$

Q: Why does the factor $dx/d\nu$ appear in this formula?

M: If you want $d\nu$ in the formula, you multiply by $d\nu$ and divide by $d\nu$.

9. Following: "The absorption maximum can be found by finding the largest term in (17) and (18), by maximizing with respect to m and to x. The $dx/d\nu$ factor is *a constant* when $k_p = k_r$."

Q: Why so?

M: I think it has something to do with the following: when you look at an energy difference you might have, say, $k_r x^2 - k_p(x-a)^2$. Then, when $k_r = k_p$, cancel the x^2 term and now you have something linear in x, you see, and so the derivative of that is a constant.

10. Following: "The remaining x-dependent factor has a maximum absorption at $x = 0$, since $G_r^s(x)$ has a minimum there."

Q: Is the absorption maximum at the minimum free energy value due to the fact that at that minimum energy there is a higher population of systems capable of absorbing?

M: Yes.

11. p. 3082 1st column top: "An additional contribution to W_a is found by *expansion* of the S-dependent factor in (18)

$$|\langle\phi_0^r \mid \phi_m^p\rangle|^2 = S^m e^{-S}/\Gamma(m+1), \quad \text{with} \quad S = \lambda_i^p/\hbar\omega_p \tag{18}$$

as a Gaussian about its maximum at $S \cong m$. It is given by

$$W_a \cong 2[\lambda_i^p \hbar\omega_p(2\ln 2)]^{1/2} \quad \text{(contribution)} \qquad (21)\text{"}$$

M: You expand the exponent there as a quadratic and now you have a Gaussian. You stop at some quadratic term. You do a Taylor expansion of the exponent. It can be done in a much more sophisticated way, that happens in statistical mechanics a lot where you have the average of an exponential and you express that in term of what is called a cumulant expansion that amounts to getting an average and going up to the quadratic term of the exponent. I didn't realize in the early days that is what I was actually doing. I was always amazed at the quadratic thing that seemed to be working and I didn't realize that I was doing a cumulant expansion.

12. p. 3086 1st column middle: "Further examination of the widths... Indeed, it would also be interesting to study CS reactions... by studying... fluorescence quenching ETs... or by studying these quenching reactions in intramolecular systems, particularly under conditions of weak electronic coupling (*small CT-LE state mixing*)"

Q: What is the "small CT-LE state mixing" in parenthesis?

M: I think it is equivalent to weak coupling. If there is small coupling, then those states will be two distinct states, representing different electronic states. There is weak coupling between the A and the B that are involved. Then one state is going to $A^* + B$, another state is A^+ and B^-, in one case the excitation is localized at the A, in the other case the charge has been transferred. They go from one to the other because of electronic coupling between A and B. The *localized excitation state* LE is where you excite the electron on A. And the charge transfer state CT is where you have an A^+ and a B^-. Now, if that localized state there, that excited A^*, had zero coupling to B, you could never go to $A^+ + B^-$.

M249 Reorganization Free Energy for ETs at Liquid–Liquid and Dielectric Semiconductor–Liquid Interfaces

R. A. Marcus

NOTES

1. p. 1050 1st column middle: "Earlier, using a *charging path* to produce a system with a nonequilibrium dielectric polarization, we obtained a classical statistical mechanical expression for the free energy of a system having *longitudinal* polarization fluctuations (**M45**)."

M: Polarization can be transversal and longitudinal. When it is due to a charge is typically longitudinal. To understand this you should know about the electrostatics of longitudinal polarization. What is due to charges is typically longitudinal. There may be something about it, I don't know, in Russian books, maybe Ulstrup says something about longitudinal and transverse polarization. I know that some of the Russians have done something about that... I don't know about the book of Pekar from 1954. There are other articles written by Russians, so there is a lot of work on that.

2. Following: "The principal assumptions used were (1) linearity of the response of the medium to a change in electric field, (2) *a static treatment of the low frequency motions*, and (3) instantaneous response..."

Q: What do you mean by "static treatment of the low frequency motions"?

M: I mean that you are not using dispersion of frequencies that you are setting at zero frequency. You are using typically the static dielectric constant instead of the frequency dependent dielectric constant.

3. p. 1051 1st column middle: "The expression for Ψ at any field point (x, y, z), denoted by P, is given by (3) and (4) when these boundary conditions are satisfied:

$$\Psi(x, y, z) = \frac{q}{D_1}\left[\frac{1}{R_1} - \frac{D_2 - D_1}{(D_2 + D_1)R_1'}\right], \quad z > 0 \tag{3}$$

......

Here, R_1 is the distance *from the ion* to the field point P..."

Q: (1) Should it be "from the center of the ion"?

(2) At the beginning of the 1st column of p. 1051 you define the optical and static dielectric constants D_i^s and D_i^{op}, then in Eq. (3) D_1 and D_2 appear without superscript specification. Are they the static dielectric constants? See following question 5.

Erratum: In note (18) you refer Eq. (3) to Jackson's "Classical Electrodynamics" pp. 110–112, which is OK for the first edition of the book published in 1962, not for the 1975 second edition, the one you mention. There the same equations are on p. 148.

M: (1) Oh, yes, sure, it is the center of the ion.

(2) It depends, depending on what you are doing it may be one or the other.

4. p. 1051 1st column middle: "Equation (8) was derived... and Eq. (7) becomes"

Typo: Misprint: it is: "and Eq. (8) becomes."

5. p. 1051 2nd column middle: Eq. (8)

$$\lambda_0 = \frac{(\Delta e)^2}{2a}\left(\frac{1}{D_1^{op}} - \frac{1}{D_1^s}\right) - \frac{(\Delta e)^2}{2R}\left(\frac{D_2^{op} - D_1^{op}}{D_2^{op} + D_1^{op}}\frac{1}{D_1^{op}} - \frac{D_2^s - D_1^s}{D_2^s + D_1^s}\frac{1}{D_1^s}\right) \quad (R = 2d) \tag{8}$$

Was derived for the case of two dielectric media. It is useful to see what it reduces to in the case where phase 2 is a classical metallic conductor. In this limit both D_2's are replaced by infinity and Eq. (8) becomes

$$\lambda_0 = \frac{(\Delta E)^2}{2}\left(\frac{1}{D_1^{op}} - \frac{1}{D_1^s}\right)\left(\frac{1}{a} - \frac{1}{R}\right) \quad (R = 2d) \qquad (11)\text{''}$$

Q: The tables of dielectric constants do not (normally) list constants of metals while you write that such constants are really infinite. Can you explain this point?

M: I can try... We want the electric field in a metal to be zero. OK? Now, the dielectric displacement, which is directly related to any charge you put there, is equal to the dielectric constant times the electric field, is finite and if the electric field is zero the dielectric constant has to be infinite.

6. p. 1052 1st column bottom: "A dielectric continuum expression for w^r ... this result is found from Eq. (14) to be:

$$w^r = -\left(\frac{(e_1^r)^2}{4d_1 D_1^s} - \frac{(e_2^r)^2}{4d_2 D_2^s}\right)\left(\frac{D_2^s - D_1^s}{D_2^s + D_1^s}\right) + \frac{2}{R}\frac{e_1^r e_2^r}{D_1^s + D_2^s} \qquad (20)\text{"}$$

Q: Eq. (14) is

$$G^e = \frac{q_1^2}{2a_1}\left(\frac{1}{D_1} - 1\right) + \frac{q_2^2}{2a_2}\left(\frac{1}{D_2} - 1\right)$$
$$- \left(\frac{q_1^2}{4d_1 D_1} - \frac{q_2^2}{4d_2 D_2}\right)\frac{D_2 - D_1}{D_2 + D_1} + \frac{2q_2 q_1}{R(D_1 + D_2)}. \qquad (14)$$

Is there some reason for using the D's with s superscript in Eq. (20) and without it in Eq. (14)?

M: Maybe in Eq. (14) I wanted flexibility. In another paper, I express a nonequilibrium state in terms of an optical constant and a static constant. I wanted generality, so, if it is not specified, the meaning is that if I wanted an optical constant it would be an optical constant, if I wanted a static constant it would be a static constant. The problem arises when you express a nonequilibrium state in terms of some hypothetical states, you may remember a paper from 1955, that's what I did, I expressed the nonequilibrium state in terms of some hypotheticals.

7. p. 1052 2nd column top: "For distances greater than about 3 Å the interaction energy G^e of an electron and the surface is approximated by

$$G^e = -\frac{e^2}{4(d - d_0)} \quad (d > 3\,\text{Å}) \qquad (21)$$

.....

To fit some LEED intensity line shapes Eq. (21) was *joined linearly* to a given value at $d = 0$.[22b]"

Q: **Typo**: Misprint: There is no reference 22b.

M: Look a nearby references, maybe 22 or 23.

8. p. 1052 2nd column bottom: "It would appear that this type of treatment would also lead to *an image repulsion of an approaching ion in solution, instead of image attraction*"

M: Well, if the object on the other side is a metal, then the image reduces the electrical energy, the field is reduced and so there is an attraction because of the image forces. If on the other hand what is on the other side has a higher dielectric constant, then it is the reverse. The electric constant does not allow to go into it, so there is a repulsion.

9. p. 1053 1st column top: "**Reaction Rate Constant** (1) *Two immiscible Phases*... a rate constant k_r for reaction between two reactants, one in each of the two immiscible phases....

$$-\frac{dN_1}{dt} = k_r n_1 n_2 A \tag{25}$$

....

If κ is some Landau-Zener factor for the ET in this region of R (=distance between reactants) and ν is *some relevant frequency for the molecular motion*, then k_r can be written, approximately, as

$$k_r = \kappa \nu \upsilon e^{-\Delta G_r^{\neq}/k_B T} \qquad (26)"$$

Q: When one speaks of frequency one normally refers to a vibration. Is it the case here? Or is the "frequency" to be intended here with a different meaning from the usual? Can you say something about the possible "relevant" frequencies?

M: It is a sort of model that Sutin is using, a molecule goes into a little cage and then "vibrates," that's a model, there are other models. It is a sort of a half motion. I you think of a molecule coming in, it's like a half frequency.

10. p. 1053 2nd column top: "One question which arises is the rapidity of the response of the electronic and *hold* charge distribution"

Typo: Misprint for "hole."

11. p. 1054 1st column top:"

$$\tau \sim H_{12}/\nu(\lambda_i k_B T)^{1/2} \tag{32}$$

Thus, the smaller the electronic interaction energy H_{12}, the shorter the time τ spent by the system *in the vicinity of the potential energy curve crossing*"

Q: Is it so because the greater H_{12}, the greater *the region* "in the vicinity of the potential energy curve crossing" through which the system is to pass?

M: When you think of the region as being one that goes from where the diabatic curves are within H_{12} of their crossing point, if you think of defining the region that way just for the sake of argument, then when H_{12} is large the region is large, when H_{12} is small... yes.

12. Following: For example, if τ is relatively small, the *Eq. (9)* and 16 may be appropriate for λ_0."

Typo: Misprint: Eq. (9) refers to λ_i. It is Eq. (8), see below.

13. p. 1054, 1st column middle-bottom: "(*ii*) '*Undoped*' *Semicoductor-Liquid Interface.* As an example of Eqs. (8) and (11)

$$\lambda_0 = \frac{(\Delta e)^2}{2a}\left(\frac{1}{D_1^{op}} - \frac{1}{D_1^s}\right) - \frac{(\Delta e)^2}{2R}\left(\frac{D_2^{op} - D_1^{op}}{D_2^{op} + D_1^{op}}\frac{1}{D_1^{op}} - \frac{D_2^s - D_1^s}{D_2^s + D_1^s}\frac{1}{D_1^s}\right) \quad (R = 2d) \tag{8}$$

$$\lambda_0 = \frac{(\Delta e)^2}{2}\left(\frac{1}{D_1^{op}} - \frac{1}{D_1^s}\right)\left(\frac{1}{a} - \frac{1}{R}\right) \quad (R = 2d) \tag{11}$$

we compare them for the case that $D_1^{op} = D_2^{op}$, $D_1^s = 10$ *(semiconductor), and* $D_2^s = 40$."

M: 10 is about right, dielectric constants for semiconductors vary between 5 and 10. Something around there. Semiconductors don't have a lot of heavily polarizable molecules. A semiconductor doesn't have a lot of polarizable molecules like water.

14. p. 1054 2nd column middle: "(*iv*) *Nonlocal Dielectric Response.*

Q: What is it meant by "nonlocal dielectric response"?

M: That is discussed in Russian literature, you see a lot of it in some of the papers I think of Kuznetsov. That runs something like this: when is local, the dielectric response is the dielectric displacement times electric field, when it is nonlocal then the dielectric displacement is some integral of some functions of r and r', times electric field at r' integrated over r′. That's all nonlocal. And I think in the way you treat the Fourier transform for the nonlocal you get a k-dependence. You get additional r-dependence... if you look at the Russian literature, then you see that the Russians have a function of ω and k, and the k is the Fourier variable corresponding to r, so if you have something which depends on r, then you get a k-dependence, essentially, when you transform. So, it would be good to look at the Russian literature... they look at the evolution with time, where the ω-dependence appears, and how the k-dependence is and, for example, if you have a dielectric constant that depends on distance, then I think you get a k-dependence, you know. Normally I assume a constant dielectric constant, independent of space, but when you take into account the spatial dependence of the dielectric constant and you do all the Fourier analysis, then you get dependence on the Fourier component associated with r, which is k, instead of just the Fourier component associated with time which is ω, but for that is best to look at Russian literature.

15. p. 1055 1st column top: "For the case where there is more than one phase some assumption for the nonlocal dielectric constant $\varepsilon(\mathbf{r}, \mathbf{r}')$ in the vicinity of the interface is needed. One assumption in the literature, a '*specular electron reflection ansatz*,' is that it correlates only points $\mathbf{r}$, $\mathbf{r}'$ in the same phase and is zero otherwise."

M: If the polarizability doesn't extend beyond the surface separating the phases, only ions in one phase polarize in that phase, the polarization is then specularly reflected.

Q: Is there then no polarization in the other phase?

M: Not due to an ion in the first phase. If there is an ion in the second phase it polarizes in the second phase. Apparently some people in the literature assume that.

16. Following: "This approach has been used for systems where one phase is a solid.... For a system of two dielectrics, the same approach yields an obvious inconsistency in the limit where the two dielectrics have the same properties. In this case the result does not reduce to the correct limit for a single-phase behaviour, for which $\varepsilon(\mathbf{r}, \mathbf{r}')$ correlates all pairs of points, not just those on the same side of a now only imagined boundary (*While the important property is* $\varepsilon^{-1}(r, r')$, *it too would be incorrect in the cited limit, since* $\varepsilon(r, r')$ *is presumed to have a unique inverse.*)"

M: If you have something at r' that affects something at r then something at r would affect something at r', that would be the inverse. In general, one would have a tensor. So maybe $\varepsilon(r, r')$ is just some kind of tensor... when you are relating two vectors, then what you have is a tensor.

17. p. 1053 1st column top: "*(i) Two Immiscible Phases*. We calculate first a rate constant k_r for reaction between two reactants, one in each of the two immiscible phases. Its units can be defined via

$$-\frac{dN_1}{dt} = k_r n_1 n_2 A \tag{25}$$

.... The units of k_r are seen to be cm^4 molecule^{-1} s^{-1}. We let υ ... be a 'volume' (units of cm^4) such that *in a unit area of interface* the center-to-center distance of the pair of reactants lies in $(R, R + \Delta R)$, where ΔR, defined in the Appendix, is the region over where there is a significant contribution to the ET process, and where reactant 1 lies wholly in phase 1 and reactant 2 in phase 2... then k_r can be written, approximately, as

$$k_r = \kappa \nu \upsilon e^{-\Delta G_r^{\neq}/k_B T} \tag{26}$$"

p. 1055: "**Appendix: Derivation of Eq. (25)**

We calculate here the "volume" υ per unit area of interface defined as in Eq. (A1), such that the reactant 1 lies wholly in phase 1 and reactant 2 lies wholly in phase 2. Coordinates z_1, θ, Φ (Φ is an angle, not shown, about the z_1 axis), and R are introduced, as in Figure 1a...Calculation of the "volume" υ involves *finding all configurations of the pair, per unit area of interface*...At fixed z_1 and R, the area element $R^2 \sin\theta d\theta d\Phi$ is integrated from...The final result is independent of the (x, y) position along the plane *and so is a constant in the unit area*.... We thus have (A1):

$$\begin{aligned}\upsilon &= \int_{R=a_1+a_2}^{\infty} \int_{z_1=a_1}^{R-a_2} \int_{\theta=0}^{\theta_{\max}(z_1,R)} \\ &\quad \times \int_{\Phi=0}^{2\pi} R^2 \sin\theta d\Phi d\theta dz_1 dRk(R)/k(R_{\max}) \\ &= \pi \int_{R=a_1+a_2}^{\infty} (R-a_1-a_2)^2 Rk(R)dR/k(R_{\max})\end{aligned}$$

M: Imagine that you have a fixed unit area there, imagine that you have particle 1 on one side and particle 2 on the other. With respect to that fixed unit area consider all possible configurations of particles 1 and 2, and consider the line joining 1 and 2 that passes through that surface of unit area. So, you are looking at all possible configurations joining 1 and 2 such that line of intersection with the plane crosses that unit area. That's the basic idea.

18. p. 1054 2nd column middle: "*(iv) Nonlinear Dielectric Response*...Apparently there is, as yet, no direct experimental measurement of nonlocal dielectric response parameter for polar solvents (e.g. a determination of an *orientational analogue* of the slow neutron scattering determined structure factor).

M: The neutron scattering diffraction gives you the k-dependence, a kind of wavelength dependence of correlations, and here we would be looking at the wavelength dependence on orientations, you see, and we don't have that or we didn't have that then and may we don't have that even now.

REVIEW M250 Theory and Experiment in Photosynthetic ET

R. A. Marcus

NOTES

1. p. 4 bottom: "Some of the reactions observed in the bacterial reaction center include . . . (3.1)–(3.9) . . .

$$\mathrm{BChl_2^*BChl\,BPh} \xrightarrow{k_1} \mathrm{BChl_2^+BChl^-BPh} \tag{3.1}$$

$$\mathrm{BChl_2^+BChl^-BPh} \xrightarrow{k_2} \mathrm{BChl_2^+BChl\,BPh^-} \tag{3.2}$$

$$\mathrm{BChl_2^+BChl\,BPh^-} \rightarrow \mathrm{BChl_2^*BChl\,BPh} \tag{3.3}$$

$$\mathrm{BChl_2^+BChl\,BPh^-} \leftrightarrow \mathrm{(BChl_2^+BChl\,BPh^-)}^T \tag{3.4}$$

$$\mathrm{(BChl_2^+BChl\,BPh^-)}^T \xrightarrow{k_T} \mathrm{BChl_2^T BChl\,BPh} \tag{3.5}$$

$$\mathrm{BChl_2^+BChl\,BPh^-} \xrightarrow{k_S} \mathrm{BChl_2BChl\,BPh} \tag{3.6}$$

$$\mathrm{BChl_2^+BChl\,BPh^-}Q_A \xrightarrow{k_3} \mathrm{BChl_2^+BChl\,BPh\,Q_A^-} \tag{3.7}$$

$$\mathrm{BChl_2^+BChl\,BPh\,Q_A^-} \xrightarrow{k_4} \mathrm{BChl_2BChl\,BPh\,Q}_A \tag{3.8}$$

$$\mathrm{BChl_2^+BChl\,BPh\,Q_A^-} \rightarrow \mathrm{BChl_2^*BChl\,BPh\,Q}_A. \tag{3.9}$$

. . . . . . Reaction (3.4) is *treated in a coherent manner (hyperfine induced) rather than with the usual chemical kinetics rate constant formalism*"

Q: (1) No rate constant is shown for reactions (3.3) and (3.9). Is it so because the rate constants of those steps are unknown?

(2) Please explain this "coherent manner" way of treating Eq. (3.4).

M: (1) In Eq. (3.3) *the separated ions* go back to form the excited state, and that is an uphill reaction and maybe it will occur at very slow rate. If the reaction is, for example, 2.5 eV uphill, its rate constant would be proportional to $e^{-2.5/0.025}$, proportional to e^{-100}. So, going back would be uphill, and that probably would make it, you know, very slow. Equation (3.9) would be even more uphill, because there is an extra quinone and that would be vastly uphill. Reaction (3.9) would take a long time to occur. *It could occur if there is another quinone. If there is something else, if you block off a competing reaction by removing that second quinone then you would have to consider it if you are looking at very long time scale.* I didn't consider the reactions because too slow.

(2) If you have something going from A to C by a B, and B is not an actual intermediate but B is simply such that intermixes the wave functions of A and C and that's a coherent transfer: it is superexchange.

The first step is OK, that's right, Eq. (3.1) is OK, and Eq. (3.2) is OK, that'd be the second step, if you had a two-step process, but I don't know why we had the other steps, was it part of a more complete thing or what? Equation (3.3) is discussing the possible back reaction, but I think that's supposed to be very small.

"*the separated ions*": You can say just the ions . . . the word separated is unnecessary and there is no actual separation that occurs during the process, that's just a way to say that the ions are separate, but it's an unnecessary word.

"Reaction (3.9) would take a long time to occur." Yes, that's because it's a long ET, and it's also uphill, you know, it would be, both thermally and in distance, adverse. I mean one can neglect its occurrence, it's so uphill energetically and besides the ET is so far.

If you have another quinone, if you have something which blocks . . . if you block off the competing transfer which is transfer from one quinone to the other, if you block that off, then the process has no alternative but to go backward, but it might take a century to do something. Because it's uphill.

2. p. 5 bottom: "The only systematic type of study of varying ΔG^0 on a k_r has been through the effect of external electric fields ... The interpretation of those effects (the coupling to the field is via the dipole change accompanying reaction) is not simple. *The modification of the external field by the system itself* ("*the internal field*") is imperfectly known."

M: You have an external field acting on the dipole plus there is the effect that the field produces on everything else there. Everything else gets polarized and effects the local field at that point. One should also consider the internal field of the molecule.

3. p. 7 middle: "... the suggestion that ... the k_T for reaction (3.5) and the ΔE_{ST} of reaction (3.4) occur via a superexchange (**M223**, **M145**). Being *a second order phenomenon*, the latter would yield a relatively small k_T and ΔE_{ST}."

Q: "Second order", I guess, in the sense of perturbation theory?

M: Yes, that's right.

4. p. 7 middle-bottom: "Equation (2.1)

$$k(r) = \frac{2\pi}{\hbar}\frac{|H_{DA}|^2 e^{-\Delta G^{\ddagger}/k_B}}{(4\pi\lambda k_B T)^{\frac{1}{2}}}, \quad \Delta G^{\neq} = (\Delta G^0 + \lambda)^2/4\lambda \quad (2.1)$$

can be used for k_T with $\lambda \cong -\Delta G^0$, since k_T depends very little on temperature."

M: Maybe that's at low temperatures.

M251 ET in a Dynamically Disordered Polar Medium

W. Nadler and R. A. Marcus

NOTES

1. p. 69 2nd column top: "The rate constant k_e used in Eq. (1)

$$k_{ET}(\tau) = \frac{k_e}{1 + I k_e \tau} \quad (1)$$

is the *semiclassical equilibrated* ET rate constant

$$k_e = \nu_q \exp(-\Delta G^* / k_B T \quad (3)\text{''}$$

p. 77 Ref. (29): "Solvent relaxation time-dependent rate constants of the form of Eq. (1), or its limiting form for $k_e\tau$ large . . . They differ in what approximate expression is employed for the *equilibrated rate* (semiclassical, quantum-mechanical) . . .

Q: What do you mean by semiclassical or quantum mechanical "equilibrated" rates? Are they maybe the rates when the relaxation time τ of the solvent is equal to zero?

M: You consider the equilibrated rate when you pretend that the solvent relaxation is very very fast.

Q: So τ equal zero, basically.

M: Yes.

2. p. 69 2nd column middle: "It has been demonstrated theoretically that a non-vanishing contribution λ_i . . . can lead to considerable *deviations from a single-exponential* time behaviour . . ."

Following, bottom: ". . . a *nonexponential* time behaviour . . . may be due not only to a competition between fast intramolecular and slow solvational degrees of freedom . . ."

Q: Do you mean that "nonexponential" really means non-*single*-exponential, that is, *multiexponential*? Or a function different from an exponential function?

M: Yes, that's nonsingle exponential. He found that there is a power low, I remember. I think Nadler found it to be a power law. Actually you can make a power law multiexponential. And a multiexponential that covers a certain range you can make a power law.

3. p. 70 2nd column top: "In many complex systems, particularly ones that show glass-like behavior, the autocorrelation function of a quantity x, with a *time scale* τ *of the relaxation,*

$$C(t/\tau) = \langle x(t)x(0)\rangle / \langle x^2\rangle \qquad (4)\text{''}$$

Following: "An example for the functional form of $C(t/\tau)$ is the Davidson-Cole (DC) form

$$C_{DC}(t/\tau_0) = \Gamma(\beta, t/\tau_0)/\Gamma(\beta) \qquad (5)$$

.... β and the time scale τ_0 are the only parameters..."

Q: (1) On p. 69 you define the τ appearing in Eq. (1) as "longitudinal *dielectric relaxation time* of the solvent."

(2) Are the terms "time scale of the relaxation" and "relaxation time" equivalent? Is τ really a *nonlocal dielectric relaxation time scale* valid throughout the system?

(3) Does the parameter β have a physical meaning? It appears also in the following Davidson-Cole form of the frequency-dependent dielectric constant

$$\frac{\varepsilon_{DC}(\omega) - \varepsilon_\infty}{\varepsilon_0 - \varepsilon_\infty} = \int_0^\infty e^{-i\omega t}\left[-\frac{d}{dt}C_{DC}(t/\tau_0)\right]dt = \frac{1}{(1 + i\omega\tau_0)^\beta} \qquad (6)$$

M: (1) That τ is one relaxation time. If you have the Davidson-Cole, I guess the behavior is more complicated than with just a single relaxation time.

(2) They are more or less equivalent except that relaxation time might imply a single time whereas the time scale of the relaxation could imply a broad time. When you have a Debye-like behavior you have a single exponential. Time scale is not rigorous, you might look to the broad range of relaxation times.

Q: So it is in the same sense in which you speak of femtosecond time scale.

M: Yes, except that in this case the time would be much longer.

(3) I don't think so. There are various sorts of empirical times of relaxation that are not single exponential, and I don't think that a single parameter would appear in some of the β. It is just giving you a range of time scale. If $\beta = 1$, that would be single exponential. This is just a way of representing something, I think empirically.

4. (i) p. 70 1st column top-middle: "*Static* models, based on inhomogeneity of the medium, as well as *dynamic* models describing complex *local dynamical processes*, have been both employed as possible descriptions of such relaxation behavior of fluctuations";

(ii) p. 70 2nd column bottom: "In the static disorder approach, it is assumed that in a medium which shows anomalous relaxation the *local relaxation behavior of the correlation function* $\langle x(t)x(0)\rangle$";

(iii) p. 70 2nd column bottom: "However, . . . the *local relaxation time* τ' varies within the medium according to a distribution function $g(\tau')$."

(iv) p. 71 1st column middle: "the *local relaxation of x*"

Q: Apparently the adjective "local" may refer to three different things: a local point in the medium, a local value x of the polarization or a local relaxation time.

M: Just to a local region. You may have a set of local regions each of which has a different relaxation time.

5. p. 71 1st column middle: "Denoting by υ the deviation of this new degree of freedom from its average value, we can now write down a simplified stochastic model for the *local relaxation of* x, the Fokker-Planck equation

$$\frac{\partial}{\partial t} P(x, \upsilon, t) = \left[\frac{1}{\tau_x(\upsilon)} \mathbf{L}(x) + \frac{1}{\tau_\upsilon} \mathbf{L}(\upsilon) \right] P(x, \upsilon, t) \qquad (9)$$

for the probability distribution $P(x, \upsilon, t)$ of fluctuations. The time scale τ_υ of the fluctuations of υ could be the time scale of the *free volume* or hydrogen bond *fluctuations*; $\tau_x(\upsilon)$ *is the* υ*-dependent time scale of the fluctuations of* x."

p. 72 1st column bottom: "In the dynamic disorder model, the polarization fluctuations are coupled to the υ-fluctuations via Eq. (9), and a treatment of the reaction problem then involves Eq. (A5).

$$\frac{\partial}{\partial t} P_r(x, \upsilon, t) = \left[\frac{1}{\tau_x(\upsilon)} \mathbf{L}(x) + \frac{1}{\tau_\upsilon} \mathbf{L}(\upsilon) - k(x) \right] P_r(x, \upsilon, t) \qquad \text{(A5)}$$

In the present paper, the simplified solution is given in which the approximation Eq. (1)

$$k_{ET}(\tau) = \frac{k_e}{1 + I k_e \tau} \qquad (1)$$

is used to describe the effects due to polarization fluctuations on the *local ET processes*. The *local rate* for the ET is given again by

$k_{ET}(\tau)$, but now with τ replaced by $\tau_x(\upsilon)$. The υ-fluctuations then give rise to the *reaction-diffusion equation*:

$$\frac{\partial}{\partial t}P_r(\upsilon,t)=\left\{\frac{1}{\tau_\upsilon}\mathbf{L}(\upsilon)-k_{ET}[\tau_x(\upsilon)]\right\}P_r(\upsilon,t), \qquad (15)$$

which describes the time behaviour of the reactant distribution $P_r(\upsilon,t)$."

p. 75 2nd column middle: "**APPENDIX** A model for the coupling of ET processes to polarization fluctuations of Debye form with *relaxation time* τ that also allows for any (assumed rapid) vibrational reorganization is given by the reaction-diffusion equation

$$\frac{\partial}{\partial t}P_r(x,t,\tau)=\left[\frac{1}{\tau}\mathbf{L}(x)-k(x)\right]P_r(x,t,\tau) \qquad (\mathrm{A1})$$

for the reactant distribution function $P_r(x,t,\tau)$. Equation (A1) describes a polarization-dependent (i.e. x-dependent) ET reaction process, given by the x-dependent rate constant $k(x)$, which is coupled to x-fluctuations..."

Q: (1) In Eq. (A1) there is a dependence on τ while in Eqs. (15), (A5), and (9) there is no probability dependence on τ_υ and $\tau_x(\upsilon)$.

Q: (2) Which kind of fluctuations are the "free volume fluctuations"?

M: "free volume fluctuations": If you think of the molecules, there is a little space between them, right? Let me call that the free volume, the molecules aren't standing still, so that free volume is fluctuating.

Now, as far as the Eq. (1), what is the dimension of L? According to that, L should be dimensionless, is it? Yes it is, the way L is defined, at least looking at Eq. (15), it had to be dimensionless. There has to be a τ in Eq. (A1) because L is dimensionless. In fact, it has to be, if you go back earlier, if you look at Eq. (9), then you see that if you look at the left hand side and to the right hand side L is dimensionless.

I'd have to see what L(v) and L(x) are. He's introduced then an additional coordinate besides x, so one not only has x but one has fluctuations in something which we would call disorder, and so you have an extra term there beside L(x). So, somehow there is a diffusional spatial term associated with the disorder, as well as a diffusional spatial term associated with the original thing. So, he is just generalizing the case where you don't have the v in there,... an analogous equation where you have a diffusional motion of disorder, so to speak, as well as the system itself. He's made the probability distribution a function of the coordinate and time, which we had before, plus some disorder coordinate. So what he's done is just introduce a disorder coordinate. And he has to it consistently for everything, which has done in Eq. (9). It's a kind of phenomenology. The τ in Eq. (A1) doesn't have a v in it, the τ in Eq. (A1) is related to τ_x in Eq. (9). Because Eq. (A1) doesn't have a v, doesn't have those kind of disorder fluctuations in it. If you look at A1, I think it would be dimensionally incorrect if you remove the τ. The L, we decided on looking at the initial equations is dimensionless, so you have to have a τ in there. You see how Eq. (9) is an add off generalization of Eq. (A1), has the same sort of functional form, except that it doesn't have the rate constant in it. So Eq. (A5) is the add off generalization of Eq. (A1) to include additional disorder, a disorder coordinate.

6. p. 72 1st column top: "These local fluctuations, in turn, determine the dynamics of the ET process. From a *phenomenological* point of view, a mathematically (*but not necessarily physical*) correct description of the local fluctuations of a quantity is sufficient for the analysis of the effects that these fluctuations have on reaction processes that are coupled to them."

Q: Do you mean here that a good mathematical model may be able to reproduce the phenomenon but can be physically wrong?

M: Yes, it can be without a physical basis.

7. p. 72 1st column middle

$$\text{“}Q(t) = \frac{\sin(\pi\beta)}{\pi}\int_0^1 dx x^{\beta-1}(1-x)^{-\beta} \times \exp\left(-\frac{k_e t}{1+Ik_e\tau_0 x}\right) \quad (14)$$

From this equation, it can be readily seen that, *using* k_e^{-1} *as time scale*, the reaction process depends on the reaction barrier parameter $\Delta G^*/k_BT$ and the time scale τ_0 of the fluctuations only via the product of parameters $Ik_e\tau_0$.”

M: That $\Delta G^*/k_BT$ is kind of redundant if you use k_e^{-1} as a time scale. You see, k_e depends on $\Delta G^*/k_BT$. The main point is that if time is given in units of k_e^{-1}, $k_e t$ becomes a new time, let's call it θ, so the survival probability in terms of that new time θ depends only on $Ik_e\tau_0$. The other way of phrasing it is that if you introduce a new time scale in units of k_e^{-1} then it depends on that new time only via that... OK, fine.

8. p. 72 2nd column bottom: “For small values of $k_e\tau_0$ the ET is single-exponential in the dynamic model... Only for large values of $k_e\tau_0$ does the time behaviour of the ET become *nonexponential* (as in Fig. 1a). There is *a noticeable difference* in the behaviour of static and dynamic models and for different values of $k_e\tau_0$...”

Q: The difference among the curves representing $Q(t)$ vs. t is noticeable but don't all the curves in Fig. 1 look pretty much exponential or exponential-like?

M: There is only one curve in Fig. 1 which is exponential, the others show some deviation, at long time they go over to a straight line, but there is some transition period in which they are not.

Q: Basically they are just deviations from exponentials

M: Yes, that's right, in other words if you look at the longer times on all of them, then they look like straight lines, so what you have then at very long times is exponential, yes, that's right. If you look at the lowest curve, on the first figure, the upper figure, then that's pretty much an exponential ... then it begins to get deviations from the exponential, you go to some transition period when you go over to a different exponential, and, if he didn't comment on which different exponential was, that was too bad, he should have said ... *Deviations near zero* That's right.

At shorter time you have deviation from exponentials, at longer times you have exponentials. If you look at the longer times, if you put your ruler along them you see they look like straight lines. But you see that at shorter times you have deviations from that, so you really have some function times an exponential, it's the way it's described, and where the exponential isn't changing much at short times, that other function is. But anyway, the main point is that asymptotically the solution is such that it goes over to exponentials and maybe Nadler commented on that in the paper. Deviations near zero that is the deviations from the exponential at short times.

9. p. 73 2nd column top: "Two natural quantities for such an analysis are the short-time rate constant

$$k_s = -\left.\frac{d}{dt}\right|_{t=0} Q(t) \qquad (18)\text{''}$$

Q: Is the general definition of a time dependent rate constant $k(t) = -\frac{dQ(t)}{dt}$?

M: Yes, that's a good definition of a time dependent rate constant.

That formula is OK. That's fine, good. Normally I would have used $k(t) = -\frac{1}{Q(t)}\frac{dQ(t)}{dt}$. You see, at $t = 0$, $\frac{1}{Q(t)}$ is unity, so the formula in Eq. (18) is a special case.

10. Following: "and the mean reaction time

$$\tau_a = \int_0^\infty dt\, Q(t). \qquad (19)\text{''}$$

Q: Intuitively I understand that the time intervals are weighted by the unreacted fraction $Q(t)$. But can you give a rigorous demonstration that Eq. (19) gives the average reaction time?

M: It depends on the definition of average reaction time. In other words, if a reaction time is a *qualitative* statement and you can't prove a rigorous definition, you can't assign a rigorous definition to it. You can only assign some formula and then you call it that. Supposing the $Q(t)$ was a single exponential, then you would find that τ that would be related to $1/k$, where k is a rate constant, it would be rigorous when it is single exponential. In other words, it gives you just a rough time scale, it is not a rigorous statement. It is OK if the reaction time is defined as 1/rate constant. You see, a question of definition comes in.

11. p. 73 2nd column middle: "We note that for any functional form $g(\tau')$ of static disorder the mean reaction time τ_a has the general form:

$$k_e\tau_a = 1 + Ik_e\langle\tau'\rangle_g, \tag{22}$$

where $\langle\tau'\rangle_g$ is the *mean relaxation time* of the fluctuations.... Since $\langle\tau'\rangle_g$ is the *time scale* of the fluctuations..."

Q: Here we have *time scale* = *mean time*. Is it always so for the definition of time scales?

M: You see, they are both loose words, if you say that a process is occurring in a certain time on the average, so that's a time scale, loose words.

12. p. 73 2nd column bottom: "It can also be seen in Fig. 2 that the short-time rate constant from the dynamical model obeys a power law

$$k_e/k_s \propto (Ik_e\tau_0)^{\beta''} \tag{24}$$

for large values of the product $Ik_e\tau_0$."

Following next page, 1st column top: "...the asymptotic behaviour of the short time rate constant..."

M: "asymptotic behavior" means behavior "for large values."

13. p. 76 2nd column top-middle: "We note that the approximation in Eq. (A8)

$$\begin{aligned}Q(t) &= \int_{-\infty}^{\infty} d\upsilon \int_{-\infty}^{\infty} dx \\ &\quad \times \exp\left[\left\{\frac{1}{\tau_x(\upsilon)}\mathbf{L}(x) + \frac{1}{\tau_\upsilon}\mathbf{L}(\upsilon) - k(x)\right\}t\right] P_0(x)P_0(\upsilon) \\ &\approx \int_{-\infty}^{\infty} d\upsilon \exp\left[\left\{\frac{1}{\tau_\upsilon}\mathbf{L}(\upsilon) - k_{ET}[\tau_x(\upsilon)]\right\}t\right] P_0(\upsilon)\end{aligned}$$

is rigorous only in the case that the υ-fluctuations are much slower than the reactive transitions along the x-coordinate...Since $k_{ET}[\tau_x(\upsilon)]$ is largest in the *half-plane* $\upsilon < 0$, the main contribute to the reaction comes from that half-plane."

Q: Here you consider an (x, υ) plane not previously defined.

M: That's something that Nadler wrote, I didn't give much attention to it, that is something that I should have looked at more carefully, it looks so new to me, I didn't look at that part of the paper carefully.

M252 Dynamics of ET for a Nonsuperexchange Coherent Mechanism

R. Almeida and R. A. Marcus

NOTES

1. p. 2973 1st column top-middle: "...but B and A are so strongly coupled electronically that the entire transfer occurs *coherently*, rather than *incoherently, in two successive steps.*"

Q: Which is the difference between "coherent" and "in one step"?

M: In one step is a little bit dated because if you say "in one step" it can mean that the electron goes from the start to a finish without ever knowing that there is an intermediate thing there, not even *sensing* it. Now, that would be a one step process. But as soon as we say "coherent" it immediately implies that the wave function *senses* that intermediate there, but of course it is not standing any real time there, when it stands some real time there then you have an incoherent transfer. The "coherent" refers to a wave function. In fact, if you have to describe the process by a wave function instead of by populations, only populations.

2. p. 2973 1st column middle: "The question that arises is whether or not for the third mechanism Eq. (1.1)

$$B^{-}_{\max} = (k_1/k_2)^{k_2/(k_2-k_1)} \tag{1.1}$$

can still approximately represent the data but with a considerably enhanced value of k_2, a value *substantially greater* than *the maximum* k_2 *for an adiabatic electron transfer.*

Q: Please give order of magnitudes intervals of rate constants for

(1) Nonadiabatic ET's;
(2) Adiabatic ET's;
(3) "Beyond" adiabatic (?) ET's.

M: For both adiabatic and nonadiabatic the potential could have a big activation energy, so the rate constant for both can be extremely slow if there is an activation energy. So let's focus on the case then where there is no activation energy, in other words on the preexponential factor. Now, for a first-order adiabatic reaction between two reactants fixed aside, it will be of the order of 10^{12}, 10^{13} per second. If it were not adiabatic, I would say that it would be of a factor 100 lower than that, then it would be clearly nonadiabatic and any process in between you may want to describe by one or the other, it would be a process in between. That brings us now to beyond adiabatic: "Adiabatic" implies that there is some nuclear motion needed, as does "nonadiabatic," *but there are some transfers where you can inject an electron from a dye into a semiconductor and there is no nuclear rearrangement necessary* so there you can get up to, you know, a few femtoseconds in other words a rate constant would be 10^{14}, 10^{15} almost, 10^{14} actually. And there are some of those. That is where you inject directly into a semiconductor, there are no rearrangements necessary.

3. p. 2973 2nd column top: "(To design a theory that in one limit would yield *two incoherently connected consecutive steps*, a dissipative term accompanying the motion *in the* D^+B^-A *stage* would also be included..."

Q: (1) When are two consecutive steps coherently or incoherently connected?

(2) Why should the dissipative term be included in the D^+B^-A stage?

M: (1) The two steps are coherently connected when you essentially don't see the intermediate, I mean the intermediate, so to speak, doesn't really have a real existence. If energy is too high it doesn't really have an existence. The wave function though is *amplified by its presence*, the wave function of the donor and the wave function of the acceptor are amplified by its presence so it is playing a role in a coherent transfer, but there is no actual intermediate, there would be too small a probability for its presence.

(2) Well, I don't know whether it has to be included, I'd have really to look at that, because I would imagine that even if you didn't use dissipation... you know, when you want to go over from a *wave function description* over to a limit where you have a *population description*, then usually you put in some imaginary object, you know... some losses, something like that, so I think a Laurentian $i\Gamma$ and then that

sort of a loss. You see, going from wave function description, where you have coherence, to the other description probably implies something that permits a loss, something to break up physically the coherence, so I'm not sure, I'd have to think more about that but my guess is that... you see, at this point you need to bring in the nuclear motion of that intermediate in between, see. So, for the coherent case the intermediate in between doesn't have to have any nuclear motion, that is serving there as a *conduit for the wave function* but when the process is incoherent, then, I mean, usually that intermediate would be vibrating, its coordinates would be changing, and so I don't know whether I'd have to think more of the equations that would involve going from coherent to incoherent, you see, and those equations are undoubtedly around. Various people have undoubtedly looked at that when they are looking at tunneling vs. hopping mechanism and trying to combine the two, that would be the where to look at. There are papers in which they are combining hopping and tunneling, a number of people have done that, I think Stuchebruchov has done that, I think that Jortner has done that, I think any number of people have done that, but I don't have references handy. See, the main thing would be to look at some concrete example, of where the theory has those two limits in it, you see, and I believe there are several people who have done work on that, they have written papers that cover the whole gamma.

4. p. 2973 2nd column middle: "... we introduce... some approximations such that *the coherent problem* can be reduced analytically from three coordinates to one."

Following below: "The treatment of the *coherent dynamics* has thereby been reduced to a one-coordinate problem"

On p. 2977 1st column middle: "In several respects a recent article by Lin on a donor-bridge-acceptor electron transfer is complementary to the present one: the case when the B^- orbital is close to D^* is included there, and the overall $D^* \rightarrow A$ transfer is assumed to be *coherent.*"

Q: Why is the present problem "coherent" and what does it mean "coherent dynamics" and when is a transfer "coherent"?

M: When the intermediate really doesn't have any finite time existence.

5. p. 2974 1st column top: "We let Φ_i denote the corresponding time-dependent *nuclear* wave function... The electronic-nuclear wave function $\Psi(\tau)$ at time τ is now given by

$$\Psi(\tau) = \sum_{i=1}^{3} \Phi_i(\tau)\Psi_i \qquad (2.1)\text{"}$$

Following, middle column: "For each of the three electronic configurations D^*BA, D^+B^-A, and D^+BA^-, a set of three equilibrium values of the q_i's is defined. When the electron is on D, namely when Φ_1 is large..."

M: You see, there are three wave functions there, and if Φ_1 is large then the others would be negligible. Let's say that Ψ_i's are normalized, so we don't speak of large Ψ_i or of small Ψ_i. Let us say that $\Psi(\tau)$ is just taking different forms at all times while the electron is going from one side to the other, when it is going from Ψ_1 to Ψ_3. When going from 1 to 2, Φ_1 becomes small and Φ_2 becomes big, the $\Psi(\tau)$, the sum, was big to start with and is big at all times. The components Ψ_i in the sum are time independent, they have fixed values, they don't vary with time. It is only the $\Phi_i(\tau)$ that are varying with time. The nuclear wave function is not only the nuclear wave function, but it is also the coefficient of the electronic wave functions, the Ψ_i's. The Ψ_i's are each individually normalized.

Q: The nuclear wave functions play then the role of coefficients of the electronic wave functions.

M: They also play that role, that's right. And they are also dependent on r, of course, on positions. I just haven't shown the exclusive dependence on nuclear coordinates, that it is implied in calling them the nuclear wave function. If the electron is first on D, then if it is not anymore on D, at that time Φ_1 and Φ_3 are zero because that's on B. So, if the electron is on B. It cannot possibly be on D or A. What makes the nuclear wave function vary are the equations of motion involving the nuclear electronic coupling. And if you solve those equations of motion, you see that Φ_1 changes with time, Φ_2 changes with time, Φ_3 changes with time. For the initial value problem that is initially in state 1, one solves the time dependent Schrödinger electronic nuclear equation.

Typo: Misprint: in Eq. (2.9) ϕ_j should be Φ_j.

6. p. 2974, 2nd column bottom: "With the change of coordinates from (q_1, q_2, q_3) to (x, y, z) it is seen from Eqs. (2.9)–(2.11) (**Typo**: misprint: it is from Eqs. (2.10) and (2.11)) that the x motion is identical for all H_{ii}'s and so is now separable from that of y and z, an anticipated result since the coordinate x was defined so as to be perpendicular to the plane containing all the geometrical changes.

Q: We know that the three points $(a_1, 0, 0)$, $(0, a_2, 0)$, $(0, 0, a_3)$ belong, by construction, to the plane (and one sees it from Fig. 1). But how can we be sure that the plane contains *all* the geometrical changes?

M: Let me tell you what I think. The three points define a plane, and when you are going from one point on that plane to another you are changing coordinates on that plane, you are not changing coordinates perpendicular to that plane. So, the coordinate perpendicular to that plane is the same for all the points that lie on that plane.

Q: So x is fixed at the value it has at the intersection with the plane.

M: That's right.

7. p. 2975 1st column middle: "Equation (3.2) is separable, and we have

$$\Phi_{N_1}(y, z) = \Phi_n(y)\Phi_m(z), \quad E_{N_1} = E_n + E_m \qquad (3.3)$$

....

$\Phi_{N_{23}}(y, z)$ becomes

$$\Phi_{N_{23}} = \Phi_{n'}^{(2)}(y)\Phi_{m'}(z) \quad E_{N_{23}} = E_{n'} + E_{m'} \qquad (3.5)\text{''}$$

Q: Why the superscript $^{(2)}$ on the above $\Phi_{n'}^{(2)}(y)$? The same symbol is used in the following Eq. (3.6)

$$\left[-\frac{1}{2}\frac{\partial^2}{\partial y^2} + {}^1\!/_2(y - A_1A_2/a_1)^2\right]\Phi_{n'}^{(2)}(y)$$
$$\equiv H_y^{'}\Phi_{n'}^{(2)}(y) = E_{n'}\Phi_{n'}^{(2)}(y) \qquad (3.6)$$

M: Maybe it is the wave function of state 2. The main point is to write it as the product of two functions, that's the main thing. Maybe it was labeled to distinguish it from something else. There is a Φ_n in Eq. (3.3), so, I guess, we wanted to distinguish it from that particular Φ_n, I imagine. It is different from Φ_{N_1} in Eq. (3.2), right? The potential energy function is different. Sure, we wanted to distinguish it from the Φ_{N_1} in Eq. (3.2). Suppose we are using $\Phi_{n'}(y)$, we may confuse it with Φ that satisfies Eq. (3.2), you see, in other words there are two equations with two $\Phi_{n'}(y)$ and they are not the same. I imagine it is something like that.

8. p. 2975 1st column top: "A Golden Rule approximation can then be used for a *first order rate constant* k_r for the disappearance of electronic configuration 1. For reactant from a particular vibrational state N_1 of D*, this k_r is given by[8,9]

$$k_r = 2\pi\omega H_{12}^2 \sum_{N_{23}} |\langle \Phi_{N_1}(y,z) \mid \Phi_{N_{23}}(y,z)\rangle|^2 \delta(E_{N_{23}} - E_{N_1}) \quad (3.1)\text{”}$$

Q: Ref. 8 is reference to Merzbacher's Quantum Mechanics. There in my 2nd edition book I read:

p. 479: "...we have for the *transition probability per unit time* (for $t > 0$)

$$w = \frac{d}{dt}\sum_k |c_k(t)|^2 = \frac{2\pi}{\hbar}|\langle k \mid V \mid s\rangle|^2 \varrho_f(E_s^{(0)}) \quad (18.107)$$

Formula (18.107) *represents* a *constant rate of transition*"

Q: Can you please make clear the relation between the concepts of first-order rate constant and of transition probability per unit time?

M: Well, let's see. First of all: the units on the right hand side are per unit time, aren't they? The units are OK. Now the probability per unit time is the probability of transition. The Golden Rule is used for a nonadiabatic process, so that only has meaning when the probability per unit time is very small. If you have a large probability per unit time, the process wouldn't be nonadiabatic, right? So that case doesn't arise when this equation is valid. The first-order rate constant has 1 over time as unit. A first-order rate constant is $k = -\frac{1}{P}\frac{dP}{dt}$, that's a probability, you are normalized with the P underneath, so it is a transition probability per unit time but normalized to that existing probability. In other words $k = -\frac{1}{P}\frac{dP}{dt}$ is a first-order expression, is a first-order rate constant, so it is *a change of probability per unit time* *per unit probability*, and you can see that our Eqs. (3.1) and (18.107) are related.

Q: So the k_r and the w are the same thing there.

M: Yes, that's right, they are the same thing. One is given in a notation of density of states, and the other assumes a constant spacing and is given in notation of sum over energies.

Q: But the probability should be a number smaller than 1.

M: Yes, but that is per unit time. If the time unit were extremely small the number would be very large or vice versa. It depends on your unit of time, what its value is. It is not just a number such as a pure probability. It has units, a pure probability doesn't have units.

9. p. 2976 1st column middle: "With a change of variable $t - \theta \rightarrow \theta$, Eq. (4.2)

$$|\Psi(y,z,t)\rangle = -i\int_0^t d\theta e^{-iH_{yz}(t-\theta)}|\Psi\rangle H_{12}\Phi_{nm}(y,z,\theta) \quad (4.2)$$

now becomes

$$|\Psi(y,z,t)\rangle = -iH_{12}e^{-(k_r/2+iE_{nm})t}\int_0^t d\theta e^{(k_r/2+iE_{nm})\theta}$$
$$\times [e^{-iH_y'\theta}\Phi_n(y)] \times [e^{-iH_z\theta}|\Psi_2\rangle\phi_m(z)] \quad (4.7)\text{"}$$

M: It is a common change of variable. Both vary between 0 and t. That kind of transformation is very common. So that's a matter of ordering. t is a constant, you see. So, $dt - d\theta$ becomes $d(-\theta)$ because t is a constant at a given time.

10. p. 2976 1st column bottom: "The "*population*" $B^-(t)$ of the intermediate electronic configuration 2 is obtained by projecting

$|\Psi(y,z,t)\rangle$ onto the diabatic electronic state $|\Psi_2\rangle$ and integrating over the nuclear coordinates:

$$B^-(t) = \iint dydz|\langle\Psi_2|\Psi(y,z,t)\rangle|^2 \tag{4.8}$$

Q: Why "population" in quotes?

M: Maybe it's a population probability, so it is not really the number of systems but it's the population probability.

11. Following: "Strictly speaking, $B^-(t)$ *does not actually exist in this mechanism*, and it would be more rigorous to treat an observable, such as the absorption spectrum in the B spectral region, rather than defining a $B^-(t)$ via Eq. (4.8). Use of Eq. (4.8) for B^- *presumes* that the quantity so defined adequately describes this depletion in the spectral region of B."

M: That comes to that same question about whether something is existing for any finite time or its concentration is negligible, you see. In other words, if B^- is *a coherent intermediate* and rather than a real intermediate, it has essentially no population at any time there.

Q: Is it not detectable?

M: No if it is a really coherent intermediate, it would be extremely small. I mean, yes, there may be a little bit, but you wouldn't not normally speak of it as an intermediate. The intermediate is described by that wave function instead of a population.

12. p. 2976 2nd column bottom: "To interpret some of the numerical results for $B^-(t)$ and for $B^-_{\max}$ given in part 2 and provide an approximate indication of what might be expected for other values of the various parameters, we obtain below a rather rough classical estimate of the time spent by the system in the 'B^- *region*'...For this purpose we approximate the length of this region by the z interval between $z^{\ddagger}_{12}$ and the intersection $z^{\ddagger}_{23}$ . . . "

Q: Why do you put B^- region in quotes?

M: My guess is that in that region there are several species and the dominant we can call B^-, there are maybe more than one species there. That region doesn't belong to B^- alone, but others are there with a certain thermal probability.

13. p. 2977 1st column top-middle: "When the wave packet undergoes negligible reflection in the region near $z^{\ddagger}_{23}$, a condition we denote by setting the $2 \to 3$ transition probability, w_{23}, to be approximately unity, the time τ_2 spent by the system in the B^- region is then, classically

$$\tau_2 \sim \int_{z^{\ddagger}_{12}}^{z^{\ddagger}_{23}} dz/v_z \quad (\text{when } w_{23} \sim 1) \tag{5.7}$$

where v_z is the local z component of velocity . . . Equation (5.7) can be written in terms of some average velocity, $\overline{v}_z$, $\tau_2 = (z^{\ddagger}_{23} - z^{\ddagger}_{12})/\overline{v}_z$.

If τ_2 is regarded as the reciprocal of an *effective* rate constant k_2, then Eq. (1.1)

$$B^-_{\text{max}} = (k_1/k_2)^{k_2/(k_2-k_1)} \tag{1.1}$$

still applies approximately for the maximum population of B^-, but now with the k_2 given by

$$k_2 \sim \overline{v}_z/|z^{\ddagger}_{23} - z^{\ddagger}_{12}| \quad (\text{when } w_{23} \sim 1) \qquad (5.8)\text{"}$$

M: The species doesn't exist if the mechanism is a superexchange there it does exist if there is the possibility of an intermediate there. So, it depends on whether if one uses a superexchange or an intermediate formula. If it is superexchange then it really doesn't exist.

An intermediate in the case of superexchange is not observable... you would not call it a *kinetic intermediate*.

14. Following: "This k_2 becomes large when the z interval in the denominator of Eq. (5.8) becomes small (*but not too small*, if the expression is to have some validity)..."

Q: What do you mean by "not too small"?

M: Well, there may be maybe an uncertainty principle...

15. Following: "... E_y, the y-mode energy in excess of the potential energy $H_{22}(y_2^0)$ after the $1 \rightarrow (2, 3)$ transition, is

$$E_y = n + 1/2 - 1/2(y_{12}^{\neq} - y_1^0)^2 + 1/2(y_{12}^{\neq} - y_2^0)^2 \qquad \text{(A1)}$$

We define the corresponding z-mode energy E_z as its value relative to that at z_2^0. Thereby

$$E_z = m + 1/2 - 1/2(z_{12}^{\neq} - z_1^0)^2 + 1/2(z_{12}^{\neq} - z_2^0)^2 \quad (H_{22} \text{ surface}) \qquad \text{(A2)}$$

......

By use of concepts employed for the intersecting parabolic surfaces, (**M30, M53**) the point $(y_{12}^{\ddagger}, z_{12}^{\ddagger})$ lying on the intersection of $H_{11}(y, z)$ and $H_{22}(y, z)$ surfaces and having the least potential energy is given by

$$y_{12}^{\neq} = y_1^0 + M(y_1^0 - y_2^0), \quad z_{12}^{\neq} = z_1^0 + M(z_1^0 - z_2^0) \qquad \text{(A3)}$$

where M satisfies the equation (**M30, M53**)

$$-(2M + 1)\lambda_{12} = \Delta E_{12} \qquad \text{(A4)}$$

Equations (A1)–(A5) then yield the energy partitioning expressions, Eqs. (5.1)–(5.2)

$$E_y = n + 1/2 + (\Delta E_{12}/\lambda_{12})(\lambda_2^2/\lambda_{23}) - \Delta E_{12} \equiv n' + 1/2 \qquad (5.1)$$

$$E_z = m + 1/2 - (\Delta E_{12}/\lambda_{12})(\lambda_2^2/\lambda_{23}) \qquad (5.2)\text{"}$$

Erratum: Something is missing in the above derivation: There are two problems: (1) There is no A5 equation in the paper; (2) In Eqs. (5.1) and (5.2) a λ_2 and a λ_{23} appear that are not present in Eqs. (A1) to (A4) from which Eqs. (5.1) to (5.2) are derived.

M: Probably there is an Eq. (A5) that got deleted. Something got chopped off. Certainly.

16. p. 2977 2nd column bottom: "In a Landau-Zener treatment the probability w_{23} of remaining on the potential energy curve $E_-(z)$ after a single passage through $z_{23}^{\ddagger}$ is given by

$$w_{23} = 1 - e^{-2\pi H_{23}^2/v_z|\Delta s|} \tag{B1}$$

where $z_{23}^{\ddagger}$ is real. Here, v_z is the velocity at the crossing point and Δs is the difference of slopes of the intersecting diabatic potential energy curves there. If we set $v_z^2 = N$ in dimensionless units (which defines N) and if we introduce the value (**M231**) for $|\Delta s|$, namely, $(z_3^0 - z_2^0)$ in dimensionless units, one finds that

$$w_{23} = 1 - e^{-2\pi H_{23}^2/(\lambda_{23}/N)^{1/2}} \qquad \text{(B2)''}$$

Q: Why can you substitute a difference of slopes with the difference between the two minima of the potential energy curves?

M: v_z is the velocity of the nuclei passing through the crossing. It is a nuclear motion corresponding to an electron jump. I wonder why we even introduced λ_2, what we used it for? I don't know. First of all, I don't know why λ_2 was introduced, I really should know that but I don't know. And secondly... do we use λ_2 elsewhere in the paper?

Q: No, you don't.

M: Somehow some of that may be related to the following: If we have two parabolas, then the difference in two slopes at the intersection, which is important in the Landau–Zener formula, can be calculated from other properties of the parabolas, such as their separation, the energy difference at the two minima and the curvatures. So you can relate, you see, the difference of slopes to some other properties. You may just try it as an exercise. Draw two parabolas, calculate at the intersection the difference of the two slopes and you find that's related to force constants in there, to the difference of positions of the minima and to the energy difference of the two minima there. So, you can take that difference of slopes and express it in terms of these other properties of the parabolas, you see.

17. Following: "Employing Eq. (B2) when H_{23} becomes very large may, however, be questionable."

Q: Why so?

M: Well, if there is an equation which was based upon some nearly nonadiabatic approximation, then if H_{23} becomes very large then that's no longer a due approximation, the position of the potential energy curves is very far removed from the crossing point, so maybe that's it.

M253 Dynamics of ET for a Nonsuperexchange Coherent Mechanism. 2. Numerical Calculations

R. Almeida and R. A. Marcus

NOTES

1. p. 2978, 2nd column middle-bottom: "In a real system, dissipation of the energy after (and during) the $\Psi_1 \rightarrow \Psi_2 \rightarrow \Psi_3$ transition occurs by redistribution of that energy among the numerous degrees of freedom. To avoid *spurious oscillations* of the wave packet..."

Q: How do you distinguish spurious from true oscillations, such as the ones described by Fig. 3?

M: You know, in chemical systems... occasionally you see some oscillation, they are rare but with a wave packet calculation you should easily get them if you haven't got something that sort of damps out the oscillations.

Q: How do you distinguish a true oscillation from a spurious oscillation?

M: Suppose you have a crossing, now a wave packet actually goes pass the crossing and hits a repulsion and then comes back and hits another repulsion and so part of it can be sort of trapped that way, those would be a real oscillation. Those oscillations can be spurious if there is very strong physical damping. I am not sure how you would tell, except if you had a wonderful physically correct model, then the oscillations come out to be correct oscillations. So, it may depend first of all on the kind of damping that you put in and if your damping is realistic or not. Unless you know whether your damping is realistic or not, and you may not know, then you can't tell whether the oscillation is spurious or not.

2. p. 2979 1st column top: "In each of these latter approximations a Golden Rule expression was used. Written here in terms of one coordinate, the rate constant k_r for reaction from a specific vibrational state m is

$$k_r = 2\pi\omega|H_{12}|^2 \sum_{m'} |\langle\Phi_m(z) \mid \Phi_{m'}^{(2)}(z)\rangle|\delta(E_m - E_{m'}) \qquad (2.3)$$

where $\Phi_m(z)$ is *the* vibrational wave function appropriate to $H_{11}(z)$, treated as a harmonic oscillator potential... and the $\Phi_{m'}^{(2)}$ in one approximation denotes a nuclear wave-function appropriate to $H_{22}(z)$."

Q: Does the index m' run over a number of degenerate wave functions with different m' vibrational quantum numbers but all of them corresponding to the same energy $E_{m'} = E_m$?

M: Strictly speaking yes, but you might approximate by going to states in a certain region . . . that's right.

3. Following: "For comparison, in a different approximation it denotes the eigenfunction Φ_A appropriate to the lower adiabatic surface $E_-(z)$ arising from the (Ψ_2, Ψ_3) pair . . . The corresponding values of k_r are denoted . . . by k_{HO} and k_A respectively. *This* k_{HO} *is, thereby, the microcanonical version of the usually calculated Golden Rule rate constant* in the literature.

M: Well, the Golden Rule is sometimes just really expressed microcanonical, in other cases one puts in the Boltzmann weighting factor and then sums over that. But the word microcanonical version didn't have to be included because if it is microcanonical that's more commonly the way the Golden Rule is stated. It's unnecessary to say microcanonical version. Because most of the times when one writes the Golden Rule one writes it microcanonically. But when one writes the thermal weighting probability, then it's canonical.

4. p. 2979 1st column bottom: "Both for the one- and three-coordinate calculation it is useful to know whether the coherent dynamics for the '$B^-(t)$,' obtained numerically, can be fitted by an *apparent* two-step kinetic equation, in which the *effective* rate constant k_2 of the second step is much larger than the *theoretical adiabatic maximum for* k_2 in a two-step (incoherent) process, *namely*, $\omega/2\pi$."

Q: Why is the theoretical adiabatic maximum equal to $\omega/2\pi$, that is, to a vibrational frequency?

M: If there is no barrier of any kind and no nonadiabatic effect, no activation energy . . . then a reaction to occur requires some nuclear motion, essentially a nuclear motion of frequency $\omega/2\pi$.

5. p. 2980 1st column bottom: "The rate constants obtained from the numerical (FFT) solution for k_r are denoted in Tables I and II by

k_1 and are compared there with those obtained by using the various approximations...A comparison with the *semiclassical results* is helpful in providing some insight on the dependence of k_r on the *initial vibrational state*..."

Q: Why do the semiclassical results provide an insight on the initial vibrational state?

M: I frequently find that the semiclassical results are useful in sort of breaking into a physical picture of what's going on.

6. (1) p. 2981, caption to Fig. 8: "Fit of $B^-(t)$ in the 3-D calculation to an *effective two-step incoherent* formalism..." cf. p. 2982, caption to Fig. 9: "Fit of $B^-(t)$ in the 3-D calculation to an *effective two-step coherent* formalism..."

(2) p. 2982 legend to Table 4: the fifth column of numbers is headed by λ_1 as the third column.

Q: (1) Effective two steps coherent?

(2) Probably the heading of column 5 should be λ_3.

M: I don't know why he says effective two steps. If it is coherent, it is one step. In the second legend I would just omit effective two steps, I would just say a coherent one. I don't even know why he says effective two steps, if the thing goes through an intermediate bridge that is not an effective two steps, that is simply a two steps. So, I would delete the word effective there. And if the whole thing is done coherently, you know, the kind of tunneling, then I wouldn't use the two steps, I would just say coherent. I would delete effective in both of those. If it is going through a bridge, I don't know why those words would be used, that's my fault for not having noticed that.

(2) OK.

7. p. 2982 2nd column middle: "Plots of $k_r(t)$ *vs* the temperature T typically had *a negative temperature dependence* when ΔV_1, given by Eq. (2.5)

$$\Delta V_1 = (\lambda_{12}/4)(1 + \Delta E_{12}/\lambda_{12})^2 \qquad (2.5)$$

was *smaller than the zero-point energy*, $\hbar\omega/2$ (50 cm^{-1} in the present case), i.e. whenever the ΔV_1 in dimensionless units was below 0.5. For all of the results in Table IV, ΔV_1 is in the neighbourhood of 0.2."

M: ΔV_1 is the barrier to go from the first to the second, to the bridge. If the intersection is occurring below the zero point level of the starting point, you wouldn't have any barrier, it can't get lower than a zero point energy.

8. p. 2982 2nd column bottom: "B^-_{max} is given by[12]

$$B^-_{\text{max}} = (k_1/k_2)^{k_2/(k_2-k_1)} \qquad (4.1)\text{"}$$

Q: **Typo**: Misprint: the number reference 12 should be 1 (reference to your former paper with Almeida), where Eq. (4.1) is referred to Marcus R.A. *Chem. Phys. Lett.* **1988**, *144*, 24.

M: OK.

9. Following: "Examination of the results for B^-_{max} in Table IV and Figures 4 and 5 shows that they are appreciably less than this value, reflecting the coherency of the overall $D^* \to A^-$ ET in the present model. Thus, if Eq. (4.1) were used to fit the observed B^-_{max}, an *effective* k_2 much larger than this maximum adiabatic value would occur for the present mechanism."

Q: If I understand correctly, what you do here is a "reductio ad absurdum," I mean you give an indirect proof that the kinetic process is a coherent one. Your argument apparently goes like this: Suppose the process is an incoherent two steps process for which $B^-_{\text{max}} = (k_1/k_2)^{k_2/(k_2-k_1)}$. One then finds that the value for k_2 is larger than the maximum allowed adiabatic value. But such a value is impossible and so the incoherent model is impossible. Is it so?

M: In other words, if there is essentially no population of the intermediate, it is a coherent type process. If something is much larger, it could be that the nonadiabatic calculation is broken down, so one would have to ask: is it much larger because the nonadiabatic calculation is broken down? Because if it has broken down, then it simply means that you can't use the nonadiabatic formalism. If k_2 is too big, it signals that there is something wrong, either because the process is coherent or because you simply can't use that kind of nonadiabatic perturbation theory. I imagine it's something like this: if one assumed that it was a two steps process, and if, after having done all of the analyses and compared with the observed rate constant, . . . the value that one came out with for the rate constant, if the value that came out for k_2 is larger than it is possible, then that means that one has to reject the two steps mechanism because if one interprets the process by two steps mechanism one would come out with too large a k_2. Then it may be possible that the process is actually a one-step mechanism, by a superexchange. In other words, the reaction would be much faster than you can imagine, so therefore for the two steps mechanism the second step is unrealistic, is unrealistically fast, therefore one has a one-step mechanism. I imagine that is probably what I was thinking about.

There are some situations, and this is probably not one of them, when the processes can occur faster than the adiabatic value. For example, if you have a dye that is attached to a semiconductor electrode, and if you excite it up to a high level so it can go directly into the conduction band, it is possible for the electron to just pop in into the conduction band, with no nuclear motion needed. In that case there isn't an upper limit of the order of $10^{13}\,s^{-1}$. Some values have been reported of the order of 10^{-15} s.

Maybe 10 fs, I think there was some time, but it's indirect, it was based on some analysis of some transfer of a charge to or from a crystal and . . . the transfer occurred in a very short time, and such a short time that there wouldn't be really time for nuclear movement to occur, so there are some reactions that apparently can occur without invoking the movement of nuclei, and it may be, for example, that if you excite a molecule and it's strongly bound to some metal electrode, the electron can simply pop over into the continuum of bands, without any nucleus having to move. So in that case you can't use then the frequency of a nuclear motion as being the upper limit of the rate constant. I saw one that claimed to be 10 fs. So there wouldn't be time for nuclear motion there.

CHAPTER 3

Matrix Elements for Long-range ET, ET across Liquid–Liquid Interfaces, Charge Transfer Spectra in Frozen Media, ET in Proteins: Calculation of Electronic Coupling, ET across Liquid–Liquid Interfaces 2. Schrödinger Equation for Strongly Interacting ET Systems

Interviews on M254, M255, M256, M259, M261, M266

M254 Comparison of Experimental and Theoretical Electronic Matrix Elements for Long-range ET

P. Siddarth and R. A. Marcus

NOTES

1. p. 2986 1st column middle: "The quantities in Eqs. (1) and (2)

$$k_{\mathrm{ET}} = \frac{2\pi}{\hbar}|H_{\mathrm{DA}}|^2 \mathrm{FC} \tag{1}$$

$$FC = \frac{1}{(4\pi\lambda k_B T)^{1/2}} e^{-(\Delta G^0+\lambda)^2/4\lambda k_B T} \tag{2}$$

dependent on R are ΔG^0, λ, and H_{DA} ..."

Q: The dependence of λ and H_{DA} on R is clear. Can you explain the dependence of ΔG^0 on R? Do you consider it for a single DA pair with R's of different lengths or do you maybe fix really R and change DA pairs in a series?

M: Is this for a charge shift reaction or for a charge recombination or a charge dissociation? If the donor has, say, a minus and the acceptor becomes a minus, that reaction is called a charge shift reaction.

Q: You don't have this in the paper.

M: Well, then we should, because for a charge shift reaction the interaction of the two reactants, before and after the ET, is sort of being about the same. If it is a recombination reaction where you have a plus and a minus and an electron transfer from the minus to the plus, then of course you have a difference in coulombic energy before and after, that is going to be R dependent, or, if you have the reverse of that, it would be R independent. So, this dependence of ΔG^0 on R doesn't apply, largely, to charge shift reactions that applies to recombination of charges, or formation of charges. That would be pretty general. It is legitimate maybe to focus on the coulombic interaction. That would be the long-range interaction. You are talking of the effect of changing the bridges, bridges of different lengths. You may trust Siddarth's papers to discuss different lengths, it is a paper I did take care of. We are talking there about changing R, that is, the work she did, involving changing R.

2. p. 2986 1st column middle: "An ET is *normally* preceded by thermal fluctuations of the various coordinates (e.g. orientation of solvent molecules, lengths of various bonds) in or near the DA pair."

Q: Are there ET cases when ET is not preceded by thermal fluctuations?

M: See NOTE 9 in the preceding paper. Another point is that, if you have one curve intersecting the bottom of the other, you need a little thermal motion to get it go. When one curve crosses at the bottom of the other curve, then if you start with the curve which is being crossed at its bottom, then the system just moves to get into the other state. There is a little motion there. Whereas in the conduction that I suggested in NOTE 9 the electron can just pop right in, popping into a continuum of levels.

3. p. 2986 2nd column middle: "... $c_{D\upsilon}$ is the coefficient of that bridge orbital υ at the *point of contact* of B with D..."

Q: How do you define the point of contact between bridge B and donor D? Is it half of the distance between the nuclei of the closest atoms of B and D?

M: Well, maybe the statement is a little bit loose, the point of contact is the atom of the bridge. I mean, if I would be precise I wouldn't say point of contact. It's intended to be the coefficient of the atomic orbitals of B.

Q: So, the last atom of B.

M: That's right. Exactly.

4. p. 2986 1st column bottom: "To calculate the electron-transfer electronic matrix element H_{DA} one procedure is to seek the *two* lowest energy many-electron wave functions of the DA pair *where*, as a result of a suitable fluctuation in the coordinates, *the extra electronic charge is equally divided between D and A.*"

Q: Two questions: (1) What do you exactly mean by "equally divided"?

(2) In the case of a DBA system in which D and A are not only bridged during the electron transfer but bound through B so that they do not separate after the transfer, isn't it possible that the final state after the transfer is, say, $D^+(BA)^-$, that is, part of the electronic density remains localized on the bridge?

M: (1) I mean that *at the crossing point* of the diabatics, then the wave functions of the adiabatics would be $a(D^+A^-) + b(DA)$, and $a(D^+A^-) - b(DA)$, and a and b would be equal. If you would imagine of *holding* fixed the system at the avoided crossing, the system could oscillate between those two adiabatic states. But you have to think of what is the experiment. Normally you can't hold the system there but if you did hold it there, if there is some magic way of doing it, depending on what you did, the system may oscillate back and forth, depending on its initial conditions. In other words, if you had all ensemble of systems, each one like that, you might find 50% in one form, 50% in the other form. But of course that's an artificial situation.

M: (2) That's certainly possible, usually the bridge is very high up, you see, so that the fraction of the electronic population of the bridges is very small, so it depends on the energy level of the bridge compared to that of the acceptor.

5. Following: "In the one electron approximation this $2H_{DA}$ is the energy difference $\Delta\varepsilon$ of the *two delocalized orbitals* which are distributed over D and A. One such orbital is, in effect, *symmetric* and the other *antisymmetric* with respect to the two centers D and A."

Q: Please check if my interpretation of your words is correct. Look at panel B of Fig. 1. There I have schematized the square of a symmetric bonding orbital and the square of an antisymmetric antibonding orbital. In both cases, we see that the transferring electron is evenly distributed over the two centers D and A but the two distributions are different from each other. The energy of the symmetric orbital is lower than that of the antisymmetric one. Is it so?

M: That's right.

Q: With the energy of the symmetric orbital lower than that of the antisymmetric.

M: Yes, usually, Yes.

6. p. 2988 1st column middle: "In order to compare the R dependence from experiment with theoretical calculations of the R dependence of H_{DA}, it is necessary to allow for any R dependence of FC. One possibility.... Another is to make the studies under conditions where a plot of k_{ET}versus ΔG^0...is at a maximum ($\Delta G^0 = -\lambda$). Then, *FC should have little R dependence.*"

Q: I guess it is so because then the exponential dependence of FC in Eq. (2) disappears and only the dependence through λ in the denominator remains.

M: The FC has that exponential dependence on $\Delta G^0 + \lambda$, then when $\Delta G^0 = -\lambda$ the dependence is essentially zero, it can't have any R dependence. The dependence of FC on λ in the denominator is probably small. It's there but in this work it should on not be allowed for, λ does depend upon distance, you know, so there is a dependence of λ on distance and the most correct thing is that you allow for that dependence, but that dependence is small when you are here on top of that inverted problem.

7. p. 2988 1st column bottom, Note (32): "Equation (6) was used and *the energies of the D orbitals alone were changed* in order to match the D and A energies."

Q: How do you change the energies of the D's orbitals? Do you move the D's nuclei until you reach a nuclear configuration at which the above matching is possible?

M: Well, it depends, you could artificially do it, just by changing, if it is an LCAO kind of calculation, the α parameter. You know, in MO you have α which is kind of coulombic and β which is exchange. So probably arbitrarily changing α's, you know, it's an exercise if you do it. In other words, you are working on a parameter in the Hamiltonian. If you move nuclei then of course everything changes.

8. p. 2989 1st column middle: "An inspection of the energies and coefficients of the bridge orbitals reveals that, for all the series presently studied, it is principally only the few lowest unoccupied energy states (virtual states) of the bridge that are responsible for the electron transfer."

Q: Might one consider these orbitals really as "channels" through which the transferring electron passes?

M: Yes, you could, sure. Usually the electron doesn't spend any time in those orbitals, yes, they would be kind of channels.

M255 Theory of ET Rates

R. A. Marcus

NOTES

About possible radiations accompanying nonadiabatic ET

Q: In the case of nonadiabatic ET don't you have some accompanying radiation?

M: Not really because you are essentially just interchanging electronic and nuclear motion. The equation that you solve for that transfer were done by Landau and Zener and Stückelberg and involve just a transfer, when the electron is transferring, during that process, there is a little change in nuclear motion, so they are coupled, without any radiation field.

Q: So $\Delta\varepsilon = 2H_{DA}$ is given by the kinetic energy of the nuclei?

M: Yes, kinetic energy of the nuclei. That's a bit a classical way of describing it, under certain conditions you need a full quantum description, that's when electronic and nuclear motion are so strongly coupled that they are inseparable in that region. I remember Stückelberg's papers, he was known for the Stückelberg oscillations, when you have that curve crossing you can oscillate back and forth. Nikitin is a very first class theotetician, so... I went to a conference in Moscow in 1963, I think it was... 1965 and the one paper that I have in Russian is number 53a in my list of publications.

1. p. 4153 1st column top: "... the rate constant k_0...

$$k_0 = 2\pi(a_1 + a_2)\kappa\nu(\Delta R)^3 e^{-\Delta G^0/kT} \tag{2.2}$$

.... ν is a typical *frequency for nuclear motion along the reaction coordinate*."

Q: Normally one speaks of frequency when there is a vibration. Which frequency is this one?

M: The molecules are sort of rattling around in a little cage, so it is kind of a vibration.

2. Following: "The κ in Eq. (2.2) is the *usual adiabaticity/nonadiabaticity* Landau–Zener factor."

M: If you look at Landau–Zener formula, then there is a preexponential factor there, and that preexponential factor is dimensionless. When it is small the process is nonadiabatic, when the factor is large, when it gets close to unity it is adiabatic. In other words, if you use the Zener form of the LZ probability that's what it is. Because Landau didn't go all the way from nonadiabatic to adiabatic, Zener using a different kind of approximation, went all the way from nonadiabatic to adiabatic, and so included Landau's result as a special case. See, it is like this: Zener considered the $1 - e^{-\kappa}$, and if you expand that, you get κ. That is what Landau got. $1 - e^{-\kappa}$ is what Zener got.

3. p. 4153 1st column middle: "Equations (2.1) and (2.2)... When each reactant can penetrate the other phase, such that the center of the reactant could even lie on the phase boundary, a larger preexponential factor was obtained, larger by a factor f:[1]

$$f = {}^1\!/\!_2(a_1 + a_2)^2/(\Delta R)^2 \qquad (2.4)\text{"}$$

Erratum: The formula is not present in ref 1.

Nonetheless, there it is possible to very easily *derive Eq. (2.4) from two equations reported in ref 1. If* we go back to ref 1, **M249**, there we read on p. 1053 1st column top-middle.

"We let υ ... be a 'volume' ... such that in a unit area of interface the center-to-center distance of the pair of reactants lies in $(R, R + \Delta R)$ where ΔR ... is the region over which there is a significant contribution to the ET process, *and where reactant 1 lies wholly in phase 1 and reactant 2 in phase 2.* If κ is some Landau-Zener factor ... then ...

$$k_r = \kappa \nu \upsilon e^{-\Delta G_r^0/k_B T} \qquad (26)$$

In the Appendix it is shown that the leading term in an expression for an effective υ is

$$\upsilon = 2\pi(a_1 + a_2)(\Delta R)^3 \qquad (27)$$

when a sharp boundary at the interface is assumed.

When the two phases are 'immiscible' liquids, some interpenetration of the two phases may occur, so that the reactants may be able to approach each other over a wider solid angle than that indicated in Fig. 1. For example, if the centers of ions 1 could each penetrate the other phase to the extent that each center could even reach the interfacial boundary, but such that the reactants would not overlap, one would obtain ... Eq. (28) instead of Eq. (27).

$$\upsilon \sim \pi(a_1 + a_2)^3 \Delta R \qquad (28)\text{"}$$

Q: As you see dividing Eq. (28) by Eq. (27) one gets Eq. (2.4).

M: The formula can be derived, OK.

4. p. 4155 2nd column top-middle: "Nevertheless, some of the assumptions made in deriving, say, 2.2, may be summarized: They include an 'ideal' (sharp boundary) interface, a *local and linear dielectric continuum theory*..."

M: When you write down the expression for the dielectric displacement in terms of the electric field, when the displacement is local there is simply a multiplication factor, ε, when it is nonlocal then $D(r)$ would be $D(r) = \int \varepsilon(r, r')E(r')dr'$ and that's nonlocal. In the nonlinear case, there would be also another term where $D \propto E^2$ and so on. The Russians did a lot in terms of nonlocal treatments, so going way back, not Levich and Dogonadze did, not they but probably Kuznetsov, maybe Dogonadze and Kuznetsov, maybe not the nonlinear, they tend to stay off the nonlinear but they surely have the nonlocal there a lot, but the trouble is that there is a lot we don't know about the nonlocal, so it is fine just as an exercise.

M256 Theory of Charge Transfer Spectra in Frozen Media

R. A. Marcus

NOTES

1. p. 4964 2nd column top: "The subsequent freezing of the solvent... presumably does little to change these orientations in the immediate vicinity of the transition point, *if the cooling is not too slow.*"

M: When you freeze slowly you allow the whole system to adapt to the new conditions, whereas if you freeze quickly then you have an all group of nonequilibrium subsystems frozen in.

2. p. 4966 1st column middle: "*Weakness of the electronic coupling* in the mixed-valence complex,

$$(bpy)_2ClRu^{II}(pz)Ru^{III}Cl(bpy)_2^{3+} \xrightarrow{h\nu_{CT}} (bpy)_2ClRu^{III}(pz)Ru^{II}Cl(bpy)_2^{3+} \quad (1)$$

and *competition with nonradiative processes*, might also make observation of the fluorescence difficult."

Q: Does the weakness of electronic coupling imply a small population of the excited electronic state and consequently a faint fluorescence emission?

M: I guess this is related to what happens after excitation, right? Yes, for example, if when you excite, the coupling is extremely weak, maybe you just have a small population, if the complex doesn't absorb there is a nonradiative process . . . the weakness of electronic coupling mixing of electronic states may lead to a very small cross section . . . I think that's all . . . I don't know why it says a competition with nonradiative processes . . . if the absorption is weak . . . so . . . now, once the system gets excited then fluorescence makes use of the coupling and if the coupling is very weak, then it may not fluoresce, so, yes, that's true, first of all the absorption may be very small, then if you look at the reverse process, the fluoresce rate may be very small . . . That seems to be OK.

M259 ET Reactions in Proteins: A Calculation of Electronic Coupling

P. Siddarth and R. A. Marcus

NOTES

1. p. 8431 2nd column middle-bottom: "To formulate a search for the 'important' amino acids or atoms of amino acids, it is necessary to have a measure of the electronic coupling matrix element V_{AB} between atoms. We use the following plausible expression:

$$V_{AB} = K \sum_{a}^{A} \sum_{b}^{B} S_{ab}(\varepsilon_a + \varepsilon_b/2) \tag{2.1}$$

where a denotes an atomic orbital on atom A, b denotes an atomic orbital on atom B, S_{ab} *is the overlap integral between a and b*...”

Q: The chemical bond between two atoms is related to the extent of overlap among their atomic orbitals. The coupling matrix element between atoms is also related to that overlap. Can we then say that a strong chemical bond is the same as a strong coupling matrix element? Are there differences between the two concepts of chemical bond and coupling matrix element?

M: That’s an interesting question. Suppose that you took the two atoms to the limit then you would have a coulombic integral for each, and if you now brought them together then you’d have, you know, an $\alpha + \beta$, an $\alpha - \beta$ for the energies, so the difference in that terminology is largely in the β, but I think that using molecular orbital for the separated thing isn’t very good, so you can have a very weak coupling element and if you have a very weak coupling element, then you normally don’t call it a chemical bond.

Probably, you know, they are not exactly the same but probably...if you have two atoms and their coulombic integrals are α, supposing the same kind of atoms for simplicity, and then the matrix element is β, and their energies are $\alpha + \beta$ and $\alpha - \beta$, those are the energies of the bonding state and the antibonding state, so the bonding state has been stabilized relative to the separate atoms, by an amount β so that β is related then to the stability of the bond that would be formed. I think that’s probably reasonable, as a rough approximation.

2. Following, last line: “...through-bond and through-space coupling...”

M: This is intended for both to be coherent, so in through space you represent it not by any hopping, that's a separated mechanism, but in this contest here you have a wave function and this wave function is describing it and when it's through space the wave function amplitude in that region becomes extremely small and so it ends up with being a relatively poor way of getting there. The wave function decreases exponentially but nevertheless is finite. And the acceptor has an exponential decrease, so the interaction depends exponentially on the distance between.

Q: Can you please explain the two mechanism vs. hopping and why are they called "coherent"?

M: By the way, "through space" doesn't refer to a hopping of electrons between molecules separated by solvent, it's hopping of electron between molecules where there is nothing in between... if it's going through solvent, then it's not going through space, that's going through solvent, and the electron is making use of superexchange. It's only when there is a hole there, essentially a gap, that's going through space, when there is nothing there, nothing between the donor and acceptor. Coherent means that when you hop you actually spend some time in the intervening sites, but when the process is coherent there is just a wave function and it's sort of a wave function which is extending all the way from donor to reactant, that mechanism is then called coherent.

3. p. 8432 1st column top: "Given a protein crystal structure, it is possible to establish covalent and hydrogen-bonded connections between atoms in the protein. As a result, one obtains the through-bond connections between the atoms in the protein. Then, for each *bonded pair* of atoms A and B, V_{AB} is calculated. The through-space *links* are obtained as follows: Through-space couplings between each atom A and any other atom B within a given radius of the atom A, *but not bonded to A*, are calculated."

Q: Here you make a difference between bonds and links: apparently the through space couplings are between atoms between which there is a V_{AB} weak enough not to be considered a bond. Is it so?

M: Yes, when it is not bonded the wave function is dying out much more rapidly than in $e^{-\beta R}$ where $\beta = 1$. β is very close to 2.7, according to some Jortner's papers. But it dies out much more rapidly.

4. p. 8433 2nd column top: "The value of H_{DA} *extracted from the experimental data* for the His-48 derivative is 0.006 cm^{-1}. Of course, it is important to note that the value of H_{DA} that was derived experimentally is itself *model-dependent* and might change if a different analysis of the experimental data were used"

Q: How can one extract the value of H_{DA} from experimental data? Just from rate constants measurements? And how and why is it model dependent?

M: Yes, there are a couple of ways. One way is to vary the ΔG^0 until essentially $\lambda + \Delta G^0 = 0$, because that means you have that systematic variation and I guess people like Harry Gray and Leslie Dutton (but he's called Les) have done it for some particular cases and then the maximum rate is essentially $\frac{2\pi}{\hbar} H_{DA}^2$ times something or other, I forget, but essentially something like that. And so you can get the matrix element from the maximum rate. All right. But another way is that you measure the activation energy and you look at the preexponential factor. The preexponential factor may depend on various factors, one of which is H_{DA}, but you see, a preexponential factor actually also has in it an $e^{\Delta S^\dagger}$ which is sort of the entropy of activation, and if the ΔG^0 is temperature dependent then there is a component of ΔG^0 which is entropy of activation, you see. So, you have to correct for that, you have to correct the exponential factor for that. So, to be careful, what you put in is going to affect what you estimate the preexponential factor from the experimental data to be. Now, maybe the best way is when one can use charge transfer spectra, and there I think Norman Sutin and before him N. Hush had expressions from the extinction coefficient, from assumptions of how you estimate the matrix element. So, there are two ways. You can see, for example, that if a charge transfer spectrum is very weak, then, among other things, it can't have much of matrix element for the coupling because the transfer involves the coupling.

Q: And how and why is it model dependent?

M: Well, supposing you want to extract the preexponential factor, suppose you are not at the maximum, so you have some ΔG^0, then part of the contribution to the preexponential factor is the matrix element, but part of it comes from the entropy of activation associated with the temperature derivative of $\frac{(\lambda+\Delta G^0)^2}{\lambda}$. In other words, if you write down $\log k$ and you differentiate with respect to $1/T$, et cetera, that gives the activation energy, put it back in and you get the preexponential factor. You see that the preexponential factor can depend on these other details, you use some sort of a model for these other details, you see, your model may have the distance in it, you know the $\frac{e^2}{R}$ that comes into λ, for example, and so that affects magnitudes you try to estimate from the preexponential factor. You see that the preexponential factor depends on details that you put in.

5. p. 8433 2nd column middle-bottom: "The present model neglects any effects due to conformation fluctuation of the protein coordinates on H_{DA}."

Q: How important can the coordinates' fluctuation be?

M: Well, suppose you had several, so to speak, paths for coupling D with A, maybe in a protein, several paths like that, then for any particular configuration you can calculate the contribution to the H_{DA} from each of those paths, and there will be an interference between them, you know, H_{DA} *will involve some coherent combination and, if there are fluctuations, that will destroy the coherence, and then* H_{DA}^2 *would be the sum of the squares by each path*, so I personally think that for paths of some length there is probably a lot of cancellation. Now, recently I was at some meeting, I think it was David Beratan who discussed the question of fluctuations over the paths and I don't remember the details but he has looked into that, made some model calculations and Alexei Stokhebrukov some years ago, before Beratan, did some work on the effects of these fluctuations on the net H_{DA}, I don't remember what he concluded. So, those are two people that have worked on it.

Q: Does the coherent combination necessarily involve the presence of cross terms?

M: Yes, that's right, if you have two paths and you connect the wave functions of the two paths, you have coherence. If you don't, then, if you treat each path separately, that means you use a square for each path, separately from what you do for the other path, then the two paths aren't coherent.

Q: Path integral calculations?

M: Not necessarily path integrals. Another paper that Siddarth and I wrote on proteins should have different proteins coupling, amino acids that are coupling...now actually you should look the whole protein but, if you want it approximated, you can look at different *protein paths*, you see, that is an approximation looking at the protein as a whole, so the path integrals don't have to be involved, just protein residues. So that's a paper with Siddarth you might look at.

M261 Theory of ET Rates across Liquid–Liquid Interfaces

2. Relationships and Applications

R. A. Marcus

NOTES

1. p. 2010 2nd column top-middle: "...the rate constant...

$$k_{12}^{ll} = 2\pi(a_1 + a_2)(\Delta R)^3 \kappa \nu \exp(-\Delta G^{\neq}/k_B T) \qquad (2)$$

...κ is a nonadiabatic factor at *the distance of closest approach* of the reactants"

Q: Is the distance of closest approach across an interface the same as for two reactants in the same phase?

M: I think it would be the same for both, it would be $a_1 + a_2$, just that.

2. Following: "$\Delta G^{\ddagger}$ is given by (**M53**)

$$\Delta G^{\neq} = (\lambda/4)(1 + \Delta G^{0\prime}/\lambda)^2 \qquad (3)$$

$\Delta G^{0\prime}$ being the 'standard' free energy of reaction 1 for the *prevailing* media"

Q: What do you mean by "prevailing"?

M: Well, if one uses the usual definition of standard free energy of a reaction, one refers to a particular salt concentration, and so on, at a particular temperature and so on. But if instead you use it to mean $\Delta G^{0\prime} = -RT \log K$ in that medium, in other words the ratio of concentrations not corrected for activity coefficients or anything, just the appropriate ratio of concentrations, equilibrium concentrations, then that gives you an equilibrium constant, an *effective* one, that gives you to an *effective* ΔG^0, in that medium.

Q: So, your ΔG^0's are not related to activities but to real concentrations.

M: Yes, that's right. It is defined as an effective ΔG^0.

3. Following: "More precisely, for the present case where the reaction occurs across a liquid-liquid interface, the driving force $\Delta G^{0\prime}$ is replaced by $(E - E^{0\prime})ne$ (**M53**), where... E is the potential drop across the interface, and $E^{0\prime}$ is the equilibrium potential drop that occurs when the forward and reverse rates are equal, at unit concentrations of all four species in reaction 1:

$$Ox_1(liq1) + \mathrm{Re}\,d_2(liq2) \rightarrow \mathrm{Re}\,d_1(liq1) + Ox_2(liq2) \qquad (1)\text{"}$$

Q: You simply extend the usual Nernstian symbolism for liquid/metal interface to the case of liquid/liquid interface.

M: It seems just the obvious thing to do...

4. p. 2011 1st column top-middle: "In obtaining Eq. (5a)

$$\lambda = \lambda_1^{el} + \lambda_2^{el} \qquad (5a)$$

it is assumed in (**M255**) and in the present section 4 that the distance from reactant 1 to the liquid-liquid interface in the transition state of reaction 1 is about the same as that in the liquid-metal electrode case and that typically *the two reactants are not 'off-center' in the transition state.*"

Q: What does "off-center" mean?

M: It means that the line of centers is not perpendicular to the interface. They are on-center when the line of centers of the two reactants is perpendicular to the interface. You can see that when the line of centers is not perpendicular to the interface the distance of the centers can be huge, so I defined it that way.

5. p. 2010 1st column bottom: "*ii. Several Layer Deep Interfacial Region Model.* As noted in part 1, one alternative limiting simple model for the interface of a pair of immiscible liquids is one in which the change of composition occurs over a distance L, perhaps of the order of a few molecules thick. In this case $\Delta G^{0'}$ becomes a varying function *of the position in this interfacial layer*, and there is also a concentration variation *of each reactant in the region.*"

Q: Are the two of the reactants in the layer? Formerly, $\Delta G^{0'}$ was "the 'standard' free energy of reaction 1 for the prevailing media" and the reactants were one in phase 1 and another in phase 2.

M: Yes, two reactants in the layer.

Q: Inside the layer.

M: Yes. I think this is what I had in mind. In other words, in one model you have an abrupt transition between the two faces, one reactant in one phase, the other reactant in the other. Another model though would be that the liquids are partially miscible over a small region and both reactants are in that region. I believe that that is what I intended. They are both inside this semimiscible layer, yes. This is a model, it may not be a good model.

6. p. 2012 2nd column top: "In the transition state of reaction 1 one expects, as a first approximation, $R \simeq d_1 + d_2$."

Q: Why $R \simeq d_1 + d_2$ and not $R = d_1 + d_2$?

M: Because they may not be exactly on centers, there may be slow fluctuations.

7. Following: "In that case the deviation from additivity in Eq. (19)

$$\lambda_0 - \frac{1}{2}(\lambda_{0,11} + \lambda_{0,22}) = \frac{(\Delta e)^2}{2}\left(\frac{1}{d_1} + \frac{1}{d_2} - \frac{4}{R}\right)\left(\frac{1}{D_1^{op} + D_2^{op}} - \frac{1}{D_1^s + D_2^s}\right) \quad (19)$$

is seen to be proportional to $(d_1 - d_2)^2/d_1 d_2(d_1 + d_2)$."

Following, mid-page: "Clearly, other things being equal, *the closer the reactants are in their radii, the smaller will be* $(d_1 - d_2)^2$..."

M: When they are exactly equal, $d_1 = d_2$, that vanishes...d_1 and d_2 are the diameters. Let say that $R = d_1 + d_2$, OK? Suppose that they go as close as they can get, so that d_1 and d_2 are the radii. They get as close as they can to reduce the reorganization.

Q: Are the d's diameters or radii?

M: I don't know, I'd have to look at the derivation.

8. p. 2013 1st column top: "The recent work of Lewis *et al.*, referred to earlier, on the ferrocene-ferrocenium couple using nanometer-sized metal electrodes...The interest in the fast systems arises since only they tend to have reorganization energies λ dominated *by solvation rather than vibrational effects.*"

M: I guess the idea is that typically when you have vibrational effects, you have some significant difference of bond length change in the reactants and that can really slow the reaction down, whereas if you have a large reactant then, when it goes from one charge to another, the polarization around it changes by a relatively little amount, so λ is relatively small for them. I guess that's probably it. You know, in other words, for molecules where there are significant vibrational effects, like ferrous compounds and some others, the processes will never be fast, whereas it is possible to make the solvent effect small just by adding more larger ions, so they can be fast. Maybe this person was more able to study faster processes using nanometer sized electrodes than was able using bigger electrodes. Maybe there is less diffusional depletion. A large reactant is something with a large radius. In λ, the larger the radius the smaller is λ. In other words, if you didn't have any inner shell contribution, then just by working with large reactants you make λ small, but if there is a vibrational contribution those large reactants or small reactants might have a large vibrational contribution, those properties are independent, so the process can never be fast. The λ's are independent. The λ has *two terms* in it and sometimes the λ_i is small, there is little vibrational contribution, but that λ_i is independent of the size of the ion, it's only λ_o that depends on the size of the ion. λ_i depends on the difference of vibrational equilibrium coordinates between reactant and product multiplied by an effective force constant.

M266 Schrödinger Equation for Strongly Interacting ET Systems

R. A. Marcus

NOTES

1. p. 1754 1st column top: "...we consider next the case where the electrons of *two* reactants (or of *one* reactant in the case of an intramolecular electron transfer) interact with each other, with the nuclei of reactant or reactants, and with the orientational, vibrational and electronic dielectric polarization of the surrounding medium. The set of coordinates $\mathbf{r}_1, \ldots, \mathbf{r}_N$ of the N electrons of the reactant(s) is denoted collectively by $\mathbf{r}_e$.The free energy of formation of nonequilibrium polarization state of the medium...is...

$$W_{rev}(\mathbf{r}_e) = -[(1 - 1/D_{op})/8\pi] \int \mathbf{D}^2(\mathbf{r})d\mathbf{r} - \int \mathbf{P} \cdot \mathbf{D}(\mathbf{r})d\mathbf{r} + 2\pi c \int \mathbf{P}^2 d\mathbf{r} \quad (3)$$

upon neglecting dielectric image effects. Here $\mathbf{D}(\mathbf{r})$ is given by

$$\mathbf{D}(\mathbf{r}) = -\nabla_r \left(\sum_j \frac{Z_j e}{|\mathbf{r} - \mathbf{R}_j|} - \sum_{j=1}^{N} \frac{e}{|\mathbf{r} - \mathbf{r}_j|} \right) \quad (4)"$$

Q1: Please check if my understanding is correct: you apparently consider two (or one) molecules of the reactants and consider the interactions of electrons and nuclei of the molecules among themselves and with the electronic and nuclear polarization of the medium. I have the following questions: Do you consider.

(1) The presence of only two reacting molecules in the whole volume of the solution? What about the interaction between the couple of molecules you are considering and the others far away reactant molecules? Can they be neglected?

(2) When you consider $W_{rev}(\mathbf{r}_e)$, you consider the contribution of the whole solution. But in practice if the distance between the reactants is R, say, which is the diameter of a sphere encircling the reacting system effectively contributing to $W_{rev}(\mathbf{r}_e)$? I mean: when you draw an orbital you typically consider, for instance, the region containing 90%, say of the charge. Can we extend this way of thinking in this case?

(3) You mention here dielectric image effect: you considered them in the case of ET through interfaces, but here there is a single phase.

M: (1): Yes, you know, in the limit of dilute solutions that's OK. If you go to concentrated solutions, then you'd have to consider this other property, but in dilute solution this is sufficient. But basically it is just the ions exchanging during the charging process, the part of the solution that is far from the reactant or reactants isn't changing at all. So, it doesn't contribute.

(2) Yes, that's right, sure, you are considering practically 100%.

(3) Well, if you have two ions and you regard them as spheres or other forms, then you really have an inhomogeneous system, then every time you have a boundary you get image effects, even in this case. The dielectric constant there is, say, around 2 and is around 80 outside. So, you have a difference of electric constant. Every time you have any surface that separates two media with a difference of dielectric constants and you look at some charge interacting with it you get what is called image effects. In other words, it is a thing you have to do to satisfy the boundary conditions. Typically, that involves an infinite series.

Q: So, outside each sphere containing an ion there is an image of that ion.

M: That's right, well, there is some image... one way of putting is that the situations are more complicated that if we just regard ions as point charges. In other words, at the boundary of the ions there are certain conditions to be satisfied, there is no free charge there, so there is a certain condition on **D** and on **E**, perpendicular and parallel and so on, on any interface when you have a charge in one of the sides of the interface there are always conditions, that's loosely called image effects.

Supposing you have some boundary between two different media, each having its own dielectric constant, then if you want, for whatever reason, the potential to be constant on the boundary, for example, one of the media is a metal, then you use images to do that or you solve the Poisson equations, but the images method is a way of getting at the constant potential on the boundary. Now, I'm not sure, it may be that the only time when you want constant potential on a boundary is if the boundary is a metal, if you have a boundary between two dielectrics then I don't think that you want the field to be zero on the boundary and so . . . the only point that's bogging me is that I'm trying to remember for the dielectric . . . for the treatment of the charge to one dielectric near another and I simply don't remember that, I'd have to look back at what I did on those things. Do I ever use dielectric images other than for boundaries when one of the materials is a metal? I don't remember what I did on images Francesco, I remember what I did when one of the media was a metal but I forgot what I did when I had two dielectrics, for example, when I had two ions immersed in a media I've completely forgotten all that, I'd have to look . . .

2. p. 1754 1st column middle: "For a given $\mathbf{P}(\mathbf{r})$, Eq. (5)

$$-\left[{}^{1}\!/_{2}\sum_{i=1}^{N}\nabla_i^2 + V_{tot}(\mathbf{r}_e) + W_{rev}(\mathbf{r}_e)\right]\Psi(\mathbf{r}_e) = E\Psi(\mathbf{r}_e) \quad (5)$$

is solved for the many-electron wave function $\Psi(\mathbf{r}_e)$ and for the eigenvalue E."

Following: "

$$E = G_i(\mathrm{i} = \mathrm{r}, \mathrm{p}) \quad (7)$$

where

$$G_i = \int \Psi_i * \left[-\frac{1}{2}\sum_{i=1}^{N}\nabla_i^2 + V(\mathbf{r}_e) + W_{rev}(\mathbf{r}_e)\right]\Psi_i d\mathbf{r}_e \quad (8)$$

G_i is the energy (really free energy of solvation plus q_i- and Q-dependent electronic energy of the solute) when the extra electron is on the original donor (i = r), or on the acceptor (i = p). Thus

Eqs. (7) and (8) yield

$$G_r(\mathbf{P}) = G_p(\mathbf{P}) \qquad (9)\text{''}$$

Q: Imagine your classical parabolas. Apparently for a $\mathbf{P}(\mathbf{r})$ corresponding to the minimum of the reactants parabola, G_i in Eq. (8) is the free energy G_r at the minimum reported on the diagram's ordinate. *Such free energy is the energy of all the electrons of the reactants (reactant).* Is it so? Changing then the m parameter, $\mathbf{P}(\mathbf{r})$ changes and G in Eq. (8) changes until at the saddle point we have the energy condition in Eq. (9) which is then the energy of *all the electrons* of the reactant or product molecule (molecules) electrons interacting with each other, with the nuclei and with the nonequilibrium polarization at the saddle point. Is this interpretation of the free energy reported on your free energy diagram correct?

M: The free energy is that of a continuum because of the solvation, that's why I use the symbol G.

Q: Is such free energy the energy of all the electrons of the reactants (reactant)?

M: Yes, that's right, and, well, you know, some of the free energy is due to the interactions of the solvent molecules by themselves, all of that's included.

Q: But $\Psi(\mathbf{r}_e)$ depends on the coordinates of the electron.

M: Right, but the electron is interacting with the solvent, and is polarizing the solvent, so G is the energy stored up in the solvent also, due to interaction with the electrons and being polarized.

Q: Which influences the electronic wave function.

M: Which influences the electronic wave function in some way or another, yes.

Q: Energy of only the electrons of the reactants, even if influenced by the solvent...

M: Yes, but look at what it says here, it says W_{rev}, that means that you are charging up those ions and that is changing the dielectric polarization of the solvent, so that change is both in terms of the interaction of the charge with the new dielectric polarization and also in terms of the new energy stored up in the polarization, those things are automatically included then in the W_{rev}. I mean, look: you have some ions. Suppose there is some of the solvent outside. Now, let us imagine of charging up the ions, the charge interacts with the solvent, the solvent gets polarized, that stores up energy in the polarization, that plus the interaction of the charge with the polarization contributes to the free energy, to the W_{rev}, but, you see, part of it is the energy stored up in the polarization, which is separate in a sense, from these ions in the reactants, although it is the electrons in the reactants that are producing the polarization, of course. So, you got to figure that in, you consider somehow that the distribution of the solvent molecules is changing during the W_{rev} and so some of that energy is stored up as polarization of the solvent.

Q: But we have an expectation value of a wave function of only the electrons of the reactants.

M: That's right, but automatically that W_{rev}, if one looks into it, it will contain an integration over some oriented polarization of the solvent. So, the solvent is certainly involved, you see. The way it is involved is in the earlier formulation that I had. I considered essentially point charges. W_{rev} in a sense would depend upon on just where the electron is, therefore the solvent is going to depend on where the electron is. You see, if the electron is in this position, it will polarize the solvent a bit differently if it is in another position. So, the solvent, in an adiabatic response, is affecting the distribution of the electrons in the reactant. In a very small way, but it's there. Otherwise I would replace the $\mathbf{r}_i$ describing the position of the electron by some single point.

Q: Please explain the "adiabatic response"....

M: Yes, I think I'd phrase it differently now... the idea is that the field due to the solvent at different points on the ions is different, and that affects the potential energy that the electron sees at different points on the ions. Now, that's a small effect, but the effective potential acting on an electron in an ion, at any point in the ion, is dependent on the field produced by the solvent, quickly.

3. Following: "The Schrödinger equation based on Eqs. (3) and (4) is given by

$$-\left[\tfrac{1}{2}\sum_{i=1}^{N}\nabla_i^2 + V_{tot}(\mathbf{r}_e) + W_{rev}(\mathbf{r}_e)\right]\Psi(\mathbf{r}_e) = E\Psi(\mathbf{r}_e) \qquad (5)\text{"}$$

Q: Do you consider here $W_{rev}(\mathbf{r}_e)$ as a potential exerted on point $\mathbf{r}_e$ by the whole solution, some kind of collective potential? Can one get *a force* from it just computing $-\nabla_{r_e} W_{rev}(\mathbf{r}_e)$?

M: $W_{rev}(\mathbf{r}_e)$ is serving as an effective potential acting on the electron. That's correct, yes. Interactions of the dipoles with each other and with the charges. So, that's how it comes in and depends on positions of the electron, it is a potential that's acting on the electron. If you differentiate with respect to $\mathbf{r}_e$ you can get a force.

4. p. 1754 2nd column top-middle: "For a *noninteracting* donor and acceptor pair, such correlative interactions reduce to those for the reactants or products interacting with the solvent molecules in a way often modelled with Lennard-Jones 6-12 or exp-6 atom-atom potentials."

Q: Does the Lennard-Jones potential model the correlative interactions *at all distances*?. What does it mean the statement "for a *noninteracting* donor and acceptor pair"? Do you refer here to interactions between a reactant molecule and another solute molecule not directly exchanging an electron with it? Is it maybe it is this sense that you mean "noninteracting donor and acceptor pair"?

M: The 1/6 part is the correlative interaction at all distances, of course at small distances that formula breaks down.

Q: For a *noninteracting* donor and acceptor pair?

M: That's when they are not forming a chemical bond, I imagine.

Q: Do you refer here to interactions between a reactant molecule and another solute molecule not directly exchanging an electron with it?

M: I think so... Yes, that's right, I mean certainly not forming a chemical bond. If you bring two molecules together, you get a repulsion because of the Pauli exclusion principle, even when they don't form a chemical bond, right? And also, because of the polarizability–polarizability interaction, you get a 1/6, at least at large distances, so those two terms are distinct from forming a chemical bond. If you form a chemical bond the Lennard-Jones formula wouldn't be adequate, You'd have to have something else.

Q: It is good for everything except when there is a chemical bond.

M: Yes, that's right. I mean, of course the exponential 6 is an approximation.

5. Following: "There are *several* frequencies describing the electronic motion of the solute. *Within* a donor or acceptor molecule, the transferring electron is of a fairly high frequency, of the order of several electronvolts"

Q: The electronic states of a molecule are characterized by their quantum numbers and their energies and by the frequencies corresponding to *spectral transitions* among states. How about the *several* electronic frequencies of electrons belonging to a certain state? They sound here like intrinsic electronic frequencies of the state. This topic is apparently not treated in standard Quantum Chemistry or Physical Chemistry books. I found two references

(1) H. E. White in his "Introduction to Atomic Spectra", p. 39 writes of the electrons moving around the nucleus in Bohr's atom where "for very large quantum numbers n the quantum-theory frequency and the classical orbit frequency become equal";

(2) In "On the Theory of Oxidation-Reduction Reactions Involving Electron Transfer. I*" on p. 976 you write: "The rate constant k_2 is κ_e multiplied by the number of times per second that the electron strikes the barrier. This number is *presumably* the frequency of motion of the valence electron in the ground state of the ferrous ion. This is of the order of the frequency of excitation of this electron to the next higher principal quantum number."

The electron striking the barrier should be the transferring electron. Is it enough for it to transfer to just go to the energy of the first electronic excited state? Or maybe this is enough to get an order of magnitude value for the frequency?

M: Yes, that's just order of magnitude. It's a loose way, you know, of describing. That's a sort of classical kind of description, but if one looks at the relation between old quantum theory and classical mechanics and semiclassical, then this sort of description emerges, most people don't go into, you know, about how the ideas of old quantum theory, of semiclassical theory, permitted one to understand the relation between classical frequencies and the frequencies of quantum transitions, those were concepts that were talked about in the 1920s, say. Most people are not aware of semiclassical implications and so on, of the old quantum theory.

Q: But you use it.

M: Yes, that's ballpark, you know, there are examples in Sommerfeld's and so on, he used in the treatments of frequencies of absorption lines before one had quantum mechanics. In fact if you look at a book that came out in nearly 1900s, I think it's a three volume set in German by Schäfer, I remember I have looked at it long long time ago, I was a student, postdoc, and I was amazed at how many electronic properties they could treat with the aid of Lorentz theory, the electronic polarization, they had the electron treated as an oscillator, in fact that's where the word oscillator strength comes in, you still use it today, its origin goes back to there, when properties were treated classically. Schäfer, I think the first name was Clemens, it was a three volumes set before the Sommerfeld's series of books, before the Landau's series of books, before the Feynman's series of books, that was a series of books, Schäfer's. It's amazing how many properties they captured, out of classical theory, optical properties of polarization, all sort of properties. The justification came later on, in the form of the correspondence principle, a justification came later on in the form of quantum theory. It is of historical interest and is of physical insight interest, but most people are just interested in calculating results. It is thing of old timers . . . with a lot of history . . .

6. Following: "*When there is also a weak donor-acceptor electronic interaction*, there is also a low-frequency *component* to the motion"

Q: How is this low frequency component related to the weak electronic donor–acceptor interaction and the nature of the motion?

M: Yes, if you have a small interaction, an interaction produces an energy splitting. That splitting divided by h gives you a frequency, a classical frequency of going back and forth if you fix a position where the splitting is.

7. Following: "a measure of lowness being the ratio of the spectral frequency corresponding to the difference between the energies of the lowest and first excited electronic state *of the solute in the transition state*, as compared with *the lowest excitation frequency of the electrons of the solvent*."

Q: Is the lowest excitation frequency of the electrons in the solvent the excitation frequency between ground and first excited electronic state of the solvent molecule?

M: The polarizability is related to that lowest frequency, among other things, of the electrons of the solvent. A measure of that frequency for the solvent is of course a refractive index, which is fairly close to unity or two... What comes into the expression for polarizability is the difference in energy between the ground state and all of the various excited states.

The polarizability is really related to the electrons of the solvent, the polarizability of the solvent is related to those differences of energy levels of the solvent, and normally the lower energy levels play a role there, I mean it's a couple of electron volts or something. I you think of the electrons in the solvent, they have a series of energy levels widely spaced, you know, there are electronic frequencies there, way up in the visible or something, far greater than a few kilocalories. Well, there are many levels and so many frequencies, but the levels involved mostly in the response of the medium of course would be just the levels of lower frequencies, which are already several volts up. There are high energy states, they're virtual states, the system doesn't actually go there, just makes use of them. The system makes use of the electronic polarizability of the solvent and what's involved in electronic polarizability is not just the ground state, it's excitations, the virtual states are excited states which are way up there. Probably what would be good to do is look up an expression for the refractive index of a material in terms of the energy levels of that material, like water, and then you will see that.

Q: Is the intermediate state used in superexchange a virtual state?

M: It is virtual as far as its processes is concerned, the system never gets in there, just makes use of it. I suppose you can call it a virtual state, it's something that is made use of, the system hasn't to spend any time on it. I don't know if it helps to call it virtual state.

Q: Could one consider the TS kind of a virtual state?

M: One wouldn't call it that, one would call in classical terms a hypersurface in a many-dimensional space. That's classical, I wouldn't call it a virtual state, the system *actually passes through* something that's sort of related to it, whereas in a virtual state you never really visit it, the system just *senses* it.

8. Following: "The first difference is perhaps of the order of 0.03 eV for weak-overlap electron transfers, and of the order of ~0.5 eV for the strong overlap ones."

Q: Such numbers don't appear in your previous papers. Where do they come from?

M: Well... I don't know where I got that from what, one may look at Carol Creutz, Volume 30 of Progress in Inorganic Chemistry, she has a lot of these interactions also in charge transfer spectra..., so that's where you can find some very strong interactions and some very weak interactions.

9. Following: "The electronic wave function may be represented *approximately* as an antisymmetrized combination of terms

$$\Psi = A\Psi_{te}(\mathbf{r}_1)\Psi_{core}(\mathbf{r}_2 \ldots \mathbf{r}_n) \tag{10}$$

Q: Why is this sum of, I believe, n terms approximate? What is missing for an exact wave function?

M: I guess you wouldn't split an exact wave function into two factors. You have just a wave function of all the variables. The approximate wave function indicates a kind of $\mathbf{r}_1$ motion that is independent of the $\mathbf{r}_2 \ldots \mathbf{r}_n$ motions.

10. Following: "For a given Ψ_{core}, we note *from Eq*. (3) that *two limiting situations*, one *preaveraged* and the other *not preaveraged* over the core electrons would have from *a single term* in Eq. (10) an energy difference proportional to

$$\Delta E = \int \langle \Psi_{core} | (\mathbf{D}^2_{core} - \langle \Psi_{core} | \mathbf{D}_{core} | \Psi_{core} \rangle^2) | \Psi_{core} \rangle d\mathbf{r}$$
$$= \int \langle \Psi_{core} | (\mathbf{D}_{core} - \langle \Psi_{core} | \mathbf{D}_{core} | \Psi_{core} \rangle)^2 | \Psi_{core} \rangle d\mathbf{r} \quad (11)$$

with

$$\mathbf{D}_{core}(\mathbf{r}) = -\nabla_r \sum_{i=2}^{N} \frac{-e}{|\mathbf{r} - \mathbf{r}_i|}$$

Q: (1) I realize that ΔE has the mathematical form of a statistical mechanical fluctuation, say $\Delta E = \overline{(E-U)^2}$, or $\Delta E = \langle (E - \langle E \rangle)^2 \rangle$ where apparently $E = \int \langle \Psi_{core} | \mathbf{D}^2_{core} | \Psi_{core} \rangle d\mathbf{r}$ and $U = \int \langle \Psi_{core} | \langle \Psi_{core} | \mathbf{D}^2_{core} | \Psi_{core} \rangle | \Psi_{core} \rangle d\mathbf{r}$ and E is the term you indicate as "not preaveraged," and U is the preaveraged term. We see here the application to quantum mechanics of a formalism I was accustomed to in classical statistical mechanics...

(2) Please explain the sentence: "we note *from Eq. (3)... two limiting situations, one preaveraged* and the other *not preaveraged* over the core electrons." Which is the physical meaning of the above "limiting situations"? Are they two physical states or two different approximations? And *which is the physical meaning of* ΔE?

(3) Which is the meaning of the energy of *a single term* in Eq. (10)? Which physical situation would a single term of Eq. (10) represent? And which one is its energy?

(4) The operator in the second line of Eq. (11) reminds me of the expression in square parentheses in Eq. (69) on p. 689 in "On the Theory of Electron-Transfer Reactions. VI. Unified Treatment for Homogeneous and Electrode Reactions":

$$\lambda_0 = \langle [U_o^r(1) - U_o^p(1) - \langle U_o^r(1) - U_o^p(1) \rangle]^2 \rangle / kT \quad (69)$$

(I just remind you that $U(1)$ depends on first power of the permanent charge density ρ_a^o of the reactants). Can you comment comparing the two fluctuations?

M: (1) The first term of the operator is not averaged and the term that follows it is averaged. It is averaged over the core. The second term is preaveraged. The first term just gives **D** of the core without any averaging over, so that **D** is a function of all $\mathbf{r}_2, \ldots, \mathbf{r}_N$ variables, what is averaged over those is no longer a function of those variables. The averages are over wave functions, instead of averages over phase space.

(2) Well, when you preaverage that core term... when one electron moves in the sea of the other electrons, there is a name for that sort of thing... and that's an approximation, to regard the interaction of that electron with all the other electrons that way, this is one way of regarding it. So there is a well-known name for that...

Q: Hartree?

M: Yes, a Hartree approximation. When you have something moving in the field of everything else, that's the preaveraged approximation, and the other possibility is when you don't do it that way, when you consider everything.

Q: The splitting refers to these two situations.

M: Yes, that's right, there is an error made by sort of preaveraging, the more correct procedure is not the preaverage. So I would assume that that's the energy difference being represents about it. No, it's not two different physical states, it's two different approximations, or two different ways of operating, and you are not supposed to preaverage,... it is related to fluctuations in fields I guess... because what you see in that second line equation 11 really is like averaging, getting the root mean square average, over a fluctuation... some difference from its average... yes, what is the name of that approximation?...self-consistent field...some sort of field, you know...you have something moving in an average field...

(3) A single term is a term that neglects the antisymmetry, you know, it neglects the identity of the electrons, their indistinguishability, neglects that. Of course we know that we have to make somehow the electrons indistinguishable, then we know that can make a big difference in calculating energies...so...a single term then is almost right when you have two almost hardly interacting systems, one interacting with just an average of another system, but if everything is sort of an all mix together with the interaction, you can't do that way. And also, I mean, first of all, there are two approximations, the approximation that neglects the A, and there is the approximation of writing the wave function as two separate factors, that already has an implication and there is another if you neglect the A. Two separate things.

(4) I have to think about it...if you have an expression for the exponential of free energy, you can write in the cumulant expansion deriving an exponential in various terms, one of the terms is this quadratic expression you see there. I don't know the answer of that, I wouldn't be surprised if there is a connection, but I just don't know the answer to it. In a sense, when you think of λ_0...if you think of the upper curve...think of two parabolas, OK? Think of the second parabola being laterally displaced from the first parabola, if you go from the first parabola vertically up to where the second parabola is, that vertical segment there really represents a fluctuation on the second parabola, it may be that that's the kind of connection when you make the bottoms of the two parabolas the same, in other words when you make $\Delta G^0 = 0$. That is a fluctuation, you see, on the second parabola. I suspect when one looked at it one would find that, but I am not sure.

11. Following: "The term in Eq. (3) linear in $\mathbf{D}_{core}(\mathbf{r})$ cancels in Eq. (11) and so does not contribute to this ΔE."

Q: You consider here the quantum mechanical equivalent of $\int \mathbf{D}^2(\mathbf{r})d\mathbf{r}$ in Eq. (3). The term $\int \mathbf{P} \cdot \mathbf{D}(\mathbf{r})d\mathbf{r}$ cancels, OK. What about the term $2\pi c \int \mathbf{P}^2 d\mathbf{r}$ in Eq. (3)? Does it contribute to ΔE?

M: Maybe that is like a constant term and is not fluctuating. In other words, for fixed **P** it is the same before and after the average. The c is $\frac{1}{D_{op}} - \frac{1}{D_s}$. It is in Pekar's book "Untersuchungen über die Elektronentheorie der Kristalle."

12. Following: "Because there is little fluctuation in the charge distribution arising from the core electrons, even in the transition-state region, one can expect *terms* such as the ΔE given by eq 11 to be small."

Q: Which other terms besides the above ΔE?

M: I don't know whether in the expression there are other things that contribute besides ΔE.

Q: If there are they are small.

M: If there are, yes.

13. Following: "In the latter, when the electron is transferred a *significant distance* in the transition state, there is a large *fluctuation inside* the wave function Ψ_{te} of that electron in the transition-state region.... Thus, for this electron, it is important to use the $\mathbf{D}(\mathbf{r})$ for electron 1 which is not preaveraged, as in Eq. (4) and in Eq. (14) below."

Q: (1) What do you mean by "a significant distance in the transition state"?

(2) Do we then have here an *ensemble of possible wave functions* for the transferring electron? And what do you mean by "inside"?

(3) Why should one use a nonpreaveraged $\mathbf{D}(\mathbf{r})$?

M: (1) From the center of one reactant to the center of the other. Yes, ballpark. On the average, that would be the transfer distance.

(2) What I mean there is... in the region occupied by the two reactants, in between the two reactants. That local region. It is not an ensemble of wave functions, for different nuclear configurations you have different wave functions, in a sense you have an ensemble of wave functions but only because the wave function is influenced by other variables. For fixed nuclear coordinates you have one wave function.

(3) Because each of the electrons interacts with, say, ... an outside electron or outside ions, each would interact and you should treat that interaction in the most general way, which involves using the wave function in some formula and not preaveraging the wave function by somehow squaring and integrating over some coordinates.

14. Following: "We consider next a common approximation in which the transferring electron 1, situated at $\mathbf{r}_1$, moves in an averaged potential $V(\mathbf{r}_1)$, the sum of the Hartree potential (due to the nuclei and the other electrons of the reactants) and the exchange-correlation potential. The problem then reduces to a one-electron one, with the $W_{rev}(\mathbf{r}_1)$ given by Eq. (3), but with $\mathbf{r}_e$ replaced by $\mathbf{r}_1$ and with $\mathbf{D}$ now given by Eq. (14) instead of by Eq. (4)

$$W_{rev}(\mathbf{r}_1) = -[(1 - 1/D_{op})/8\pi] \int \mathbf{D}^2(\mathbf{r})d\mathbf{r} - \int \mathbf{P} \cdot \mathbf{D}(\mathbf{r})d\mathbf{r} + 2\pi c \int \mathbf{P}^2 d\mathbf{r} \tag{13}$$

$$\mathbf{D}(\mathbf{r}) = -\nabla_1 \left(-\frac{e}{|\mathbf{r} - \mathbf{r}_1|} + \sum_j \frac{-Z_j e}{|\mathbf{r}_1 - \mathbf{R}_j|} + \int \frac{\rho_{core}(\mathbf{r}')d\mathbf{r}'}{|\mathbf{r} - \mathbf{r}'|} \right) \tag{14}$$

where $\rho_{core}(\mathbf{r}')$ denotes the electron charge density at $\mathbf{r}'$ due to the $N - 1$ core electrons. In Eq. (14) the major difference in the contribution of electron 1 and the core to $\mathbf{D}(\mathbf{r})$ contains an average over the positions of the core electrons but not over that of electron 1."

Typo Q: Please look inside the $\sum$ in Eq. (14): Shouldn't there be an $\mathbf{r}$ instead of $\mathbf{r}_1$?

M: Yes, that should be an **r**.

Well, you know, there are a lot of subtleties here, in fact in... I don't remember what paper it was, but there were correlations that somehow I neglected or I really didn't take into account well, then David Chandler wrote a paper with K. C. Hynes maybe a couple of years after this, where they introduced a particular model where he was very careful about some of these subtleties on fluctuations, so there are some real subtleties there, this is how you treat a self-consistent field approximation, and some of the approximations in using self-consistent fields, so it may not apply to this, but it certainly applies to something related to it. This is one area I never really did explore and where David I am sure did a better job and I may have written something in response to what he wrote. You may come across that. There is something that I know I did not treat as carefully as I should. It is interesting what you are bringing it out, there are so many different sides to this problem one brings in, maybe one is not too conscious of many different aspects that really come in.

Q: How about this idea of the frequency of an electron inside an electronic state?

M: It is a classical idea but it is the kind of ideas that are roughly OK in the correspondence principle even if one should not really think of it that way quantum mechanically.

Q: But in this new attosecond experiments, don't you catch somehow the electron moving in an orbital?

M: Oh, yes, I suppose so, but you see, there you have done an experiment where you somehow localize the wave function, but it is not correct strictly speaking to think of the electron running around in an orbit, and the frequency is just that classical representation, but the main point is that there is some ballpark similarity to that. In other words you can get results that aren't too far off if you use that concept, and the justification for it comes in into the subtleties involving the use of the semiclassical theory, the WKB theory, in relating quantum mechanics to the old quantum theory. So there is a kind of justification but one has to be very very careful because strictly speaking one shouldn't speak of the electron just circling around, I mean, that's not the right description, but the main point is that if you do it in a certain way, you get something, ballpark.

Q: In the attosecond experiments...

M: You take the wave function, when you do the attosecond experiments, you localize the wave function and then there is probably a good way of interpreting all that in a way that doesn't violate quantum mechanics, and now whether there is a semiclassical theory that sort of provides somehow a description of all that, I never really looked into it, that may well be.

15. p. 1755 1st column bottom: "We let $\mathbf{D}_i(\mathbf{r})$ denote a field derivable from a potential $\Phi_i(\mathbf{r})$

$$\mathbf{D}_i(\mathbf{r}) = -\nabla_r \Phi_i(\mathbf{r}) \tag{17}$$

$$\Phi_i(\mathbf{r}) = \int \rho_i(\mathbf{r}')/|\mathbf{r} - \mathbf{r}'|d\mathbf{r}' \tag{18}$$

where $\rho_i(\mathbf{r}')$ is a charge density, which may be a sum of terms with or without *one or more delta functions*"

Q: In your first fundamental papers you use only continuum charge distributions $\rho(\mathbf{r})$ and $\sigma(\mathbf{r})$. You consider here the charge density represented also by delta functions.

M: Well, sometimes you see treatments with point charges, and so this would be intended to include those.

16. p. 1755 2nd column middle: "The eigenvalue E in Eq. (15)

$$[-\tfrac{1}{2}\nabla_1^2 + V(\mathbf{r}_1) + W_{rev}(\mathbf{r}_1)]\Psi(\mathbf{r}_1) = E\Psi(\mathbf{r}_1) \tag{15}$$

(where

$$W_{rev}(\mathbf{r}_1) = -[(1 - 1/D_{op})/8\pi] \int \mathbf{D}^2(\mathbf{r})d\mathbf{r} - \int \mathbf{P} \cdot \mathbf{D}(\mathbf{r})d\mathbf{r} + 2\pi c \int \mathbf{P}^2 d\mathbf{r} \tag{13}$$

and

$$\mathbf{D}(\mathbf{r}) = -\nabla_1 \left(-\frac{e}{|\mathbf{r} - \mathbf{r}_1|} + \sum_j \frac{-Z_j e}{|\mathbf{r} - \mathbf{R}_j|} + \int \frac{\rho_{core}(\mathbf{r}')d\mathbf{r}'}{|\mathbf{r} - \mathbf{r}'|} \right) \tag{14}$$

is varied by varying the parameter m in $\mathbf{P}(\mathbf{r})$, q_i and, in the dissociation case, Q, so as to obtain the saddle-point value of E along the reaction coordinate";

Following: "The resulting E, minus the E when $m = 0$, is *the solvational and electronic contribution to the free energy of formation of the transition state*"

Q: Please compare with the former discussion: apparently in going from the reactants equilibrium state to the transition state at the saddle point the energy change of the whole set of all reactants' electrons of $\mathbf{r}_e = (\mathbf{r}_1, \ldots, \mathbf{r}_N)$ coordinates is equal to the energy change of the only transferring electron of $\mathbf{r}_1$ coordinates, that is, the core electrons energy remains the same in going from the reactants equilibrium state to the transition state. Is it an approximation?

M: Yes, because I guess we are looking at energy differences, and the other electrons... they would cancel, in the initial state compared with the fluctuating state. Yes, that's an approximation. As long as their configuration isn't changing, though, they are having some effect... the core electrons, because they are interacting with the polarization outside too, but if they're the same throughout, all along. You know... it's an interesting question..., I wonder if that is really the case now... I have to think about that... because if the electron transfers at different positions it of course is affecting the system and the core electrons, the other electrons are included in this...

Q: Is the energy change carried by the transferring electron?

M: Well, not necessarily, it is a reflection of change of the whole configuration, it is the whole and when the transferring electron is changing, the other electrons are changing too... scaling just what the polarization is looking like as you go along the reaction coordinate, that everything inside is changing, all the electrons are changing in some way, and their interactions with the solvent are also changing. So, it is not just the transferring electron, because everything is changing. I use m as a kind of parameter that characterizes how the polarization of the whole is changing, you see, rather than how the transferring electron is changing.

Q: Nonetheless the question remains of how much of this energy change is due to the transferring electron and how much is due to the core electrons...

M: Well, suppose you have a certain charge on the core, and of course you have nuclei in the core which are largely modifying it, but supposing you have a charge which isn't fully compensated, let say in an ion, OK? And now you have an electron that is going over, all the other electrons that are not part of the transfer are interacting, you know, the nucleus is interacting but those electrons are also interacting, so all the electrons are interacting with the solvent, not just the transferring one...

Q: But you write two different Eqs. (15) and (8)...

M: All right, Eq. (8) includes all the electrons.

Q: And Eq. (15) only one electron . . .

M: Oh . . . it might be better to write that formula involving the equation of all the electrons but integrated over all the electrons but one.

Q: It is a delicate point because it goes right to the core of your theory somehow . . .

M: Yes, yes . . . , all I can think about at the moment is that it is really an approximation, that $W_{rev}(\mathbf{r}_1)$, is sort of right for a fluctuation.

Q: From the way you write it, it looks like that in going from the ground state to the transition state most of the energy is carried over by the transferring electron . . .

M: Yes, the question is, when you have very strong interaction, to what extent can you use the kind of average mean field approximation, that's basically what has been done there, and saying that all the other electrons are not doing much, yes.

Q: It looks like for your theory, that is, the case of weak coupling . . .

M: Yes, sure, there is clearly an approximation there, and with a very strong interaction you probably can't even separate that transferring electron from the other electrons, yes, exactly, you are right, you right, that's a good point. And you see, there are other points too, . . . you see, I think the new ideas for strongly interacting systems are probably in that paper of Chandler, because Chandler wrote one where he brought in Hynes because Hynes had made a mistake and I think this is an oversimplification too, so I may have something that's fishy there. But Hynes had made a mistake and Chandler, you know, did it right, and out of courtesy and maybe to have a good discussion too, brought Hynes too into the paper, so there is a paper involving Chandler, Hynes, and two students, one of each. A paper with four authors. And indeed around this time, sometimes around this time, or a little bit later, they were really doing a much better job on the subtleties, what is going on in a sort of adiabatic or intermediate adiabatic–nonadiabatic regime.

Q: By the way, I have seen a reference to the book of Clemens Schaefer...

M: The classic set of volumes of Clemens Schaefer, I was amazed at looking at them at the time when I was a postdoc, I was amazed when I was looking at those, and to see how with the Lorentz oscillator, I mean this is before quantum theory, they captured so many optical properties of crystals and other properties, and of course what has been preserved since then is the idea of oscillator strength.

17. Following: "An improvement in this E can be made by then expressing $\mathbf{P}(\mathbf{r})$ in terms of its Fourier components c_{k} and truncating the set of components."

Q: I do not think that until this paper you mentioned the above way of expressing $\mathbf{P}(\mathbf{r})$. How does it compare with your classical $4\pi c\mathbf{P}(\mathbf{r}) = \mathbf{D}_r(\mathbf{r}) + m[\mathbf{D}_r(\mathbf{r}) - \mathbf{D}_p(\mathbf{r})]$?

M: If you take that $\mathbf{P}(\mathbf{r})$, then you can write a Fourier decomposition in terms of $e^{i\mathbf{k}\cdot\mathbf{r}}$, OK? And so you can write it as some integral or some sum over $c_{k}e^{i\mathbf{k}\cdot\mathbf{r}}$, it would be in the Russian papers, it would be in the book of the fellow from Denmark... of Ulstrup. Anyway, it should be in various places, and, of course, even though in the solvent there aren't really harmonic oscillators, normally they have been using harmonic oscillators assumptions when they do that. This is a way of discretizing, instead of working with continuum variables you can work with discretized variables and especially if you want to put in an approximate dynamical model, like oscillators, which the Russians do and some others, you know, that's convenient. Convenient for calculations, of course it is an approximation.

Q: How does it compare with your classical formula there?

M: Oh, well, the classical formula does not assume any harmonic oscillator decomposition, it has a harmonic free energy, but that is much milder than a harmonic oscillators approximation. You can have all sorts of local anharmonicities, and nonetheless get harmonic free energy. I mean, many people have done computer simulations and have found harmonic free energy curves, even though you have water molecules which are very anharmonic in motion. Part of the understanding of that may be that if one does cumulant expansion in statistical mechanics and if one goes up just to the second order in the cumulant expansion, then you run a harmonic approximation for the free energy. So, my treatment was harmonic although I never used cumulants, I used something that was related to that without realizing it, my theory then makes no *molecular* harmonic approximation. On the other hand, the advantage of the molecular harmonic approximation is that because harmonic oscillators are so simple you can go far more, you know, into details, dispersion and all that, but of course you are making a major assumption. I wouldn't say that one is more general than the other, mine is more general in a certain respect, theirs is more general in a different respect. I made use of it in a paper with Georgievski, in a later paper, that's right, and there we used the harmonic approximation and then showed that we could get certain relationships that we had gotten using a continuum description.

Yes, there would be rigor in what they do in the harmonic oscillator case. The point is that the way I treated it allows passably for anharmonic motions of the solvent, if you use harmonic oscillators for everything you don't get any anharmonic motion but *the actual behavior in the solvent molecules is anharmonic*. But, for example, when you go over to quantization you can quantize a harmonic oscillator very easily and that's what they did, but you can't quantize what I did, you'd have to use much more sophisticated methods, like path integral methods, that's just one way.

18. Following: "Using the functional two-state form for E obtained in ref 5 or often used before, we would have

$$E(P, q, Q) = {}^1\!/\!_2[G_r(\mathbf{P}, \mathbf{q}, Q) + G_p(\mathbf{P}, \mathbf{q}, Q)] - [{}^1\!/\!_4(G_r - G_p)^2 + H_{rp}^2]^{1/2} \quad (24)$$

where now the dependence of the *diabatic* eigenvalues G_r and $G_p \ldots$"

Q: From a physical point of view, does this mean that your classical G's for weak-overlap electron transfer, that is, for $H_{rp} \simeq 0$, are diabatic curves?

M: Yes, they are based on a diabatic approximation of electronic motion. That's right.

19. p. 1756 2nd column top: "For *strong overlap ET's either some simple functional behaviour* should be formulated..."

Following: "The *simplest functional behaviour* for a *strong-overlap* system would be a two-state one, an approximation to Eq. (24): Some generalized reaction coordinates q could be introduced (**M30**) and the results of the original quantum plus statistical mechanical calculations *fitted* by an equation of a functional form

$$G(q) = {}^1\!/\!_2(G_r + G_p) - [{}^1\!/\!_4(G_r - G_p)^2 + H_{rp}^2]^{1/2} \quad (25)\text{"}$$

Q6: Do you mean that, contrary to the above, the adiabatic curves are suggested to describe free energy curves for strong-overlap ET? Is it so because the adiabatic curves are the ones usually computed in quantum chemical calculations?

M: Yes, supposing that you did the calculation in a rigorous way and their interaction was so strong that you really called it an adiabatic electron transfer, and you can calculate the free energy of the whole process in a rigorous way. Supposing you did that. That is a horrendous undertaking. But suppose you did that, and so you had then the G as a function of some kind of reaction coordinate. Then you could try to approximate that function by the one that I wrote down there. People have made calculations for some very strong interactions, for example, in some charge transfer transitions, they may have computed some adiabatic curves too. And I think it was when Hynes tried to compute the adiabatic curves that Chandler detected that there was something that was funny there, anyway... if you look at the Chandler Hynes paper plus two other papers, then you might see something.

20. Following:

$$\text{“}G_r(q) = \frac{k}{2}(q - q_r)^2, \quad G_p(q) = \frac{k}{2}(q - q^p)^2 + \Delta G^0 \qquad (26)$$

k is the curvature of the '*assumed quadratic*' G_r and G_p curves, and the *fitted* H_{rp} is *assumed* independent of q."

Bottom: "If the λ appearing in these equations proves to have the same additive property that it has for weak overlap ET's"

Q: In this paper, you are trying to extend your classical treatment for weak overlap ET to the case of strong overlap. When is this extension possible as suggested, to what an extent, and when not?

M: Yes, I can answer that a bit... I tried in a very oversimplified description, using a BEBO model, so that's far simpler than this, to obtain adiabatic curves, and then, using that, I showed that using those adiabatic curves for the cross reaction and for the self-exchange reactions, to some extent you had the additivity approximation. Then I also showed that, although there was no inverted region because when you have a strictly adiabatic reaction that disappears, for the ΔG^0, when λ is relatively small, you can still have that linear relation where the slope is 1/2 and so on, what you have for diabatic. So, certain properties carry over for the adiabatic, as long as you are not getting anything near the inverted region.

Q: Why there is no inverted region in the adiabatic case?

M: Well, if you think how the inverted region comes up, take the two parabolas, consider a weak interaction, by lowering one parabola steadily you can see how the inverted region comes up. But if you have an adiabatic, now instead of the two parabolas, take a strongly interacting system, now distort that by lowering the well of the products, and you'll see that you'll never get the inverted region. Because the curves are kept separate then one curve just slides downhill more and more, that's all. Just think of the usual skewed axis description, of an atom transfer reaction, for example, in Eyring's book and in many other places. As you lower one of the wells, things are just going roll down faster, that's all, there is no kind of topology which is similar to that for the weak overlap case. The lower curve will always stay lower because of the repulsion of the curves. You can never get a reversal.

21. p. 1757 1st column middle: "Until now, one method which has been used to approximate λ for *strong overlap* ET is to assume that it equals the value for the *corresponding weak overlap* ET, multiplied by a factor α^2

$$\alpha = H_{rp}/h\nu_a^{\max} \tag{35}$$

α^2 representing in first-order perturbation theory the factor describing the *reduction of charge* on the donor center due to the strong electronic interaction of the charged centers."

Q: (1) When and how is a strong overlap ET corresponding to a weak overlap one?

(2) Why and how does a strong electronic interaction lead to a reduction of charge on the charged centers? Is this reduction of charge on the donor center a localization vs. delocalization situation?

M: (1) I don't know the answer to that without looking at the reference. Certainly Hush and Creutz have been interested in charge transfer spectra, and so I guess that they look at the vertical transition, and they ask what happens if you take your diabatic curves and introduce some repulsion in the form of that $H_{12}^2/\Delta E$. You know, you introduce that and they maybe ask how much that is going to change the lower curve, is going to change the upper curve, when you introduce that, and so how much it is going to change the vertical transition. So, it may be that that's what they have done. Start with diabatic curves and then you add the perturbation in, so that the perturbation not only changes the splitting but it also changes the lower curve and changes the upper curve, in roughly this manner I guess.

Q: How does that all relate to the above sentence "*reduction of charge on the donor center due to the strong electronic interaction of the charged centers.*" in your paper?

M: Supposing you have two separate charges, and now supposing you have some binding of one of the charges to the other, so that the excess electron is not on the negative donor, is somewhere in between, then you no longer have such a clear charge distinction, such pure charge separation, I think that's what I meant. So that effectively reduces the charge on the donor, it distributes it more, it diffuses charge on the acceptor, it distributes it.

(2) Well, by reducing the charge on the charge center, it is delocalized in some other part of the system, it's taking part of the initial charge and spreading it over some other part of the system.

22. p. 1757 2nd column bottom: "Other strong-overlap ET's such as those accompanied by bond dissociation and separation of the fragments, would have *additional specific solvent effects* which would tend to obscure *the solvation effects considered in Eq. (5)*. Perhaps the most suitable systems for investigating *solvent effects of the type discussed in ref 3*"

Q: Which ones are the additional solvent effects?

M: You know, there are some solvent effects where you have a kind of bonding, almost a chemical bonding of sorts, between the solvent and the charge, it changes the distribution of charge on the ions, say, whereas there is a solvent effect which really doesn't do that, you have that same charge distribution on the ion, and that charge distribution is interacting with the solvent back and so on. But in some cases instead, the solvent sort of binds with the charge and changes the charge distribution. So that's different. And, for example, you'll see solvent scales which are based upon more on this latter kind. Some sort of bonding.

CHAPTER 4

ET Matrix Elements of Bridged Systems: A Molecular Fragment Approach; Theory of ET, Comparison with Experiments; Tight-binding for Semi-infinite solids; ET in Proteins: An Artificial Intelligence Approach to Electronic Coupling; Surface Properties of Solids Using a Semi-infinite Approach and the Tight-binding Approximation; Theoretical Model of Scanning Tunneling Microscopy

Interviews on M269, M270, M272, M273, M274, M275

M269 Calculation of ET Matrix Elements of Bridged Systems Using a Molecular Fragment Approach

P. Siddarth and R. A. Marcus

NOTES

1. p. 3214 1st column middle: "**Theory**—When applied to *atoms* the perturbation approximation given by Eq. (2)

$$H_{DA} = H_{D1} \left(\prod_{i=1}^{n-1} \frac{H_{i,i+1}}{E_D - E_i} \right) \frac{H_{nA}}{E_D - E_n} \tag{2}$$

poses the problem that $|H_{i,i+1}/(E_D - E_i)|$ and $|H_{nA}/(E_D - E_n)|$ may not be less than unity, as required in a perturbation expansion. For example, in the system investigated below, the *atom–atom* coupling elements $H_{i,i+1}$ and H_{nA} are typically about 4–5 eV, and $E_D - E_i$ or $E_D - E_n$ is about the same value. In contrast, when extended to *molecular fragments*, as in the present paper, the magnitudes of the coupling are typically reduced to 0.2 eV or less, while the magnitudes of $E_D - E_i$ becomes about 1 eV. In this case, therefore, the extension of Eq. (2) to fragments makes the new equation an excellent approximation. The coupling between these fragments is smaller because, unlike individual atoms in a bridge, the coefficients of the relevant orbitals at each fragment-to-fragment link are relatively small instead of being unity."

Q: Can you give a physical explanation for the coupling between molecular fragments being smaller than the coupling between atoms?

M: When you use group orbitals on one molecule and group orbitals on the other molecule, and there is a bond joining the two molecules, really in this case the two different amino acids, then in any orbital, molecular orbital, its coefficient is distributed over the whole orbital, that means that the coefficient at the atom that is going to be binding to the next one is relatively small, you know, if there are 20 atoms relevant in that amino acid, then that coefficient is $1/\sqrt{20}$, so it is small, that means that in the interaction of two such atoms one on the ith amino acid and the other on the jth amino acid is small, because there you are dealing with small coefficients, so there is a small active perturbation of the original orbital than otherwise. Do you see what I mean?

Q: Yes, when you are combining two amino acids and they are bound by two atoms, one belonging to one amino acid and the other to the other amino acid...

M: Right

Q: Then each one of these atoms has a coefficient rather small because the orbital is distributed over the whole amino acid.

M: Right, that's correct. OK, that's the answer.

2. p. 3215 2nd column middle-bottom: "A generalized coordinate q_i is introduced for each reactant $i = (1, 2)$, to include both solvent and (low frequency) vibrational contributions to the free energy fluctuations. The free energy G^r of the reactants and bridge in reaction (12)

$$\mathrm{Red}_1 + \mathrm{Ox}_2 = \mathrm{Ox}_1 + \mathrm{Red}_2 \tag{12}$$

is then written as

$$G^r = \left[\frac{k_1}{2}(q_1 - q_1^r)^2 + \frac{k_2}{2}(q_2 - q_2^r)^2\right] + G_1^{r,0} + G_2^{r,0} + E^r \tag{14}$$

where the first two terms are associated with the *fluctuations around reactants 1 and 2*, respectively, the next two are the equilibrium solvation free energies, namely the values at $q_i = q_i^r$, and *the last term* is the *electronic energy* of the *entire system* at $q_i = q_i^r$."

Q: (1) Why does one consider fluctuations around the reactants and not around the bridge?

M: The electrons don't actually exist on the bridge, exist on the reactants, the bridge is at too high energy for them to exist, so they are virtually there, there is always zero probability of being on the bridge, therefore they don't polarize the solvent around the bridge. If, by the way, the bridge were some aromatic molecules, more or less in resonance with the reactants, then the electrons could reside on the bridge, then you have to take into account the bridge to form an intermediate, that would exist for finite time, and solvate, but usually these bridges are saturated hydrocarbons, are something like that, and a couple of them are in resonance.

Q: (2) Is the "entire system" the reactants + bridge system or the reactants + bridge + solvation molecules system? My guess is that you refer only to reactants + bridge.

M: It's the latter, yes. The electronic energy is then that of the reactants, yes. Reactants plus bridge. The bridge electronic energy just doesn't change at all. Whereas in ET the electronic energy of the reactants changes.

Q: (3) Is such an electronic energy characterized by quantum numbers? We have here a supermolecule made up by reactants + bridge. In the third book of Herzberg's *Electronic Spectra*, a chapter is devoted to building up principles of molecules from atoms. Are there analogous principles to build up electronic states of the reactants + bridge system from the electronic states of separated reactant 1 + reactant 2 + bridge?

M: Yes, except that in a sense is kind of trivial because the bridge is normally just weakly coupled to the reactants.

Q: (4) The reactants + bridge + solvation molecules is some kind of a mesoscopic system. How big is its characteristic length? How many molecules (order of magnitude...) would make up the whole ET reacting system, reactants + polarized solvent molecules? In other words, supposing the system contained in a sphere, how big the diameter of the sphere would be?

M: The field falls off as $1/r$ or something like that, and so the number of molecules involved goes as r^2, but of course increasingly less influenced, you know... I imagine it goes out quite a distance, because you have a slowly varying field, you see. If you had there some ions that shielded it, then of course it would fall off exponentially, and the characteristic length there is the Debye length, but when the field is it a Coulomb field, then it goes out for quite a long length, and the number of molecules involved goes as r^2. The size... one way of evaluating it may be, I suppose, to use the Born formula for solvation, something like the Born formula but let's stop any polarization at some finite distance r, that's an electrostatic problem one can solve, I never tried to do it, but it is something one can solve and answer your question.

Q: (5) Being the system reactants + bridge + solvent molecules made up of a varying number of solvent molecules its electronic energy can be defined only as some mean energy, I believe. And moreover, it is probably very difficult to define quantum mechanically. The electronic energy is in this case to be considered as just the electronic part of the thermodynamic internal energy of the mesoscopic system?

M: Yes, I think so, that is normally so when you add on the solvation energy, and the vibration energy, that's the usual thing, in other words you consider the lowest energy of the system, then add solvation energy and . . . you know, it's a matter of definition, and everything is tied together, so you can define terms pretty much any way you want. If you have a term that you call solvation energy then you also speak of electronic energy and you want to be sure that you are not double counting. So, you just have to be careful, that's all.

Q: (6) When we have a vibrating diatomic molecule can we consider the stretching oscillation elongating the equilibrium bond length a *mechanical vibrational fluctuation* and the going back toward the equilibrium bond length a *relaxation*? Or is the fluctuation a purely *statistical mechanical* concept that cannot apply to a single diatomic molecule?

M: When I have a single diatomic molecule, then any vibration it had is independent on the surroundings because it is single, therefore the process it is not statistical. If it is isolated it won't relax except by radiation.

Q: You can then speak of relaxation only in a statistical sense.

M: Yes, that's right.

3. The statistical mechanical fluctuation defining λ_0 is

$$\lambda_0 = \langle [U_o^r(1) - U_o^p(1) - \langle U_o^r(1) - U_o^p(1)\rangle]^2\rangle / kT$$

Q: (1) Is there an analogous formula defining λ_i? I mean, you define λ_0 as $\lambda_0 = (ne)^2\left(\frac{1}{2a_1} + \frac{1}{2a_2} - \frac{1}{R}\right)\left(\frac{1}{D_{op}} - \frac{1}{D_s}\right)$, but also as above in terms of fluctuations. And you define λ_i as $\lambda_i = \frac{1}{2}\sum_j k_j(\Delta q_j)^2$. Is there even in this case a definition in terms of fluctuation, maybe of the form $\frac{\lambda_i = \langle [U_i^r - U_i^p - \langle U_i^r - U_i^p\rangle]^2\rangle}{kT}$ where U_i^r is the sum of the instantaneous vibrational potential energies of all normal modes?

M: I don't know, I'd have to think it through.

Q: (2) Are the different normal modes vibrations in some phase relation or are the phases of the different normal modes largely uncorrelated?

M: Uncorrelated. They are independent degrees of freedom. That's the purpose of having normal modes.

Q: (3) Is it possible to draw a parallel between different solvent dipoles randomly oriented around a reactant molecule, giving an instantaneously varying polarization, and different normal modes of the reactant whose sum of random potential energies give an instantaneously varying total vibrational potential energy of the reactant?

M: Well, people, some of the Russians in their work, use a harmonic oscillator treatment of the solvent, I think that's artificial and, for example, it doesn't account for large changes in entropy of reaction that can occur when you have a change in charge in the ions, so people have tried to do that, in fact have done that in the literature, but I think it's artificial.

I see what you mean. Yes, you have fluctuations in the amplitudes of the normal modes, and you have fluctuations in the solvent, so there is an analogy there, Yes.

Q: (4) How are the two different definitions of λ_0 related? Is it maybe so, that $\frac{\lambda_0 = \langle [U_o^r(1) - U_o^p(1) - \langle U_o^r(1) - U_o^p(1) \rangle]^2 \rangle}{kT}$ is the general formula for fluctuations while $\lambda_0 = (ne)^2 \left(\frac{1}{2a_1} + \frac{1}{2a_2} - \frac{1}{R}\right)\left(\frac{1}{D_{op}} - \frac{1}{D_s}\right)$ is *that* fluctuation which is needed to reach the transition state?

M: I believe that's correct, I don't think that in papers I showed the connection of the two formulas, I don't think I did, But I think it to be correct. How one would show it is another story, yes... What it is, is sort of how you relate a statistical mechanical expression to continuum expressions, you see, and so now from statistical mechanics one should be able to introduce some operations and get a continuum expression as an approximation, all right? It is just an approximation, of course, and so there is probably some way of doing that here, but you see the problem would be how do you relate the Born solvation to the statistical mechanical expression for the solvation, it is not immediately transparent how you do that, I'm sure that it has been done under certain conditions in the literature, though, that somebody said: OK, here is the formula for the statistical mechanics of this ion and solvent and if you do such and such and such, then we approximate such and such and we get the Born formula out. It involves quite an elaborate calculation, and it is only approximate because the Born formula is only approximate... The problem of showing the relationship is the same as the problem of deducing the Born formula from statistical mechanics, perhaps somebody has done that, they certainly have done something analogous in some transport phenomena, where they have shown some connections, you know, some derivation of the Stokes law and something like that, but these are approximate derivations.

Note: *Both formulas are related to the λ_o's to reach the transition state.*

4. p. 3216 1st column middle: "We consider next some omitted bridge effects. In the vertical charge-transfer spectrum *from the donor to the* *adjacent fragment of the bridge*, the electron on the bridge polarizes the electrons of the solvent, an effect which lowers the value of $h\nu_{CT}$. Again, in the case of Δ_{TS}, which is involved from the donor to the acceptor *via the superexchange mechanism*, the transferring electron in the *virtual orbitals* of the bridge interacts with solvent polarization, both the electronic and *the (static) orientational*, an effect which may tend to lower Δ_{TS}."

Q: But if one has a superexchange mechanism is the residence time of the transferring electron on the bridge long enough to affect the slow orientational polarization?

M: The answer is no. That *above* was in a CT spectrum to the bridge, that's where it is put on the bridge by light, not in the reaction.

5. p. 3217 1st column middle: "One way to correct for the variation of FC with distance is to study the temperature dependence of the rate constant, a procedure used by Isied and co-workers. From the experimental heats of activation, the variation of *the electronic factor alone* in Eq. (23)

$$k_{\mathrm{ET}} = \frac{2\pi}{\hbar}|H_{\mathrm{DA}}|^2\mathrm{FC} \tag{23}$$

and values of H_{DA} are then obtained."

M: If you measure the rate constant, then if the rate constant is equal to that quantity there, in Eq. (23), times an activation energy, $e^{-E_a/kT}$, then if you measure the activation energy you could correct the rate constant, and then extract this quantity. You have to be careful that you may also have an entropy of reaction, so you have to take that into account too. In other words, the pre-exponential factor may not be just this but there may also be an entropy term associated with, you know, some entropy of activation.

6: p. 3217 2nd column top: "It is seen that both occupied and unoccupied orbitals are needed"

Q: By "occupied" do you mean "singly occupied" or even "fully occupied"?

M: Probably I mean both. In other words, the question is the change in occupation in electron transfer.

REVIEW M270 Theory of ET Reactions and Comparison with Experiments

R. A. Marcus and P. Siddarth

NOTES

1. p. 55 top: "In the present case, this transition is one from *the electronic configuration* of the reactants to that of the products, and is the electron transfer."

Q: Is "electronic configuration" synonymous of "electronic state" or does it carry some other meaning in the case of ET? As a matter of fact, a molecular electronic state is a well-defined state characterized by well-defined quantum numbers. But here we have a complex mesoscopic state with many solvent molecules whose number is also fluctuating. Which one is then the rigorous definition of "electronic configuration" in electron transfer theory?

M: By the way, it is intended to be the same as electronic state, nothing deeper. Of course, it is true that there are fluctuations and so on but, you know, it is largely the electronic state.

2. p. 59 top: "In order to find the values of $\mathbf{P}^{\neq}(\mathbf{r})$ and $\mathbf{q}^{\neq}$ in the TS, $G_r(\mathbf{q}, \mathbf{P})$ is minimized with respect to variations in $\mathbf{P}(\mathbf{r})$ and $\mathbf{q}$, subject to the condition that $G_r(\mathbf{q}, \mathbf{P})$ and $G_p(\mathbf{q}, \mathbf{P})$ are equal at the transition state:

$$G_r(\mathbf{q}, \mathbf{P}) = G_p(\mathbf{q}, \mathbf{P}) \qquad \text{(TS)} \tag{19}$$

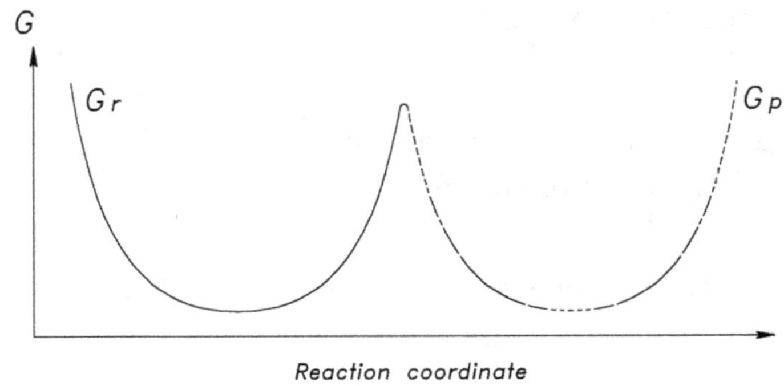

Fig. 1*.

Thus,

$$\delta G_r = 0 \tag{20a}$$

subject to

$$\delta G_r - \delta G_p = 0 \tag{20b}$$"

Q: Please check if the following pictorial illustration of the above is correct. Please look at my Fig. 1* where the schematic free energy surface for adiabatic ET is shown, with the left part of it drawn as a continuous line representing G_r and the right part of it as a dashed line representing G_p. At the TS, the two curves meet and have the same value, so $G_r(\mathbf{q}, \mathbf{P}) = G_p(\mathbf{q}, \mathbf{P})$ and, there being a maximum at TS, it is also $\delta G_r = \delta G_p = 0$.

M: Yes, except that for a weak overlap ET, it would have more of a cusp rather than a gentle slope around the middle.

Q: If the parabolas are the same the slopes at the crossing are the same but of opposed sign.

M: Ja, but only if $\Delta G^0 = 0$. If you have $\Delta G^0 \neq 0$, the slopes of course are different.

Q: If the slopes are different, then you don't have any more $\delta G_r = -\delta G_p$

M: If you go from the bottom to the transition state, yes, then, look, $G_r(\mathbf{q}, \mathbf{P}) = G_p(\mathbf{q}, \mathbf{P})$ is *well-defined* when there is essentially no interaction. When there is interaction, then G_r and G_p are first order terms, and you have a correction due to the interaction that comes in, which causes the splitting. When the force constants are different, and even when the force constants are equal but $\Delta G^0 \neq 0$, the slopes would be different at that point. Because the slope is *kx* and the farther the slope from its minimum, the bigger the slope.

Q: But in this case you don't have anymore $\delta G_r = -\delta G_p$.

M: With this rounded thing more than at zeroth order you don't have really have a G_r at that point. The G_r only refers to the zeroth order value where occurs with cross. $G_r(\mathbf{q}, \mathbf{P}) = G_p(\mathbf{q}, \mathbf{P})$ is approximate in zeroth order at the intersection, but because of its being not of zeroth order you get a splitting there. $G_r(\mathbf{q}, \mathbf{P}) = G_p(\mathbf{q}, \mathbf{P})$ is only in the zeroth order thing. Because G_r refers to the electronic configuration to the reactants, but in the rounded curve, when you are in the transition state, you know, near it, you don't have a G_r as such, *G_r refers to the diabatic curves, not to the adiabatic.*

Q: But is the condition $\delta G_r - \delta G_p = 0$ always valid?

M: Well G_r refers to the electronic configuration of reactants that is valid when you think of the unperturbed system, of the crossing, when you are looking at to the maximum there, G_r as far as referring to the electronic configuration of reactants doesn't exist. That's a diabatic concept. $\delta G_r - \delta G_p = 0$ is valid when the splitting is small, when it is reasonably valid to think of an electronic configuration having an energy that is approximately that at the top. In other words, in the case of small splitting.

3. Following: "

$$\lambda_0 = \frac{1}{8\pi c}\int (\mathbf{D}_r - \mathbf{D}_p)^2 d\mathbf{r} \tag{25}$$

$$\lambda_i = \sum \frac{1}{2} k_i (q_i^p - q_i^r)^2 \tag{26}$$"

Q: (1) Do you use here Eq. (25) for the first time? If not, where did you introduce it?

(2) The forms of the above two equations are very similar and they appear to suggest a parallelism between q_i^p and $\mathbf{D}_p$, between q_i^r and $\mathbf{D}_r$. Is it so? In this case it would be better to invert the two symbols inside the parentheses in Eq. (25), I guess.

M: (1) You mean the first time that I have written down explicitly, even though some of the things may have appeared before, is that it? In fact, I think that came later, it wasn't that simple. I think that in the 1956 paper there is a kind of simplicity when you use **D** and **E** instead of **P** and something else. And I used **D** later on, much later, because there is a kind of simplicity in the formalism, but when you put **P** in there, things are more complicated.

(2) If the transition state were near the reactants, then that $\mathbf{D}^\ddagger$ would be $\mathbf{D}^r$, if it were near the products it would be $\mathbf{D}^p$. In other words its typical value is the value where the TS is, depending on the ΔG^0, and if ΔG^0 is quite negative then that **D** would be like the $\mathbf{D}^r$, the reactant would be the TS, and so on, so that $\mathbf{D}^\ddagger$ is sort of measure there of the TS in capital **D** space, it's a reaction coordinate in the capital **D** space, in other words the reaction coordinate can go steadily from reactants to products, this $\mathbf{D}^\ddagger$ if you change ΔG^0 continuously can go steadily from reactants to products, so it's a kind of a reaction coordinate.

Q: Normally you consider m as a measure of the reaction coordinate.

M: That's also a measure, there are *multiple measures of the reaction coordinate*, if m happened to be zero, then the TS would be like the reactants, if it were like -1 it would be like the products, if it's some way in between the TS is in between. If one wanted to—although I don't recommend it—one can vary it continuously and one would generate the whole set of systems in which you move along the reaction coordinate, so, in a sense, you can relate them all to a reaction coordinate, they take a particular value at the TS for the given reaction, but they could have had any value depending on ΔG^0.

Q: So they are both OK...

M: Yes, I mean I wouldn't use them as a reaction coordinates, but I would use, as I had in the 1960 paper, the *difference of potential energies surfaces*, that *global* reaction coordinate, those are sort of quickies....

4. Following: "The ν in Eq. (14a)

$$\kappa\nu = \nu \qquad \text{(adiabatic)} \tag{14a}$$

is approximately given by

$$\nu = \left(\sum_k \lambda_k \nu_k^2/\lambda\right)^{1/2} \qquad k = (i, o) \tag{28}$$

using the notation in Eq. (24),

$$\lambda = \lambda_0 + \lambda_i \tag{24}$$

or if there are several vibrational frequencies contributing to $\lambda_i \nu_i^2$, the sum in Eq. (28) is over all of them, $\sum_j \lambda_j \nu_j^2 + \lambda_0 \nu_0^2$. The often neglected ν_0 is associated with the "*inertial motion of the solvent.*"

Q: What do you mean by "inertial motion of the solvent"?

M: The solvent has sort of slow frequencies, the usual dielectric relaxation, but there are also some high frequencies, related to the *vibrational libration*, of the water molecules for example, that's the motion that's responsible for infrared frequencies, the dielectric constant being something around 4.5 instead of about 1.8, so you can think of it as sort of the solvent having oscillators, that would correspond to whatever the infrared frequency is associated with.

Q: Which ones are the librational motions and in particular the vibrational librations?

M: Anything that isn't just a hindered rotation is basically a vibration,... *librational motion is a motion of rocking back and forth, it means not fully rotating,* as you have in hydrogen bond situation, in water the vibrational polarization associated with it is responsible for the dielectric constant of water in the infrared being about 4.5 instead of being 1.8. When you consider the dielectric response function of water, you know, at zero frequency the real part is 78 and then gradually it goes down as you go to higher frequency, it goes down a lot where you have a major change, like where the frequency becomes too fast for the librating molecule to follow, and then the only thing that can follow it are the vibrations, and then, when you get to a frequency which is like the bending frequency, if you go higher than that, the dielectric constant drops from 4.5 down to 1.8 or something like that. So, in other words, what this is all about is interpreting the dielectric function of solvent or of anything else in terms of the various kinds of dynamical motions that occur and different dynamical motions have somewhat different characteristic frequencies. One shouldn't say vibrational libration, that's really not right, one could say associated to libration...typically libration is just sort of a vibration with a large amplitude, in other words it is quite a rotation but it doesn't quite go round, but it has large swings, sometimes they call that a libration, a libration is a particular case of a vibration and a libration will certainly be very anharmonic.

Q: Why do you call it inertial?

M: Oh, inertial because it's vibrational, instead of relaxation. If you think of motion in a system, at very high frequencies the system will oscillate, and then at longer times things have damped out, you know, and the system is not oscillating. And that's referred to as inertial because to get an oscillation you have a mass times acceleration, and that term is inertial. Whenever you have vibrations you have a mass times acceleration, when you're only talking about slow motion you neglect that mass times acceleration and you have sort of *diffusive Langevin equations* describing the motion.

5. p. 62 middle: "If the reactant were bound to the electrode, instead of being free to move in solution, Eq. (36)

$$\lambda_{el} \cong \frac{1}{2}\lambda_{11} \tag{36}$$

would remain applicable only *if the dielectric aspects were similar in the two systems* and if the R in Eq. (27)

$$\lambda_o = \left(\frac{1}{2a_1} + \frac{1}{2a_2} - \frac{1}{R}\right)\left(\frac{1}{D_{op}} - \frac{1}{D_s}\right) \tag{27}$$

were equal to the R in Eq. (35)

$$\lambda_{o,el} = \left(\frac{1}{2a_1} - \frac{1}{R}\right)(ne)^2\left(\frac{1}{D_{op}} - \frac{1}{D_s}\right) \qquad (35)\text{"}$$

Q: What does it mean "if the dielectric aspects were similar in the two systems"?

M: Well, if the solvent, water say, were the first layer, were sort of bound to the electrode, it wouldn't have the same rotational property as it does in the bulk liquid, and so the dielectric constant would be effectively small, there would be less dielectric response to the field, that's what I meant. Similar because a water bound to the electrode, as part of the solvent, is different from the water not bound in solution.

Typo Misprint: p. 63 bottom: "This result has the effect of distorting somewhat the $\ln k_r$ *vs* ΔG^0 curve in Figure 7"

It is really Figure 6.

6. p. 64 top: "The *inverted behaviour* is sometimes also seen in the preferential formation of an electronically excited product"

Q: (1) Can you list the different ways (experiments) in which the inverted behavior may be seen?

(2) Is a maximum observed in a ln k_r vs. $-\Delta G^0$ plot a sure sign of existence of the inverted effect?

M: (1) The ways to observe the inverted effect: One is the way Bard followed when, in some experiments in the sixties, he did some electrochemiluminescence by rapidly changing the electrode's polarity between positive and negative so that he had the formation of aromatic cations and aromatic anions. Their recombination was heavily exothermic and apparently there were a couple of cases in which he got 100% formation of excited state, the ground state wasn't directly accessible because of the presence of the inverted effect. So, that's one way of observing it. Then of course the Closs–Miller experiment was another, and then there was quite a bit of work of Mataga, of the Japanese, and also some other people in Paris had sort of an inverted effect plot, and there have been some people in Rochester . . . there is a plot in radiationless transitions in aromatics due to Siebrand, in excitation transfer, something like that, or maybe radiationless transitions, and there you have what is called the energy gap law . . . which suggests the existence of the inverted effect in ET reactions.

Q: Was Bard's experiment the following:

$$A^+ + A^- \rightarrow A^* + A \rightarrow 2A + h\nu \quad (1)$$

and what the others studied was

$$B^+ + B^- \xrightarrow{\text{inverted effect}} 2B? \quad (2)$$

M: The first step in Eq. (1) is correct, then the excited state would emit fluorescence and so that second step is also right. It is by detection of the fluorescence that they deduced the inverted effect. In other words, you normally don't get luminescence in the chemical reactions, but in this one you did and in order to explain it they invoked the inverted effect. They couldn't go to the ground state because of the inverted effect, so they went to an electronic excited state. The other case is that of Miller and Calcaterra, a totally different experiment, where they really observed the inverted effect. The case (2) is what everybody else studied, that was studied by the people in Rochester, there is a group of people in Rochester which published papers and a group in Japan that published papers and they are of the type where you recombine two ions and they got the inverted effect all the times.

Q: So Bard did not really see the inverted effect.

M: He saw evidence of the inverted effect.

Q: How did he see the evidence? Basically he saw only the excited state.

M: That's right and you may ask why he didn't see the ground state that was because of the inverted effect.

Q: But he went to the ground state after the emission of the light.

M: That's a separate process, but the main point is that he saw luminescence in a chemical reaction.

Q: But he saw luminescence after he went to A^*.

M: You saw luminescence of the A^*, that A^* can luminesce. Actually, the steps are a little bit more complicated, I think you got singlets first, then you got triplets, one would look at the papers to see, but this is the essential idea. And remember though that you first form singlet states and then I think from two singlet states you may form the triplet state, you see its emission, in other words there is a long story associated with that, one would look at the papers and see what it is.

(2) Yes, I think so.

7. p. 65 middle-bottom: "In the case of a semiconductor...the inverted effect...becomes possible"

Q6: After this prediction of yours (1992) was it experimentally observed?

M: I believe so, I believe there are some examples where for a narrow banded semiconductor that's true. I know that often in a number of times in semiconductors the inverted effect has been involved, but I don't remember any details.

The prediction of the inverted free energy effect was *implicit* in the equations in 1956 but I don't know that I spoke of it until the paper in 1960 and then there is a paper of Raphy Levine that said that there couldn't possibly be the inverted effect. Then many people tried to find it.

8. p. 65 bottom: "...there may be some nuclear tunnelling through the barrier in Figure 3"

Q: In Fig. 3 the potential energy U is reported as function of the nuclear coordinates. In Fig. 4 the free energy is reported as a function of the reaction coordinate q. Would it make sense to speak of nuclear tunneling referring to Fig. 4?

M: The potential energy should really be reported as a function of about 10^{23} coordinates, so in Fig. 3 there is only a profile of it. I think it would probably be wrong because Fig. 4 is a free energy and you know...some of the high frequency vibrations might be tunneling through a potential energy barrier but you wouldn't have tunneling through the free energy barrier, tunneling is through some local barrier, and that means that you are starting at some tunneling point, so to speak, that part may be roughly carried over to the free energy barrier...but you know, it wouldn't be strict.

9. p. 71 middle: "Reactions which are required to occur over large distances, include those where the donor and acceptor in solution are separated by a long rigid saturated bridge, or are far apart in frozen media, or when they are far apart in proteins, or in an electrochemical system described below. *All of these systems have a small matrix element* H_{rp} and hence are clearly nonadiabatic"

Q: (1) Is there a class (or classes) of reaction which are *typically adiabatic*?

(2) Being adiabatic reactions the ones for which $\kappa \cong 1$ and nonadiabatic the ones for which $\kappa \leq 1$, the adiabatic reactions appear as a *limiting case of the nonadiabatic* ones and—consequently—the number of nonadiabatic reactions should be much bigger than the number of adiabatic reactions. Is it so?

(3) What if $\kappa = 0$? Does this case exist or is it unphysical? Or is there some kind of selection rule for ET completely forbidding it in some cases?

M: (1) Well, if you get away from ETs, *all atom transfers or group transfers are adiabatic*, I imagine that a number of adiabatic ETs exist if the overlap is strong enough. If you think of other things above ETs, and I think of dissociation, the photodissociation of N_2O to N_2 and O, if you look at the potential energy curves, they present sort of an intersection or a crossing, you go from one curve to the other and the splitting is small for those two curves. In that case then that reaction is nonadiabatic because there is only a weak coupling going over from the reactants to the products curve, the reactants curve is that of N_2O and that of the products is that of $N_2 + O$, there is only a weak coupling, so that reaction is nonadiabatic.

(2) Yes, I think in the case of ETs that's probably true. But maybe just weakly nonadiabatic, you know, differing by a factor of ten or something from adiabatic as far the coupling goes.

(3) I think that if one plotted potential energies as a function of rotation of an angle, in the case where you are trying to transfer from an orbital

to another, that there may be some angle such that the orbitals are sort of being orthogonal, and you might get zero electronic interaction, that may be a point of measure zero, so to speak.

10. p. 71 bottom: "For slower reactions the estimation of $\kappa\nu$ from the pre-exponential factor of the bimolecular rate constant tends to be masked by other effects, such as *the entropy associated with the work terms w_r and w_p*"

Q: If I remember well, you treat this point in one of your earlier papers.

M: The w_r is associated with the ions coming together and if there is repulsion, then that's changing the polarization of the solvent and it may have some entropy effect. When the two ions come together, the polarization at any point of the medium is also changing because the field is changing . . . if two ions are coming together, supposing consider the case where the charges are the same sign, then of course when they are near each other you have some varying intensity at any point of the solvent, more intensity than you had when the ions were separated because both ions are contributing to an electric field at any point of the solvent, and if the two ions are of the same charge then of course that's a stronger field than if you had by either ion alone, so that is going to try to orient the molecules more, so the molecules are going to be more restricted, so there is an effect on entropy.

11. p. 74 bottom: ". . . the back ET from BPh^- to form ground state $BChl_2$ appears to be extremely slow, *perhaps due in part to an inverted effect*"

M: The main evidence for the inverted effect is just in those chemical reactions that we discussed above, that was the main evidence for it, the work of the Japanese and the group of the University of Rochester on one hand, the work of the group at Argonne, John Miller for example, on another, and the chemiluminescence results of Bard, those are all sort of the key things. Now, it is often the case in a photoexcitation, not just in a photosynthetic reaction, that the forward reaction after the light has been absorbed, will be faster than the back reaction, but that can be for *two reasons*, one the inverted effect, and the other is that the two orbitals that are interacting are different. In a back reaction you are interacting the product with the ground zero orbital, in the forward reaction you are interacting the product with an excited state orbital, and those are different interactions with different strengths. When you're going to the ground state, you're using a different orbital of the $BChl_2$ than you did when in the excited state of $BChl_2$, and so there's going to be less overlap because the ground state orbital doesn't extend quite as far. So, there could be two reasons why there is this almost universal behavior, universal as far as I know, that the forward reaction after an excitation is substantially faster than the back reaction. Now, the back reaction is very downhill, so part of the reason could be, as the Nobel Committee suggested, or believed, the inverted effect. About the other results, the results of Bard, Miller, there is no question but, you see, here, in the case of photoexcitation, there are two reasons why the forward reaction could be faster than the back reaction, one is the inverted effect and the other the difference in orbitals.

Q: Bard basically found that instead of going through the inverted effect he went through an excited state. But then how could he say that there is an inverted effect? He could say only that he went to an excited state, that's all.

M: Like in one of the stories of Sherlock Holmes, the hound that didn't bay one night, here it was actually an *indirect* evidence, the fact that it did fluoresce instead of nonfluorescing. Yes, *there is no explanation why it would fluoresce other than there had to be something which prevented the system from going to the ground state, and that is the inverted effect.* The natural thing for reaction is to produce products in the ground state, but here it couldn't. He couldn't go to the ground state of the products because it was too exothermic and there was the inverted effect, so the net result was that they went to the excited state of the products. End of story.

Q: The fact that he couldn't go directly to the ground state but he is forced to go through an excited state was an *indirect* demonstration of the existence of the inverted effect.

M: Right. And actually, you know, often people say that the first demonstration of the inverted effect was the work of the people at Chicago, at Argonne, you know, that famous paper of 1985 or 1986, and Bard always said that he had evidence of it back in 1965 or 1966, which is true, but the evidence of Miller and Closs was very *direct*, you can see the turnover of a parabola, Bard saw it *indirectly* by the fact that he had 100% formation of excited state, not just a few, a little bit, the whole went that way in the two reactions that he studied. Normally you don't get any, if there had not all that excess energy he disposed of, he wouldn't have got any, that's the usual behavior. Normally in a chemical reaction in those case you don't form excited states, you form ground states, there is not enough energy to form excited states. The excited states are way up, they are too high, you only form the ground states of the products. He could go to the ground state indirectly by first forming the excited state, then that could form the ground state. The reaction in Eq. (1) in question **6** above where the first step is OK, formation of an electronic excited state, but the second step is more complicated, one had to read the original articles. The second step is made up of a few steps.

12. p. 76 top: "When the wave function is *constrained* to reside on the donor, this U becomes the surface U_r."

Q: Do you refer here to the wave function of the transferring electron?

M: Yes. Because the other electrons are pretty much constrained by energetics.

13. Following: "The frequency with which an electron would oscillate *between two reactants whose nuclei were fixed* is shown below to be the Bohr frequency, $2H_{rp}/h$, where $2H_{rp}$ is the splitting in Figure 3 due to the electronic interaction of the two reactants."

M: Now, of course, that would be the case if you *fix* the nuclei so that they are right at the avoided crossing, and then you would have such oscillations going on, but of course life doesn't work that way, so you normally have the nuclei that *pass through* there, so, you know, you don't see the oscillations, but under that other condition you would have free oscillations, but they don't actually occur. Because you don't have things just existing at that point, you have other motion going on.

Q: Is there the possibility for an electron of oscillating in a solvent cage?

M: No, because the electron isn't in the solvent, the electron is in the ion, and if you want to use semiclassical motion, where an electron is sort of moving in an ion and going over, that would be localized on the ion and not much leaking out to the solvent. If you want to think classical... the electron would be oscillating each time when would be sort of coming near the barrier and, seeing if it would tunnel through or not, then you can interpret somehow that way but you have to be careful. Anytime you speak of an electron oscillating it's largely a classical description, I mean there are some quantum mechanical analyses... anyway, the electron has to ionize from the reactant and this is too high in energy. You can have electrons, you know, when you excite with x-rays or in some other way, you can produce electrons in water. There is all that work that was done on electrons in water, but that's at much higher energy.

In quantum mechanics you have, say, an electron with different energy levels. Now, associated with the difference of the energy levels is a frequency, in old quantum theory that frequency was actually an actual frequency, in this case of the electron a going back and forth and so on, or a frequency of vibration if you're talking of vibrational levels, so in QM the difference of levels can correspond semiclassically to a classical motion actually undergoing kind of oscillations, or a more complicated motion than an oscillation, but somewhat periodic as a quasi-periodic in semiclassical parlance.

14. p. 77 bottom: "It is the *difficulty of intersection* of the two parabolas in Figure 8 at a conveniently small height which leads to the inverted effect"

Q: Where is the difficulty? It is not just that the intersection happens to be on the other side of the parabola's vertex?

M: Of course, if you lower one parabola relative to another there is no difficulty involved, . . . but under most conditions it is unlikely that there will be an intersection *on the back side* of the curve rather than on the usual side . . . that's really not a well phrased statement.

15: p. 78 top-middle: "6. *Are proton transfers expected to display an inverted effect?*

M: I know of no such case. In the case of electronic levels you really have to go to a fairly excited state and then you can bypass the inverted effect by going to a higher state, just like Bard did. The analogous thing in the case of proton transfer would be that the reaction would be so exothermic that you would go into an *excited state of the proton* in a reaction and then that proton would maybe emit a radiation. But there a very small probability for it, it's more likely for it to degrade to the ground state just thermally, so if it ever occurred in a proton transfer reaction, that maybe would be difficult to see it because of competing thermal processes that would compete with infrared emissions. Infrared emission intensities are low by a huge factor compared with a visible light process, it goes as a high power of frequencies, so the chance of observing it is small and besides the levels of the proton, they are only, what? 3000 wave numbers apart, something like that, *they are really pretty close together compared with electronic levels*, so the chance that you can be able to observe something like that is probably not large. In other words, supposing that you could do the same kind of experiment that Bard did, and you found that you had a 100% of reaction occurring by an infrared emission, in other words you form an excited state, in this case an infrared state, and then emit it, that would be the analogous thing to Bard, but emission intensity by infrared is smaller by orders of magnitude than the intensity at the visible, you know, it goes as the power of the frequency, so the chance of observing it would be very small, and, besides, the energy levels of the proton especially when you bring in the coupling, are relatively close comparing with those of the electronic levels, so you don't have quite the same opportunity and I don't know any example where there has been an emission from a proton transfer and I think for understandable reasons. If the reaction would form an excited state of a proton, I don't think that I would see that emission because of the low emission intensity in the infrared.

16. p. 78 middle: "7. *Is it meaningful to speak of the splitting of the free energy curves, as well as the PES?*

The splitting $2H_{rp}$ of the two PES depends mainly on the separation distance R and so, for a fixed R, is essentially constant along the intersection surface. Because of the relation between the free energy curves in Figure 4 and the PES in Figure 3, the splitting of the $G_r(q)$ and $G_p(q)$ curves is in that case essentially equal to $2H_{rp}$ also. *In other cases*, if one wished to depict a splitting in Figure 4, that splitting would have to represent the value of H_{rp} averaged over the intersection surface using a Boltzmann weighting factor"

Q: (1) Which other cases? Cases in which the splitting depends on other parameters beyond the separation distance?

(2) Would the Boltzmann factor be something like $\exp(-H_{rp}(R, .?./k_BT)$?

M: (1) It can depend on angle rotation of two species, in other words there are some selective effects I think, where if you have, say, two orbitals one on one species and one on another, and if they interact in a certain configuration, maybe you have a π orbital, say, you rotate one of the species and then that interaction should fall off. In fact, two my students and I wrote a paper on orientations...on orientational effects, electronic orientational effects, and I think that Closs did experimental observations related to that. So, it can depend upon electronic and symmetry effects on the electronic state.

(2) If you are in a 10^{23} dimensional space you write $\exp(-H_{rp}(R, .?./k_BT)$ as a Boltzmann weighting factor in that many-dimensional space. Potential energy curves are in 10^{23} or so dimensions, it's hard to draw them, so all you can do is to draw profiles, on the other hand you can take those potential energy functions, you can use statistical mechanics and calculate free energies as a function of reaction coordinate, and the place where... the point where the free energy curves of reactants and products intersect is close to where the major parts of those 10^{23} dimensional potential energy curves intersect and so *the dynamics of course is not on the free energy surface but is on a potential energy surface*, so when one tries to put the dynamics on a free energy surface and talk about the tunneling and this and that, that would involve a large averaging over the potential energy in the 10^{23} dimensions. So you are tunneling from one potential energy curve to another, formally it looks like tunneling from one free energy curve to another but of course the free energy curve is composed of many components. Each pair of PESs has its own splitting, and if you are drawing a free energy curve you're drawing a whole ensemble of potential energy curves, so if there is any splitting that you draw for the free energy curves, it has to be some sort of complicated average of the splittings for individual curves, especially since individual pairs of curves have their own splittings. Free energy is an ensemble of many different potential energy curves, so you have an enormous number of pairs of curves, really you have a many-dimensional surface, so

any splitting is some complicated average over all those dimensions, in other words in $N - 1$ dimensional space you have a whole seam of intersections and if there is 10^{23} dimensions then you may have 10^{23} H_{RP}'s, in effect the H_{rp}'s vary with position, over that $10^{23} - 1$ dimensions.

Q: Practically then you don't do it anyway.

M: No, no.

17. p. 80 bottom: "$Q^{\ddagger}$ is the partition function for the transition state:

$$Q^{\ddagger} = \int \cdots \int \exp[-H(q_1 = q_1^{\neq}, p_1 = 0)/k_B T] \times dq_2 \cdots dq_N dp_2 \cdots dp_N / h^{N-1} \tag{A4}$$

The H in Eq. (A4) is now the sum of the potential energy of the TS, measured relative to the lowest value on the TS, *plus the kinetic energy of the TS*."

Q: Why $p_1 = 0$? Is then the kinetic energy of the TS due to the contribution of the other degrees of freedom?

M: It's because if you have the Hamiltonian containing all terms but the p_1 term, and write the p_r terms as $\frac{p_r^2}{2\mu}$, then if you start using transition state theory, that p_1 disappears, and so you end up with the Hamiltonian for a system which has one degree of freedom less than the actual system, as describing the TS.

Q: But $q_1 = q_1^{\neq}$.

M: Because you have integrated over the q_1 in writing down the transition state theory. That $q_1 = q_1^{\neq}$ is the value of q along the reaction coordinate at the TS. So, the actual transition state does have translational energy as the system is sweeping through it, but when you describe the properties of that transition state and you have got the flux and multiplying by the velocity to get a current, anyway you get an $\dot{x}dp$. $\dot{x}$ is a velocity, and *dp* is part of *dpdx*, which is the volume in phase space. In other words, if you have a two-dimensional phase space, you have coordinate and momentum, so x is the coordinate and if you want to get a current you need to get a density, and you get the density by dividing by *dx* and that gets rid of the *dx*, you multiply that by a velocity and you get the result. That's a way of getting a transition state theory and I think that way is described in Eyring, Walter, and Kimball, not in Glasstone, Laidler, and Eyring. That is, instead of using the way that Eyring did in his 1935 paper where he considered a small integral and he used the particle in a box formula, and so on, you use the Wigner approach, you have a linear density... you calculate the probability of being in *dxdp*, you divide by *dx* to get the density along the TS coordinate, and you multiply by a velocity in order to get a rate, that's the standard procedure in all sorts of flow problems, having nothing to do with reaction rate theory. In other words you really use the derivation that Wigner had and that I occasionally used in some of my papers. Glasstone, Laidler, and Eyring did it the old-fashioned way, and in Eyring, Walter, and Kimball, probably as the effect of Kimball, they did it in the way Wigner did. They may not even refer to Wigner but basically that's what they did. There is a book by Steinfeld that gives both derivations, a textbook, this is the second edition, maybe the first edition does too, they should give the two kinds of derivation of TS theory.

18. p. 81 top-middle: "For bimolecular reactions, three of the coordinates in Eq. (A1)

$$k_r = \int \cdots \int \dot{q}_1 \gamma \exp$$
$$\times (-H(q_1^{\ddagger})/k_B T) dq_2 \ldots dq_N dp_1 \ldots dq_N / h^N Q \qquad \text{(A1)}$$

can be chosen to be the *center of mass of the two reactants.* The contribution of these coordinates and their momenta to the integral in Eq. (A1) just cancels their contribution in Eq. (42)

$$Q = \int \cdots \int \exp(-H/k_B T) dq_1 \ldots dq_N dp_1 \ldots dp_N / h^N \quad \text{(A2)''}$$

Q: (1) Who was the first to give such a general expression for k_r?

(2) Do you consider as reactants the "naked" reactants or also the closest solvent molecules dragged along with them? Of course in this last case wouldn't the center of mass be kind of "fuzzy"?

M: (1) The first guy who came close to it was killed in 1915 in the first world war, I'm trying to remember his name, I referred to it, frequently the Chemistry textbooks refer to it, René Marcelin, that was his name. He had the idea of moving in phase space . . . the key idea of Eyring's was to do something about what's happening at the top, like that there is a little translation. Marcelin didn't put that in and, if he had put it in, he would have had transition state theory, a classical form, about 20 years before Eyring. I remember I read Marcelin's thesis, I got a hold of his thesis, long time ago, long long time ago.

Q: But about that specific form you write down in Eq. (A1) . . .

M: I have something a bit more sophisticated than that in some papers around 1964 where we talked about curvilinear coordinates, and there are all sort of effects there that come in. You see, here in getting this formula they pretended that there was just a little Cartesian coordinate crossing the barrier, but there can be temperature effects, so I have all of that in there, I think it was rediscovered by Truhlar, he may have referred to it, Truhlar made a point of it but that was many years later.

(2) Yes, sure, that's right, that would be a bit fuzzy, but normally one just takes the reactants, you see. Surely you have solvent molecules probably bound to like a ferric ion . . . but it doesn't affect very much.

19: p. 82 bottom: "Physical insight into Eq. (A7)

$$\kappa(\upsilon) = \frac{2\pi}{\hbar} \frac{|H_{rp}|^2}{\upsilon|s_r - s_p|} \tag{A7}$$

in terms of the 'frequency' of the electronic motion in the TS, obtained as $2H_{rp}/h$ in Section 4, namely the separation of the adiabatic energy levels in the TS in Figure 3 divided by h, and of an *effective frequency* for the nuclear motion in the vicinity of the crossing point is given in ref 90"

Q: Why "frequency" in quotes for the electronic motion and why "effective" frequency for the nuclear motion in the vicinity of the crossing point?

M: You shouldn't think of an electron as oscillating back and forth with a frequency, that's a classical picture, that's not what's meant. The nuclear motions have a whole group of frequencies, some slow, some fast, and so on, so this is sort of a typical one which is relevant for the reaction. But of course it is not a single one that is relevant, but several and you sort of average over the several. The frequencies that are relevant are the ones that have to change over the course of the reaction.

M272 Tight-binding Approximation for Semi-infinite Solids. Application of a Transform Method and of Delta Function Normalization

R. A. Marcus

NOTES

1. p. 5608 2nd column top: "We turn next to the question of *bound surface states* for a one band system."

Following middle-bottom of column: "Equation (5.5) provides the condition for a *bound surface state below the energy band*."

Following bottom: "We consider next the case of *a surface state with an energy above the band*..."

Q1: (1) Does a bound surface state for a solid's surface have some similarity with a bound state for a molecule?

(2) Please explain this story of bound surface states and their energies relative to the energy band...

M: (1) It could be bound in several ways, you could have a surface state which is stable on the surface, you can have another one whose wave function sort of goes into the solid... it could be that you have a surface state that's imbedded in a continuum... in the case of a molecule when you are above the dissociation you probably have more than one potential energy curve and then have a coupling to a second potential energy curve, so the state as a whole might be unstable...

(2) I should have been more specific there because you can have an energy band and, you know, for example, it could be a conduction band or a valence band... most bands, all bands I guess, really have a finite width, that's certainly true, maybe of the order of 5 eV or something like that... I should have said whether we are talking of valence band or conduction band but the point is that if one considers the case of a semiconductor, where one has valence band and conduction band, then you can have a state, a surface state, just below the conduction band and you can have a surface state which is above the valence band.

2. p. 5609 2nd column bottom: "One reason for using a box type normalization, I am told, is to assist in the counting of the electronic states in treating various phenomena."

M: I think I probably referred to papers when I was talking about the normalization of the states... and although a box normalization is commonly used, I used a delta function, as others have too. When you are dealing with electron transfer into solids where maybe you regard them as semi-infinite, in other words they start at a fixed place at the interphase between the solid and the liquid and then you just go on to infinity. Now, you have various wave functions of the solid and you have to normalize them in some way. Well, if they were in a box there would be an obvious normalization. When something goes off to infinity there is what is called the delta function normalization, that's what it is. In other words, you don't normalize to unity, you normalize to the delta function. Infinite or semi-infinite space wave functions.

M273 ET Reactions in Proteins: An Artificial Intelligence Approach to Electronic Coupling

P. Siddarth and R. A. Marcus

NOTES

1. p. 2400 1st column middle: "These *long-range electron-transfer reactions* in proteins are nonadiabatic."

Q: (1) Why are they nonadiabatic?

(2) Are there examples of long-range adiabatic electron-transfers?

M: (1) Because the coupling between the reactants is weak. And if the coupling between the reactants is weak, it doesn't matter whether it is long range or short range, it is nonadiabatic. In other words, if you had a bridge that was on resonance with the reactants, then the electrons could go through that bridge easily and there wouldn't be a nonadiabatic reaction.

(2) Well it may be, I haven't look that closely, there is some work of Wasielewski where he did some work with a bridge almost on resonance with the reactants, I don't remember the reference but he did it, and in fact it may be that the long bridge may be more on

resonance and the reaction was actually faster. That was present in one of the papers. When he had a longer bridge, it was such that the reaction actually went faster because when you lengthen the bridge you change the energy levels and it turned out that in this case it brought the energy levels more into resonance with the reactants, so that the electron didn't have to go over a barrier. In other words, the closer the energy levels of the intermediate are to those of the reactants, the more the electronic levels match those of the reactants, the less there's sort of a tunneling barrier.

2. p. 2400 2nd column top-middle: "The path integral method of Kuki and Wolynes...includes electron tunneling and neglects *hole transfer*"

Q: How are electron transfer and hole transfer related with bands in the solid?

M: If you have, say, conduction bands and valence bands, then you can have a hole transfer by the valence band, and an electron transfer by the conduction band. Different sorts of orbitals are involved.

3. p. 2401 1st column bottom: "In the electron-transfer problem, for each intermediate node I, the *true measure of electronic coupling* from the starting point D to the node I is estimated from the product of the appropriate H_{MN}'s divided by an energy denominator. Let this quantity be denoted as V_{DI}:

$$V_{DI} = H_{DI}\frac{H_{12}}{\Delta E}\frac{H_{23}}{\Delta E}\cdots\frac{H_{I-1I}}{\Delta E} \qquad (4)\text{"}$$

M: The formula is a result of perturbation theory and if you have a wave function which extends on to a neighbor, then that bit of wave function extends to another neighbor and you have a product of those small extensions.

4. Following: "It has been well documented in the literature that the rate of falloff of electronic coupling with distance is approximately exponential,

$$H_{MN} = H^{o}_{MN} \exp(-\beta R/2)\text{''}$$

Q: Is this formula theoretically founded or is it simply an experimental result or both?

M: It's both, it is theoretically founded in other words, if you use perturbation theory, you'll get that sort of result. Chao Ping Hsu, a former student of mine, wrote a nice paper on that... well, in fact McConnell gets it by perturbation theory, the rate varies as the number of nodes, exponentially, and the distance varies as the number of nodes.

5. p. 2402 1st column top-middle: "The transition state for electron transfer occurs at the 'intersection' of the *energy surface* for the reactants with that for the products, i.e., where there is zero change in *electronic energy* upon electron transfer. (These surfaces are those of *electronic energy vs* nuclear coordinates.) On this intersection, *the energy of the donor orbital* E_χ *equals that of the acceptor orbital* E_μ"

Q: (1) You write here "energy surface" and qualify the energy as "electronic energy." In view of the previous equations

$$G_i = \int \Psi_i * \left[-\frac{1}{2} \sum_{i=1}^{N} \nabla_i^2 + V(\mathbf{r}_e) + W_{rev}(\mathbf{r}_e) \right] \times \Psi_i d\mathbf{r}_e \quad E = G_i \quad (i = r, p) \qquad (7, \mathbf{M266})$$

where

$$-\left[{}^{1}\!/_{2} \sum_{i=1}^{N} \nabla_i^2 + V_{tot}(\mathbf{r}_e) + W_{rev}(\mathbf{r}_e) \right] \Psi(\mathbf{r}_e) = E\Psi(\mathbf{r}_e) \qquad (5, \mathbf{M266})$$

and where

$$W_{rev}(\mathbf{r}_e) = -[(1 - 1/D_{op})/8\pi] \int \mathbf{D}^2(\mathbf{r})d\mathbf{r} - \int \mathbf{P} \cdot \mathbf{D}(\mathbf{r})d\mathbf{r} + 2\pi c \int \mathbf{P}^2 d\mathbf{r} \quad (3, \mathbf{M266})$$

can we say that the *free energy surfaces* are the same as the *electronic energy surfaces* once we modify the Schroedinger equation that would be valid for the gas phase

$$-\left[{}^1\!/\!{}_2 \sum_{i=1}^{N} \nabla_i^2 + V_{tot}(\mathbf{r}_e)\right] \Psi(\mathbf{r}_e) = E\Psi(\mathbf{r}_e)$$

inserting inside the square brackets of the above operator the potential energy operator $W_{rev}(\mathbf{r}_e)$?

(2) If the above is true, can we say that the free energy of the reactants + bridge + solvation molecules system is nothing but the electronic energy of the reactants + bridge system when their electrons are under the effect of the field produced by the potential energy $W_{rev}(\mathbf{r}_e)$?

(3) In the above Eqs. (7) and (5) you use the symbols $V(\mathbf{r}_e)$ and $V_{tot}(\mathbf{r}_e)$. Are they the same object? I believe so. In **M266** you define $V_{tot}(\mathbf{r}_e)$ as "the Coulombic interaction of all the electrons in the reactants with each other and with the nuclei of the reactants, *and of the nuclei with each other*." With the subscript *tot* for "total" you probably want to stress this last point. Is it so?

(4) In going from reactants to transition state does the whole nuclear framework of reactants + bridge + solvent change?

(5) If I understand it correctly the energies of the donor and of the acceptor orbitals, E_χ and E_μ, vary along the reaction coordinate depending on the solvent polarization. Is it so? Do they depend also on the nature of the solvent? If different solvents would have the same

polarizing effect, would the two above orbital energies be the same, or different solvents would influence electron transfer in different ways even if the polarization effects were the same?

M: (1) The energy surface is the electronic energy plotted vs. the coordinates. That G_i . . . that's giving you the free energy surface, the whole integral is the free energy surface. It's just that there are subtleties because that $V(\mathbf{r}_e)$ has coordinates and so depends on the values of those coordinates, and that $W_{rev}(\mathbf{r}_e)$ has already been averaged over all of those individual coordinates, so there is a difference there. In other words, it may well be that the equations are approximations, in fact I think there is a tacit approximation, a tacit approximation that the electronic frequencies of the solvent are very fast and can adapt to this sort of slow moving electron that we are solving the Schrödinger equation for. There is some assumption about time scales involving all the other electrons compared to this electron, this is a mean field approximation, you know, it's an approximation, if there are electronic frequencies that are comparable with these frequencies then it is a bad approximation. This is actually assuming that this electron is a kind of slow moving object, compared with the electrons in the solvent, you see, and that may not be a good approximation. Otherwise you have to take into account correlations of the motions of this electron an those in the solvent.

(2) Yes, in the approximation we were talking about.

(3) That can well be. Yes, sure, it comes in as a constant term. It is there, yes.

(4) What you mean by nuclear framework? All of the nuclei? Then yes because the vibrational polarization, rotational polarization, involve the different statistical distribution of the nuclei compared with the unpolarized system.

Q: Also nuclei of the bridge?

M: Well, it depends on what the bridge is doing, if the electron resided on the bridge for some time that would be a factor, if the bridge was just a virtual intermediate, the bridge may not really make a difference.

(5) Yes, they depend on the nature of the solvent, they go up and down, and they go into resonance with each other, and so on. If the only effect is polarization it would be the same. Only if the electron resided on the solvent as a solvated electron then other effects come in.

6. p. 2402 1st column middle-bottom: "We have used the term electron transfer, but it is intended to include, in Eq. (8),

$$H_{DA} = \sum_{\alpha} \sum_{m_D m_B m_A} \frac{(C^*_{\chi m_D} H_{m_D m_B} C_{m_B \alpha})(C^*_{\alpha m_B} H_{m_B m_A} C_{m_A \mu})}{E_\chi - E_\alpha}$$

transfer by any occupied orbitals χ of the bridge as well, i.e. hole transfer."

M: If for example, there is a hole in the valence band, then that hole can serve as a conduit, if there is no hole there, then of course it is all saturated and the electron can't transfer.

7. p. 2402 2nd column bottom: "The values of H_{DA} extracted from experimental data, as well as their ratios, are also given in Table II. (The absolute values of the "experimental" H_{DA} may be also model dependent and might change somewhat if a different analysis of the experimental data were used.)

Q7: Why "experimental" in quotes?

M: I put it "experimental" because it is something that you infer from experiment, from an activation energy measure, *this part* of the preexponential factor is something that you infer from the preexponential factor, there may be other terms in the preexponential factor, there is an entropy of activation associated with some entropy change in the transition state.

8. p. 2404 first column bottom: "**Appendix. Estimation of the Donor (Acceptor) Orbital Energy, $\mathbf{E_{\chi}}$, at the Transition State** In the expression for the electronic matrix element H_{DA}, Eq. (8), a key quantity is $E_{\chi} - E_{\alpha}$, where E_{χ} is the donor energy level... and E_{α} *the energy of a molecular orbital of the bridge*... we estimate it using a cytochrome *c* charge transfer spectrum."

Following, p. 2404 2nd column top: "The electron-transfer reaction of interest is

ET: Fe(II)–Cyt *c*–Ru(III) → Fe(III)–Cyt *c*–Ru(II) (A1)

It has been reported in the literature that ferricytochrome *c* exhibits a *porphyrin to metal charge-transfer absorption*... This process can be depicted as

LMCT: Fe(III)–Cyt *c* → Fe(II)–Cyt c^{+} (A2)

The free-energy surfaces of these two processes can be schematically plotted as in Figure 5. Δ_{TS} *denotes* the vertical energy difference from the transition state of the electron-transfer reaction to *the products of the charge-transfer excitation* process and $h\nu_{CT}$ is the absorption maximum associated with this charge transfer."

Following: "Δ_{TS} *represents* the energy difference $E_{\chi} - E_{\alpha}$ where E_{α} *is the energy of the HOMO of the porphyrin from which the charge is transferred in the LMCT.*"

Q: (1) Apparently a first intermediate step in going from reactants to products in reaction A1 is

Fe(II)–Cyt *c*–Ru(III) → Fe(III)–Cyt c^{-}–Ru(III)

in which the transferring electron originally on Fe(II) goes to the LUMO orbital of the porphyrin in the heme of 4 *c*. The LUMO becomes then the HOMO wherefrom the electron transfers to Ru(III). This interpretation appears to agree with the last meaning you give of Δ_{TS}.

(2) Is this same MO the one implied in the LMCT of reaction (A2)?

(3) I have a last general question. When an electron transfers from a donor D to an acceptor A through a bridge made up by a chain of molecular fragments like in

$$D\text{–}MF_1\text{–}MF_2\text{–}\ldots\text{–}MF_i\text{–}\ldots MF_n\text{–}A$$

don't we have an all *series of transition states* when the transferring electron is successively on the different molecular fragments? Don't we have a set of activation barriers, one in going from D to MF_1, then from MF_1 to MF_2, and so on, and maybe we have to consider only the one with the highest activation energy? Please look at my Fig. 2*.

M: (1), (2). OK, yes. If you want to know something about its property in the transition state, you have to use a vertical transition from there and not the vertical transition starting from the reactant. You see, I even draw that line there.

(3) The bridge should be consider immobile in the first approximation, in a more accurate approximation the bridge has some vibrational motion and that will affect the ET some way or another. We do not take that effect into account.

Q: I don't understand if this thing is valid always and then in the case of superexchange you can forget completely?

M: No, no, well. . . in all of these cases you have to take in account vibrational motion, in any problem, but in some approximations you neglect some properties and you focus on the electronic motion.

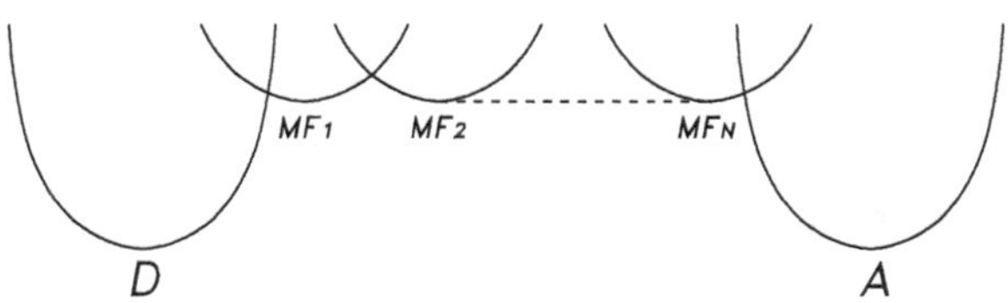

Fig. 2*.

Q: And the approximation can happen in superexchange?

M: It can happen in superexchange that is even apart from the approximation due to superexchange, this is an additional approximation that is made typically besides superexchange. You have superexchange but you also neglect some coupling of the vibrational and electronic motions. Two separate things.

Q: I was asking if you have a series of TSs to go through.

M: Not of TSs. I wouldn't call TSs the intermediate thing, when you have superexchange you actually have thoroughly wave functions, you can't speak of going from one state to the other literally, the uncertainty principle would prevent you from doing that when you have superexchange. In conditions where you have superexchange the electron classically can't be allowed inside that bridge, so it's only a purely quantum mechanical phenomenon, you don't speak of there being TS, that would imply that the electron is eventually spending some time in those sites, but it doesn't.

Q: But in case where you don't have the superexchange?

M: Then you can have a hopping, that's a separate question sure, then you can.

Q: Then you can have a series of TSs...

M: Yes, right, but there are sections where you have hopping and there are sections where you don't have hopping.

Q: So when you have hopping you have a series of TSs...

M: You have a series of states... normally when you have a series of states you try to find the bottleneck, and that one is the TS.

Q: Because there is one that is the most important...

M: Yes, I mean in this case where there are so many equally important steps, if you are hopping, then you refer to each step having its own TS. You do it step-by-step. Every step has single TS. In other words, if you go from one side to another by a hopping, you have to go through a TS.

Q: So my question was not stupid...

M: No, not at all...

Q: How about polarization and excited electronic states?

M: The theoretical expression for the polarizability involves the excited electronic states and the ground state. I imagine if you look at the Pauling and Wilson's Quantum Mechanics and you look up "polarizability," if they give a theory for polarizability you'll see it there, otherwise any Quantum Chemistry book that talks about polarizability if they have a formula there you'll see it in the formula.

M274 Surface Properties of Solids Using a Semi-infinite Approach and the Tight-binding Approximation

H. Ou-Yang, B. Källebring, and R. A. Marcus

NOTES

1: p. 7405, 1st column top: "*...electrochemical inverse photoemission.*"

M: That was a paper that Shachi Gosavi and I wrote, a later paper, it's where by chemical reaction you can eject an electron high up into the empty conduction band of the metal and for most of the time the electron just simply relaxes to the Fermi level, but in about one millionth of the time emits light, so that's what that is, it was studied experimentally for a while, maybe by electrochemists.

M275 A Theoretical Model of Scanning Tunneling Microscopy: Application to the Graphite (0001) and Au(111) Surfaces

H. Ou-Yang, B. Källebring, and R. A. Marcus

NOTES

1. p. 7565 1st column top: "...the real tip structure in STM is unknown..."

Q: How so?

M: They don't know if the tip consists of one atom, several atoms, they don't have that kind of resolution, you know, you have something sticking out, some sort of whisker, but at the end of the whisker how many atoms are there? It's not known. Or it wasn't at that time. One of the edges is closest to the film are you looking at.

2. p. 7565 2nd column middle: "the present treatment is next specialized by making use of a simplified tip model in which the tip is assumed to be a semi-infinite linear chain of atoms"

Q: Why this semi-infinite chain, why not a finite one?

M: Semi-infinite because it stops at the end-atom. So, that's why the semi. And infinite is a wire going off to infinity.

Q: Why is it not a finite thing?

M: Well, the finite thing brings an end effect at the other end. If the other end is a long distance away then presumably everything will converge.

3. p. 7567 1st column middle: "The net current from tip to sample is then given by (14a).... In the limit of low voltage υ, Eq. (14a) reduces to (15a)..."

M: I think it goes something like this: you write that $\delta(\varepsilon_s - \varepsilon_t + ev) = \delta(\varepsilon_s - \varepsilon_t) + ev\delta'(\varepsilon_s - \varepsilon_t)$ and then you integrate by parts that δ', and $\delta'(\varepsilon_s - \varepsilon_t)$ will operate on those f's, the f is like a step function, the derivative of a step function is a delta function, so I think it's probably something like that I would have to go from one equation to the other.

Q: The Fermi–Dirac function looks like a Heaviside function.

M: That's right, when you differentiate a Heaviside function you get a delta function.

4. p. 7567 2nd column top-middle: "... we consider a chain with one electron per atom and one orbital per atom, the *hopping integral* H_{ij} between nearest neighbors being denoted by β"

Q: You denote here for the first time the coupling as "hopping integral." Is there some reason for that? I mean, can the general donor–acceptor coupling H_{DA} be called "hopping integral"?

M: Why did we call hopping integral? That may not be a great term, but, you know, when you have sort of a group of atoms like that and you are using perturbation theory, the overall matrix element will come out to be proportional to that hopping integral, that β, to some nth power, n is the number of intervening atoms and so it is not as though it's really hopping from site to site, although it would be so in a certain limit.

Q: Which limit?

M: The limit of... if you work in the superexchange regime, in other words if the electron really didn't hop, which isn't doing here, because this is more of a tunneling process, then that would be something which characterizes the strength of the hop, but that's probably not a great designation anyway, calling it the hopping integral, because there is no need to even involve that image. The hopping is valid when the electron really resides on that atom, but it doesn't really reside on that atom, so it's not as though you go from a wave function that is concentrated on one atom to a wave function that is concentrated on another atom, so that term hopping integral is not a great term, you know, it clearly introduces confusion. In a certain limit, when the electron would actually reside on that intermediate atom, that you can really call it the hopping integral.

Q: So in the general case H_{DA}?

M: Well, you can use the word hopping integral, but I don't think people do, they may call the interaction matrix element.

Q: You never did until now.

M: No, no, my guess is that one of the chaps put it in there and I didn't notice, you know, there is no real merit, you know, of course you can call as you want but there is no advantage to call it that way. It's usually called the exchange integral, in the case of tight binding or LCAO.

5. p. 7567 2nd column bottom: "...as an approximation, only the matrix element between the first tip orbital and the sample has been included"

Q: Do you mean that atom 1 interacts with the sample and with atom 2 and atom 2 interacts only with atom 1 and 3 and so on?

M: Well, I think what it is, is that if you have a tip and you pick an atom at the end of the tip, normally it's a little bit more complicated because there are maybe several atoms that are trying to decide who is the king of the castle... or you will focus let's say on one of them which is sticking out a little bit more than the others... then you would talk about its interaction with the other material.

6. p. 7569 1st column top-middle: "We consider a *hydrogen tip* (a semi-infinite chain of hydrogen atoms) first"

Q: Is this a purely theoretical tip, I imagine, some kind of simplest of all possible tips?

M: Absolutely, absolutely, in a realistic case you may put in a platinum tip...

Q: By the way, I saw that in the last edition of the book of Kittel, he introduces the bands considering a ring of 20 hydrogen atoms.

M: Yes, yes, also when one is considering a sort of a long linear problem, to simplify treatment one uses a ring and essentially the properties of the interior of a long linear chain would be similar to those of a ring, you know, in terms of electron density, the idea of using a ring to represent a long polymer, that can be used in other cases too, it is an approximation. It is to simplify the treatment and may capture some of the essential details but, of course, if you would have an end the approximation wouldn't catch it, but something about density of states it may catch, you see.

CHAPTER 5

ET in Proteins: Electronic Coupling in Myoglobin, Quantum Corrections for ET Rates. Polarizable vs. Nonpolarizable Description of Solvent, ET in Proteins: Electronic Coupling in Modified Cytochrome *c* and Myoglobin Derivatives, Free Energy of Nonequilibrium Polarization Systems. 4: A Formalism Based on the Nonequilibrium Dielectric Displacement, ET in Ferrocytochromes

Interviews on M279, M283, M286, M289, M293

M279 ET Reactions in Proteins: Electronic Coupling in Myoglobin

P. Siddarth and R. A. Marcus

NOTES

1. p. 6111 1st column middle: "Ruthenium-modified myoglobin derivatives"

Q: What does the word "derivative" mean in this case? Is it a protein derived from myoglobin modifying it by insertion somewhere of Ruthenium atoms?

M: It says chemically from it, the derivative is something that has been changed.

2. p. 6111 2nd column middle: "ET systems that are not structurally homogeneous"

Q: Can you give examples of structurally homogeneous systems?

M: Well, I'm not sure that I can, if you had a solvent and you didn't have any of your ions in there, that'd be structurally homogeneous, in other words one part statistically is the same as another part. But in the case of a protein that's not the case, a part of a protein is not statistically the same as another part.

3. p. 6112 1st column top: "Second-order perturbation theory can be used to determine this bridge-mediated electronic coupling. The electronic properties of the bridge, namely, the molecular orbitals and their eigenvalues, are calculated separately, and the interaction of the donor/acceptor with the bridge is treated as a perturbation."

Q: Does the superexchange theory consist of the above theory? If not, where to look up for a detailed treatment of superexchange theory?

M: Yes. Superexchange is a term that's popped in, I think probably one thing to do would be to look on books on ET that have some quantum mechanics in them, and then look up superexchange and, you will see, it's second order perturbation theory.

4. Following: "The above formulation includes hole transfer as well as electron transfer, since the summation over α in

$$H_{DA} = \sum_{\alpha} \frac{T_{\alpha}^{D} T_{\alpha}^{A}}{\Delta E_{\alpha}} \tag{4}$$

runs over both occupied and unoccupied orbitals of the bridge."

Q: Is the hole synonymous of unoccupied orbital?

M: The hole is in the valence band, so you have something transferring out of the valence band, to somewhere else... If you are making use of the conduction band that's sort of an ET. The hole is synonymous with a vacancy in an orbital.

5. p. 6112 middle: "The search starts from the donor site D in the protein and *attempts to locate the acceptor site* A by traveling via the intermediate atoms I of the amino acid residues..."

Q: From the above sentence one has the false impression that the A redox center is not known and that one somehow tries to locate it. Please explain...

M: Maybe that's not a good way of phrasing it because the A redox site is known, so it would really reach that site, probably by intermediate atoms. Rather that locate it'd be better say "one attempts to reach the acceptor site, the known acceptor site."

6. p. 6112 2nd column middle: "Three mutant human myoglobins were modified by replacing the heme by Zn mesoporphyrin (ZnP) and coordinating a pentaammineruthenium complex to a surface histidine residue (RuHis$X$Mb, where $X = 70$, 48 and 83)."

Q: What is X and what do the numbers after it mean?

M: That's a numbering, in proteins you number the various amino acids and, you know, that's what they are, that's a convention, they're numbered along a chain and so the number 70 means that there is a histidine there, there is a histidine in position 48, so it's a numbering scheme for numbering amino acids in proteins, I imagine when you go along the chain... I don't know how they do it.

7. p. 6113: Table 1, $k_{\max 1}/k_{\max 2}$

Q: Please look at Table 1: what do the subscript 1 and 2 mean?

M: Yes, that would be the value where $\Delta G^* = 0$, if you look at Eq. (5), that's a maximum. This is the maximum rate constant you could have, in other words the value it would have for the most favorable ΔG^0 in that step. k_{max1} for system number 1 and k_{max2} for system number 2, maybe. The definition of k_{max2} is simply k_{max} for component number 2 which is the His48, so you can see that from the table.

Q: But there is also a third thing.

M: Number 2, that's the standard. You see, if you put k_{max2} you get unity. The square root is there because it is proportional to the matrix element. Because your errors are in calculations of the matrix element, if you show errors in a matrix element squared in the calculation of the matrix element, you are exaggerating the errors.

8. p. 6113 2nd column bottom: "the calculated value of the energy of the donor orbital at the transition state."

Q: How do you determine when the donor is at the transition state and how do you determine then its energy?

M: The superexchange is occurring at the TS for the reaction, I mean the TS is where everything is occurring, so you reach the TS and in this case instead of being a simple exchange there is a superexchange. That means that all energies that appear in the superexchange expression have to be calculated at the energy of the TS, not at the energy of the reactants, so you have to look for the vertical energy change that's appropriate for the superexchange but at the TS of the reaction, not at of the initial state of the reactants. The various energies of the energy levels depend somewhat on the reaction coordinate, which means that it's the vertical energy that counts, the vertical energy's changes that are involved in the superexchange, they will vary with the reaction coordinate, so you pick the value of the reaction coordinate that is at the TS and it's that vertical energy at that configuration that you use.

9. p. 6114 1st column top: "A B-O pathway is a single chain of atoms connecting the donor and the acceptor and the electronic coupling provided by a pathway is estimated by the number of links where each bond, hydrogen bond and *through space link.*"

Following: "via through space connections"

Q: What is it a "through space link"? What a "through space connection?

M: If you have two atoms that are connected by a chain, the chain is the through bond link, and the direct is the through space, without paying any attention to the chain, in other words you look at the orbital on the donor atom, you look at the orbital on the acceptor atom, you don't perturb them by the chain and you look just at their direct through space interaction. Through space connection is the same as through space link.

10. p. 6114 1st column bottom: "the need for theories more sophisticated than the existing pathway analyses and *homogeneous barrier models.*"

Q: What do you mean by "homogeneous barrier models"?

M: It is if you just use the square of a barrier between the reactant and the product, that is not distinguishing at the molecular level.

M283 Quantum Corrections for ET Rates. Comparison of Polarizable versus Nonpolarizable Description of Solvent

X. Song and R. A. Marcus

NOTES

1. p. 7768 1st column top: "By classical and quantum simulation methods they observed the parabolic behaviour of the *free energy* *surface* with respect to the solvent polarization coordinate, a behavior which plays an important role in the theory developed by one of us."

Q: Have you been the first to use free energy surfaces in the study of ET in the Discussions paper of 1960? Have you been the first to discover their being parabolic?

M: I don't remember exactly when, when I had free energy and, yes, I was the first, to my knowledge.

2. Following: "...the rate is expressed in terms of the spectral density (the dielectric response) of the system."

Q1: I read in the book of Papoulis (Probability, Random Variables, and Stochastic Processes) that the spectral density $S(\omega)$ of a process $\mathbf{x}(t)$ is the Fourier transform of its autocorrelation

$$S(\omega) = \int_{-\infty}^{+\infty} e^{-j\omega\tau} R(\tau) d\tau$$

where

$$R(t_1, t_2) = E\{\mathbf{x}(t_1)\mathbf{x}(t_2)\} = \int_{-\infty}^{+\infty} x_1 x_2 f(x_1, x_2; t_1, t_2) dx_1 dx_2$$

and the mean $\eta(t)$ of the process $\mathbf{x}(t)$ is

$$\eta(t) = E\{\mathbf{x}(t)\} = \int_{-\infty}^{+\infty} x f(x; t) dx$$

My questions are the following

(1) Which one is here the process $\mathbf{x}(t)$?
(2) Which is the physical meaning of its spectral density $S(\omega)$?
(3) Why and how is the spectral density related to the rate?
(4) You gave me the reference to Eyring, Walter, and Kimball for TS. Have you a reference for this way of getting the rate constant in terms of the spectral density of the system?

M: I know Papoulis very very well. We were good friends. The process $\mathbf{x}(t)$ here can be a coordinate, a random variable changing as function of time, it can be a bond length, it could be sort of generalized normal coordinate of the solvent.

What you've written down there, that

$$S(\omega) = \int_{-\infty}^{+\infty} e^{-j\omega\tau} R(\tau) d\tau$$

is a theorem actually, the spectral density is normally not defined that way, that is a theorem. The spectral density is defined differently, in a related way, and you prove that theorem. Normally the spectral density is defined in terms of the square of some amplitude that is fluctuating and then you show how that is related to a correlation function by a theoretical method argument. It is defined in basic books on time dependent statistical mechanics like McQuarrie which may have it, there are quite a few books on statistical mechanics that have some time dependent statistical mechanics in there.

In the present case of ET, I suppose that $\mathbf{x}(t)$ to be the polarization in any particular volume, you know, and fluctuations in it. Or it could be the dielectric displacement, or the electric field, and fluctuations in it. Typically, in this case it is mainly the polarization as a function of time at any particular point or is the dielectric displacement, one can see what Ulstrup does in his book that is on his thesis.

Probably what I meant there is that describing the whole process in terms of normal coordinates for a solvent is a kind of poor method to do but lots of people have done it. Treating the solvent as a collection of harmonic oscillators... a lot of people have done that, especially the Russians did a lot but the trouble is that doesn't take into account all of the hindered water molecules and *how they change depending upon the state of the ion, whether it is +2 or +3 charge, all of that's ignored* when you use harmonic oscillator description.

Q: Here we have ET between two iron–water complexes.

M: Yes, the random variables are the vibrations, the Boltzmann motions of the solvent, these are all random variables.

Q: And the physical meaning of the spectral density?

M: I think, but I'm not sure, that if you analyze the motion of a particle or mode, you know, anything, any combination of coordinates, then if you did a Fourier analysis of it.... I think if you look up McQuarrie Statistical Mechanics it has some description and basically if you have some sort of a noise and you look at the fluctuations in that noise, you can get a spectral density out of it but I don't remember the details.

Q: But a spectrum of what?

M: A spectrum of the frequencies of motion, I should know the answer to that precisely but I don't, I should know it exactly. It might be polarization as a function of time, at any particular site, it might be the electric field as a function of time, at any particular site, it might be the dielectric displacement, I'm not sure which one they use... but it might be one of those things.

Q: Why and how is the spectral density related to the rate?

M: It's related to a correlation time, if you take the Laplace Fourier transform of the correlation time, I am not sure if you get the spectral density out or not. For the moment, let's assume that you do but that one would have to check, I don't remember. The rate constant for crossing a barrier can be related to a correlation function, I never did that, but Chandler has done it and maybe Yamamoto may have been the first to do that around 1960, in a paper I think in JCP, so there are expressions for the rate in terms of a correlation function, and I think is a crossing the barrier, and maybe you can see that... if you look at the coordinate in the vicinity of crossing the barrier... first of all there is a probability of being there, and the velocity of crossing the barrier sort of carries you across the barrier... I am not sure if there is, but maybe if one looked at Chandler's small book... does he have an expression of the rate in terms of a correlation function? There would be other references though, but Chandler is a key reference, it is related to sort of velocity of crossing the barrier, and being at the barrier and a correlation function related to them, but anyways way back in the 1960s Yamamoto was the first to give an expression for the rate constant. Chandler and also Bill Miller used that quite a bit in various papers, you know, in the bulk theory, and so did some others.

Chandler related the rate constant to certain correlation functions. Before that I guess either Levich and Dogonadze or one of the Russians did, I don't remember, maybe all of this is in Ulstrup's book, related that to correlation function of I don't know of whether dielectric displacement or what have you. Now how's that related, the details of how it's related, the physical insight, that I really don't know, I'd have to think about it, but at the moment I don't know since I never really derived anything like that. This is really the heart of the Russian work and is the heart of Chandler's work and although occasionally I've been involved in a paper connected with it, like a paper with X. Song, I've never really done it on my own, so you know, I don't have a real feeling for it.

"Is a crossing the barrier" Yes, I think in the Chandler's formulation he had a correlation function in the vicinity of crossing the barrier, so you should look up at Chandler's paper.

3. p. 7768 2nd column bottom: "If the solvent and the nuclear motion of the reactants and products are described as a *harmonic bath*"

p. 7769 1st column bottom: "A key assumption... is the use of a harmonic approximation for the *bath modes*"

p. 7770 1st column top: "If the polar medium is treated as *a collection of harmonic oscillators* characterizing the dielectric polarization..."

Q: Can you explain more in detail this description of the solvent as a harmonic bath? Should one consider the solvent as a crystal lattice with normal modes like a big molecule with related collective motions? Or should one consider the absorption of electromagnetic radiations of varying frequencies ω and imagine that to each absorbed frequency corresponds a particular collective motion of the solvent molecules? Are there in this case infinite normal modes, one for each frequency?

M: First of all it is a model that some would use to treat some problems. I don't think it is a great model, maybe the first ones to use that model for electron transfer effectively were Levich and Dogonadze, in a paper in 1959. They used a collection of harmonic oscillators . . . a solid is acting harmonic, which of course a solvent doesn't, but it is a way of proceeding, and so a lot of people have done that and the modes are actually supposed to be collective modes of motion of the solvent, you take the solvent and treat it almost as a crystal lattice broken up into normal modes. That's basically what it is. It is just a model, but it's a model which permits you to make all sorts of calculations, and there are some classic papers by Caldera and Leggett in Annals of Physics, I don't know if it was in the 70s, or 60s, or 80s, and they are enormously referred to. . . . So, it's just a model, in a paper David Chandler credited me with spin-boson Hamiltonian but I never knew what it was, I may have been using it without realizing it, I think I was using a dielectric continuum equivalent without knowing anything about it. So I didn't use it in the full dynamic sense.

Q: When an electromagnetic wave goes through a liquid and you have a dielectric constant depending on the frequencies of these waves, are these frequencies related to the modes?

M: Yes, it is connecting with the so called normal modes. But that's just an approximation, an approximate description, because strictly they are not normal modes, *they are highly damped* and so on, they are a convenient vehicle for developing theories, so it's fantastically widely used. But everybody recognizes, I think, that it's not really the solvent, but it's a convenient description. I mean, for example, recently, on a totally different topic, I was looking on some results on some frequency generation and the model we were using turned out to have some bad properties in some ways, it was OK in others, they gave us too wide a spectrum but if I was using an area of the spectrum and using the experimental width together, then we would get the right answer, and so I wanted to try to understand it and I thought I could understand what are called sum rules and, after thinking about it through, maybe a month or two, I finally realized: yes, I can use a spin-boson Hamiltonian for this specific problem to obtain the sum rule.... Now, I think that in a number of these problems what's important is sort the *quadratic structure* of some magnitude. Now, for the polarization, for the free energy, the quadratic structure of that is probably in good shape, you see, so based upon decomposition of the polarization into Fourier components and using a finite Fourier decomposition, that's sort of like equivalent to normal modes, but they are not really great modes, I think I could rephrase the whole theory in terms of some dielectric properties, just like what I did, maybe with Georgievski in some paper, in order to compare with people who do spin-boson Hamiltonians, we did something of that... some analogy to continuum.

I think one of the points I was trying to get at, but I'm not sure, I've to look at it, was that you can have a quadratic expression for the free energy in terms of the fluctuations, and that can be pretty good, that's essentially a cumulative expansion dropping higher terms. A much more specialized assumption is that you have essentially dynamics associated with harmonic oscillators, that's the model that Levich and that Caldera and Leggett are using, in other words that's a collection of harmonic oscillators, so, I mean, that's convenient but is not accurate, because the system really does not consist of harmonic oscillators, and is not assumed in the quadratic expression for the free energy that the system involves harmonic oscillators, instead you just use the cumulant expansion, and doesn't have to be harmonic oscillators. So, treating the, so to speak, statics by that free energy where you just end up with harmonic terms, that's much more general than a model which is based on harmonic oscillators, but the dynamical model that some people use, like Levich and later Leggett that assumes that you have harmonic oscillators dynamically, that's a much more restricted model.

Q: Did David Chandler use this model too?

M: David Chandler... I'm not sure if he did or not, he may have used... I don't think he did, I think he used the general method and determined the actual frequencies from computer calculations. So I don't think that he explicitly assumed harmonic oscillators. In fact, what was remarkable about his work was that he came up with a very harmonic looking expression for the free energy curves, so he didn't assume that.

4. p. 7768 2nd column bottom: "This kind of system has been described by the spin-boson Hamiltonian:

$$H = \frac{H_{AB}}{2}\sigma_x + \frac{\varepsilon}{2}\sigma_z + \frac{\sigma_z}{2}\sum_{i=1}^{N} 2c_i y_i + \sum_{i=1}^{N}\left(\frac{p_i^2}{2m_i} + \frac{m_i\omega_i y_i^2}{2}\right),$$

.... The y_i is a polarization coordinate, p_i is the canonically conjugate momentum..."

M: In this model for each polarization coordinate there is a normal mode. But often when you speak of normal modes you use some sort of analysis where the masses are unity. I don't know why it wasn't done here, normally with normal modes you steer your momentum, and so on, so that the masses are unity. And the momentum then has units of square root of energy. N is the number of vibrational normal modes.

Q: What do you mean by "you steer your momentum"?

M: You scale your momentum. In expression for kinetic energy you have $\frac{p^2}{2m}$, you can adjust the unity so you have a new $\frac{p^2}{2}$, but of course that implies changing other units too.

Q: Apparently here one should consider internal vibrational modes of the coordination shell of the iron complex and solvent modes...

M: Yes, all of them. N is the number of the solvent modes. In this usual spin-boson Hamiltonian the conventional thing it is usual to refer to the solvent, but I guess that because that c_i is going to be related to dielectric properties, I guess you could say: among those N I'm going to include the internal vibrational modes of the coordination shell. So N would be all of the modes.

Q: Is mass associated to bath vibrations?

M: If you analyze in terms of normal modes, if you do a normal modes analysis, you choose your momentum in such a way that the mass is unity. In normal modes you will see an expression $\frac{p^2}{2} + \frac{\omega^2}{2}$. With bath normal modes you can do that too. Even with bath normal modes the masses would be unity, the way you normally do with normal modes. When you do that, then you usually have in front of your $\frac{p^2}{2}$ an $\hbar\omega$, because you insert down something to get rid of your masses but you want to have an energy unit. I guess $\hbar\omega$ has been absorbed into the p. The masses associated with bath normal modes would be a collective property. In other words, take a crystal, and then consider a normal modes analysis, the various masses come into the description of normal modes in different ways, for example, if you have one particle vibrating against some others, then it is the mass of that particle that is involved, if instead you have two particles that are vibrating with respect to one in the middle, that would be another kind of normal mode for triatomic molecule, a different effective mass comes in, so a lot depends on the nature of the normal mode.

But all of that comes out when you do a normal modes analysis because then you take into account all the masses, and in a way end up with normal modes, and when you look at their description they depend on masses in the correct way. It can be probably found in maybe Wilson, Decius, and Cross, sort of, they do normal modes analysis, you'll see they end up with $\hbar\omega$, and then the normal mode energy... You may find that book easier than Born and Huang.

If you look at the statics we're just calculating the free energy that relies on this cumulant expansion, that's pretty good and is not assuming the actual dynamical motion as harmonic oscillators.

There is something here, that ω^2 should be multiplied by q^2, you need to have a potential energy term there. You have $\frac{p^2}{2}+\frac{\omega^2 q^2}{2}$. If you look at the quantum mechanical expression often you'll see having an $\hbar\omega$ and then you have creation and annihilation operators, in this case for a harmonic oscillator. Those creation and annihilation operators are related to $p \pm iq$, so you'll see a quantum mechanical counterpart of that and in front of the creation and annihilation operators you'll probably see an $\hbar\omega$.

5. p. 7769 1st column "For the nonadiabatic case, the ET rate constant is

$$k = \frac{2\pi}{\hbar Z_b}\left|\frac{H_{AB}}{2}\right|^2 \sum_i \sum_f \exp(-\beta E_i^0)\left|\langle \chi_i^0 | \chi_f^0 \rangle\right|^2 \delta(E_f^0 - E_i^0 + \varepsilon) \tag{2.2}$$

Q: Where does the solvent appear in the above expression? Only in ε which is equal to $-\Delta G^0$?

M: I guess the χ's are probably the vibrational modes. Overlap of the wave functions. Initial and final vibrational wave functions. This is an expression that was originally used by Levich and Dogonadze in 1959 in their model of electron transfer theory, they relied I think quite heavily on some earlier work, maybe on F centers, and they quote what they relied on, and this formula here, the delta function, has an energy difference, so when they finally got the classical expression they instead of having a ΔG, they had a ΔE, an energy of reaction. In 1970, I met Levich at a meeting in a summer school, in Yugoslavia, and talked to them about this and I told them that yes, their expression is right but that going over to ΔE from that, in the exponential, is wrong and I explained to them why, and then Levich agreed and said he is going have to tell Dogonadze to go wash his hair, apparently some Russian expression when somebody is wrong, but the point is that when you use spin-boson Hamiltonian the frequencies of the solvent are the same before ET and after ET, that means that if a reaction produces a large change in the solvent in the form of organizing it a lot more, for example, when considering two neutral molecules producing positive and negative ions, you have an entropy of reaction which their treatment doesn't have. It can't because it assumes the same solvent bath, and there is no chance that it is going to change the solvent frequencies, for example, so that's a drawback of the spin-boson Hamiltonian. The quantum mechanical expression has a ΔE like that, and the quantum mechanical treatment doesn't take into account the way you usually do of the entropy excess. The solvent is much more highly organized around high charges, that's going to affect usual solvent frequencies, you change the solvent. So that's a shortcoming of this spin-boson Hamiltonian. It doesn't really handle reactions with large entropies of reactions, large standard entropies of reaction. Now, some place I wrote an article in which I tried to bring this out and show how one goes from this to a free energy, and I don't remember what I did now, because it was a long time ago, maybe... I have a series of articles 1, 2, going up to 6 I think, I am not sure, it would have to be after 1970, maybe what I did was maybe I treated vibration... in a solvent... large entropy, I don't remember what I did, maybe a paper related to absorption spectra reaction rates, certainly not in the last 10 years... I forgot what I did. Oh, yes, where I probably said something about the large error that you can make in entropy and so on.

That meeting in Yugoslavia was really wonderful. Actually if you look at the frequencies around an ion, *there's going to be a different distribution of frequencies depending on the charge of that ion.* Because if you have, say a +3 ion that's going organize things more, there'd be a higher representation of higher frequencies in some Fourier decomposition and, on the other hand, in the model that Levich and Dogonadze use you have fixed frequencies, they don't depend upon the charge state of the ion. So physically there's something missing. I mean, not only is the motion not strictly harmonic because you have these big vibrational motions of the water molecules, but also the *distribution of frequencies of the solvent really change depending on the charge state of the ion.* All of that is not present in their model. In fact you can't use that model.

6. Following: "... Upon using the usual δ-function expression

$$\delta(E_f^0 - E_i^0 + \varepsilon) = \frac{\beta}{2\pi}\int_{-\infty}^{+\infty} \exp[-i(E_f^0 - E_i^0 + \varepsilon)\beta R]dR \tag{2.3}$$

and the overlap integral of the wave functions assumed to be harmonic, the following formula has been obtained

$$k = \frac{\beta}{\hbar}\left|\frac{H_{AB}}{2}\right|^2 \int_{-\infty}^{+\infty} dR\exp\left[-\left(\frac{\beta}{2} + i\beta R\right)\varepsilon - \frac{2}{\pi\hbar}\int_0^{+\infty} d\omega \frac{J(\omega)}{\omega^2}\frac{\cosh(\beta\hbar\omega/2) - \cosh(iR\beta\hbar\omega)}{\sinh(\beta\hbar\omega/2)}\right] \tag{2.4}$$

where $J(\omega)$ is spectral density of the system

$$J(\omega) = \sum_{i=1}^{N} \pi\delta(\omega - \omega_i)\frac{c_i^2}{m_i\omega_i} \tag{2.5}$$

Q: (1) Which is the meaning of R?

(2) Are the ω's the frequencies upon which the dielectric constant $\varepsilon(\omega)$ depends?

(3) Are these frequencies the same as the harmonic frequencies of the bath?

(4) Which is the physical meaning of $J(\omega)$?

M: (1) It is a dummy variable. That's all it is. You can put anything in there. It is an identity.

(2) Those are the frequencies of the solvent's normal modes.

Q: Which are the same upon which the dielectric constant depends?

M: (3) Yes, the c_i^2 that appears there is eventually related to $\frac{1}{\varepsilon_r} - \frac{1}{\varepsilon_s}$, see the Ulstrup's book.

(4) I think it is the contribution of each mode to the coupling, I think that's what it is. See that c_i^2 that appears there, is related to coupling to the reactants, to changes going on due to reactants.

There are others who understand this much better than I do, this was not part of the original formalism that I worked with, so my knowledge is largely derivative, but since I really never derived that formalism I sort of don't have an immediate insight into it, but somebody like... what's his name? The fellow of Iowa State who used a fair amount of it when he was a student with me, X. Song, he'd be the one to ask about that, in many of these problems he is really the student who has done more than I, they can give you more specific knowledge.

That ε there if you use a spin-boson Hamiltonian is really ΔE^0.

7. p. 7769 2nd column top: "For the ferrous-ferric system ε is zero. The saddle point of R is zero in this case. The rate constant can be expressed in the following simple form using *the saddle point approximation*

$$k \approx \frac{2\pi}{\hbar}\left|\frac{H_{AB}}{2}\right|^2 \left[4\hbar \int_0^\infty d\omega J(\omega)\cosh\left(\frac{\beta\hbar\omega}{2}\right)\right]^{-1/2} \times \exp\left[-\frac{2}{\pi\hbar}\int_0^\infty d\omega \frac{J(\omega)}{\omega^2}\tanh\left(\frac{\beta\hbar\omega}{4}\right)\right] \tag{2.6}$$

In the classical approximation for bath $\beta\hbar\omega \ll 1$ Eq. (2.6) reduces to the classical nonadiabatic expression..."

Q: (1) Is the saddle point approximation justified by the fact that the saddle point gives the most important contribution to the reaction rate?

(2) Why is the classical approximation for bath determined by the relation $\beta\hbar\omega \ll 1$?

M: (1) The saddle point approximation... if you look at that function as a function of ω, that's a peaked function..., I don't know where it's peaked at the moment, but the integrand there is highly peaked and you evaluate by expansion, that's a saddle point approximation. The first who invented this approximation was Debye, around 1901 or something, but I first learned about it when I was a postdoc at Carolina. I had a certain integral I wanted evaluate and I had been sitting in a course of Nathan Rosen who was teaching physics there, he told me about the saddle point approximation. I used it in unimolecular reactions theory, that's how I discovered what the saddle point approximation is. Apparently that function is peaked at some frequency I suppose; I don't know, I don't know where it's peaked the integrand, how it's peaked, that's a saddle point approximation. Well, if that function is a function of ω like a Gaussian, you know, highly peaked some place, then you can use the method of the saddle point approximation to evaluate integrals like that.

(2) We broke it up into two regions, something about unity, but really in a real classical motion is $\beta\hbar\omega \ll 1$, because for the high frequency

modes you have to take into account that they are quantized, and for low frequency modes they compare to kT.

8. p. 7770 1st column top: "If the polar medium is treated as a collection of harmonic oscillators characterizing the dielectric polarization, *the changes in electric field of the ions, where the products are compared with the reactants, shift the equilibrium positions of oscillators describing the polarization... the shift of the dielectric polarization oscillator of the medium* q_{i0} is related to the dielectric constant in the following way, neglecting spatial dispersion:

$$\frac{1}{2}\sum_i \hbar\omega_i q_{i0}^2 = \frac{\int d\mathbf{r}|D^f - D^i|^2}{4\pi^2}\int_0^\infty \frac{d\omega}{\omega}\frac{\mathrm{Im}\varepsilon(\omega)}{|\varepsilon(\omega)|^2} \tag{2.8}$$

Q: How is the story of these shifts of the dielectric polarization of the medium?

M: You have to distinguish between dielectric polarization and the dielectric polarization of oscillators, you have to distinguish the focus. Yes, how to put it... I'm not sure if the word shift is the best one... well, if you use the model of an oscillator, then if you apply an electric field the oscillator adopts a new position, in other words, if you now have $\frac{kx^2}{2} + ax$, where a is the electric field, essentially, that function has another minimum, that function has shifted the position. Now, in actual fact of course you don't really have oscillators, that's just a model that people have used to simulate, it has advantages and has physical disadvantages, especially in the case of ions in solution, it doesn't take into account reorientation effects, entropy effects, but it is a very simple model and it carries some physical meaningful consequences but it can be in big error in terms of entropy. You represent the electric polarization by oscillators and if you apply an electric field, like due to an ion, the equilibrium position of those oscillators shifts. The point is that dielectric polarization consists of more than just harmonic shifts of oscillators, there are reorientations of molecules which the oscillators don't exactly take into account.

So, it's a useful model because you can do so much with it, but when you come down to certain issues it can be horribly wrong, it depends on the issues, but it is a very useful model.

Erratum: Wrong citation: You refer the following formula

$$\frac{1}{8\pi}\int d\mathbf{r}|D^f - D^r|^2 = \left(\frac{1}{2a_1} + \frac{1}{2a_2} - \frac{1}{R}\right)(\Delta e)^2 \tag{2.9}$$

to your review with Sutin, but the formula is not there.

M: I surely have it in some place.

9. p. 7770 1st column middle: "For the spin-boson model this reorganization energy can be written as

$$\lambda_o = \sum_i \frac{2c_i^2}{m_i\omega_i} = \int_0^\infty \frac{2}{\pi}\frac{d\omega}{\omega} J_o(\omega) \tag{2.10}$$

Following: "...the outer contribution $J_o(\omega)$ to the spectral density can be written as

$$J_o(\omega) = \frac{\int d\mathbf{r}|D^f - D^i|}{8\pi}\frac{\mathrm{Im}\varepsilon(\omega)}{|\varepsilon(\omega)|^2} \qquad (2.11)\text{"}$$

Q: At this point we have *four* different ways in which λ_o can be written, that is

$$\lambda_o = \sum_i \frac{2c_i^2}{m_i\omega_i} = \int_0^\infty \frac{2}{\pi}\frac{d\omega}{\omega} J_o(\omega) = \frac{1}{8\pi}\int d\mathbf{r}|D^f - D^r|^2$$
$$= \left(\frac{1}{2a_1} + \frac{1}{2a_2} - \frac{1}{R}\right)(\Delta e)^2$$

(1) Can you comment on the four different forms?

(2) Is there a physical interpretation of the evident close relation of λ_o and $J_o(\omega)$?

M: (1) If you express your D's and your **P**'s and so on in terms of discrete decomposition . . . normal modes, so to speak, then you have a transformation between them and those various forms there they are all related, they can be all derived from the transformation.

(2) The λ_o refers to the solvent and the reorganization, and $J_o(\omega)$ refers to some harmonic description of the solvent. The point is that if you put in the appropriate values for each of these magnitudes, if you use some harmonic oscillators model of the distorted field when you have ions in solution you might well go over from, say, Eq. (2.10) to the equation for λ_o involving $(\Delta e)^2$, but, you know, there is dipole reorientation and it's not easily taken into account . . . in other words, I'm sure that when one introduces a certain harmonic oscillator of a solvent that had two ions in it somehow you can go over from one λ_o expression to another, but they're totally different formalisms. I'm thinking that . . . I'm sure that one can make a model of solvation of an ion that gave rise to that 1/2a using harmonic oscillators, in other words one can probably model these properties in an approximate way but they're different languages.

I was never familiar with this detailed model in terms of collections of harmonic oscillators, I didn't derive it, we used it in papers, in various papers, but that is not something that I feel at home with.

10. p. 7770 2nd column top: ". . . the spectral density of the inner contribution can be written as

$$J_i(\omega) = \left(\frac{\pi}{2}\right)\omega\delta(\omega - \omega_0)\lambda_i \tag{2.12a}$$

where

$$\lambda_i = 6m\omega_0^2 d_0^2 \tag{2.12b}$$

using the fact that there are six ligands for the symmetric breathing mode in $Fe(H_2O)_6^{2+}$ and in $Fe(H_2O)_6^{3+}$."

Q: Same question as the preceding last one, even considering the parallelism between $\lambda = \lambda_i + \lambda_o$ and $J(\omega) = J_i(\omega) + J_o(\omega)$?

M: That may be the following: that supposing you had N harmonic oscillators, and you introduce new coordinates as has been done where one of the coordinates is perpendicular to the TS line, then the rest of the harmonic oscillators under certain conditions sort of are the same for reactants and products, they don't change in distances, they don't contribute, the only oscillator that changes is the oscillator that moves perpendicular to the TS, and that reduces to one frequency.

11. p. 7772 1st column middle: "For comparison with previous work, we give next an approximation to the outer contribution in Eq. (2.13)

$$\begin{aligned} k \approx \frac{2\pi}{\hbar}\left|\frac{H_{AB}}{2}\right|^2 \left[4\int_0^\infty d(\ln\omega) J_o \hbar\omega \cosh\left(\frac{\beta\hbar\omega}{2}\right)\right. \\ \left. + 2\pi\hbar\omega_0\lambda_i \cosh\left(\frac{\beta\hbar\omega_0}{2}\right)\right]^{-1/2} \\ \times \exp\left[-\int_0^\infty d(\ln\omega)\frac{2J_o(\omega)}{\pi\hbar\omega}\tanh\left(\frac{\beta\hbar\omega}{4}\right)\right. \\ \left. - \frac{\lambda_i}{\hbar\omega_0}\tanh\left(\frac{\beta\hbar\omega_0}{4}\right)\right] \end{aligned} \quad (2.13)$$

by dividing the complete frequency range of the dielectric response in two parts, a 'classical' part and a 'quantum' part, where a separation frequency ω_{cl} is defined by relation $\beta\hbar\omega_{el}/4 = 1.0$."

Q: When and why is a particular frequency to be considered classical and when quantal?

M: When it is much more than kT is quantal, when it is much less than kT is classical.

12. p. 7772 2nd column middle-bottom: "These two types of modes tend to play different roles in ET, the former giving a *nuclear tunnelling* effect and the latter generating an *activation barrier*"

M: The quantum mode gives a nuclear tunneling effect, in other words, in the case of a high frequency mode... when the frequencies are high you have the biggest deviation from classical mechanics, so the high frequency mode is the one that can give rise to tunneling, and the low frequency modes are the ones whose net result gives similar results to classical, and the relation to activation barrier is classical.

If there is weak overlap, that's the situation where you have nuclear tunneling. Yes, probably a lot of the barrier is due to that, the high frequency modes can tunnel through, so they don't have such a high barrier, because they can tunnel through it, but the low frequency modes can't, so under bigger barrier they suffer, they can't tunnel through.

13. Following: "From the tunnelling factor expression (3.4)

$$\sigma = \exp\left[-\frac{2}{\pi\hbar}\frac{\int |D^f - D^i|^2 d\mathbf{r}}{8\pi}\int_{\omega_{cl}}^{\omega_{op}} d\omega \frac{\mathrm{Im}\epsilon(\infty)}{\omega^2|\epsilon(\omega)|^2}\right] \tag{3.4}$$

it is clear that the really high frequency modes (say higher than $7.2 \times 10^{14}\,\mathrm{s}^{-1}$) do not make a significant contribution to the tunnelling effect due to the negligible imaginary part of the dielectric constant."

Q: Mathematically clear. Can you give a physical explanation?

M: I guess physically the very high frequency modes of the solvent... whether they make a contribution or not would depend on how much different they would have to be in sort of equilibrium position, if the the equilibrium positions of a high frequency mode were identical before and after, of course they wouldn't make any contribution, but I'm just trying to remember why they make very little contribution... it may be that those modes of the solvent are really essentially unchanged by the reaction... See, if they are unchanged by the reaction, that would be the case, now I guess that there is some change but maybe they are less shifted, possibly, one would have to look at the solvent mode, a high frequency mode, for example, that could be an H–O–H bending motion, you know, vibrational polarization, and that makes some contribution but the contribution that makes probably is small because it makes very little contribution to the energy of activation, but it makes some contribution to the preexponential factor, in other words to the tunneling, but maybe it turns out that... I don't know, one would have to look at it in detail, that the difference in sort of equilibrium position of those modes in the initial state and in the final state for reactants and products are not very different, it could be, I just don't remember.

If that displacement of the oscillators of reactants compared with products, if that's a small displacement for a particular mode then that mode is not in to reorganize there.

The main point is that you can use the classical expression up to a certain frequency, beyond that frequency you have quantum effects, you can't use classical, my expression was classical, so beyond that if you want to make the quantum corrections, you can't use my expression, you have to use some expression like Eq. (3.4), so that's how it starts with the classical, in other words that ω_{cl} is the point where you can... after that higher frequency you can no longer use classical, that's the key point. And so I think that at thousand wave numbers, for that kind of frequency you can't use classical, it is far greater than kT, so... and that 1,000 wavenumbers, when you convert using the velocity of light 3×10^{10} used for converting per seconds to per centimeter wavenumbers you have 3×10^{13}, so that's really a better starting point when you start using the quantum, in other words that ω_{cl} is closer to 3×10^{13} than it is to 7×10^{14}.

Q: But this frequency is 7×10^{14}. Isn't it an electronic frequency?

M: Oh, they are really high frequency modes, yes, that's right, they are not present in the solvent to speak of, yes, that's right, no. You are right, that is not that vibration I was talking about. What it means, I guess, is that there is very little imaginary part there in fact they are not doing anything, so they don't make any contributions. That frequency is the electronic frequency of the solvent and they are pretty much the same, you know, they are not changing, you know, they are sort of like bystanders.

Q: And which is the relation with the fact that the dielectric constant is imaginary?

M: It's related to the absorption and you remember there is a relationship between absorption spectra and something else that may be the connection why the imaginary part comes in there. I know that it does, and of course you see it in many places, that it comes in, but I'm just trying to remember whether... you know, if you look at absorption spectra, and if you look at the free energy barrier, you remember I wrote something about it, Jortner has written something about it, there is a certain parallelism and in fact the one side of the absorption peak corresponds to the inverted region, so there is a parallelism between the barrier and the absorption, and the absorption is related to the imaginary part, that may be it.

You know, the polarization can be expressed in terms of the dielectric displacement and the electric field, and if you have a dielectric constant partly imaginary and you look at the polarization you'll find it will decay with time,... if you have some oscillating field, if the dielectric constant is imaginary that means that something decays with time, that means that some energy has been absorbed, so often the imaginary part that's responsible for the susceptibility function and dielectric constant are related to absorption spectra. In fact, if you take what's called the dielectric function, which is a complex value function, defined in terms of relation between the dielectric displacement and electric field, then you find that the imaginary part of that dielectric function is responsible for the absorption.

14. Following: "In this sense the electronic polarization does not make large contribution to the ET rate. However, the electronic polarization by creating a *shielding effect*, does influence *the other aspect* of the $\varepsilon(\omega)$ behaviour"

M: Yes, you know, for example, it comes into the $\frac{1}{\varepsilon_{op}} - \frac{1}{\varepsilon_s}$, that $\frac{1}{\varepsilon_{op}}$ is sort of where it comes in and it is, I guess, shielding the solvent from the changing charge of the reactants so it reduces sort of the changes that have to occur first.

Q: How does this shielding come about?

M: Well, if you have a charge and if you have a dielectric that will reduce the field at some point, in other words it partially shields the charge, ϵ_{op} instantaneously provides a shield, reduces the effect of the charge, in fact it lowers the field immediately by $\mathbf{D}/\varepsilon_{op}$ where **D** is the dielectric displacement. **D** *is the field directly due to a charge*.

Q: Apparently the electric constant has two different aspects in its behavior.

M: I guess the other aspect of that behavior is that related to the $\frac{1}{\varepsilon_{op}}$, it is the fact that you have $\frac{1}{\varepsilon_{op}} - \frac{1}{\varepsilon_s}$ instead of $1 - \frac{1}{\varepsilon_s}$.

15. Following: "Furthermore, the atomic polarization (in the vibrational resonance region) does contribute to the electron transfer, both *directly* and *via shielding*, indirectly."

M: We have that $\frac{1}{\varepsilon_{op}} - \frac{1}{\varepsilon_s}$, because the optical constant really responds so quickly, so you have that instead of $1 - \frac{1}{\varepsilon_s}$. Song's analysis, sort of broke that up, just like the electronic constant would be acting as a shield, if you have $\frac{1}{\varepsilon_{ir}}$ instead of $\frac{1}{\varepsilon_{op}}$, then it would be acting as a shield. You see, the best expression to use is what is given in Eq. (3.4)

$$\sigma = \exp\left[-\frac{2}{\pi\hbar}\frac{\int |D^f - D^i|^2 d\mathbf{r}}{8\pi}\int_{\omega_{cl}}^{\omega_{op}} d\omega \frac{\mathrm{Im}\epsilon(\infty)}{\omega^2|\epsilon(\omega)|^2}\right] \quad (3.4)$$

but of what is given in Eq. (3.4) you can make an approximation, and one approximation is given in the paper with Song, and is where you break things up into two parts. Now, if that break up were rigorous, then you would have a $\frac{1}{\varepsilon_{ir}}$, but it isn't very rigorous, and so... I forgot what we did in detail but those are the talks that come in.

Q: So you have three regions of dielectric constant.

M: You have two regions, one from $\frac{1}{\varepsilon_s}$ to $\frac{1}{\varepsilon_{ir}}$ and one from $\frac{1}{\varepsilon_{ir}}$ to $\frac{1}{\varepsilon_{op}}$. Now that $\frac{1}{\varepsilon_{ir}}$ is, you know, a little bit arbitrary, that was introduced to get a simplified expression, but there is no need to break it up if you use that imaginary expression. I mean, once you get at the infrared regime you are already getting into quantum phenomena, so that *ir* is a rough division between the one side where doing the classical treatment is OK, and the other side where you better do something that is more quantum-like. That *ir* is sort of a transition area, I mean for example... *kT* is 200 wavenumbers, so if you have 1,000 you are already widely spaced compared with *kT*, you are already well into the quantum area. With 1,000 you are already in a region where classical partition function would break down very badly. In other words, think of how well the quantum partition function, which for vibrations, one knows, is approximated by $kT/h\nu$, the classical value. But once you get well above *kT* that approximation of using classical instead of quantum breaks down. But you can see, it's not a very sharp line, it's sort of a very approximate line because in that intermediate region it's not just a question of classical versus quantum, it's not a sharp division, so I have to use the full formula that would bridge the two regions, you see, here we are using an approximate formula where, up to a certain point, we use a quantum treatment and beyond that we use classical, or below that frequency we use classical. That's an approximation, it's not an uncommon approximation to take something which is continuously varying and break it up into two parts, use something for one part, something for the other part, but it's an approximation.

M286 Correlation between Theory and Experiment in ET Reactions in Proteins: Electronic Coupling in Modified Cytochrome *c* and Myoglobin Derivatives

P. Siddarth and R. A. Marcus

NOTES

1. p. 13078, beginning: "Electron-transfer reactions in proteins have been investigated in detail by several experimental groups in recent years in order to understand the factors that control the rate of these *nonadiabatic* reactions"

Q: Why are these reactions nonadiabatic?

M: Because the distance is so long and automatically that makes the coupling weak.

2. p. 13078 1st column middle: "It is known from *semiclassical ET theory* that the rate for a nonadiabatic ET reaction depends on a nuclear factor, which in turn depends on the driving force of the reaction ($-\Delta G^0$) and the extent of nuclear reorganization (λ) accompanying the electron transfer, and an electronic factor which depends on the distance and the medium separating the electron donor and the acceptor."

Q: (1) Are there *purely quantal* ET theories?

(2) Did ever somebody try to formulate a *purely classical* theory only based on classical physics, at least limited to some systems?

I mean, suppose you have very small conducting spheres in suspension in some liquid and they transfer very small charges from one to another... to which limit could such a system be treated only with classical electrodynamics?

M: Well, you see, if you got to put the electron in the medium itself, then you got big complications, you can have an infinite self-energy..., I mean you wouldn't be using classical mechanics there. Essentially the electron would go from some place to another by some tunneling. You know, maybe if there is an aromatic system you might do it differently, but by and large the electron has to tunnel. The electron going into the medium would have to have substantially higher energy because it would be *less solvated, so less bound to the systems* there, and so that's the reason why if you have an electron going from one reactant to another it doesn't really spend any actual time normally in between the two reactants, it's really tunneling through there.

If you have, say, an aromatic chain between two reactants, there may not even be a barrier at all, because their energy levels can be in resonance with those that are in circus, the two reactants that are touching the aromatics, and for example, in some experiments of Wasielewski at Northwestern, for a longer system he had a faster charge transfer rate, because... with a longer system the energy levels in that particular case became more in resonance with the Fermi level of the donor.

3. Following: "The rate constant is given by...

$$k_{\mathrm{ET}} = \frac{2\pi}{\hbar}|H_{DA}|^2(FC) \tag{1}$$

where H_{DA} is the electronic coupling matrix element...

p. 13078 bottom: "The electronic matrix element, H_{DA}, is a measure of the coupling of the interaction between *the orbitals* of the donor (D) and the acceptor (A)."

Q3: Please check if my understanding is correct:

(1) The ET reaction is DA $\rightarrow$ D^+A^-. In order to compute H_{DA} one can compute the PES's relative to the DA and to the D^+A^- systems in the neighborhood of the distance R between D and A, look at where the splitting is, and see the amount of it, and from that one gets H_{DA};

(2) But there is the simpler method of directly computing $H_{DA} = \langle \Psi_{DA}(\mathbf{R})|H|\Psi_{D^+A^-}(\mathbf{R})\rangle$ at the equilibrium distance R of D and A. Could you comment on these two possible methods that come to mind?

M: (1) That (1) would a way of doing it, maybe the whole potential energy would involve two pure adiabatic surfaces, and you subtract the difference between the two adiabatic surfaces where the transition state would be.

Q: So that would one method.

M: Yes, probably. I don't know if they have done that way, but certainly in principle one could. They certainly have done the other way (2). Somebody may have done the other way (1) also, I don't know. You compute in the neighborhood of the transition state, then... The main thing is that you want really the two adiabatic surfaces. If you had two diabatic surfaces and the splitting is small, then from those diabatic surfaces you can get the two adiabatic surfaces in a simple way. At the intersection the diabatic surfaces are, say D_{11} and D_{22}, and if there you calculate some interaction then your adiabatic surfaces would become $D_{11} + D_{22}$ and $D_{11} - D_{22}$, in other words if you want the adiabatic surfaces you can either try to calculate them directly or indirectly starting off with the diabatic surfaces and calculating some perturbation, the effect of some perturbation.

4. p. 13078 1st column middle-bottom: "In the classical limit, FC is given is given by

$$FC = \frac{1}{(4\pi\lambda RT)^{1/2}} \exp\left(-\frac{(\Delta G^0 + \lambda)^2}{4\lambda RT}\right) \tag{2}$$

When the driving force equals the reorganization energy, the rate constant for the ET reaction is at a maximum"

Q: (1) If I have two parabolas equal and with minima at the same energy, $\Delta G^* = \lambda/4$. Where should one correctly draw λ? From the minimum of the R curve upward or upward from the minimum of the P curve?

(2) What if one has two different parabolas? Here we have two different λ's and the correct meaning of them is important.

(3) Apparently k_{ET} is function of four variables, that is, $k_{ET} = f(H_{DA}, T, \Delta G^0, \lambda)$. Are they always independent?

M: (1) If you have two equal parabolas and $\Delta G^0 = 0$ one can draw λ starting from the minimum of the R curve up to the crossing with the P parabola or from the minimum of the P parabola up to the crossing with the R parabola and that implies by the way that one symmetrizes the parabolas.

(2) Then the equation simply becomes more complicated. You don't have a single λ like that, you just have a more complicated thing, you got two different force constants, you have two different λ's, it's more complicated. First of all... in the 1965 paper even if the force constants differ by a factor of three, I symmetrize the λ's, or the k's, I had an asymmetric term of the force constant and if one doesn't include that, as one typically does, you then get the equation with a single curvature.

Q: But if the parabolas are not equal...

M: Then you have to use a much more complicated formula. If you are dealing with a symmetrized case then by definition one has a single λ.

If one is not dealing with a symmetrized case then by definition one doesn't have a single formula with a single λ. In the symmetrized case you get the same answer. If there is a symmetry the lambda's from the bottom of the R curve or from the bottom of the P curve they are equally correct.

(3) They are independent. Completely. For example, you can have a reaction with a particular ΔG^0 if the reactants are far apart you'll have a small H_{DA}, if they are close together you'll have a large H_{DA}, but they still have just one ΔG^0. If you are thinking of the λ, the λ is going to be independent of ΔG^0, so the H_{DA} is going to be independent of the λ. The temperature is related to the usual probability of being near the intersection, that's all.

5. p. 13078 2nd column middle: "The evaluation function...

$$EF = \frac{V_{DI} T_{IA}}{\Delta E} \tag{4}$$

....

The first factor in the EF, V_{DI}, is a true measure of the electronic coupling from the donor D to the intermediate atom I since the atoms involved in the path from D to I (i.e., D, 1, 2, 3, ..., I − 1, I) are all known. Presently, we calculate V_{DI} as the product of the *atom–atom couplings* from D to I, divided by an energy denominator,

$$V_{DI} = H_{DI} \frac{H_{12}}{\Delta E} \frac{H_{23}}{\Delta E} \cdots \frac{H_{I-1,I}}{\Delta E} \tag{5}$$

...we believe it is adequate to replace the exact ΔE's by some mean value, except when *the actual* ΔE_{IJ}..."

p. 13079 first column top-middle: "The couplings H_{MN} between any two atoms M and N in eq 5 are themselves estimated by a modified form of the Wolfsberg–Helmholtz resonance integral between two atomic orbitals:

$$H_{MN} = \max \left| KS_{mn} \left(\frac{\varepsilon_m + \varepsilon_n}{2} \right) \right| \tag{6}$$

where m denotes an atomic orbital on atom M, n denotes an atomic orbital on atom N, S_{mn} is the overlap integral between m and n, ε_m and ε_n are the orbital energies, ... The max in eq 6 indicates that *only the mn orbital pair which gives the maximum contribution to an atom M-atom N interaction is used in the estimation of an* H_{MN} ..."

Q: (1) The ΔE_{IJ} energy difference is not defined in the paper. What does it mean?

(2) Eq. (5) looks like a perturbation expression.

M: Eq. (5) is just being used as an evaluation function that has been used as a guide. Then, after the evaluation, one uses the correct function. In other words, each atom has a number of orbitals so there are many possibilities for m and n, out of all those you just pick the *mn orbital pair which gives the maximum contribution to an atom M-atom N interaction.*

(1) Well, think of donor and acceptor at the transition state. There donor and acceptor have the same energy. You have an all group of orbitals, you pick an orbital of the donor that has an energy equal to the energy of an orbital of the acceptor but then you are not going to satisfy the condition that all the orbitals in the donor equal the energy of those in the acceptor, just the most important orbital satisfies the equality of energies condition, so donor and acceptor have the same energy but the orbitals, some orbitals, have a different energy.

The evaluation function was used to see how strong, *how interactive a particular path* was, in a simple way, I guess.

"The meaning of the ΔE_{IJ}" What that would be would is the following: supposing that you have a *donor level* and an *acceptor level*, and you had the TS, so they have the same energy, now look at the levels in between and, let say, one of the levels has an energy E_i, then E_i minus the energy of the donor which is equal to the energy of the acceptor, is ΔE, so in the TS the donor and the acceptor have the same energy but if in superexchange in the material in between there is some particular part that has an energy E_i, you subtract that from the energy of the donor and the acceptor, the donor and the acceptor have the same energy in the TS.

Q: So the intermediates in between can have different energies.

M: They can be different, yes. You pick an energy of an orbital and you calculate the tunneling to that orbital. You could use all orbitals but the approximation would reduce to involving the orbitals that are nearest in energy to those of the reactant and the product in their TS.

(2) If you think of first and second order perturbation theory, just to start it out you have an H_{12}, that's first order perturbation theory, second order is $\frac{H_{12}^2}{\Delta E}$ ... well, it's really a product $H_{12}H_{23}$, for second order perturbation theory, if you are linking orbitals 1 and 3.

6. Following: "The second factor in EF, namely T_{IA}, imparts knowledge about *the unexplored domain of the protein space* to the search and thus transforms the search from a *blind* or *uninformed* search to a true AI or *informed* search."

Following: "This factor, by definition, cannot be exactly calculated but can only be estimated. A possible estimate of this factor can be obtained by making use of the well established *empirical* finding that the electronic coupling decays exponentially with distance *in an average sense*. Therefore, we choose to estimate T_{IA} as

$$T_{IA} = C \exp(-\beta' R) \tag{7}$$

where C is a constant which represents the electronic coupling between I and A at contact distance and R is the actual distance between node I and the acceptor A."

Following. "The AI search thus uses the EF (eq 4) to proceed from one redox center to another (from D to A as well, with $V_{\mathrm{DI}}T_{\mathrm{IA}}$ replaced by $V_{\mathrm{AI}}T_{\mathrm{ID}}$ from A to D). At each step of the search, this equation is applied, and the most promising of the nodes (atoms) is selected."

Q: (1) Is the exponential decay only an empirical fact, as asserted here, or does it also have a theoretical foundation? Or does it have a theoretical foundation when the interaction is through space but not through bond?

Q: (2) Why do you use this method instead of using an evaluation function made up of two T terms, I mean $EF = \frac{T_{DI}T_{IA}}{\Delta E}$, or by two V terms, that is, $EF = \frac{V_{DI}V_{IA}}{\Delta E}$? Which is the reason for your mixed and two ways (forth and back, from D to A, and back and forth, from A to D) method?

M: This is Prabha's doing,...it is just an evaluation function,...it is just for selecting some amino acids, so I'm pretty sure she was probably pretty loose on that, the main point is just that you have 150 amino acids, you are trying to pick 10 to 15, so the main thing is to use some method that is pretty simple because you are just using it to get an idea to what amino acids to work with, and then you better had to diagonalize using them.

7. Following: "The search does not necessarily proceed along the backbone of the protein but instead chooses the atoms that couple in the most optimal way the donor and the acceptor, based on both overlap *and energy* considerations."

Q: The overlap is clear. Which are the energy considerations? Maybe how much does a Δ_{IJ} contribute to the above average ΔE?

M: Well, if she is using the same ΔE, then there are no energy considerations, if she is using a ΔE which is dependent on orbitals, you see, the point is that ΔE going from an atom to another atom that

ΔE doesn't mean much anyway, it is really more a ΔE of orbitals that really counts, you know, the orbitals of the amino acids.

8. Following: "The interaction of the donor D with the bridge B is given by

$$T_\alpha^D = \sum_{m_D m_B} C_{\chi m_D}^D H_{m_D m_B} C_{m_B \alpha}^B \tag{9}$$

Similarly, T_α^A can be written as

$$T_\alpha^A = \sum_{m_B m_A} C_{\alpha m_B}^B H_{m_B m_A} C_{m_A \mu}^A \tag{10}$$

where χ and μ denote the molecular orbitals in D and A that are involved in ET, ; $m_{\mathrm{D}}, m_{\mathrm{B}}, m_{\mathrm{A}}$ are the atomic orbitals..."

Q: The C's are undefined in the paper. Are they the coefficients of the LCAO representation of the molecular orbitals?

M: Yes, I think so, that way it makes sense.

9. Please look at Fig. 1*: I have drawn two different parabolas with two different λ's. Is λ_2 the λ appropriate to the ET in the direction

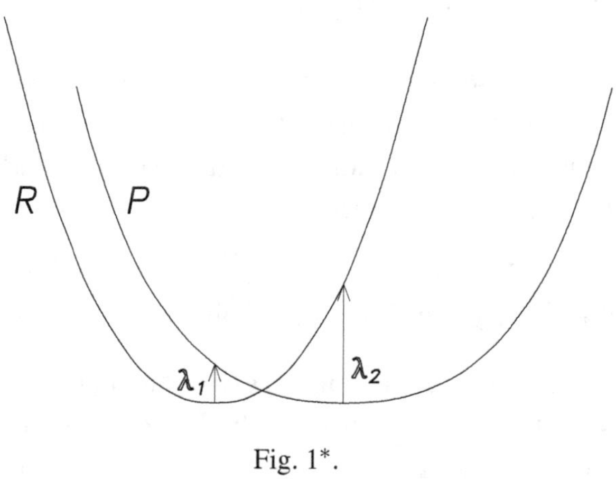

Fig. 1*.

$R \rightarrow P$ and λ_1 the λ appropriate to the reaction in the direction $P \rightarrow R$?

M: I have to think about that... I'm not sure if to call them λ, maybe it would, but you see, what I did when the two λ's for the reactants are different, and remember λ refers to both things, not just one, to both reactants, it characterizes both reactants in their initial state, λ refers to both 1 and 2, to the starting initial configuration of each. But you see, ... the formulas in that case become extremely complicated, and in the 1965 paper I may have them in there, but then rather than use those which I have left in the complicated form, I symmetrized things, so I introduced symmetric and antisymmetric combinations of the λ's, really of the force constants, so I just have *one* λ, which is a combination of the two λ's in a symmetric combination.

You would use the same λ and you would use the symmetric and antisymmetric combination. For any further discussion, you wouldn't use one λ for the reactants and one for the products, that wouldn't be correct.

Q: You consider parabolas that are the same.

M: Yes, but that comes after an approximation that I make and I get corrections in Appendix IV of the 1965 paper (**M53**). In other words I've taken differences of parabolas, differences of force constants within each one, I have introduced symmetric and antisymmetric combinations of those force constants, then got an expression, then found out how big the effect of the symmetry was. And that's in Appendix IV.

Q: Does this figure mean anything?

M: What that figure would mean, is that if you start off in the equilibrium state of the two reactants, and you just let the electron transfer, and don't change any of the coordinates, just let the electron transfer, that λ would be the free energy difference. Similarly if you start off at the products that would be the free energy difference. But you have different λ's and the overall formula is pretty complicated. So I didn't want to use that, so I made a symmetrization, and I showed it in the 1965 paper. You can draw that figure but you don't get any simple relation coming out of it. The figure is all right but the formulas that would be associated with the calculation of the free energy barrier would be more complicated, and they have not been used. The two λ's... you want to symmetrize before applying any formula. Otherwise it is too complicated. The chance that $\Delta G^0 = 0$ is negligible unless you have the kind of reaction where the two λ's are equal.

NOTE: Meanings of λ's in the Figure: (1) For λ_1: Imagine an ET induced by absorption of radiation, with fixed nuclei during the photon absorption. One passes from the minimum of the R curve to the point in which the perpendicular from there meets the P curve; (2) To that same point of the P system one may get at by a thermal fluctuation starting from the minimum of the P curve; (3) For λ_2: Fluctuations starting from the minimum of R until one reaches the point of R on the perpendicular at the minimum of P.

10. p. 13080 1st column bottom: "The results of the present study of modified cytochrome *c* derivatives, together with those given earlier, are presented in Figure 1. They are compared there with the maximum rate constants k_{max}^{expt} inferred from experimental data. The corresponding results obtained earlier (but represented in a different way) for modified myoglobin derivatives are given in Figure 2. In both cases, there is seen to be a smooth correlation between calculated and experimental k_{max}'s, though *the slopes of the log–log plots differ from the value of unity*"

Q: Why so?

M: By working with the maximum you are getting at the electronic factor, that's the purpose of working with the maximum. And assuming that all coulombic effects are being taken care of, and they may have been or may not have been, because I think Sutin had to make corrections for coulombic effects. So, I don't know what the coulombic situation is here, but I know that in some cases Sutin was treating he had a correction of the maximum for the coulombic effects. So, I don't know whether that comes in, but it can well be . . . tight binding theory is an approximate theory, there is no reason why it should be in great agreement here. What is the slope?

Q: The slope is a little bit smaller than 1.

M: Yes, because when we calculated for simpler systems slopes, and we compared with experimental, there was a perfect agreement, maybe they differed by 0.1 slope, maybe 0.2. So, if you treat simple systems with tight binding there wasn't a perfect agreement there, and there may not be a perfect agreement here.

Q: The line really passes through the points.

M: Yes, but that's slope is unity. That could be because the theory isn't perfect, the data not be correct for coulombic effects. I think in an earlier paper Sutin corrected for coulombic effects, you know, if there is a coulombic interaction.

They are just approximations, in other words the approximation that we were using to make the comparison with the experiment is just an approximation, anytime you make big approximations God knows what the origin is . . . we were assuming for example certain energetics . . . in other words the slope depends upon details of the energetics and the coupling and so on, and of course the whole formalism of using some sort of the equivalent of tight binding approximation or whatever approach have you, so anytime you have approximations you may have deviations, God knows what, why. I mean, with approximations first of all we pray and then if they work we thank God.

Coulombic effects may not be perfect for a variety of reasons.

11. Following: "The *intercepts of the plots are also not equal to the ideal value of zero*; the intercepts are *a measure* of *both the variation of the rate constants within the protein* and the agreement of the absolute values of the experimental and theoretical rate constants."

M: Yes, it may be that the absolute value was off a tight binding approximation, and the absolute value may be more in error than the slope.

12. p. 13081 1st column middle: "The atom-by-atom path method of Beratan and Onuchic.... The atom–atom interaction is...too large to be treated literally by a path method, if the justification used for the path method is usual one of perturbation theory."

M: Well, it means that the interaction of one atom with the next atom is very large, that the path method was a perturbation method, but if the interaction is large you can't choose a perturbation method. One of the things that Prabha did is that she was using the single amino acids in the path. Because there the coefficient where the two amino acids join is very small because everything is diluted. So there you can use perturbation theory. We should have kept up on it more, we have been sending it to those who have applied a lot more than we did. We just let it go and we shouldn't have done that. In other words Siddarth really had something and probably we should have applied to more cases because the method was so much better than what Beratan was doing, Beratan later on went to more complete description, but that single chain sort of thing, that was really bad, you know, it gave the right qualitative result but we should have done more, if you don't write a number of papers on something especially when there's competition, then you may lose out, so we should have written more papers on it.

Q: But are there biochemist who are using your method?

M: I don't know, I mean, nowadays people may just being diagonalizing the matrix, there isn't much of the analytical done, you know.

M289 Free Energy of Nonequilibrium Polarization Systems. 4. A Formalism Based on the Nonequilibrium Dielectric Displacement

R. A. Marcus

NOTES

1. p. 7170 1st column top: "...in the polaron system the electron is distributed over the nearby solvent molecules"

Q: Isn't this different from your theory where, I believe, the electron remains on the reacting ions?

M: In the polaron system the electron is distributed over molecules, that's not the case in the ET, what the two theories have in common though is that in polaron theory when the electron is moving through the fluid it's dragging with it the polarization which means the polarization isn't an equilibrium polarization, that is what the theories have in common, otherwise they're different.

2. p. 7170 1st column middle-bottom: "The derivation itself is much simpler than that in part I, and the resulting equation, eq 2.17,

$$\Delta G^{\ddagger} = \frac{1}{8\pi} \int \left(\mathbf{E}_{\mathrm{op}}^{\ddagger-\mathbf{r}} \cdot \mathbf{D}_{\mathrm{op}}^{\ddagger-\mathbf{r}} - \mathbf{E}_{\mathrm{s}}^{\ddagger-\mathbf{r}} \cdot \mathbf{D}_{\mathrm{s}}^{\ddagger-\mathbf{r}} \right) d\mathbf{r} \qquad (2.17)$$

is more compact, though equivalent (for longitudinal $\mathbf{D}(\mathbf{r})$) to the previous one."

Q: What does the "longitudinal $\mathbf{D}(\mathbf{r})$" mean?

M: When you have a polarization and it's oscillating, then the oscillation can be along the direction of the field, along the direction of the dielectric displacement or can be perpendicular...in a certain system you may get the vibrations at right angles to **D**...maybe it requires an anisotropy of the material, under the most common condition, the phonon would be longitudinal, along the direction of **D**, but if the material is anisotropic you can get phonons produced that are perpendicular to that, that's my guess.

3. p. 7170 2nd column middle: "$\varrho_\lambda(\mathbf{r})$ is any bulk charge density."

Q: Is it the charge inside the volume of the ion considered, for instance, as a sphere, or the charge distributed like in the above polaron case? And, are there surfaces in the polaron case?

M: There are surfaces of the ions but if you just have an electron in a medium there is no surface.

4. p. 7171 1st column middle-bottom: "Equation 2.6

$$w_{\mathrm{I}} = \frac{1}{8\pi}\int (\mathbf{E}^{\ddagger}\cdot\mathbf{D}^{\ddagger} - \mathbf{E}^{\mathbf{r}}\cdot\mathbf{D}^{\mathbf{r}})d\mathbf{r} \tag{2.6}$$

is valid regardless of whether or not interfaces are present."

Q: Why so?

M: I think it was the nature of the derivation... yes, that expression comes from potential times the charge density and you can write down the energy in terms of the potential times the charge density, integrating both over the entire system. Then you can relate the charge density to the divergence of **D** and once you do that you can use the divergence theorem and after you have carried that out you'll get a derivative, a gradient of the potential, that will give you **E**, so the whole derivation there, if you start out with ψ, ϱ, dV, and the integral of that, one half of that, that's the energy of the system, the free energy, where if you do certain electrostatic operations you'll get this regardless of whether there are surfaces or not. So it's a matter of looking at the derivation, to go from the basic one, starting from the potential times the charge density at every point and integrating over all that and then doing certain operations which are standard.

5. p. 7171 1st column bottom: "The final value of $\mathbf{D}^{\lambda}$ in step II, denoted by $\mathbf{D}^{\mathrm{r}}_{\mathrm{II}}$, is chosen to be the **D** appropriate to the initial charges. It differs, however, from $\mathbf{D}^{r}$, since the polarization field at the beginning of step I differs from that at the end of step II, even though the ionic charges are the same. In a medium with interfaces there are usually dielectric image effects, which cause $\mathbf{D}^{\mathrm{r}}_{\mathrm{II}}$ and $\mathbf{D}^{r}$ to differ."

M: In (**5**) you can have the same charge density corresponding to different dielectric displacements, because the charge density is only related to the divergence of the dielectric displacement, so if you have some vector that's divergence-less you will still get the same charge density. So, in other words, ρ doesn't determine **D** uniquely, it just determines the divergence of **D** uniquely.

6. p. 7172 middle:

$$\text{“}\delta\Delta G^{\ddagger} = \frac{1}{4\pi}\int (\mathbf{E}^{\ddagger-\mathbf{r}}_{\mathrm{op}} - \mathbf{E}^{\ddagger-\mathbf{r}}_{\mathrm{s}}) \cdot \delta\mathbf{D}^{\ddagger}_{0} d\mathbf{r} \qquad (3.2)$$

Since $\delta\mathbf{D}^{\ddagger}_{0}$ is arbitrary, $\mathbf{E}^{\ddagger-\mathbf{r}}_{\mathrm{op}} - \mathbf{E}^{\ddagger-\mathbf{r}}_{\mathrm{s}}$ vanishes, and since $\epsilon_{\mathrm{s}} \neq \epsilon_{\mathrm{op}}$, $\mathbf{E}^{\ddagger-\mathbf{r}}_{\mathrm{op}}$ and $\mathbf{E}^{\ddagger-\mathbf{r}}_{\mathrm{s}}$ vanish separately. Thereby, the nonequilibrium field $\mathbf{E}^{\ddagger}$ reduces to the equilibrium field $\mathbf{E}^{\mathbf{r}}$, when the variation of $\Delta G^{\ddagger}$ to find a minimum is unconstrained."

Q: A physical explanation?

M: If you have a fluctuation, and what's is represented there is really a fluctuation in some free energy, then when the field goes to equilibrium the fluctuation goes to zero.

7. p. 7172 2nd column top-middle:

$$``\lambda_o = \frac{1}{8\pi}\int (\mathbf{E}_{\mathrm{op}}^{\mathrm{r-p}} \cdot \mathbf{D}_{\mathrm{op}}^{\mathrm{r-p}} - \mathbf{E}_{\mathrm{s}}^{\mathrm{r-p}} \cdot \mathbf{D}_{\mathrm{s}}^{\mathrm{r-p}})\mathrm{d}\mathbf{r} \qquad (3.10)$$

If in eq 3.10 for λ_o the reacting systems were treated as a pair of spheres....

$$\lambda_o = (\Delta e)^2\left(\frac{1}{2a_1} + \frac{1}{2a_2} - \frac{1}{R}\right)\left(\frac{1}{\epsilon_{\mathrm{op}}} - \frac{1}{\epsilon_{\mathrm{s}}}\right) \qquad (3.14)$$

when dielectric image effects are neglected.

Q: (1) Is the usual formula (3.14) usually a good approximation or are there cases where it would be wrong to use it?

(2) How would Eq. (3.14) modified if image effects were not neglected? And is Eq. (3.10) comprehensive of image effects?

M: (1, 2) If you didn't have spheres, supposing you didn't have spheres, you'd have ellipsoids or big squares or something like that, this expression obviously is inappropriate to that. And in fact there are expressions in the literature, I may have written one of them, for ellipsoids. There is a question of the radii, you know this is a dielectric continuum theory, so at what point will you use discreteness for the first solvent layer maybe and use dielectric continuum theory outside of that? And also this involves a very approximate description of image charge effects when you have two particles, there are no image charge effects in Eq. (3.14), whereas normally there would be, this expression came from a treatment that still didn't exactly satisfy the boundary conditions. To satisfy the boundary conditions, one way of doing that is to bring in image charge effects, so there are a number of approximations.

M293 Theoretical Study of ET in Ferrocytochromes

A. A. Stuchebrukhov and R. A. Marcus

NOTES

1. p. 7581 2nd column top: "In the experimental studies, a very fruitful idea has been used to choose a Ru-acceptor complex so as to compensate the reorganization energy of the reaction by the driving force. As a result, electron transfer in these systems is almost activationless."

Q: (1) Your classical formulas to estimate the reorganization energy $\lambda = \lambda_o + \lambda_i$ are $\lambda_i = \sum \frac{1}{2} k_i (q_i^p - q_i^r)^2$ and $\lambda_o = \frac{1}{8\pi c} \int (\mathbf{D}_r - \mathbf{D}_p)^2 d\mathbf{r}$ or, for two spherical reactants, $\lambda_o = (\Delta e)^2 \left(\frac{1}{2a_1} + \frac{1}{2a_2} - \frac{1}{R}\right)\left(\frac{1}{\varepsilon_{op}} - \frac{1}{\varepsilon_s}\right)$. It is clear how to apply them to a reaction of, say, two iron–water complexes. But how does one estimate it in this case of very complex molecules like proteins?

(2) How do you operate, theoretically and practically, to individuate, to pick, such complexes? Is there some "a priori" method to pick them? Or maybe one estimates λ and one measures ΔG^0 experimentally?

(3) Are there some criteria by which one would be able to make a reasonable guess of a reaction's λ and ΔG^0 just at first sight?

M: (1) Yes... actually probably one might not be applying those. One might be going back to first principles to get, you know... with a statistical mechanical formula, what the actual formula looks like. And this description here is for some system but when you have a system which has a whole set of different dielectric constants, you have local paths and so on, that's more complicated and it doesn't mean it can't be done but people do.... *I mean one way of doing it,* *is that you calculate the free energy of the system, then you exchange the charges and with those you calculate the free energy once again but with the same set of nuclear configurations that gives you that vertical height of the two parabolas.* If you know or calculate the standard free energy of reaction, then the vertical height is equal to $\lambda + \Delta G^0$, you just subtract the ΔG^0 and that gives you the λ. In other words, people like Warshel calculate for proteins these parabolas in one way or another... that's a very difficult question. In the proteins that have been treated there one of the reactants was a Ruthenium, in other words they're not real proteins, and one of the reactants might be an Iron complex, you treat the inorganic complexes the way you would do in water, you know, whatever their size is, you have to decide whether to put the coordination shell in that size, probably you would, because much of the polarization is done outside of the coordination shell of the ion. In other words, these are typically systems where Harry Gray has done research, they typically have been inorganic complexes, so you treat them as you do in water, that's what the radii are.

Well, that way one doesn't use a dielectric continuum, you're probably using statistical mechanics. Whenever you are looking at an energy difference in a many-dimensional space then you're talking about statistical mechanics. The way Warshel does it, for example.

Q: So doesn't Warshel use your theory?

M: Well, he uses the basic idea, for the reaction coordinate he uses the energy difference, but I think he came about it independently, he makes statistical mechanical calculations, he doesn't use dielectric continuum theory for the central part of it.

(2) You systematically vary ΔG^0 and, you see, if you are changing ΔG^0, you vary it, you may get $-\Delta G^0 = \lambda$.

(3) No, just whether it is going to be small or large. That's all. For example, in a step in the bacterial photosynthesis, there is not much

water around there, and the reactants are large, so you know that λ is going to be very small. And it is about 0.25 eV's, it is very small, but then it's hard to estimate the ΔG^0's, there are differences in solvation, difference in electronic energy ... unless you do something like what Warshel does and then interpret it ... you only have a complicated medium around there, you can calculate λ for simple media but here it's complicated with dielectric constants varying all over the map, so it is not all that easy, so really for it probably one has to end up doing a statistical mechanical calculation of λ, or really the whole statistical mechanical calculation, which is what Warshel does for other systems.

2. Following: "The matrix element $H_{DA}(E)$ ...

$$H_{DA}(E) = \sum_{\alpha} \frac{V_{A\alpha}V_{\alpha D}}{(E - E_{\alpha}^{B})} \langle A|V|\alpha\rangle \frac{1}{(E - E_{\alpha}^{B})} \langle \alpha|V|D\rangle \qquad (2.1)$$

E is the *tunnelling energy* (energy of the donor and acceptor orbitals at the transition state)."

Q: You have taught us that thermal ET can only happen at the energy at which $U_r = U_p$. Here for the first time you denote this energy as the tunneling energy. Is there something new attached to this extra characterization?

M: Well, that E happens to be for transition, an energy where $U_r = U_p$, right? And these other energies are the energies of other orbitals relative to that, so, I mean, that's all it is, the E is related to where the two potential surfaces intersect but the E_α's are related to energies of intermediate orbitals of the system. Well, the point is that the tunneling happens where the energies are equal. There is no tunneling when you are not at the transition state, you are not tunneling then. Really tunneling happens when you get to the intersection. But you got to be careful, one doesn't want mix up nuclear tunneling with electron tunneling. When you are using the parabolas and you are going along some reaction coordinate, above or below, if you're going below the intersection that's *nuclear* tunneling, it has nothing to do with electron tunneling. Electron tunneling affects the splitting between the two parabolas, nothing else.

The electronic energy at the transition is different from that of reactants in their equilibrium position and different from that of products, so the energy of the system when one is sort of doing the transfer is really the energy of the TS, so one compares that energy with the energy of I guess some orbital . . . in other words the E that comes in there is not the E of the reactants, is not the E of the products, but is the E of the TS, and I have discussed that, I know, in some paper where I talked about the three sites description of bacterial photosynthesis, I had a few papers, in one of them I go into in some detail, and that's where that description would be found.

If you are using that expression (2.1), that's really the equivalent of a tunneling expression, and that's getting the H_{DA}, the splitting associated with the tunneling. It's an *electron* tunneling, it's not a nuclear tunneling. That Eq. (2.1) tells you something of the energy splitting at the TS and basically the electron is not residing on each of those sides in between the reactants and the products, in other words it's not there in real time, so it's there in sort of in an ephemeral object, it's not actually there, so an equivalent of saying that is saying that it is tunneling, it's not there in for any length of time.

Q: In high energy physics one considers particles living for extremely short times . . .

M: In the chemical literature what counts is the time of about 10^{-13} seconds, there are different time scales associated with nuclear motion and associated with electronic motion, and once you are talking about times of electronic motion you use the BO approximation, the nuclei don't move, so they can't tunnel. The ephemeral object is the electron between the two sides that is tunneling, it's not spending any actual time in there, you wouldn't know the energy that well, if you talked about the time there, this time is unknown, it doesn't make sense to talk about it.

3. Following: "It is usually assumed that *there are no resonances between states in the bridge and the donor–acceptor states*"

Q: Do you mean that states of the bridge and donor–acceptor states don't have the same energy?

M: Oh, usually they are higher or lower by 1 or 2 eV's. Now, if you have a conjugated bridge you can arrange it so that it is essentially on resonance. In that case the wave function hardly decays with distance, the decay is very small, instead of decaying exponentially.

4. p. 7583 1st column middle-bottom: "In the reaction of present interest, $Fe^{2+/3+}$(heme)-cyt-(HisX)-$Ru^{3+/2+}$, the *transferring electron* in the initial state and final state occupies the 3d shell of Fe ion and the 4d shell of Ru respectively."

p. 7583 2nd column middle: "Thus the *tunnelling electron* in the initial and final states is localized on the t_{2g} subshells of the metal ions."

Q: I believe that here the transferring electron is the same as the tunneling electron. Are there examples of transferring electrons that are not tunneling through some barrier?

M: *The tunneling electron and the transferring electron is the same thing*.

5. p. 7585 2nd column middle-bottom: "The overlap matrix S_{ij} was calculated with a *variational cutoff distance* between pairs of atoms"

M: Well, what I imagine is this, but I have to think back. In the big formulation you have matrix elements... all the atoms are coupled, so you have matrix elements between all the atoms, but if the atoms are quite separated in space their direct matrix elements for coupling can be neglected, so you have to decide on some cutoff distance that includes all matrix elements that will be very very small not worthwhile to consider their contribution.

6. p. 7586 1st column bottom: "In a first series of calculations, we explored how the direct long distance interactions (distances longer than the typical bond lengths) between individual atomic orbitals of the protein matrix affect the net calculated tunnelling probability. Typical data of this series are shown in Figure 1 where the cutoff distance for direct atom–atom interactions was varied. If the cutoff distance is approximately the same as typical bond lengths, then only a network of transitions over chemical bonds in the protein is involved in the superexchange coupling."

Q1: (1) Look up the graph in Fig. 1. Which is meaning of a "cutoff distance" dependence compared with a simple "distance" dependence?

(2) Are the interatomic distances frozen during the superexchange coupling?

M: (1) When you cutoff at certain distances then you get a certain overall matrix element, I guess, for your interaction, then if you gradually increase that cutoff distance you build up your matrix element until eventually further distances won't make any contribution, you reach some saturation.

(2) You calculate the superexchange coupling in the transition state, I mean you calculate the interatomic distances that are appropriate to the transition state. *Now, when one would be rigorous one has fluctuations about that, because the transition state is not a single configuration, it's a whole hyperplane of configurations.*

7. Following: "We find that the matrix elements for such calculations are several orders of magnitudes smaller than the observed ones and that *the coupling increases dramatically if the interaction distance is increased beyond the typical bond lengths*"

Q: But doesn't the coupling exponentially decrease with distance?

M: Well, yes when you are going in a straight line. But if you are connecting all paths . . . you know, instead of just one path, you have an all set of paths, maybe that is not falling off exponential first. If you'd have only one path it would fall off exponentially, but if you are riding a whole set of paths and you have more and more paths that you are adding in, when you have larger cutoff distances, because you are not cutting off the paths, then it will fall off not as rapidly. And you can see that there, in that diagram, where the actual matrix element increases when you increase your cutoff distance. Because of all those extra paths coming in.

8. Following: "These calculations confirm that the conclusion that Beratan et al. reached earlier that through space transitions are important for electron tunneling through protein media. These transitions serve as *bottlenecks for the overall tunneling.*"

Q: Do you mean that the through space transitions are important but that anyway the through space tunneling is smaller than the through bonds tunneling?

M: Yes, it normally would be smaller, but it may be that when you are looking at the effect of cutoff, some of that interaction there is kind of through space, you see, because you add the matrix element of two rather distant things directly, you also have it indirectly, that's through bond, but clearly here you are also including the indirect interaction, so let's say that through space has some importance.

9. p. 7586 Figure 2

Q: The ordinate should show the density of states but apparently this density is simply represented by the density of vertical lines in the figure. Is it so?

M: Yes, I suspect that's right.

10. p. 7587 2nd column bottom: "One particularly remarkable figure of the results shown in Table 2 is that *the coupling for different components* can vary by an order of magnitude for the same *tunnelling energy*"

Q: I know that tunneling is related to the height and width of the barrier. But which one is the relation of tunneling to coupling?

M: Oh . . . well, the matrix element coupling two components could be related to the tunneling to go from one to the other, because the matrix element is essentially a product of two wave functions and some sort of potential energy, and the overlap of the two wave functions if it's exponential could be described, if you describe it in a semiclassical way, in terms of tunneling through a barrier. It is kind of semiclassical, except that there is a potential energy term there that complicates the situation.

CHAPTER 6

Tunneling Matrix Elements in Ru-modified Blue Copper Proteins: Pruning the Protein in Search of ET Pathways; Gaussian Field Model of Dielectric Solvation Dynamics; Symmetry or Asymmetry of k_{ET} and i_{STM} vs. Potential Curves; A Sequential Formula for Electronic Coupling in Long-Range Bridge-Assisted ET: Formulation of the Theory and Application to Alkanethiol Monolayers; Time-dependent Stokes Shift and Its Calculation from Solvent Dielectric Dispersion Data

Interviews on M296, M299, M300, M301, M303

M296 Tunneling Matrix Elements in Ru-modified Blue Copper Proteins: Pruning the Protein in Search of ET Pathways

J. N. Gehlen, I. Daizadeh, A. A. Stuchebrukhov, and R. A. Marcus

NOTES

1. p. 272, 2nd column bottom: "Within the Hückel approximation, the matrix $(E - \hat{H}_B)^{-1}$ can be diagonalized and the matrix element reduced to

$$H_{DA} = \sum_{\alpha} \langle A|\hat{V}|\alpha\rangle \frac{1}{E - E_\alpha} \langle\alpha|\hat{V}|D\rangle \qquad (2)$$

The states $|\alpha\rangle$ are the *one-electron eigenstates of the bridging protein with the corresponding eigenvalues* E_α."

Q: Are these eigenstates the molecular orbitals of the bridging proteins? If yes, are they like the molecular orbitals of, say, a bridging aromatic molecule, I mean are there bonding, nonbonding, and antibonding protein molecular orbitals?

M: Yes.

2. p. 275 2nd column middle: "The convergence results are similar for all tunnelling energies *within the gap between the occupied and unoccupied orbitals*"

p. 281 2nd column middle: "No adjustable parameters for individual molecules, apart from *the uncertainty in the tunnelling energy...*"

Q: (1) This gap is apparently the one between the HOMO and the LUMO. Why does ET happens *within* this energy gap?

(2) Apparently ET can happen only at one energy within the gap but one considers the whole energy gap interval because of the uncertainty of where the tunneling energy is?

M: (1) It's only in that region that you will have tunneling, if your reactants or products are at an energy level in either bands there wouldn't be any tunneling there, there would be beautiful conduction.

Q: So that's why the ET happens within this energy gap.

M: Oh, if there is tunneling, yes.

M: (2) The orbital of, say, the donor might be some place between the two orbitals, typically between the highest orbital of the valence band and the lowest orbital of the conduction band. That E may be in the band gap there. Only when the bridge is aromatic might it be more or less about the same, there might be very little band gap.

Q: This energy gap is so important because the electron transfer happens within this energy gap.

M: Yes, it happens wherever the two curves intersect, and they are intersecting near the orbitals of the reactants, say, or orbitals of the products, each of them modified a little bit, and that typically could be in the band gap of a bridge unless the bridge is an aromatic bridge, then it might be almost in resonance.

Q: There is a band gap but apparently there is only one energy which is important within the gap...

M: The energy is wherever you get the intersection, that's the energy to put in.

Q: So, the intersection is within the gap.

M: Usually it is within the gap, usually there is an orbital that is lowest, an empty orbital but is lowest the donor or the acceptor, that's usually the case, yes.

Q: Why it is not exactly in resonance?

M: If the bridge orbital would be of the same energy as the donor orbital, the electron could easily delocalize onto the bridge, but typically it doesn't. There will be aromatics with energy levels that are closer to those of the reactants, are more accessible, the bridge might even be completely accessible to reactants, to the donor and acceptor, and so could be almost in resonance, I mean if the energy is exactly the same it would be exactly in resonance, the main point is that they are very accessible.

"if the bridge orbital would be of the same energy as the donor orbital, the electron could easily delocalize onto the bridge" That's an electron in resonance, yes, well, the smaller the bridge, the larger the difference between the energy levels, and, you know, you have a lot of spacing, *when you make the bridge longer the levels come close together*, the main thing is that the levels change their position depending on the length of the bridge, and it may even happen, if can happen, then did happen, that a longer bridge happens to have levels closer in resonance with the donor and acceptor than the shorter bridge, and so with the longer bridge the ET actually went faster, that's the work of Wasielewski.

Q: So, it is not necessary that they are exactly the same energy, if they are the same there is delocalization, if they are not the same there is some kind of tail of the wave function . . .

M: It is the tail of the wave function that is able to overlap the acceptor, yes.

3. p. 276 2nd column bottom: "Though by changing the curvilinear axis of the tubes, the truncation method serves as a valuable probe of the *inhomogeneity of the tunnelling mechanism* . . ."

Q: What do you mean by "inhomogeneity of the tunnelling mechanism"?

M: Well if we are doing all this without an almost continuous medium where all parts of medium would be identical, but there are patches, you may prefer to go down one patch, or down another patch, anyone that has a homogeneous medium.

M299 Gaussian Field Model of Dielectric Solvation Dynamics

X. Song, D. Chandler, and R. A. Marcus

NOTES

1. p. 11954 2nd column top: "Where standard approximation to dielectric continuum theory would predict a single *longitudinal* relaxation time, τ_L, the exact analysis of the column top.

Q: (1) What is it meant by "longitudinal"? Is there something like a "transversal" relaxation time?

(2) What is the Debye relaxation time? It seems that τ_L and τ_D define the extremes of an interval of times.

M: Longitudinal is when the polarization is along the direction of the local electric field. When the polarization is perpendicular to that.... For that you'd had to have an anisotropic medium.

2. p. 11955 1st column bottom: "The dielectric susceptibility, $\tilde{\chi}^{(m)}(\mathbf{r}, \mathbf{r}'; s)$, is the response function as *modified* by the presence of the solute (hence the superscript m). The modified response function generally differs from that of the pure solvent, $\tilde{\chi}(\mathbf{r} - \mathbf{r}'; \mathbf{s})$."

Q: Why in the case of the pure solvent appears the difference of **r** and **r**'?

M: See, that's a nonlocal situation, yes, **r** and **r**' instead of just **r**. $\mathbf{r} - \mathbf{r}'$ is probably due to translation symmetry, if it's pure solvent then you can start one place and you go with this to another place and you're going to get the same result if you started a different place, so that's if it's solvent. If, on the other hand, there is solute the response function depends on how close you are to the solute.

3. p. 11959 1st column middle: "In the absence of the solute, the normal modes of the Gaussian model are the Fourier components of the polarization field."

M: Then the Gaussian model is related to the whole quadratic model... you know, when you discuss polarization and electric field and there are no cubic terms in the polarization, only quadratic or linear, and all of that is of course an approximation, but if you stay within that, that's probably what is meant by the Gaussian model. It is another word for quadratic, although Gaussian will normally refer to a distribution, the distribution probably contains just quadratic forms so... like the free energy, the exponential of the free energy, maybe a Gaussian-like function and all of that is based on the free energy being a quadratic function. My guess is that the Gaussian model is something which expresses the free energy as sort of a functional of a quadratic description of the solvent, and then when you say normal modes you use quadratic in energy or free energy...

Q: Are normal modes always related to quadratic?

M: Well, yes, the Hamiltonian that generates them is quadratic.

M300 Symmetry or Asymmetry of k_{ET} and i_{STM} vs. Potential Curves

R. A. Marcus

NOTES

1. p. 3905 1st column middle: "In ET reactions, energy conservation for ET is assured, *regardless of the intrinsic affinity of the electron or hole for each site*, by fluctuations in the environment."

Q: The above specification appears for the first time in your papers. Is it the energy to extract the electron from the site bringing it to infinite distance from the site in vacuum? I mean some kind of a generalized electron affinity?

M: I think so, yes. It is just one way of phrasing an energy level, the energy of a site relative to some standard.

2. Following: "Outside the "normal region" ($|e\eta| > \lambda$ in those equations) the electron or hole goes into levels distant from the Fermi level, *to avoid the "inverted effect*"

M: Yes, the point is that if you think of a single parabola intersecting many parabolas, one of those parabolas is essentially the Fermi level, and all the rest are not. Now, if the Fermi level is on the wrong side of the first parabola, in other words if going into the Fermi level would require an inverted effect, the electron will go into a level *higher* above the Fermi level, an unoccupied level in the metal, because that would intersect the bottom of the first parabola. The i_{STM} is related to the Fermi level, the current depends upon how close the state is to the Fermi level, if the state is at the Fermi level, essentially at the exchange current where the forward and the reverse currents are equal to each other, then your electron is going in at the Fermi level. Now, each level has its own parabola, and so supposing that the parabola for the Fermi level was on the wrong side of the parabola, then a higher energy state of the metal or of the semiconductor would permit one to go through the bottom of the parabola, then you wouldn't be in the inverted region. Think of the following: have a parabola for the reactants, now for products including the electrode you have a whole group of parabolas, one of which corresponds to the Fermi level, and that'd be the one where one electron would normally go into, say, if there is no overpotential, but if the Fermi level is not intersecting near the bottom, it will go to another level, which is there, and so you never reach the inverted region, there is a level that permits you to intersect near the bottom.

3. Following: "On the other hand, in STM, energy conservation at a substantial bias potential is assured by the electron or hole always going into the distant levels."

M: Yes, anyway if the hole goes into the Fermi level from the regions we just described, depending upon whether you are above or below, it will either make use of an unoccupied level or of occupied level.

Q: A level distant from the Fermi level.

M: That's all it means, yes.

4. p. 3907 1st column top: "… the energy of the LUMO of the bridge … relative to vacuum, the work function of the electrode … relative to vacuum …"

M: There is an energy relative to infinity, if you put an electron in there, then there would be some energy relative to infinity, if you brought an electron from infinity to it, you see. If one talks about extracting an electron from the metal to infinity, then one has to talk about taking that electron from infinity and bringing it to the unoccupied orbital.

5. p. 2907 2nd column top-middle: "As we discussed elsewhere[6] [6: R.A. Marcus, unpublished data] if the individual $H_{\beta t}$ matrix elements differ mainly in the phases associated with modulation factors such as $\exp(i\mathbf{k}_{||} \cdot \mathrm{r}_i)\sin k_z z_i$ at each electrode site i in the state $|\mathbf{K}_t\rangle$, $[\mathbf{k}_t \equiv (\mathbf{k}_{||}, k_z)$, where $\mathbf{k}_{||}$ and k_z are components of $\mathbf{k}_t]$ then the $|H_{\beta t}|^2$ may be relatively insensitive to the difference in the ε_t samples at the two biases."

M: That $H_{\beta t}$ has in it an $e^{ik\mathbf{p}\cdot\mathbf{r}t}$ etc., right? As a factor in it. What it might be is that if one has a sum of a group of terms that have exponentials and there is a factor in each of those terms, then when you take the sum and you square, if you have a whole group of phases, then the average of the phase terms may be zero. They cancel each other.

6. p. 3907 2nd column bottom: "...a possible asymmetry in the reorganization of the system, e.g., due to a difference in vibrational force constants of corresponding vibrational modes of the oxidized and reduced forms, was omitted in Eqs. (1) and (7)

$$k_{\text{rate}} = \frac{2\pi}{\hbar}|H_{\text{DA}}|^2 \frac{1}{(4\pi\lambda k_{\text{B}}T)^{1/2}} \exp[-(\lambda + \Delta G^0)^2/4\lambda k_{\text{B}}T] \quad (1)$$

$$k_{\text{rate}} = \frac{2\pi}{\hbar}\frac{1}{(4\pi\lambda k_{\text{B}}T)^{1/2}} \int [V(\varepsilon)]^2 \exp[-(\lambda - e\eta + \varepsilon)^2/4\lambda k_{\text{B}}T] \times \frac{\exp(\varepsilon/k_{\text{B}}T)}{1+\exp(\varepsilon/k_{\text{B}}T)} d\varepsilon \quad (7)$$

...Any such effect would cause the electrochemical transfer coefficient at $\eta = 0$ to differ from the value of 0.5. At large values of η, both positive and negative, however, the effect of the asymmetry would disappear. *The limiting rate constant does not involve any reorganization*, and so in this region any difference in limiting rate constant at $+|\eta|$ and $-|\eta|$ would only reflect a dependence on the $|V(\varepsilon)|^2$ in Eqs. (8)–(10) on the sign of η."

M: There are several aspects, it's a long statement. In the 1965 paper, I looked in the question of what happens when you have different force constants. Before that I assumed that they are essentially equal, you know, I assumed that the curvatures of the two parabolas were the same.

Q: You did symmetrize.

M: Yes. Now, in the 1965 paper I symmetrized the force constants and showed that to first order around 0.5 for the transfer coefficient, for $\Delta G = 0$, if you use a symmetrized force constant then there were no asymmetric contributions. Only when I got away from that 0.5 transfer coefficient, for $\Delta G = 0$, did the effect of the difference of force constants appear. Now, when you get to very high driving forces, then eventually the barrier due to reorganization simply disappears, you go into a level where there is no inverted effect, crossing the bottom of the parabola where you started with, and so reorganization doesn't come in there. What does come in is the matrix element, the electronic matrix element, the reorganization doesn't come in.

Q: Because you cross the parabola at the bottom.

M: Yes, that's right.

7. Following: "Another potential source of asymmetry is the difference in the local electrostatic field around the site D due to the electrolyte, at $+|\eta|$ compared to $-|\eta|$. Any such effect should again not affect the limiting k_{rate} values at large $|\eta|$."

M: The first sentence is certainly true, if you have a large η there is an asymmetry in the ion atmosphere around the electrode depending on the potential and so if you have a large η in one direction you may have a very different ion atmosphere than you will have in the other direction, so that creates an asymmetry. Now, is this a donor that's attached to the electrode or an ion in solution or what?

Q: I think you are considering here an electrode with a layer and then you have the donor.

M: Is the donor sitting on top of the layer or in the layer? If so we are not affecting migration or anything like that, we are not affecting the probability of D being next to the electrode, to the monolayer. I that's the case, if you change the potential then if you have that D is part of the monolayer, then the assumption is being made tacitly that change in the potential isn't changing that, otherwise everything would be changing. Maybe you are changing the monolayer but the assumption implicit there is that you have the same monolayer.

Q: Yes, the donor is somehow fixed to the monolayer.

M: Yes, that's right, the only ions that you have are outside the monolayer. Now this is an STM setup, so there is no liquid there? In the STM setup, there is just a tip that's touching the monolayer that's attached on the metal. So, if you don't have any ions in the layer, you don't have changing. Some experiments there are certainly with no liquid, now they have done in STM, I believe, some experiments where you do have a liquid.

M301 A Sequential Formula for Electronic Coupling in Long-range Bridge-Assisted ET: Formulation of the Theory and Application to Alkanethiol Monolayers

C.-P. Hsu and R. A. Marcus

NOTES

1. p. 584 1st column middle: "Theoretical studies on molecular wires... show exponential dependence of the conductance with the length of wire when the electron is at an energy *outside of the wire's energy band* and large conductance is obtained...for the energy of an electron *inside the wire's band*"

M: If the electron is in the band gap it won't be metal-like, in other words if the energy level of the starting electron is inside the band gap then you have exponential decay, It's like a square well barrier then, you have an energy, there is no way it can be inside there other than by exponential decay of a wave function. So, it decays exponentially. If you start up with an energy that is in the band gap, the electron has a negative kinetic energy, that means that it has an imaginary momentum $e^{\int pdx}$, then p is imaginary and there is exponential decay. If on the other hand the energy were, say, above, in the conduction band or below, in the valence band, ... then you would get a very easy transfer across it. You can get conduction by both bands, conduction or valence, depending on what the energy levels of the donor and acceptor are. If the donor and acceptor had their energies overlapping at the conduction band you get easy transfer of electrons, if they at both their energies are overlapping at the valence band you get easy transfer also. When, on the other end, as it's usually the case, they have their energy in the band gap, then you have an exponential decay. Instead of energy band you can say energy bands, there can be overlapping bands, I suppose, yes, there are also gaps between bands. The thing is that if you start changing the principal quantum numbers of an electron, you get a different band, different bands often probably have the same principal quantum numbers, you have interactions between the atoms which sort of broadens a single energy into a band of energies, the degenerate energy into a band of energies.

Q: I was asking you if by "above the conduction band or below" you mean the valence band?

M: If it's below the conduction band then it could be in a band gap, unless it's a metal, the metal doesn't have a band gap, if it's above it depends whether it is a semiconductor or a metal, a metal doesn't have a band gap, the valence band and the conduction coalesce in a sense, you know, there is just one band, you wouldn't call it valence or conduction band, I think, probably would be whole conduction band, you wouldn't have a band gap... you have only one band in that case. Although in principle you could have several overlapping bands, they are all conduction bands with different principle quantum numbers, there's no gap.

2. Following: "... $k = \frac{2\pi}{\hbar}|H_{DA}|^2(FC)$ ... where H_{DA} is the *effective* electronic coupling"

Following: "The effective coupling element can be defined as the coupling between the eigenstate $|\psi_D\rangle$ and the *zero order state* $|\phi_A\rangle$, *namely*, $\langle\psi_D|H|\phi_A\rangle$"

M: Of course you want to have as close an eigenstate as you could for either of those, for both of those, what you have is the donor and then if the electron can sort of going into the bridge and in between certainly it will be promoted along, the wave function will not die off rapidly if you don't have the vacuum there. If you have the vacuum it dies off as an exponential involving imaginary momentum and if the imaginary momentum would be infinite there would be nothing, then it would die off immediately. If on the other hand you have some level, you know, the conduction band or what have you in between, then it dies off at a slower rate, but the main point is that that *wave function is extended, by perturbation* if you want, by the material in between and of course in a way you really want to have that extended, you want to have your acceptor also extended. Now, in lowest order theory, or first order theory, I suppose that if you use the product of the two wave functions, that would be a second order wave function, if you use one and the extended version of one of them, you know, first order, then you got first order perturbation theory, so really you want both but in first order it might suffice just have the tail of one wave function going...I have not thought that through for a long time, so I don't know.

If you think of semiclassical theory, or WKB theory, a wave function is like $e^{i\int pdx}$. Now, if p contains some imaginary part, like in the classically forbidden region, then that p becomes an $i|p|$, so now you have e^{-kx} which is dying off exponentially. Whenever you are in a classically forbidden region the wave function dies off exponentially, from the point where you entered the classically forbidden region. If you have one atom whose orbitals overlap another atom, and then the first orbital was part of something connected to your donor, then this by that interaction extends the donor wave function one atom further and the next one extends it one atom further, further, further, so you get an exponential drop off but you get things coupled in that way and the wave function can finally reach a distant atom. If you look for the expression for the coupling matrix element, it's a product, it's a set of factors, you would have $\frac{1}{E-\alpha}$ or something like that in the bottom, and the β on top and you raise to the n'th power if there are n atoms...If you just look at the unperturbed wave function the coupling would be the coupling of the unperturbed wave function, but both wave functions are perturbed and that extends the coupling and so the material in between comes into that whole coupling element. This is the *effective* coupling, you know, I am not sure what exact coupling would be since you are not solving all equations exactly, so it's an approximate coupling, but it is a pretty good approximation. You see, the direct coupling would be that in which you use the orbitals on D and the orbitals on A, and you just couple them directly and that would be very small, but when you put that material in between, then you have sort of appendages on each of them and that provides extra coupling and, you now, you sort of label that whole complex coupling by the word "effective."

3. p. 586 2nd column top: "here Δ denotes the diagonal $m \times m$ matrix

$$\Delta = E\mathbf{1} - \mathbf{e} \qquad (14)\text{"}$$

Q: The **e** matrix is given in Eq. (9) p. 586 and its elements are, I believe, the energies of the m orbitals belonging to a bridge unit. E was defined on p. 584, 2nd column bottom, as the energy of the electron to be transferred, referring to McConnell's formula

$$H_{DA} \propto \left(\frac{\beta}{E - \alpha}\right)^n . \qquad (3)$$

Is the E in Eq. (14) the same as the one in Eq. (3)?

M: Yes, I guess it is the same E except that in McConnell's formula there is really just sort of one orbital bridge. Normally a bridge with quite a few atoms would have quite a few energy levels, E is the energy of the starting orbital. If you have a number of orbitals. . . I think in Chao-Ping Hsu's case she had a repeating atom, so all those E's are the same, if you have *different* atoms strung along in the intermediate you would have different E's.

4. p. 586 2nd column bottom "Since there is negligible direct coupling between the donor and acceptor states in the long-range electron transfer, the *effective* coupling is calculated from the second term in Eq. (2)

$$T = V + VGV \qquad (2)$$

Q: Does this mean that you neglect the V term?

M: You neglect in that case the first V term that V term is smaller than the VGV term. Because it is smaller, then when you put in the wave functions, left and right, the overlap of those two is so small whereas in the other case the overlap is made larger because of the G.

5. p. 586 2nd column bottom: "Since there is negligible direct coupling between the donor and acceptor states in the long-range electron transfer, the effective coupling is calculated from the *second term in Eq (2)*. Also because the tight-binding model is used, only one block of the Green's function $G^{(n)}$ is needed in that expression, namely, the block relating transition *from the first bridge unit to the nth one*."

p. 567 2nd column top-middle: "It should be noted that that the *bridge Green's function* was used here, while in obtaining Eq. (4)

$$H_{DA} = \frac{\beta_A \beta_D (-\beta)^{n-1} 2^{n+1} \zeta}{(E - \alpha + \zeta)^{n+1} - (E - \alpha - \zeta)^{n+1}} \tag{4}$$

the full tight-binding Hamiltonian, including donor and acceptor states, is used instead."

M: If you diagonalize the whole system, without breaking it up into donor, bridge, and acceptor, that's one case. If on the other hand you take the donor, and the bridge, and the acceptor, then treat the system in some way, you get this expression. That's different from diagonalizing the whole system. That first term V in Eq. (2) only comes in if you are looking at the direct overlap without any intervening material, just the overlap of the donor and acceptor, of course that is negligible.

Q: So you completely neglect direct coupling between the donor and acceptor states.

M: Yes, completely.

6. p. 589 1st column bottom: "Curve III corresponds to the superexchange state denoted by $D^+ + B^-$ (unsolvated) + M, if it is an electron transfer (For a hole transfer, a curve representing D + B^+ (unsolvated) + M(e) should be used instead.)"

Q: In the last case of hole transfer should it not be $D^- + B^+$ (unsolvated) + M?

M: The main point is that in both cases there it's totally neutral because you have D^+ B^-, and you have B^++ M(e), OK. So, in one case on the B you have a hole, it's making use of a hole of the bridge, the other case is making use of an electron of the bridge. For the hole transfer you would look at a hole in the bridge, and for an electron transfer you'd look at the conduction band of the bridge, so the − there and the + there, that's OK. Now, if you start off with an electron at D and now put the electron on the bridge, then that gives you B^-. If you start off with a neutral at D, and you put the electron on the bridge, that gives you the D^+, if you start off with the D and you want to have a hole transferred, certainly you have a B^+ in there.

Q: Is the − on D or on M?

M: Well, there may be no stable D^-, in fact D may be more stable than D^-, so in that case the electron would have to be on M. In other words, you may have a D that can possibly form a D^+ but not D^-, D^- may be an unstable species, then the only place the electron can be is on the metal.

7. Following: "The bridge B can become a B^+ (or a B^-) in the virtual state which occurs in the superexchange mechanism, but because of the *off-resonance condition* this *supertransient* B^+ or B^- can be regarded as unsolvated."

Q: (1) How long does the "supertransient" state live?

(2) Is there a relation (uncertainty principle?) between the state being supertransient and being off-resonant?

M: (1) It doesn't live long enough to get solvated. If you say the solvation takes a certain time t, and if it is off resonance by the amount ΔE, and you divide ΔE by h and then you compare that t with the time required to be solvated...

Q: Is it something related to an uncertainty principle?

M: Well, it is something related to... well, not exactly that, but it takes a certain time to be solvated, whatever that time is, and... yes... in a way the time that is spending in the bridge is that $\hbar/\Delta E$, and if it is spending very little time in the bridge compared with the solvation time... it is like the uncertainty principle, using the uncertainty principle would give a measure of how long that transient state would last. And if that transient state lasts too short to become solvated then it's unsolvated. Normally when one's talking about a superexchange one's not giving any lifetime, in principle one could say something but one is not giving any lifetime, if you want to speak of a lifetime you would take $\frac{\hbar}{\text{off} - \text{resonance energy}}$. Well, it's sort of related but that gets away from the sort of quantum mechanics, I don't know if it helps to bring in the uncertainty principle.

(2) Yes, when it is off-resonance it means that the ΔE is very large.

Q: And Δt is very small.

M: Yes, and if you did a perturbation you'd have a very very small probability of finding it there.

8. p. 592 2nd column middle: "We also calculated the coupling strength through hole transfer mechanism by doing the direct summation only over the *filled* states of the bridge part. Our result shows that the hole transfer scheme provides the major pathways of the coupling, and it yields more than 89% of the total coupling strength."

M: Probably because the valence band edge is closer to the energy of the donor, acceptor, than the conduction band edge.

9. p. 593 1st column top: "Since the energy of the electron being transferred lies between the conduction band and the valence band of the polyethylene"

Q: In the paper you consider single bridge bands

M: That means you can neglect the other band then, that's what it means. Either that or I'm using a whole class of wave functions to form the conduction band, if I use a different class of wave functions I form another conduction band, so I'm not sure what I mean there. Single band can refer to either neglecting the other band, because there is always to be another band, or just considering one pair of class of wave functions that gives you one conduction band, one valence band, and then you want to consider another you have another and then another, another... and you are just considering one of those. So, it means one or the other. A single band case could be that you neglect the other band, that's all. In other words, by a single band I could mean... but I'm not sure, I'd have to look at it, I could mean that there is one pair of conduction and valence bands, it could mean that you're considering one class of orbitals. Or it could mean that you are considering that but you are forgetting about one half of those, it depends. In any case, the basic idea is that you always are going to have a conduction in a bridge band, that if you have a kind of an insulator there you might neglect one of them, but if you bring in another orbital you have another band. If you have a conduction band and a valence band it's certainly very possible that you have, say, some dye or something else whose energy level that you're interested in, is in some place in the band gap. It depends though, if it is an excited state of a dye then it would be in the conduction band, so it varies, it's all a matter of individual energy levels but certainly it is not uncommon to have the ground state of the dye with an energy between a conduction band and a valence band, in the band gap, but each case has to be examined individually. Often in the solid state... they consider a number of classes of orbitals, they may have in a common model eight bands, so if you just have a single pair of conduction and valence bands it means that if you are using a very restricted class of orbitals, maybe p orbitals, but a common model in semiconductors which is called, I think, the Kane model, has eight bands, it considers s, p, d, s*orbitals...

Q: When you are considering these bridges that are used for ET, how would their structure differ from that of a normal semiconductor like silicon?

M: Well, one way they differ is that the bridge is largely one-dimensional, and then it is only a finite length, so you don't have a $\boldsymbol{k}$ wave vector that is really good, that's about it, I guess. But, you see, a tight binding can be used for both, so... you can use the same method for both.

M303 Time-dependent Stokes Shift and Its Calculation from Solvent Dielectric Dispersion Data

C.-P. Hsu, X. Song, and R. A. Marcus

NOTES

1. p. 2547 1st column top: "The reaction field $\mathbf{R}(t)$ can be obtained from linear response theory:

$$\mathbf{R}(t) = \int_{-\infty}^{t} dt\mathbf{r}(t - t')\mu(t') \tag{4}$$

where $\mathbf{r}(t - t')$ is a response function ....

The Fourier transform is introduced

$$\bar{f}(z) = \int_{-\infty}^{+\infty} dt f(t)e^{-izt} \tag{5}$$

....

Using the convolution theorem, Eq. (4) yields

$$\bar{R}(z) = \bar{\mathbf{r}}(z)\bar{\mu}(z) \qquad (7)\text{''}$$

Q: Why do you use the variable z?

M: Well, it's really ω and the reason why I use z sometimes is to make some integrals converge. For certain integrals, you want to go a little off the real axis, just to make them converge.

2. p. 2547 1st column bottom: "In linear response theory,

$$\mathbf{P}_i(\mathbf{r}, t) = \int_{-\infty}^{t} dt' \alpha_i(t - t')\mathbf{E}_i(\mathbf{r}, t') \qquad (9)$$

where it is assumed that the polarizability of the region α_i *is local in space*. For a two phase system, $i = 1, 2, \ldots$"

Q: Apparently α_i depends on a phase index i but not on $\mathbf{r}$. Is then α_i local "in space" or in phase?

M: If you look at some dielectric property, it may depend on ω, then it may also depend upon the variable $\boldsymbol{k}$, meaning then that the material is inhomogeneous, the dielectric constant depends on $\mathbf{r}$, so really in Fourier space it depends on $\boldsymbol{k}$, you may have some $\mathbf{r}$ dependence in each of the phases. Local implies that you don't have something constant within a phase, but it's a function of position. Then of course one phase is different from another phase.

3. p. 2547 2nd column top: "Using the standard reaction field expression, one thus obtains for the case of a point dipole in a sphere,

$$\bar{r}(z) = \frac{2}{a^3}\frac{\varepsilon(z) - 1}{2\varepsilon(z) + 1} \qquad (14)$$

where a, the radius of the cavity, represents the size of the solute. Since *a normalization appears in calculating* $S(t)$ *in Eq.* (2),

$$S(t) = \frac{\Delta E^{solv}(t) - \Delta E^{solv}(\infty)}{\Delta E^{solv}(0) - \Delta E^{solv}(\infty)} \qquad (2)$$

any constant factor cancels, and so the final response function is independent of a.

M: Well, consider that $\bar{r}(z)$... you have something like that in the numerator and something like that in the denominator of Eq. (2), so the $\frac{2}{a^3}$ would cancel... both in the numerator and denominator...

Q: What about the normalization?

M: You have a $\frac{2}{a^3}$ in the nominator and a $\frac{2}{a^3}$ in the denominator, that would cancel.

Q: And is that what is meant by normalization?

M: Maybe normalization wasn't the right word to use, maybe I should have said ratio.

4. p. 2549 2nd column bottom: "... the rapidly dephased intramolecular modes of water,"

Q: I imagine that they dephase because of collisions with other water molecules. Is it so or are there other reasons?

M: Well, I'm trying to think what the motions are... now, the intramolecular modes of water... what's the property that we are looking at? In the experiment I think they just had a 100 fs resolution and so they really couldn't follow in detail the vibrational relaxation, as they didn't have that time resolution. Nevertheless, it does occur, and so it occurred in our calculations too, but we couldn't see then the experiment other than maybe where the dephasing started, you know, I'm not even sure about that, but I do remember that in the experiment they couldn't really tell what was happening in the first 100 fs. We could calculate what was happening but they couldn't tell, so, you know, that part we couldn't really compare with experiment, I guess, I just don't remember if their experimental correlation function dropped from unity to 0.5 or so, in 100 fs, I don't know how reliable that was, I don't remember. I'm talking about the drop in a time where you can't... they couldn't have time, their technique was not fast enough. There is a vibrational relaxation, the bending vibration for example, that's what makes the difference between a dielectric constant of 4.5 and a dielectric constant of 2.0 in the infrared region, and I don't think they could follow that. We could calculate that but we couldn't really compare with their experiment because they couldn't really measure that.

CHAPTER 7

ET Model for the Electric Field Effect on Quantum Yield of Charge Separation in Bacterial Photosynthetic Reaction Centers; Theory of Rates of S_N2 Reactions and Relation to Those of Outer Sphere Bond Rupture Electron Transfers; Source of Image Contrast in STM Images of Functionalized Alkanes on Graphite: A Systematic Functional Group Approach

Interviews on M304, M305, M308

M304 ET Model for the Electric Field Effect on Quantum Yield of Charge Separation in Bacterial Photosynthetic Reaction Centers

S. Tanaka and R. A. Marcus

NOTES

1. p. 5032 2nd column middle: "Also, it is shown that a *steady or quasi-equilibrium* state realized in the primary ET processes, $^1P^*BH \rightleftharpoons P^+B^-H \rightleftharpoons P^+BH^-$, ..."

M: In chemical kinetics one speaks of a steady state for a species whose concentration is low, it is an intermediate. So, for example, suppose you have a series of reactions, and you have some intermediate in very low concentration, simply because as soon as it is formed it quickly reacts again, OK?, then you write down the steady state for the intermediate, $dx/dt = 0$ and you write down the equation of formation... and so on. That's the usual procedure for steady state in chemical kinetics. For example, in a lot of these free radicals chain reactions for the concentrations of the free radicals, not for the concentrations of the starting molecules, but for the concentration of the free radicals, you introduce the steady state assumption, namely that for each one the rate of change of concentration is zero. And really... if you look at the concentrations of the intermediates, you see that they never build up in comparison with the other concentrations, these concentrations always remain small, so therefore the rate of change of concentrations has to be small, approximately zero.

If you start the reaction, then you have a certain transient period before the steady state, and then after that you get the steady state and then eventually everything disappears to form the products.

Only species are intermediates whose concentration is always small. It's small because it is used up so quickly in a reaction step and so that's what permits to have that steady state.

2. p. 5033, **Table 1: Rate Constants k_{ij}** $\cdots$

Q: Why do you present in the table the rate constants through by inverses $1/k_{ij}$?

M: One reason might be that sort gives you a half-life, so you think in terms of reaction times, picoseconds, because normally you think in terms of picoseconds, or whatever, you see...

3. p. 5034 1st column middle: "We proceed next to the full treatment of the rate Eqs. (2.1)–(2.3). To this end, it is necessary to evaluate the electric field dependence of several ET rate constants explicitly. In addition, the effect of including a *static heterogeneity* associated with the distribution of free energy gaps..."

p. 5035 1st column middle: "…there may exist experimentally a distribution of free energies of the ion pair states P^+B^-H and P^+BH^- due to any *static heterogeneity* in RCs."

Q: What is it this "static heterogeneity"?

M: I'm not sure, but if you have different molecules…yes, it could be that your sample isn't completely uniform, you know, if you have slightly different species or something like that, in fact there is free energy associated with one set of molecules of one species, free energy associated with a set of molecules of a slightly different species…

4. p. 5035 2nd column middle: "In Figure 3a the calculated electric field dependence of the ET rate constants k_{12}, k_{21}, k_{23} and k_{32}, is depicted for $-1.5\,\text{MV/cm} \leq F_{ext} \leq 1.5\,\text{MV/cm}$, when the mean values of the free energy gaps, $\overline{\Delta G_{12}}$ and $\overline{\Delta G_{23}}$ are introduced, for comparison, into Eq. (2.19). The relations $k_{12} > k_{21}, k_{23} > k_{32}$, and $k_{12}, k_{23} > (5ps)^{-1}$ are obeyed in this parameter region. *The first two inequalities imply that $\overline{\Delta G_{12}}$ and $\overline{\Delta G_{23}} < 0$ for all values of F_{ext}*."

M: If you take the ratio of rate constants you have an equilibrium constant, that just gives you the sign of ΔG for the equilibrium constant, so if $\frac{k_{12}}{k_{21}} > 1$, then it is related to an equilibrium constant and to a free energy change, so that's how, from taking the ratio, I'm writing the ratio in terms of ΔG. Just the definition of equilibrium constant in terms of those rate constants.

5. Following: "The last inequality, which implies a relatively weak dependence of the forward ET rate constants on the applied electric field…"

M: If some very fast reaction, it is hard to make it faster.

6. Following: "…explains why it was difficult to describe in earlier work the extent of the experimental charge separation QYE (field-induced quantum yield effect) *by considering only the competition between the internal conversion plus fluorescence rate constant* $k_{10}(= (190ps)^{-1})$ *of* $^1P^*BH$ *and the forward ET constants* k_{12} *and* k_{23} (cf. Figure 2)."

p. 5039 "**4. Conclusions**: In the present paper a theoretical model is proposed for the electric field effect on the charge separation quantum yield in bacterial photosynthetic RCs. In previous models of the effect, the explanation for experimental results was based on a competition between the internal conversion plus fluorescence rate constant and the *forward* ET rate constant of $^1P^*BH$, i.e., on Eq. (1.1), $Y_Q = \frac{k_{et}}{k_{10}+k_{et}}$, and led to too small a calculated effect (dotted line in Figure 2). In the present paper, the effect of the electric field on *both the forward and backward rate constants* is incorporated, leading to a larger effect."

M: The effect of the electric field on the forward reaction is probably small because it is *fast*, but the back reaction may be slow and maybe the electric field has a bigger effect on it or vice versa, so there is some connection with that... if you have a fast rate constant you can't change it very much with an electric field, but if, as part of the whole process, you have a small rate constant some place, then you can change it a lot.

7. p. 5040 2nd column middle: "As seen in Figure 4b, after a steady or quasi-equilibrium state among $^1P^*BH$, P^+B^-H, and P^+BH^- is realized at $t \sim 20$ ps, these states decay into the PHB and P^+HBQ^- states with the rare constants k_{10}, k_{20}, k_{30} and k_{34}."

Q: Please look at Figure 4b: the quasi-equilibrium state is, I believe, the one corresponding to the maximum of P_3, to the contemporary decreasing of P_1 and P_2 and to the contemporary picking up of P_4, P_5, and P_6. Is this way of finding out the point of quasi-equilibrium just by inspection of the populations correct?

M: Going from P_1 to P_3 I guess you go by P_2, do you? So, the species decreasing to small concentration is P_2, so maybe approximately, although it is not exactly correct, dP_2/dt is always small compared with the other derivatives, it may not be usually small if you look at it and, it is true, there is only one point for dP_2/dt actually $= 0$, but this is not the result of making a steady state approximation, it is the result of solving all the equations. In a steady state approximation, you would neglect the P_2 concentration relative to the others. The quasi-equilibrium state would be not where P_2 is growing up.

Q: But where P_3 is a maximum...

M: Certainly 20 picoseconds is where P_3 is a maximum, but I hope I didn't say that the steady state is where P_3 is a maximum because that's confusing the whole steady state business. The idea is that a steady state occurs when you have a species in very small concentration, but there P_3 has a big concentration. It is true that if you think about P_3, then briefly you have $dP_3/dt = 0$, but that's just briefly, and so P_3 is not normally called a steady state.

Q: So, it is the population that gives an idea of where the steady state is P_2.

M: Yes, the small concentrations. And in fact throughout, and I mean here in drawing that picture none of the species is said equal to zero, no rate is set equal to zero, you don't really have there a steady state as such, but if you look at the picture then you might be able to roughly get the concentrations of P_1, P_2, and P_5, as a function of time, using the steady state for the other species, but to say there that if at one point $P_3 = 0$ then P_3 is a steady state, no, that's not right. There can be some intermediate region of maximum concentration, that's all. The essence of the steady state approximation is simply missing from there.

8. p. 5043 1st column middle: "Thus, if we incorporate the field-induced enhancement of k_{nr} into the calculations, the agreement between calculation and experiment for $Y_f(F_{ext})/Y_f(0)$ is improved slightly. In addition, some dependence of k_r, k_{20} and k_{30} on F_{ext} and other '*hidden*' *pathway rates* may also be responsible for the disagreement."

Q: Can you expand on that? Apparently by claiming the influence of something "hidden" you can explain everything, much better than the adjustable parameters!

M: Yes... hidden means maybe some other pathway that we just not even think about...

Q: It could be the Holy Ghost...

M: That's right.

9. p. 5044 top: "On the other hand, the coupling constant *V* used in the present model calculations ($V_{12} = 32$ cm^{-1} and $V_{23} = 59$ cm^{-1}) in the main text should be regarded as a '*renormalized*' *quantity* which effectively takes account of the higher-order contributions through the *enforced fitting* to experimental ET rate within the nonadiabatic expression."

Q: (1) What does "renormalized" mean in this case?

(2) Isn't a renormalized quantity obtained by enforced fitting a close relative to an adjustable parameter?

M: (1) One possibility is this: you have a direct sort of matrix element between A and B, say. OK? But now, if you got some intermediate there, and you have a superexchange mechanism, then your effective matrix element contains several of these matrix elements, V_{12}, V_{23} or... etc., there are several matrix elements involved then, divided by an energy denominator, so maybe that's what I was talking about renormalized. In other words, you have something that's complicated and you write it in some place simple, but it really represents something more complicated, for example, if you have coupling by some superexchange, then the matrix element instead of being H_{12}, maybe you have $H_{12}H_{23}$ over some energy difference, that's renormalized, maybe that's what I meant, an effective coupling, corrected by way of a superexchange. If you have an effective coupling due to a superexchange, not a direct coupling, instead of having an H_{12}, you have something like $\frac{H_{12}H_{23}}{E_1-E_2}$, so you have an effective coupling.

(2) V_{12}, V_{23}... how are they related to V? Or are they just examples of V?

Q: These V_{12} and V_{23} are obtained by fitting. Aren't they then adjustable parameters?

M: Yes, they are adjusted to fit. When you determine something from data, they are effectively adjustable parameters. I don't know why we call them "renormalized." Unless it's really much more complicated and there are maybe a whole set of similar species that are slightly different. So, you know, instead of taking into account the individual different species, they have some effective average species... I don't know if that was meant or not.

M305 Theory of Rates of S_N2 Reactions and Relation to Those of Outer Sphere Bond Rupture Electron Transfers

R. A. Marcus

1. p. 4072 2nd column middle: "Stereochemistry has played a significant role in studies of the reaction mechanism, inasmuch as 100% inversion implies an S_N2 reaction mechanism only, while *partial inversion* can imply the operation of both mechanisms, as in stereochemical studies of the reaction of anthracene radical anion with optically active 2-octylhalides."

p. 4073 2nd column top: "The S_N2 reactions have a larger steric effect than the outer sphere electron transfers."

M: I imagine that if one looked at that reaction with anthracene radical anion, probably experimentally there wasn't a 100% inversion, you can tell what percentage inversion is, you know, in other words what structure, what percentage of the products corresponds to inversion and what percentage doesn't and I gather from that in that particular case the inversion wasn't 100%.

Q: It looks like that there is more steric effect in S_N2 than in electron transfer.

M: For S_N2 you need to orient, the reactants have to be oriented with respect to each other. Now, it is rarer in electron transfer that this is the case, people have been looking at ETs between different isomers, optical isomers or stereoisomers, and for outer sphere electron transfers in very special cases you can require some orientation, but the stereoselectivity is nothing like the one you have to have for S_N2 reactions. Anytime you sort of break a bond you really have to orient a lot and, depending on the isomers you use, you get a different structural fitting together of the two reactants... in other words, there is a local structural difference.

2. p. 4073 1st column bottom: "In the present paper we focus on the process leading from R to P and then calculate the bimolecular reaction rate constant by assuming, in effect, a *pre-equilibrium for R*"

M: Supposing that the essential reactant was R, that the actual reactants were $A+B$, then $A+B$ is in pre-equilibrium, one might assume, with R and then R reacts,... does anything precede R?

Q: Yes, please look up p. 4075 Figure 2

M: I see that the reactants go to R and it may be that one may assume a pre-equilibrium, it depends on how fast R disappears. All right, it is common to assume a pre-equilibrium, that's not a big deal because if one did not assume a pre-equilibrium... one would get the same results, it depends on how restricted the pre-equilibrium were. The pre-equilibrium may or may not actually exist, it's just a convenient way of reducing everything to almost like a first order reaction . . . but whether it actually exists long enough so that you actually have a pre-equilibrium, that's a question . . . I think the first who started to do that was Norman Sutin, because I didn't do that in my early papers. It is a convenient way of visualizing things. Whether it exists or not I don't know.

3. Following: "If the *diffusion* from the separated reactants to form R, or from P to yield the separated products, becomes *sufficiently slow,* the calculation below also provides a *unimolecular* rate constant for the R → P process, upon *dividing* the bimolecular rate constant by the equilibrium constant for forming the encounter complex R"

Q: That is, if $A+B \rightarrow R, \upsilon = k[A][B], \frac{[R]}{[A][B]} = K$, then $\upsilon = \frac{k}{K}[R]$?

M: Yes, I think the above equations are right. You see, that $\frac{k}{K}$ becomes an effective first-order rate constant.

4. p. 4073 2nd column top: "(iii) Solvent effects typically *increase* the S_N2 reaction barrier relative to its value in the gas phase."

Following: "Some partial desolvation, with an accompanying *increase* of in energy barrier, is expected to accompany the formation of the TS . . ."

Q: Isn't there a contradiction with the preceding statement? There the barrier in solution is higher than in the gas phase, here desolvation (i.e., going to a situation similar to the gas phase) increases the barrier . . ..

M: One possibility is this: that the solvation increases the barrier, that means that the reactants are more favorably solvated than they are in the transition state. If the solvent increases the barrier, that's one way of doing it, it stabilizes the reactants more than the transition state, so that means that, let's say, there is some partial desolvation to reach the transition state. And to make a statement like the one you mentioned, I must have surely been looking at some data comparing gas phase and solvent reactions. I should have cited.

5. Following: "When the charge distribution in the TS is dominated by *two very different contributions*, a nonequilibrium polarization of the solvent may occur..."

M: Yes, like in a weak overlap electron transfer, the transition state isn't the state of the reactants, it isn't the state of the products, it's a combined object, and the combined species corresponds to a strange kind of species where you have two very different distributions that are making up the transition state, so the solvent can't possibly accommodate easily, so that it adopts an intermediary configuration, in order to accommodate that kind of *split transition state*.

Q: I believe you refer to your peculiar TS made up of the states X* and X of your 1956 paper. The two states have the same energy and are thermodynamic states. May they be considered degenerate macroscopic quantum mechanical states?

M: Yes, they are degenerate macroscopic quantum mechanical *diabatic* states. They are the ones when you don't put in the coupling, so they are diabatic states. When you put in the coupling then you get the adiabatic states. They are *hypothetical* because there is no coupling. It is OK if you neglect the coupling. Then you use the Landau–Zener theory, you put in the coupling and then you ask what the effect is of the coupling.

6. p. 4073 legend to Figure 1: "Schematic contour plot...two electronic states...In one of these *(the shaded region) the extra charge is localized on* A_1^-, and the other on A_2^-."

Q: Please look at part (a) of Figure 1 ("Lower Surface (a)"):

(1) A_1^- is in the nonshaded region contrary to what is written in the caption (and in the text);

(2) You see an arrow crossing the ordinates axis indicating the shaded area where the charge is localized on A_2^-. Now look on the same figure where, down to the right, the system $A_1B + A_2^-$ is localized in a small nonshaded area. Why so? Here A_2^- is once in the shaded and once in the nonshaded region.

M: Everywhere below the dashed line or dotted line the charge should be localized on the A_2^-. If the small part with A_2^- doesn't have the same kind of shading that the main part has, then that is wrong.

7. p. 4074 1st column top: "... a two-interacting state model ... is less valid for some S_N2 reactions, those whose resonance energy lowering is so large that other zeroth-order states are almost certainly mixed in."

Q: Is the situation similar to that of Pauling's resonance where the resonance energy increases with the number of resonant forms?

M: Yes, I suppose there is a similarity there.

8. p. 4074 1st column middle: "The dotted borderline region ... In the dotted line region there is a large splitting (avoided crossing) of the two potential energy surfaces when both A_1B and A_2B distances are small. The splitting becomes *small*, presumably exponentially so, when either of those distances becomes large."

M: *By the splitting I mean the difference between the adiabatic and the nonadiabatic curves.* If you have a two states approximation, then one has to distinguish between the splitting part and the energy difference, by the splitting we mean the difference between the diabatic and the adiabatic corrections, the H_{12}. When those reactants are very far apart it may be that there is so little interaction of the two electronic configurations that the H_{12} is very small. I'm talking about the H_{12}.

9. p. 4075 1st column middle: "For ΔG_r, the free energy of formation of the state (X_1, X_2, Y) from R, excluding the Q and $q_{\text{vib}}^{(1)}$ terms (cf. Figure 2) we write

$$\Delta G_r = D_i(X_i - 1)^2 + D_2 X_2^2 + \lambda_0 Y^2 \tag{5a}$$

where

$$X_i = \exp(-a_i x_i), \quad \lambda_0 = \lambda_0(X_1, X_2) \quad (i = 1, 2) \tag{5b}$$

and x_i refers to the $B - A_i$ bond displacement coordinate. In Eq. (5) there is a solvent fluctuation term $\lambda_0 Y^2$, in which a generalized fluctuation coordinate Y is introduced whose equilibrium value is 0 for the reactants' state and 1 for the products. The λ_o in Eq. (5) will later prove to have its usual significance as a reorganization energy. *The* λ_o *in* $\lambda_0 Y^2$ *may be a function of* (X_1, X_2), *since the geometry of the solute depends on* (X_1, X_2) *and is assumed to be slowly varying.*"

Q: Until now I had been accustomed to completely separate the λ_i and λ_o contributes to λ but here I see that, more in general, λ_o depends somewhat on λ_i. At this point the question comes natural: can λ_i be influenced by λ_o?

M: That's right, up until to a point, yes, but up until that point the vibrations really didn't have much effect on the solvation, but the vibrations could be very large, and so there can be major effects on the solvation. So, now, the whole size of things is changing...

Q: So, λ_o can be influenced by λ_i.

M: Yes, that's right.

Q: And λ_i from λ_o?

M: You know, if you have coupling, then everything is affected, but it is a question of how you phrase it, how you put the effects in. For example, one has solvation and, if the solvation is affected by the stretching of the molecule, in this case you might say the stretching of the molecule affects the solvation. So, everything is sort of connected, you need just be aware of it. Well, you know, the contribution for the vibrations affects the contribution for the solvent outside of the vibrations, these two terms aren't entirely independent, so if you want to do everything correctly you have to include the interaction terms, you can't do it exactly with single terms, but approximately that's what we did. You make whatever approximations you can, but whenever you have vibrational contribution and contribution outside, of course there is some interaction to them, so strictly speaking they're not independent, so if you want to do everything right you put in some interaction term, of course that complicates everything. In a full molecular dynamics treatment, you don't distinguish between inner and outer, you do everything.

10. p. 4075 1st column middle: "In Eq. (5)

$$\Delta G_r = D_i(X_i - 1)^2 + D_2 X_2^2 + \lambda_0 Y^2 \qquad (5a)$$

where

$$\lambda_0 = \lambda_0(X_1, X_2) \quad (i = 1, 2) \qquad (5b)$$

there is a solvent fluctuation term $\lambda_0 Y^2$, in which a generalized fluctuation coordinate Y is introduced whose equilibrium value is 0 for the reactants' state and 1 for the products."

Q: Please look at Figure 1*: (1) is it OK? By the way such a figure with equal parabolas refers, I believe, to electron exchange reactions, that is, like the classical

$$^*\mathrm{Fe(OH)_6^{2+}} + \mathrm{Fe(OH)_6^{3+}} \rightarrow {}^* \mathrm{Fe(OH)_6^{3+}} + \mathrm{Fe(OH)_6^{2+}}$$

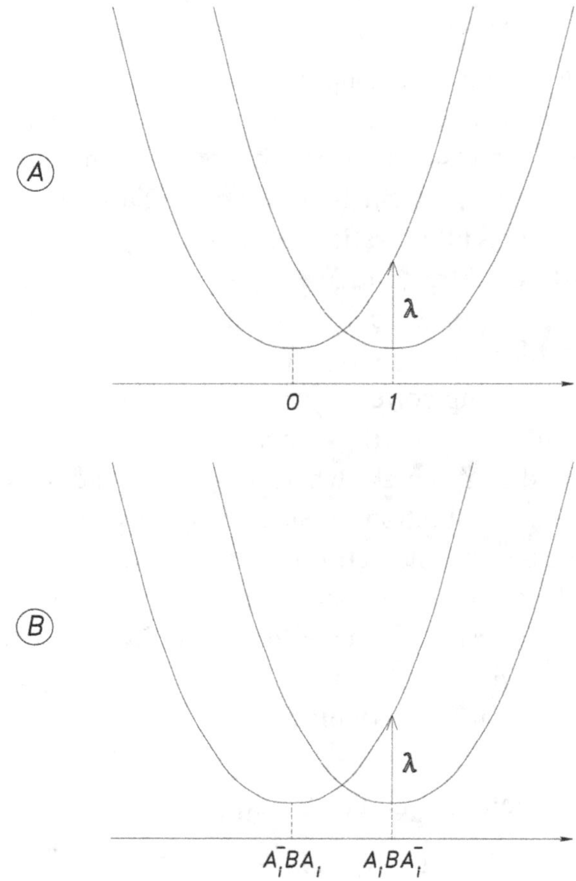

Fig. 1*.

(2) Are there other nonexchange reactions for which one has such symmetrical parabolas?

M: (1) Yes, that's fine. The answer to your question is yes.

(2) There is certainly to be some deviation, *any kind of dielectric saturation will produce some deviation* . . . but how great will be the deviation one can only decide on the basis of some computation. In the 1965 paper, I did introduce symmetric and antisymmetric combination of

force constants, so even if the properties are not symmetric I said: All right, let's look at the symmetric combination and *unless*ΔG^0 *or* λ *is not too large* then we can forget about the asymmetric combination, but actually there would be some asymmetry. One would have to start to calculate for each case, with some molecular dynamics, for some solvent. But even when the force constants of the two reactants are not equal I showed that *when* ΔG^0 *is small* you can forget about the nonequality, but only by using certain reduced combinations of force constants. This approximation I called symmetrization, as described in Appendix IV of the 1965 paper.

The linear response is the linear response to a quadratic free energy curve, it may be not exactly quadratic, you know, there is a little bit of nonlinearity, but in the existing calculations that I know of, one due to David Chandler and Tom Miller, the quadratic curve was pretty good.

If you have dielectric saturation that means the polarization is no longer proportional to the electric field, that means that the free energy is no longer quadratic in, say, the polarization. Everything is based on the validity of the quadratic approximation and if you have dielectric saturation that's no longer the case. So, for example, if you have dielectric saturation in one approximation you have to take the dielectric saturated part and put that separately, so it doesn't change and then work with the remaining part, but you can't have systems that are partly dielectric saturated and still be in the regime where the polarization is proportional to the local electric field. But that linearity is essential for a quadratic free energy approximation. *Linearity means response proportional to the activating force, polarization proportional to the electric field.* Anytime you have a response that's proportional to some imposing field, that's called a linear relation, but the response may depend on the quadratic field, you know, the power of the field, then it's no longer linear, then the free energy has now a more complicated expression. You can't use parabolas, you'd have to use modified parabolas. You only get parabolic, quadratic free energy if your response is proportional to your imposing force. And λ is defined by means of the parabolas.

Q: So, in this way you did generalize...

M: Exactly, and that was one of the strengths of the 1965 paper.

11. Following, bottom: "There will also be some *solvent caging effect,* but from an energetic point of view its effect on Eq. (5) is expected to be relatively minor."

M: Yes...in other words, that ΔG_r deals with fluctuations in X and Y and the cage is probably roughly the same for the reactants and the products, so the caging is not affecting too much that ΔG_r. In other words, always fluctuations are occurring in the cage system and the size of the cage isn't changing much, presumably.

12: Following: "Its effect should be mainly on the *diffusion* aspects of the problem, when diffusion to R from ∞ or from P to ∞ becomes slow."

Q: Please explain the caging effect in relation to diffusion...

M: The caging effect is related to diffusion in that if everything is caged, sort of, it's hard to get out, hard to move, the diffusion constant is slow, so that would come in to making the rate partially controlled by the diffusion rate. I'm not sure if it is the matter of the caging being large, as much as if it is A being out of the cage, A being out of the cage depends on how tightly the other molecules are sort of, you know, bound to each other and so it is not so much the size of the cage itself that is important.

Q: When did first the caging effect appear?

M: It first came up in photolysis, I think, of some alcohol, halide in solution, and the quantum yield was very small, I think that is work of E. Rabinovich, not the Seymour but E. Rabinovich and I think he found that the quantum yield of that photolysis was small when it was in solution, and he described that caging, that's where that effect first came in. Since then, there have been probably statistical mechanical discussions about caging but I haven't followed that area, a number of people must have written about it from time to time, but I don't know of any sort of long paper on caging. One might perhaps, but this is going too far back, look at Dick Noyes' paper in 1960 in Progress in Reaction Kinetics, more than one, he has something on diffusion, maybe he has something on caging in there, but the best thing would be to sort of intersecting diffusion and caging...to see if anybody has written any detailed article, I just don't know if they have or not.

13. Following: "The TS should be located by minimizing $\Delta G_r - k_B T \ln Q^\dagger$,...A variational parameter would be introduced into $Q^\dagger$ for this purpose...For the moment we treat the variation of $Q^\dagger$ along the reaction path as '*slow*'...The G_r and G_p surfaces in the (X_1, X_2, Y) space intersect, and we find the lowest point in the intersection by minimization of ΔG_r, in Eq. (5a), subject to the constraint imposed by Eq. (7), and treating $\lambda_0(X_1, X_2)$ as a *slowly* varying function of (X_1, X_2)."

Q: Why "slow" and "slowly"?

M: Maybe "small" may be a better word. A slowly varying function means a small varying function. So, I really mean small. In other words, the variation of that function is small. A function is slowly varying because it's not changing rapidly. You know, if you plot something versus x, if there is big changes than you call it a rapidly changing function, even though time is not involved, slow not in time but slow in distance.

14. p. 4075 2nd column bottom: "The λ_0 in Eq. (12) is the change in solvation energy accompanying a vertical transition $A_i^- BA_i \rightarrow A_i BA_i^-$"

Q15: (1) Please look at Fig. 1* panel (B). Here I have identified the above 0 and 1 positions with $A_i BA_i^-$ and $A_i^- BA_i$. I have also indicated the λ_o. Is the diagram OK?

(2) Which is the relation between solvation energy and the reorganization energy λ_o?

M: (1): Yes, that's fine.

(2): The solvation energy is associated with, say, one of those curves, the reorganization energy is associated with two. The reorganization energy is the difference, as you got here, between two curves, the solvation energy is the difference between one curve and the curve in vacuum. The two are very different. λ_o is associated with something about the properties of two states, not one. I said: "a change in solvation energy for a vertical transition." That means that λ_o is associated with two electronic states, not one. A true solvation energy would be associated with only one electronic state in the liquid compared with in the vacuum. But here we are taking a vertical transition that means it is associated with two electronic states.

15. Following on p. 4076 1st column: "The *distorsion of the free energy* by the β_{ij} interactions may cause the TS not to lie exactly where the free energies for *the two undistorted states* are equal."

Q: The above statement reminds me of adiabatic and diabatic potential energy curves. Can one speak of *adiabatic and diabatic free energy* curves?

M: Yes, I think one could, I haven't seen that in the news but I think one could, sure, absolutely.

16. p. 4076 2nd column middle: "One question is whether, when $\Delta G^{0\prime}/\lambda$ approaches -1 and the TS becomes, *energetically, reactants-like*, the $Q^\dagger$ (apart from the Z_{S_N2}/Z) approaches $Q(0)$.

Similarly, when $\Delta G^{0\prime}/\lambda$ approaches $+1$ and the TS becomes, *energetically, products-like,* does $Q^{\dagger}$ approach $Q(1)$?

Q: Why the Q's don't simply follow the energies?

M: Now...what is $Q^{\dagger}$? Is it a ratio of partition functions?

Q: It is the partition function for the transition state.

M: When the parabolas become shifted, the TS becomes like the state of the reactants, and that would become the Q for the reactants, the Q follows the intersection.

Q: But you write a question mark as if it was not always so.

M: Oh, I see, I leaved as a question...now, it may be that... Is this a case where there is a finite splitting? If there is a finite splitting, then it can never be the same as the reactants, there will always be some tint with the products state. So, there is some splitting, there is no exact crossing.

The TS depends on the reactants site. When the two reactants approach, one coordinate in the partition functions of the reactants is removed, remember that the partition function of the TS has one less coordinate than that of the reactants and of the products, so that $Q(0)$ there may refer to the partition function of the reactants where somehow you exclude one-dimensional partition function along the reaction coordinate there. $Q^{\dagger}$ won't be a ratio of partition functions, it will be the partition function of the TS and it has one dimension less than the full partition function, the $Q(0)$ probably is the same, probably defined in a certain way, one dimension less that the full partition function of reaction, then we remove the partition function along the reaction coordinate.

"does $Q^{\dagger}$ approach $Q(1)$?" The answer is yes. Remember that $Q(1)$ is not the partition function of the reactants, is that instead divided by the partition function along the reaction coordinate there.

It refers to a system with one coordinate less than the reactants. There shouldn't be a question mark there.

"there will always be some tint with the products state." Maybe what this is all about is that if there is a finite splitting then of course there'll never really have $Q^\dagger = Q(1)$. There, by the splitting, the energy is a little bit lower. If you neglect the splitting then there is no question but if you include the splitting then the $Q^\dagger$, even when the intersection becomes near the bottom of the products, never exactly equals that, because there is a splitting.

17. p. 4077 1st column top: "One useful experimental quantity is the λ_o for a single ion $\lambda_o(A_i^-)$ obtained from the threshold energy E_t of the photoelectron emission by the ion in solution"

Q: Can you shortly describe this kind of measurement?

M: Paul Delahay did some experiments a long time ago where he photoexcited some anions in solution and the electron would go off, ejected, and he inferred some reorganization energy from them, and that's a vertical transition, so you can get something related to that.

M305 Theory of Rates of S$_N$2 Reactions and Relation to Those of Outer Sphere Bond Rupture Electron Transfers

(Continuation 1)

R. A. Marcus

1. p. 4077 1st column middle: "Using a linear response approximation, we also note that the λ_o for any geometry and any initial charge distribution can be calculated from *a difference of equilibrium solvation* free energies,

$$\lambda_o = G^{e,sol}_{1-0} - G^{e,op}_{1-0} \tag{22}$$

Where the 1 denotes the final charge distribution after a vertical transition (in the present case after loss of the electron) and 0 denotes the initial charge distribution, so that 1-0 denotes the difference of the two charge distributions. The *e*, *op* superscript denotes the equilibrium solvation but when only the electronic polarizability of the solvent enters rather than the total (electronic plus the nuclear) contributions."

Q: Please check if the following statements are correct looking at Figure 1*:

(1) Let us try to understand the above formula referring it to the ET process $(A_1^- BA_2)_s \rightarrow (A_1 BA_2^-)_s$ (where $A_1 = A_2$ and the s subscript reminds us that the systems are solvated. The subscript is dropped in the following for simplicity) and in my figure with parabolas. There, the $A_1^- BA_2$ system is represented, when in its equilibrium nuclear and charge configuration, by the 0 position on the "generalized fluctuation coordinate Y" (p. 4075 1st column middle) axis. Let us now change the nuclear coordinates of the $A_1^- BA_2$ system (those of the supermolecule $A_1 BA_2$ and those of its associated solvent molecules) along Y, going gradually toward the coordinates characteristic of $A_1 BA_2^-$ sitting in the 1 position.

(2) While the nuclear coordinates are gradually changed, *let the charge distributions adjust to them* so that one goes smoothly from the charge distribution appropriate for $A_1^- BA_2$ to the one appropriate for $A_1 BA_2^-$. Likewise, the solvent coordinates adjust gradually to the inner coordinates of the system and to its charges. Does one in this case move along the *lower adiabatic curve*?

(3) In this case, one passes smoothly through the transition state and the solvent is in equilibrium with the transition state which is characterized by a charge distribution intermediate between that of the initial and of the final state. Is it so?

(4) If one now changes the nuclear coordinates as described above *but one keeps the charge distribution of the system in the* $Y = 0$ *position*, the charge distribution is fixed and isn't any more in equilibrium with

the successive nuclear configurations. Does the energy of the system now move along the *diabatic R curve*?

(5) When the representative point of the $A_1^- BA_2$ moving along R is just above the 1 point on the Y axis (representative of the $A_1 BA_2^-$ in its nuclear and charge distribution in equilibrium), the energy of this point on R is equal to λ_o, panel B in Figure 1*. Is it so?

(6) At the crossing point of the R and P diabatic curves the solvent isn't in equilibrium with both the initial and final charge configurations.

(7) If one considers, as in the case of weak overlap electron transfer, R and P diabatic curves slightly split at the crossing, and if one considers the lower curve how can one define this curve? It is apparently made up of two pieces of R and P diabatic curves except in the small region of the neighborhood of the crossing where it has the character of a lower adiabatic curve. Is it so?

(8) In the above case, the solvent is not in equilibrium with the charge distribution before the crossing and after the crossing. At the crossing and in its immediate neighborhood the solvent is in equilibrium with a charge distribution intermediate between the initial and final charge distributions. Is it so?

(9) What does exactly mean the $1 - 0$ symbol?

M: (1, 2) But the way, if something is slowly varying that means that for certain equations you don't differentiate with respect of it, it's too small an effect.

Now, are all the coordinates changing? I don't know, I have a feeling no, because if one tried to reduce it to the case where you didn't have the bond stretching, it wouldn't be right. In other words *when you don't have the bond stretching, the electron transfer is not occurring over a whole range of distances, it's occurring very sharply at one point.* Here from the way it is drawn it seems to me occurring over the whole range of distances, so I have a gut feeling that the answer is no. In other words, what we have done is really to reduce the bond breaking to a kind of quadratic, and within that framework you sort of translate the whole coordinates then, you know, from exponential to this and that. Within that framework, certain things should be slow retained, namely *the electronic structure changes abruptly at a point, not gradually changing*. If you have a bond stretching and if it stretches, something happens to the coordinates of the material around it, and so strictly speaking one should take that into account, *always a bond stretching affects the coordinates around it.* The question is this: with that formula there you're focused on basically one coordinate, the stretching coordinate, OK? Now, if you're just focused on that, you are making some assumption one way or the other, about what the other coordinates are doing, so you choose whatever assumptions you want to, the better your choice, the better your whole treatment. So, basically when one uses that one-dimensional bond stretching, one is making a tacit assumption that what one is doing is sort of using a mean field for the influence of all the other coordinates around it.

Q: In quantum chemistry you have kind of a supermolecule, the supermolecule has a certain distribution of charge, you change slowly all the coordinates going from one system to the other and the charge adjusts itself, somehow.

(2) While the nuclear coordinates are gradually changed, *let the charge distributions adjust to them* so that one goes smoothly from the charge distribution appropriate for $A_1^-BA_2$ to the one appropriate for $A_1BA_2^-$. Likewise, the solvent coordinates adjust gradually to the inner coordinates of the system and to its charges. Does one in this case move along the *lower adiabatic curve*?

M: If one would vary nuclear coordinates very slowly, one would always move along the below adiabatic curve. Let me just say what the essential is. The idea when one is using a weak interaction and a nonadiabatic interaction is that one is going from one charge distribution, the reactants' charge distribution, to the products charge distribution. *At the crossing point one is doing that abruptly*, conceptually one is doing that abruptly, and the nuclei don't have time to respond to it, so during that transformation at the intersection the nuclei retain their position, so the vibrations don't change the position, the orientation of the solvent molecules doesn't change. What does change because its response is so fast, is the electronic response of the entire system, you change, you transfer the charge, all the electrons would respond to that, very quickly, so that does change, so the electronic polarization is distinct from everything else, the nuclear polarization of any kind, atomic or vibrational polarization, whatever you. So, you have a separation of variables in that sense, that the electronic coordinates respond quickly to any sudden change, like going from reactants to products electronic configuration whereas the nuclear configurations don't change during that instance. And that is an essential part of the whole ET theory.

(3) But now it's a different kind of theory.

(4) The answer is yes.

(5) That could well be. If you keep $Y = 0$, so that you have the solvation of the reactants, no. If you think of an XY space then . . . OK.

(6) That's correct.

(7) Yes, that's correct, and of course we can't say anything about that little splitting until we put something for the splitting in there, which we don't have.

(8) You mean just at the right near the crossing, yes, that's correct. Well, it may depend, if you were in the inverted region it wouldn't be in between but at one side. The actual charge distribution is sort of *split charge distribution* in the case of weak overlap, and the energy is sort of independent of just where you are there, in a very short

distance the energy hardly changes when you have small splitting, and yet you change the charge distribution completely. Split charge distribution means that *in the TS you have one charge distribution that abruptly changes to another charge distribution in a Landau–Zener fashion.*

Q: But along that small thing you are in equilibrium . . .

M: That's an interesting question, *the point is that in that small distance in between the energy is hardly changing even though the charge distribution is changing a huge amount*, so one could say in equilibrium . . . I have to think about that, I'm not quite sure what that means . . . The statement that I make is correct, in a Landau–Zener theory at the intersection the energy is essentially not changing, but the nature of the electrons is changing dramatically.

Q: It is a delicate point . . .

M: Yes, it is a delicate point, the trouble is that the charge distribution is largely a split charge distribution, then when the splitting is zero the system is not in equilibrium, but to have an infinitesimal splitting and to say that it is an equilibrium . . . that seems a bit drastic . . . I don't know the answer to that. I suppose that the position of, so to speak, exact resonance is when the two charge distributions, the two wave functions are to contribute equally and so in a sense maybe there is *a kind of local equilibrium*, but that isn't a useful way of calculating the free energy, you see. In other words, if you say I'm going to use charge distribution on each part you wouldn't get the right answer, you need to have *something which is very close to nonequilibrium, and an equilibrium calculation never gives you something close to nonequilibrium*, so the point is that maybe it's true something after the fact . . . but you can't reverse it, you can't go the other way. That's my gut feeling, it could be wrong. The point there is that in one case before the electron's jump you have a distribution of coordinates which is really a fluctuation from what it would be at the base of the parabola, just like the kx^2 where the $x = 0$ is the equilibrium point, when you have a finite x you have a fluctuation, that's what it is here. Now you can have fluctuations in one coordinate and equilibrium, so to speak, with respect to everything else, in this case the coordinate is the energy difference of the two electronic states, it's a generalized coordinate. But for any value of that coordinate you can have, given that value of the coordinate, sort of an equilibrium distribution involving the other coordinates appropriate to that value of the coordinate. The point is, you see, that you may not be able to calculate the interaction of the electronic wave function with the rest of the medium using conventional methods to do that, you want to make use of the nonequilibrium polarization, in other words if the average charge is slowly changing you may not get the right answer, you may have to use that strict personality of the wave function, if that's possible, I'm not sure. If the process were completely diabatic you would not end up with your desired final state. If it is diabatic you stay on the original curve. In a completely diabatic way you reach the upper curve but you don't reach the down curve. How do you calculate the energy of the system? I don't believe, my gut feeling is, that you can't just use the average charge distribution to calculate it. Yes, you don't give the personality of the wave function, it's a strict property of the wave function, I'm not sure I would describe in that way. I think I'm just saying that when you look in the, say, reactants, electronic reactant state at the intersection, you have that charge distribution and for that calculate what the properties are for that charge distribution, but at that value of the reaction coordinate there is a kind of constrained equilibrium distribution, it's a distribution which is clearly not in equilibrium is the electronic state, that equilibrium would be at the bottom of the parabola, and way up there it's a fluctuation that is in equilibrium subject to a constraint, extremely different from equilibrium, in other words, you reach the fluctuation but given that constraint, huge constraint, on one coordinate, the so-called reaction coordinate, now you can treat it as an equilibrium subject to a constraint, but you have to impose that constraint and that's a very severe thing in calculating and people have done that, Chandler has done that, Miller has done that, so people have done that. You can't use the average charge distribution, you have to calculate for one charge distribution and with the same distribution in coordinates you have to calculate for the other charge distribution and those two have to be equal in energy. In other words, the system is not in equilibrium with an average charge distribution, it's in a constrained equilibrium with the reactants distribution, same constrained equilibrium with the products charges, but is not in equilibrium with an average charge. If you put in the average charge distribution and calculate you get the wrong result, because you wouldn't know what to put in for the distribution of coordinates. In order to get the distribution of coordinates you have to find the intersection, but that means you have to make two separate calculations, one with the charge distribution of the reactants, one with the charge distribution of the products, you cannot get away with just one calculation.

Q: I should ask some very clever quantum chemist if it is possible even to do such calculation, if you can move the coordinates of the molecule and also the solvent molecules distribution and the charges, is it possible to do that?

M: Yes, well, I think that *if one tried to do in a conventional way you might find out that when the splitting is very small there is essentially simply a breakdown of the Born–Oppenheimer approximation,* then you can't do in a conventional way, that's my gut feeling.

Q: But if the splitting is very large it is possible...

M: Yes, if the splitting is very large, sure, but then the diabatic theory isn't valid.

Q: Your theory appears somehow to be partly diabatic before and after the crossing.

M: Yes, yes...

Q: And then there is a small region where it is adiabatic, somehow...

M: Yes, that's right, that's correct, the question is, you start up with the diabatic calculation, then you correct it, and basically that is what I have done although I haven't put in the correction, you see, and that is what Warshel does, he corrects it in the simplest possible way, an adiabatic correction, but the whole theory is based on the diabatic formalism. Warshel essentially has what I had, he takes into account the resonance energy at the intersection and the effect of that on lowering the value of the energy a little bit at the intersection. It's the usual thing that you would expect if you want to make the correction for the resonance at the intersection.

(9) The 1 and the 0 in each place is a matter of subtraction, in other words, that 1 − 0 means the charge of 1 minus the charge after it, which is zero, so the charge on A is 1, and for the other one the charge at the start is 0 and ends up at 1 so the change in charge there is −1, so 1 − 0 is equivalent to a +1 and −1 at each site.

Q: You see how many delicate points there are in your theory...

M: There are, there are... I know.

2. p. 4078 1st column middle: "Once again, the minimization should be of $\Delta G_r - k_B T \ln Q^\dagger$, subject to the constraint imposed by Eq. (7)

$$\Delta G_r - \Delta G_p = \Delta G^0_{RP} + k_B T \ln Q(1)/Q(0) \equiv \Delta G^{0\prime} \qquad (7)$$

However, to avoid introducing at this point variational parameters into $Q^\dagger$, we minimize ΔG_r subject to the constraint imposed by Eq. (7)."

p. 4078 2nd column top: "A question arises now concerning the factor $Q^\dagger/Q(0)$ in Eq. (33)

$$k_{ET} = Z[Q^\dagger/Q(0)]\exp(-\Delta G^*_{ET}/k_B T) \tag{33}$$

The correct value of $Q^\dagger$ would again be obtained from a minimization procedure that included $-k_B T \ln Q^\dagger$ in the minimization. However, one simplifying approximation is to assume that the ET/bond rupture reaction is

(i) *fully outer sphere* and that
(ii) $Q^\dagger/Q(0)$ equals the ratio of the large moment of inertia of the TS to that which appeared in R and hence in Z, and
(iii) to suppose that the new rotations that may eventually appear in P (in $q^{(2)}_{\text{int}}$) are still, in the TS, the vibrations that they were in the reactants. In that case we have

$$k_{ET} = Z_{ET}\exp(-\Delta G^*_{ET}/k_B T) \qquad (35)\text{"}$$

Q: For convenience in our discussion I have written your words in form of a list. Please explain the above points . . .

M: (i) *"fully outer sphere," that means you are not taking into account any bonding between the two reactants.*

(ii) If you consider a different problem, consider a bimolecular collision, and use TS theory, and don't put in anything that is equivalent to steric factors, then if you want to get the usual collision frequency expression, you did just as I said there, the ratio of the largest moment of inertia and so on, so that's just taking for what I and probably others, many others, have done *to get the collision theory expression from transition state theory*. In other words, you know, the collision theory expression for the gas phase has a $\pi\sigma^2$, σ is collision diameter in there, then you can get that out of transition state theory by doing just that operation in (ii), that I showed there, and I have certainly done that in lectures, during my classes, I probably evidenced some notes, some place, some notes from the Yugoslavia summer school, I don't know, I've certainly done it in various places, so... *if you consider two balls colliding and think of a transition state, the balls rotate around the line of centers, there is a small moment of inertia, the other one is the largest moment of inertia,* that moment of inertia is a reduced mass times the collision diameter, you see, so that's how you go from TS theory to collision theory, if you wanted to do that, so in (ii) I'm just doing sort of a standard procedure for getting a collision frequency out of it. *If in the transition state the rotations associated with the individual particles, not the overall rotation, are the same as in the reactants, if you assume that, then of course such partition functions will cancel.* You have to consider the partition function of the whole system, part of which involves hindered rotation, part of which involves vibration and the hindered rotations of the molecules in the reactants, in the whole reactant state. It differs of course from that in the products, the electric fields are different, so the restrictions on the rotations are different and when you get out up to the TS you only have one set of rotations that's common for both reactants and products because the nuclear configuration is the same for both in the TS, it's only the electronic configuration that's different in the two states, but everything else is the same.

3. p. 4079 1st column middle: "In general, as in all applications of TS reaction rate theory, some estimate must be made for the behavior of the coordinates in the TS, namely whether they are vibrations hindered rotations, or rotations, based on a description of the relevant parts of the potential energy surface for the solute in the TS"

M: *In the TS are the two reactants oriented especially toward each other? Or is any orientation of one with respect to the other? That statement, what's the actual situation, that affects the partition functions*, and if, for example, the potential energy of the system is independent of the orientation of each of the reactants, then the TS has the same rotational partition function for reactants and products, and similarly for the vibrations. You see, if the frequencies of the vibrations aren't changed, then the two partition functions will be the same. So, normally you start up with reactants, which have various properties including rotations and vibrations, in the TS they have properties such that *their location may or may not be hindered. If it is hindered that gives rise to the equivalent of a ratio of vibrational partition functions over rotational partition functions, in the TS that is equivalent to a steric factor.* That all describes how you can go from TS theory to collision theory, in an approximate way. So, that concerns a model of how you can go, when you make appropriate assumptions, from TS theory to a collision theory.

4. Following: "Equations (20) and (35) are limiting cases

$$k_{S_N2} = Z_{S_N2}(q_{\text{vib}}/q_{\text{rot}})^3 \exp(-\Delta G^*_{S_N2}/k_B T) \qquad (20)$$

$$k_{ET} = Z_{ET} \exp(-\Delta G^*_{ET}/k_B T) \qquad (35)$$

When most trajectories cross the TS, instead, in some region between those two paths (arrows in Figure 1a), the three coordinates to the q^3_{vib} in Eq. (20) may be, instead, hindered rotations, and we simply write the partition function as $q^3_{\text{int}}(X_2)$. The latter reduce to q^3_{vib} for the pure S_N2 path and to q^3_{rot} for the pure ET/bond rupture path. We have

$$k = Z(q_{\text{int}}(X_2)/q_{\text{rot}})^3 \exp([-\Delta G^*_r(X_2) + \beta_{ij}(X_2)]/k_B T) \quad (37\text{b})\text{"}$$

Q: Please explain how from the potential energy surfaces and from the paths you deduce the different partition functions for the three cases above.

M: What is the two paths? I have to see what the two paths are.

Q: I think S_N2 and ET.

M: Typically, in solvent you have hindered rotations, the central ions are not really rotating freely, they are commonly hydrogen bonded to the others. It is only when the ions and their coordination shells aren't hydrogen bonded that they'd be rotating freely, but of course there may be a big difference if it's ET or if it is pure S_N2, there is a whole nuclear change, I mean the reaction coordinate is different, the nature of the solvation, how it changes is different in the two cases. You are going to have hindered rotations of the solvent that are going to be more restricted near the more highly charged ion, and the most highly charged ion is different in the two electronic states.

5. p. 4080 1st column middle: "We compare the expressions for k_{S_N2} and k_{ET} ... From the equations for S_N2 reactions and those for ET reactions we obtain, in *the linear expansion* ($|\Delta G^{0'}_{RP}|/\lambda \ll 1$) *regime,*

$$\frac{k_{S_N2}}{k_{ET}} \cong \frac{Z_{S_N2}}{Z_{ET}}\left(\frac{q_{vib}}{q_{rot}}\right)^3 \exp\left(\frac{\lambda_0^{ET} - \lambda_0^{S_N2}}{4k_BT} + \frac{D_{A_2B}}{4k_BT} + \frac{(\gamma_1 D'_{A_1B}\gamma_2 D'_{A_2B})^{1/2}}{k_BT}\right) \quad (42)\ldots$$

(*If the two $\gamma_i D'_i$ differ considerably*, the approximation of replacing their geometric mean by an arithmetic mean becomes poor.)"

Q (1) Please look at Figure 2*: I have drawn three P parabolas: (a) for $\Delta G^0 = 0$, (c) for $\Delta G^0 = -\lambda$ and a (b) dashed line, just below the (a) line to represent a linear regime situation. Is it OK?

M: (1) Yes, it looks right, that looks right. As long as the dashed line is close to the solid line... that's a linear regime, it could be below or above but it's certainly in the linear regime....

(2) Please check if my explanation for your last statement above is correct: the geometric mean of two numbers a and b is that number c_g such that $a \cdot b = c_g^2$; the arithmetic mean is that number c_a such that $a + b = 2c_a$. If $a = b \Rightarrow c_a = c_g$.

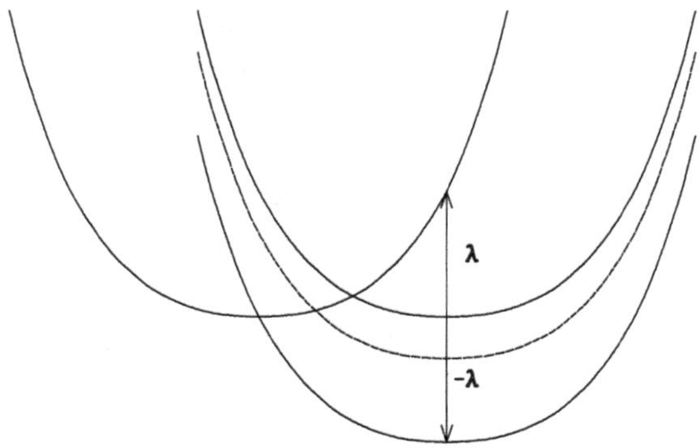

Fig. 2*.

M: (2) That's correct.

6. p. 4080 2nd column middle: "They also noted that there are *two opposing effects* influencing the k_{S_N2}/k_{ET} ratio: (1) *a more restricted transition state for the S_N2 reaction and (2) a lowered energy activation for S_N2, because of the bond formation. The two effects are present in Eq. (42).*"

Q: Which are the terms in Eq. (42) that describe, respectively, the two effects?

M: Well, the $\frac{q_{\text{vib}}}{q_{\text{rot}}}$ that's one of the terms, because that's present in the S_N2 reaction but is not present in the ET reactions, for the S_N2 you have to have that orientation of the reactants and that gives rise to that $\frac{q_{\text{vib}}}{q_{\text{rot}}}$ term. And for the other part... the three terms in the exponential reflect the second effect, in other words, the bond formation is reflected in the ratio of the qs, and also in the ratio of the Zs, you probably need a closer collision for the S_N2, so that preexponential factor certainly is related to the item number 1. And I suspect, but I have to look at it, that all three quantities are related to item number 2, low energy of activation. Because the exponential affects the energy of activation. But one would have to look at the derivation of

Eq. (42). So, it looks that the preexponential is associated with 1 and the exponential is associated with the other. That's my guess.

7. p. 4081 1st column top: "We next use *results from some gas phase* S_N2 *reactions to estimate the preexponential factor* in Eq. (20).

$$k_{S_N2} = Z_{S_N2}(q_{\text{vib}}/q_{\text{rot}})^3 \exp(-\Delta G^*_{S_N2}/k_B T) \qquad (20)\text{''}$$

Q: Does this mean that the preexponential factors are the same in gas phase and in solution? Is it always so, or it is so only sometimes? And when in this case? Or is it so that the preexponential factor in the rate constants in solution is the same as for the reaction in the gas phase and the effect of the solvent on the activation energy is taken care by ΔG^0 and λ_o in the exponential factor?

M: If you write it in the form that I have it there, and refer to preexponential in that form, yes. If you wrote it as an Arrhenius expression, and instead of ΔG^* you have an entropy of activation, then the answer is no. If one writes $k_{\text{rate}} = A\exp(-E_a/k_B T)$, where A is a preexponential factor, that preexponential factor is not the same as the preexponential factor in Eq. (20), there the ΔG^* can have some chemistry dependence, an entropy term associated with it. If you write it in this way and if you wrote the reaction in solution in that way, and wrote the reaction in the gas phase in that way, yes.

Q: So the preexponential factor is not the same in solution and in the gas phase.

M: *Not the Arrhenius preexponential factor. With the Arrhenius preexponential factor you don't write a* ΔG^*, *you write an activation energy, they differ by entropies.*

Q: They also differ by the solvent....

M: Yes.

8. Following: "For such reactions when $\Delta S^0 \cong 0$, we can write

$$k \cong Z(q_{\text{vib}}/q_{\text{rot}})^3 \exp(-\Delta U/k_B T)$$

where ΔU is a term essentially independent of temperature."

Q: Is it so because ΔU is the energy difference of two electronic states and the energy of the electronic states is essentially independent of temperature?

M: *Yes, it's only when the exponential has a free energy, the free energy may have a temperature dependence.*

9. p. 4081 1st column middle: "When the dependence of the λ on T is neglected (*the dependence is expected to be weak* as in the usual ET's)"

Q: Even in this case, is the weak dependence on temperature due to the fact that λ is a difference in solvation energy between two electronic states whose energies are weakly dependent on T?

M: Yes, I think so, generally these energies... the energies themselves, the thermodynamic energies, are weakly dependent, I mean their dependence on a heat capacity is relatively weak. But free energies, if there are some strong entropy changes, can depend strongly on temperature.

10. p. 4081 bottom: "...if we neglect the $\partial w/\partial T$ terms, which are expected to be minor *unless both reactants are charged*"

Q: Does this mean that the most important contribution to w is the coulombic one?

M: It depends... sometimes I have been using w to be reorientational effect, so I don't know what I did here, it depends on to what these w's are related to, if they included reorganizational effects. I think in the 1968 paper I included the reorganizational effects in w. So, it depends on what this w is here.

Q: So it is not the normal work term.

M: Well, *it's intended to be the part of the free energy change term which is not affected by the standard free energy of reaction, so it refers to what occurs beforehand, when there is some assembling*, then goes along on to forming the transition state. If the system has to orient itself then sometimes I have included that, I have included that in the w term sometimes, it depends. When the reactants are charged you have the coulombic interaction, the dielectric constant depends on temperature. And from that there is in principle quite a change of entropy.

11. p. 4081 2nd column middle: "We have already noted the limiting situations, often referred to in the literature, of a pure S_N2 reaction having a very negative $\Delta S^\dagger$ and a relatively low $\Delta H^\dagger$, and a pure outer sphere ET reaction having a much less negative $\Delta S^\dagger$ and a relatively high $\Delta H^\dagger$."

Q: Why are there different values of $\Delta S^\dagger$ and $\Delta H^\dagger$ for S_N2 and ET?

M: In the S_N2 reaction you have a negative $\Delta S^\dagger$ because you need a lot of orientation to form the bond, and that corresponds to negative ΔS, reactants are not rotating freely, they orbit with respect to each other, so you can get steric factors in some reactions, which, like in gas phase reactions... 10^{-3}, 10^{-4}, can be huge. When you have an S_N2 reaction you form a bond, so the two reactants can't rotate freely, they're stopped by the bond, so that corresponds to quite a negative $\Delta S^\dagger$ and a relatively low negative $\Delta H^\dagger$ because of your bonding that is going to help the reaction. In the electron transfer case you don't have by and large, except for some very special reactions, any reorientation, any orientation due to transfer of the electron.

Q: But you did write some paper on orientation with Bob Cave.

M: That was for the electronic effect. Well, it's true that if you have a weak overlap ET reaction, if it is extremely weak, that corresponds to another basis for negative entropy of activation, so the above statement would apply to reactions that are not with very weak overlap. Because the weak overlap type introduces a low preexponential factor, the probability of reaction is small and that corresponds to a very negative $\Delta S^{\dagger}$, if you put it in the transition state theory terms.

12. p. 4081 2nd column bottom: "A related aspect of this temperature or $\Delta S^{\dagger}$ behavior concerns *the linearity of the* ln *k vs.* $1/T$ mentioned earlier for the reaction between anthracene radical anions and n-butyl bromide over the wide temperature range of -50 to 50°C. There was no indication of a possible change of mechanism, i.e. no change of slope..."

Q: Is the change of slope always an indication of a change in mechanism? Is it possible for the mechanism to change while the slope remains the same?

M: Well, one point that we are not talking about here is tunneling, where you can get a change in slope and essentially you go from a nontunneling to a tunneling mechanism. Here there is no change in slope, so there is no change in activation energy. When there is a change of mechanism there should be a change in activation energy. In his case yes. But I added the other thing, which is not relevant here, that you can get a change in slope without a change in mechanism if you go over to tunneling, but we are not doing that here. But there may be slow changes, you know, another change in mechanism where you can have a small change in slope, that you hardly notice, because it is a heat of activation, and a heat of activation has a temperature dependence, because the heat capacity changes and so on. So, any heat of activation or energy always has some slight temperature dependence, so slight that you don't normally see it.

13. p. 4082 1st column top-middle: "We conclude this section with some remarks on the entropy term associated with the $\Delta G^{0\prime}$ appearing in Eq. (32)

$$\Delta G^{0\prime} = E^0_{A_2^*/A_2^-} - E^0_{A_1^*/A_1^-} + D_{A_1 B} + w_p - w_r - k_B T \ln Q^2_{\text{int}} Q^1_{\text{vib}} \tag{32}$$

and which would make a contribution $1/2 \Delta S^{0\prime}$ to the entropy of activation $\Delta S^\dagger$ in the linear regime ($\Delta S^{0\prime} = -\partial \Delta G^{0\prime}/\partial T$)."

Q: Please check my explanation for $\frac{1}{2}\Delta S^{0\prime}$. It comes from Eq. (34)

$$\Delta G^*_{ET} = w_r + \frac{\lambda}{4} + \frac{\Delta G^{0\prime}}{2} + \frac{\Delta G^{0\prime 2}}{4\lambda} \tag{34}$$

neglecting the square term and deriving with respect to T.

M: Yes, that's right. And, of course it corresponds to λ being hardly temperature dependent, so when you differentiate with respect to temperature you just do a differentiation of the $\Delta G^{0\prime}$ with respect to temperature.

14. Following: "*If the $q^{(2)}_{\text{int}}$ in Eq. (32) is largely $q^{(2)}_{\text{vib}}$, with small amplitudes, as we have tacitly assumed..., then the last partition function term in 32 will make relatively little contributions to $\Delta S^{0\prime}$. If, however, the $q^{(2)}_{\text{int}}$ is $q^{(2)}_{\text{rot}}$ or involves very floppy hindered rotations, then the term will contribute a positive contribution to $\Delta S^{0\prime}$ and hence to $\Delta S^\dagger$.*"

M: Yes, that's right, typically you can form a partition function, you can calculate the entropy associated with it, you calculate the free energy, you calculate everything from it. And considering the entropy change associated with vibrations... the temperature dependence is relatively small, you know, there is some entropy change but is a relatively small thing, frequently vibrational partition functions are close to unity, they can't change much with temperature.

Well, if you have a rotation, then there is an entropy associated with it, you can be in many rotational states, they are all close to each other, if you take the partition function that will give you a free energy, if you differentiate with respect to T, you can get an entropy,... vibrational motion is fairly stiff, no matter what, but rotation gets broader and broader you know, it goes to many more states and so on, there are lots of states associated with it, so there is an entropy associated with it, whereas a vibration typically is in the lower state, there is hardly any entropy associated with it. Rotation can be in many states so there is intrinsically a large entropy associated with it.

305 Theory of Rates of S_N2 Reactions and Relation to Those of Outer Sphere Rupture Electron Transfers

(Continuation 2)

R. A. Marcus

1. p. 4074 1st column top: "It is useful to consider first what the potential energy surface might look like for an S_N2 reaction,

$$A_2^- + BA_1 \rightarrow A_2B + A_1^- \tag{3a}$$

$$\text{or } A_2^{*-} + BA_1 \rightarrow A_2B^* + A_1^- \tag{3b}$$

and for an ET reaction

$$A_2^- + BA_1 \rightarrow A_2^* + B^* + A_1^- \tag{4a}$$

$$\text{or } A_2^{*-} + BA_1 \rightarrow A_2 + B^* + A_1^- \tag{4b}$$

Q: The second reaction is a concerted ET/bond rupture reaction. Can we define an S_N2 reaction as a concerted ET/(bond rupture + bond forming) reaction?

M: Let's see... suppose that your reactant instead of an ion is a neutral, both are neutrals, then you have an S_N2 reaction and there is no electron transfer.

Q: But in this case?

M: In this case yes, OK.

2. p. 4082 2nd column middle: "We turn next to the barrier for the *gas phase identity reaction*, which is $(0.5 - \gamma_i)D_{\mathrm{A}_i\mathrm{B}}$ according to Eq. (16)

$$\Delta G^*_{\mathrm{SN2}} = w_r + \Delta G^*_{\mathrm{r}} \cong w_r + \frac{\lambda_o + D_1 + D_2}{4} - \beta_{ij} + \frac{\Delta G^{0\prime}}{2} + \frac{\Delta G^{0\prime 2}}{\lambda} \quad (16)$$"

Q: Please check my derivation: It is $\Delta G^*_{\mathrm{SN2}} = (0.5-\gamma_i)D_{\mathrm{A}_i\mathrm{B}}$ because being the reaction in the gas phase $\lambda_o = 0$, being the reaction an identity one $D_1 = D_2$ and $\beta_{ii} = \gamma D_i$.

M: I converted from the exponentials over to quadratics and then this kind of equation is following from that. All of this was based on a transformation which converted the exponentials and that's how I got the λ, that part of the λ, was a D/4, so the derivation of this equation comes about just straight after that transformation... But you kind of derived in a different way, is that it? But I don't see β_{ii} any place in here.

Q: You have an identity reaction.

M: Oh, I see, all right, in other words, we take 16 as given and we see what happens when we have an identity reaction and gas phase. Yes, then what you have there is OK.

3. p. 4082 2nd column bottom: "It has been pointed out that terms such as *leaving group ability* and *nucleophilicity* are replaced, in the language of 7b and 11, by other terms: *intrinsic barrier, thermodynamic driving force.*"

Q: The 7b and 11 references refer to your papers. Can you comment on the two different ways?

M: *Intrinsic barrier* and *driving force*, that's sort of the results of the leaving group ability leading to nucleophilicity. In other words, those two concepts affect both of these quantities, so both have provided maybe physical insight into each of these quantities, I mean just saying intrinsic barrier doesn't give the chemical aspect, and talking about nucleophilicity leaving group ability brings in kind of chemistry, you see. And so, if you can apply that, the group ability and nucleophilicity to, for example, an exchange reactions, then you would have a direct comparison with the intrinsic barrier in terms of those concepts, because in that case the other concepts, barrier and driving force are absent. In other words, intrinsic barrier is the result of those two quantities. And, of course, the driving force is in effect related to those two quantities too.

Q: So, each one of your two terms refers to both of the preceding concepts.

M: Both, yes. Both of those will be involved when you try to understand intrinsic barriers for different systems.

4. p. 4083 2nd column top-middle: "In a system with many coordinates, one task is that of finding the transition state hypersurface, and thereby a suitable reaction coordinate. In the case of weak overlap electron transfer reactions, the *energy difference* ΔU *between the products' and the reactants' potential energy* has been a useful *reaction coordinate* and has permitted the definition of the transition state as a particular hypersurface in the space."

Q: How about the dimensionality of ΔU vs. ΔG?

M: ΔG *is not defined in a* 10^{23}*-dimensional space*, so only ΔU is.

Q: But $\Delta G = \Delta U - T\Delta S$

M: ΔU *is a function of* 10^{23} *coordinates.* ΔG is the average of $\Delta U - T\Delta S$. That already does a lot of thermal averaging, over many coordinates. Well, ΔG is not defined in 10^{23} ... it's related to something that is defined in the 10^{23} dimensional space, in other words, *we don't plot G vs.* 10^{23} *dimensions, we plot G vs.* 1 *dimension, we can plot potential energy vs. 10*^{23} *dimensions, but we can't plot G vs.* 10^{23} *dimensions. So, G is not in the* 10^{23} *dimensional space although it describes what is happening in the* 10^{23}*-dimensional space.* ΔG has the average of $\Delta U - \cdots$

Q: And the $T\Delta S$?

M: Well, you don't speak of the average of ΔS, but you can speak of the average of ΔU. That already does a lot of thermal averaging. You do that in the conventional way, when you want a free energy of an ensemble that has 10^{23} coordinates, you write it appropriately in terms of an integral involving an exponential and so on, it's an average in a Boltzmann-like fashion. It's that kind of an average. For any thermodynamic property that's true. Really in the sense of Gibbs, which is more general than Boltzmann's, but it is same basic idea.

5. Following: "However, in atom or group transfers particular care is needed. This energy difference can be a misleading coordinate, not so much near the TS, where the ΔU between the zeroth-order states is zero, but away from the TS, for example if the shape of the ΔG_r plot *vs* the reaction coordinate is being investigated, as it often is."

Q: Question similar to the preceding one: which are the merits and the informations of a ΔU plot vs. reaction coordinate compared with that of the ΔG_r plot? When should one consider ΔU and when ΔG_r?

M: They have different purposes. The ΔU plot is used to try to define a method for getting theories statistically mechanically, you need a reaction coordinate for that. This ΔU is a function of 10^{23} coordinates, so if you want to discuss processes in terms of the 10^{23}

coordinates, one way or another you have to consider ΔU and do the *complete averaging*. When you have done that you can get ΔG at that value of the reaction coordinate. So, one is the consequence of the other. And *you don't have a monotonic behavior in the difference* ΔU *of the potential energy surfaces*. If you look at the potential energy of this kind of system, it has a lot of bumps and deeps, it's not a monotonic function in one direction, it goes up and down, in anyone of the 10^{23} directions, it's not like a harmonic oscillator. The surface itself in this many-dimensional space is like a set of sand dunes, a lot of local minima on the way that is not true for a harmonic oscillator but is true for a real solvent. I'm referring to a real solvent, and its interaction with the reactants. For a collection of interacting harmonic oscillators you wouldn't have a set of sand dunes.

6. Following: "Such a difficulty can occur when the contribution ΔU_e *of only the* (x_1, x_2) *terms*[$U(x_1, x_2)$ for products minus $U(x_1, x_2)$ for reactants] is *not monotonic* along the expected reaction path in the (x_1, x_2) space. (Such a situation appears to have occurred at small A_1B distances in Ref. 33, perhaps arising from the difficulty noted there of accurately describing the $A_2B^- \cdots A_1$ repulsion.)"

Q: How about this nonmonotonicity problem?

M: Oh, well, for the weak overlap ET if you use the diabatic curves, then you have... you can always use as reaction coordinate something which is monotonic.

Q: Why so?

M: Yes, because the coordinate is the energy difference between the reactants and the products at any given... at each nuclear configuration, so *you collect all the nuclear configurations for which that energy difference has a certain value* and now you collect all and everything which has a lower value, and you collect all the things which have a higher value, that means you can do the plot as a monotonic function. Even though an individual coordinate label is not monotonic, the coordinate defined in that way can be a monotonic coordinate, just by the definition of it. Even when the coordinates are highly anharmonic, if you use this definition you can get a reduced coordinate that always goes in one direction.

Q: Which definition?

M: At any nuclear configuration you take the potential energy of the products and subtract from that, at the same nuclear configuration, the potential energy of the reactants, there is a certain difference, now you can order those differences in a monotonic way. You have the bumps in the real space but if you have *an ordered energy coordinate*, there are no bumps there. There are no bumps in the coordinate itself. At each value of the coordinates you calculate the free energy. The coordinate itself is monotonic because by definition it goes monotonically, from a small difference to a large difference, and in fact *it changes sign too, at the intersection.* It is so because you have chosen it that way. You say: I'm going to take an energy difference so I'm going to use as the coordinate an energy difference, that's automatically something that's monotonic. The single generalized coordinate, which happens to be the energy difference, is monotonic. That's a monotonic curve, you go from one minimum at a point to another minimum at a different point, on the x axis.

7. Following: "To avoid this problem, and to learn about the λ_o in the various ΔG_r and ΔG_p expressions away from the TS, one can introduce there a new coordinate ΔU_s, the difference in U due to all but the above (x_1, x_2) terms, and calculate $G_r(x_1, x_2, \Delta U_s)$ and $G_p(x_1, x_2, \Delta U_s)$ as functions of ΔU_s, x_1 and x_2."

Q: Is ΔU_s the contribution to G_r given by the solvent? Is then ΔU_e the gas phase ΔU? I mean, do we have $\Delta U = \Delta U_e + \Delta U_s$?

M: Yes, I think so, but it is not so suitable for the interaction term, so if there are no interaction terms then you can write $\Delta U = \Delta U_e + \Delta U_s$

but, if there are interaction terms, then you can't write something that way. ΔU_e is function of a certain number of coordinates of, I don't know, 10, 15 coordinates, and ΔU_s is function of 10^{23} coordinates, so you can write something as a sum but when you see something written as a sum like that, then you neglect the interaction of those coordinates...

M: In ΔU_e in ΔU_s is e the electronic and s everything else?

Q: Yes, I guess so.

M: I have to assume what it is...let's say is that, OK? What is (x_1, x_2)?

Q: Those are bond lengths.

M: I think you can define ΔU_s to include all the interactions within the solvent and the interactions of the solvent with the bonds that are breaking but when you write it that way, $\Delta U = \Delta U_e + \Delta U_s$, it's assuming no interaction between the solvent and the breaking system, which isn't right. There has to be some interaction, an ion may be forming another ion, there is going to be some interaction, that ΔU_s is really everything that isn't s_1 and s_2, the s_1 and s_2 can be identical with the gas phase but the ΔU_s is everything that's additional, including interactions, with x_1 and x_2, so that ΔU_s probably depends on x_1 and x_2, so it is an approximation to neglect that, and so what it means is that if one wants to use this then strictly speaking for the dependence of $G_r(x_1, x_2, \Delta U_s)$ and $G_p(x_1, x_2, \Delta U_s)$ on x_1 and x_2 one shouldn't use the gas phase values, one can certainly retain that functional form, those Morse curves and so on, but not the actual values, the constants that we are using in the gas phase, they should be modified somehow. The functional form is probably OK, but the constants would be somewhat different in the solution phase compared with the gas phase.

Q: This paper is impressive.

M: I hope we didn't exhaust you...

Q: You almost did...

8. Following: "From this information λ_o can be obtained: at $x_1 = 0$, $x_2 = \infty$, $G_p - G_r = \lambda_o + \Delta G^{0\prime}$, so yielding λ_o for the reactants, while at $x_1 = \infty$, $x_2 = 0$, $G_r - G_p = \lambda_o - \Delta G^{0\prime}$, so yielding λ_o for the products. These λ_o's in the reactants' region and in the products' are unambiguously known, since the two distinct charge distributions are known. However, in the TS region questions such as the validity of a two-states approximation arise and complicate a *two-charge distribution* description...particularly when β is large."

M: In a vertical transition from any point on the lower curve you go vertically upward. In the lower curve, the curve of the reactants, you have A_1^- reacting with HA_2, and on the upper curve you have A_1H instead of A_1^-, so that means *the vertical transition is not the usual vertical transition which involves no nuclei moving*, in this case a proton has jumped over. So, this is treating the proton more or less like an electron, it's treating the case where... I'm not sure it's a good way of doing it though, but is treating the case where you are looking at a state where the H is initially part of A_2, and in the upper state it is part of A_1, in other words, the two parabolas here whatever they are, the two potential energy surfaces, on one to the left you have $A_1^- + HA_2$, in the one to the right you have $A_1H + A_2^-$, if you extend the one on the right upward, so that now is over the $A_1^- + HA_2$, then upward it becomes $A_1H + A_2^-$, or maybe I have in reverse, so that's what that's about, in other words I am treating the proton in that way, I'm not saying it's the correct way, it is treating the proton as though it's like an electron, sort of it can jump like an electron, and change the state of the system, so what that is done is translate into electron transfer language what would happen if you treated the proton jump as a jump, you know, the proton is either here or there. I am not sure that that's the correct way of doing it, though, because you have a different number of coordinates, you know, because now you have a proton up there, so it's a little bit hazy, but basically that is what I am doing and what I phrased there, I don't know if it's a good way of doing it or not, is that portraying that proton like an electron, and so the upper state instead of having a 1− it has H^-. And the lower state has $A_1^- + HA_2$, the upper state has $A_1H + A_2^-$. Basically what it would be is that if you took the potential energy surfaces for $A_1^- + HA_2$, and the one for $A_1H + A_2^-$, I write them as two parabolas, and that's how you can have a vertical jump, you see, you go from one to the other, it's the kind of a jump though where the proton is like another electron essentially and you look at the energy of interaction of the upper state there with the surroundings, and similarly for the bottom state, in other words you treat it like an ordinary ET, in that crude model. You know, a number of people, in the early days Levich and Dogonadze, treated the proton transfer the way they and we treated the ET, so essentially they had two diabatic curves where in one case the proton was on one reactant, in the other case the proton was on the other reactant, just like when you have two diabatic curves in one case the electron is on one reactant, in the other case on the other reactant, so it's just a way of treating proton transfers in an approximate way that has some analogy in ET, so in some treatment, like one of Dogonadze and Kuznetsov, that's a kind of model they use.

Q: And in the reaction $A_2^- + BA_1 \rightarrow A_2B + A_1^-$, where x_1 is the bond distance?

M: OK, I see, all right, yes...this is a different kind of system...there is the difference between G_p and G_r, but this is a different kind of system and we didn't define λ_o in between...

Q: So, you have different systems.

M: Yes, we didn't define, it's just defined in the two ends... but doesn't say what they are...

Q: In between.

M: Yes... and now the transition state we don't have a way to.... That's a very different situation.

Q: You already mentioned the problem of determining the form of the free energy surface in the neighborhood of the crossing. And how about the validity of the Born–Oppenheimer approximation near the crossing? Is it possible that strictly speaking near the crossing one should not draw a potential energy surface at all?

M: No, you can, you can draw adiabatic surfaces and as an approximation.

Q: Always?

M: Always.... For adiabatic surfaces... about a two state approximation and two diabatic states... The Born Oppenheimer approximation is really valid if the system is moving slowly, the problem is that when the system is going in the region where the splitting is small then you can't say that the system is moving on the adiabatic surface and you have all of those corrections evolved in the 1930s by Landau, Zener, and Stückelberg.

9. Following: "Some quantum chemistry calculations have been performed seeking, for a gas phase system, specific transition state configurations for S_N2 and ET reactions... *The TS for the ET reaction, however, has a much broader definition than a saddle-point, because of the 'looseness' of the TS.*"

M: Yes, well, when you have bond breaking, bond forming, the reactants are usually pretty ordered with respect to each other, when you have an ET that ordering is much less. ET is different for different orientations only in some electron transfers.

10. p. 4084 1st column top: "The slow-moving solvent molecules cannot be appropriately aligned to the instantaneous position of the electronic charge in $A_1BA_2^-$. Thereby, the set of nuclear configurations adjusts, instead, to some averaged charge distribution of the electrons in $A_1BA_2^-$."

Q: Please look at Figure 3*: Here I have drawn the diabatic and adiabatic curves for the process. Please check if the following reasoning is correct: When one is at the crossing of the diabatic curves the charges of the system are equal to the charges of the system at the minimum of the *R* curve and are also equal to the different charge of the system at the minimum of curve *P*. This means that at the crossing the system isn't in equilibrium with either charge, there is *no equilibrium* at all. At TS, the representative point is not at the crossing of the diabatic curves but is lowered at the maximum of the lower adiabatic curve where there is an *unstable equilibrium*.

M: Chandler and Hynes' did work on this kind of point, there are subtle points.

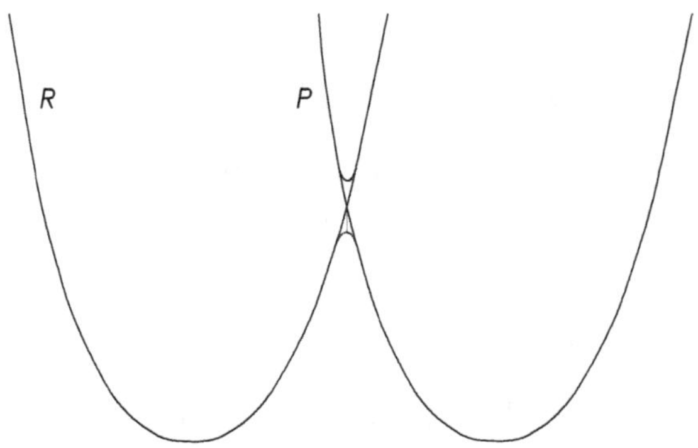

Fig. 3*.

M305 Theory of Rates of S_N2 Reactions and Relation to Those of Outer Sphere Rupture Electron Transfers

(Continuation 3)

R. A. Marcus

1. p. 4074 2nd column middle: "**Two-Interacting—States Model**: In an extension of electron transfer theory to ET reactions accompanied by bond rupture Savéant employed a Morse potential energy function $D_1[1 - \exp(-a_1x_1)]^2 - D_1$ for the rupturing bond RX in reactions 1a or 1b

$$A^- + RX \rightarrow A^* + R^* + X^- \quad (1a)$$

$$A^{*-} + RX \rightarrow A + R^* + X^- \quad (1b)$$

Here x_1 denotes the bond distance displacement from its equilibrium value in RX. He also assumed for the repulsion term between R^* and X^- the quantity $D_1 \exp(-2a_1x_1)$."

p. 4075 1st column middle: "For ΔG_r, the free energy of formation of the state (X_1, X_2, Y) from R... we write

$$\Delta G_r = D_1(X_1 - 1)^2 + D_2X_2^2 + \lambda_0Y^2 \quad (5a)$$

where

$$X_i = \exp(-a_ix_i), \quad \lambda_o = \lambda_o(X_1, X_2) \quad (i = 1, 2) \quad (5b)"$$

Following: "The $D_1(X_1 - 1)^2$ term in Eq. (5) describes the interaction between A_1 and B in A_1B, and the $D_2X_2^2$ term denotes the repulsion between A_1B and A_2^-. As noted earlier, the D_1 and D_2 include the effect of the changes in bond angles."

Q: Apart from the mathematical elegance of your representation, are there physical reasons for this mapping of exponentials into parabolas?

M: Incidentally, this you might find amusing that a Savéant's paper came to me for review and so I reviewed and then I said: "Why don't you make a transformation that looks simpler?" I did offer it free, gratis, but he didn't want to use it... so I let him go and then I wrote the transformation in connection with ET *where bond is breaking*. Under certain conditions depending on what the PES looks like you can consider a reaction where the reactants are bound but the product is unbound, in other words there is dissociation, and under certain conditions when you can write each of those PECs as a certain combination of exponentials, one a Morse potential and the other an exponential dissociation, under those conditions with certain exponents you can introduce a change of variable such that the exponential of, let say, $e^{-a(r-r_0)}$, is then changed to something called Y and if you do that, everything then looks like the ordinary quadratic expression for the free energy of the reactants or the free energy of the products and you can use the standard theory.

Q: So, you use your theory in this case of bond breaking.

M: That's right, this is where the reactant is a stable molecule and the product is a dissociated species.

2. p. 4084 1st column top: "The slow-moving solvent molecules cannot be appropriately aligned to the instantaneous position of the electronic charge in $A_1BA_2^-$. Thereby, the set of nuclear configurations adjusts, instead, to some average distribution of the electrons in $A_1BA_2^-$."

Q: (1) Please look at Figure 3*: Here I have drawn the diabatic and adiabatic curves for the process. Please check if the following reasoning is correct: When one is at the crossing of the diabatic curves the charges of the system are the charge of the system at the minimum of the R curve and the charge of the system at the minimum of curve P. At the crossing the system isn't in equilibrium with either charge.

(2) But at TS the charge must be in an unstable equilibrium with a charge intermediate between the charge appropriate to the *R* curve and that of the *P* curve. Due to this unstable equilibrium the representative point of the TS system is not at the crossing of the diabatic curves (where there is no equilibrium at all) but is lowered in the neighborhood of the maximum of the lower adiabatic curve (where there is an equilibrium but an unstable one). Is it so?

M: (1) Yes, that's right.

Q: Does it make sense my distinction between unstable equilibrium and no equilibrium at all? Normally at a potential energy maximum there is an instability but maybe not in the sense that the electrons distribution there is unstable....

M: (2) There is a real subtlety here Francesco, that I personally haven't looked through because the point is that one of the diabatic curves has the electronic distribution of the reagents, it responds to one charge, and the other curve has the electronic distribution of the products and so one can't say that what one has is an average of some charge distribution, it's just that they are related to different charge distributions, so one can't use some conventional ideas there. Now, to treat the problem correctly one would have to sort of associate a time scale with the electron that is being transferred and associate a time scale with all the other electrons, the time scale of the other electrons is probably very short, but one has different time scales and one should do the dynamics of the whole thing with the appropriate time scales and I haven't done that, and to really answer your question one would really have to do that. Now, it is possible, but I'm not sure, that in one of his papers David Chandler did that more fundamental treatment, I never have but it is possible that he did, he has written very few papers on ET but they have been significant and this was one of them. So, one would have to consult him and that paper to really get a final word on this. The main point is that the system doesn't simply have the average charge distribution of the reactants to the products, because the electronic polarization is not appropriate to that, it's a combination with its own time scale, to do a very fundamental treatment and to answer your question in detail one would have to really do the full dynamics of the electron population.

3. Following: "One main problem is then how to treat the correlation between the electrons in $A_1BA_2^-$ and those of the solvent.[36] The treatment of that correlation together with finding the nonequilibrium distribution of positions of the nuclei of the solvent, was the main focus in the electron transfer theory.[11b]"

Q: Here the reference 36 refers to a paper of Chandler and Hynes and others and reference 11b refers to your theory. I shall read David's paper but can you tell now something about how your approach compares with that of the Chandler and Hynes?

M: I don't remember all details of the paper of Chandler and Hynes, but I think the basic idea is this: that you have to treat both the electron dynamics and the nuclear dynamics at once, you have to take into account a combined dynamics of the electron and the nuclear motion. They may have done it using the so-called spin boson Hamiltonian, the main point is that you have to treat electrons and nuclei on the same level, on the same par, which I didn't. That is the way to treat all situations correctly.

Q: So, basically they didn't use the Born–Oppenheimer approximation.

M: No, not at all. You treat dynamically on the same level the electron and nuclear motion and not the way I did it.

4. Following: "One electronic frequency *in the TS* of $A_1BA_2^-$ is that associated with the splitting $\Delta\varepsilon$ of the two states there and is the frequency $\nu = \Delta\varepsilon/h$ for the electronic oscillation that would occur in *a time-dependent electronic oscillation problem*, $A_1^-BA_2 \leftrightarrow A_1BA_2^-$ or $A_1^{+\delta-1}B^{-\delta}A_2 \leftrightarrow A_1B^{-\delta}A_2^{\delta-1}$, where δ may be positive or negative."

Q: Please check if my understanding is correct:

(1) We are now at the nuclear and solvent configurations corresponding to the crossing of the diabatic curves. At this point the ET represented by $A_1^-BA_2 \leftrightarrow A_1BA_2^-$ becomes that of an electron going from one potential well to another separated by a wall moving along a purely electronic coordinate;

(2) Aren't we now in a typical superposition state of two states spatially separated? Isn't the transferring electron at the same time and with equal probability in both wells? How different is this situation from that in which the two electrons are either in one well or in the other and oscillates between the two with a certain frequency?

(3) Does the symbol $A_1^{+\delta-1}B^{-\delta}A_2 \leftrightarrow A_1B^{-\delta}A_2^{\delta-1}$ mean that instead of considering the electron localized on A_1 or on A_2 you consider also the possible localization on A_1B or on BA_2, that is, you admit a less localized (or, if you want, a more delocalized) situation?

M: (1, 2) First of all: for oscillations, refer to Kauzmann's Quantum Chemistry, here there is an interpretation of the Landau–Zener formula, so I suggest that you look at Kauzmann's book in Quantum Chemistry... he has a description of a physical interpretation of the Landau–Zener. Let me tell you what the possible issue is: it is a hypothetical thing. *Pretend that the nuclei are fixed at the transition state, they are not, but pretend that they are*... then the electron at one of the items oscillates back and forth, only if you did that it oscillates back and forth. The relationship of that splitting to that kind of a motion, *it's a hypothetical situation where you have fixed nuclei. If* *the nuclei are not fixed... If you don't do that you don't have oscillation.* You have oscillation when you fix the nuclei at the intersection. In the initial state the electron is on one of the nuclei,... If you fix the nuclei make sure that the electron starts at one of the atoms. You have the oscillation when the nuclei are fixed. You fix them at the cross point. If you fix the nuclei make sure the electron starts on one of the atoms, so it is forced, OK? Then you get the oscillation.

(3) Yes.

5. Following: "The '0' in Eq. (22)

$$\lambda_o = G_{1-0}^{e,sol} - G_{1-0}^{e,op} \tag{22}$$

refers to the *charge distribution of the solute which determines the distribution of nuclear coordinates of the solvent in the TS.*"

M: I think the 0 referred to the initial charge distribution, 1 refers to the final charge distribution, and 1 − 0 is a hypothetical charge distribution... in one of the 1963—I think—papers I showed how you express the usual formula in terms of a hypothetical charge distribution.

Q: Aren't normally the nuclear coordinates that determine the charge distribution?

M: Yes, right, effectively that difference in charge distribution determines—I proved—I didn't assume, I proved, determines the distribution of the nuclear coordinates of the transition state, that's right. I'm making use of the result of 1965. They only determine the coordinates states. It's an intersection that determines it... in other words it's a condition that determines the transition state. What I did was to show that the expression I had before could be expressed simply in terms of this hypothetical charge distributions, 1 − 0 is a hypothetical charge distribution, the charge distribution of the products minus that of the reactants. And if you refer the coordinates distribution to that, then that would give you the first term on the right hand side, if instead you consider the equilibrium solvation but where only the electronic polarizability of the solvent enters rather than the total (electronic plus the nuclear) contributions, that would give you the second term on the right hand side. In other words, I expressed the whole thing in terms of a certain equilibrium distribution, a hypothetical distribution.

6. Following: "This ρ_0 is, *in the TS state*, approximately equal to $^1/_2(\rho_{A_1} + \rho_{A_2})$ in a two state approximation if electronic overlap is neglected; ρ_{A_i} is the charge density if the electron were concentrated on A_i^- or on A_iB^-; that is, this ρ_0 is *distributed equally* between the A_1 and A_2 regions."

Q: Does this $\rho_0 = \frac{1}{2}(\rho_{A_1} + \rho_{A_2})$ refer to a -1 charge delocalized over the whole system in the TS?

M: Yes, that's right, in other words, *in the TS*, if there are strong interactions, things will be distributed equally between the two sides and also some place in the middle, OK? That for adiabatic, and so that would be this $\frac{1}{2}(\rho_{A_1} + \rho_{A_2})$, and then there would be corrections for being in the middle, you see, but we neglect what's in the middle, neglect the electronic overlap, then you have this sum here. So that would be an approximation to the adiabatic charge density in the TS.

7. Following: "The '1' in Eq. (22) refers to the *actual charge distribution*, which, in the present instance, *for the TS* is ρ_{A_1} when the electron is in the vicinity of A_1 and is ρ_{A_2} when in the vicinity of A_2."

Q: Apparently, whereas you have first considered a possible delocalized charge distribution in the TS, now you consider the possibility of localized charge distributions, always in the TS. So, you consider first a delocalized charge distribution and now an *actual* charge distribution.

M: Yes... well I'm just wondering about that... I don't know that I would agree with that statement, I don't think that statement is correct, I think that's a misstatement.

8. Following: "Thus, ρ_{1-0} is approximately $1/2(\rho_{A_1} - \rho_{A_2})$ in the vicinity of A_1 and $1/2(\rho_{A_2} - \rho_{A_1})$ in the vicinity of A_2, for this case, where both A_1^- and A_2^- bear a single negative charge in reaction 3.

$$A_2^- + BA_1 \rightarrow A_2B + A_1^- \quad (3a)$$

$$A_2^{*-} + BA_1 \rightarrow A_2B^* + A_1^- \quad (3b)$$

Thereby, 1-0 corresponds to there being *a hypothetical charge of* $-e/2$ *on* A_1 *and* $-e/2$ *on* A_2."

Q: (1) Is $\rho_1 = \rho_{A_1} = -1$ near A_1 and $\rho_1 = \rho_{A_2} = -1$ near A_2 with $\rho_{A_1} + \rho_{A_2} = -1$?

(2) Is $\rho_{1-0} = \rho_1 - \rho_0$?

(3) Is then $\rho_{1-0} = \rho_{A_1} - \frac{1}{2}(\rho_{A_1} + \rho_{A_2}) = -\frac{1}{2}e$ near A_1?

(4) What do you mean by "hypothetical"? Is it a classical electrostatics way of describing the delocalization of a -1 charge?

M: The $1 - 0$ refers to the difference between the two charge distributions, of the reactants and of the products. Why do I have that? (a hypothetical charge of $-e/2$ on A_1 and $-e/2$ on A_2) Let me think about that now... The key point is that you take the charge distribution of the products, that's 1, and you subtract from that the charge distribution of the reactants, that's 0. In the $1 - 0$ there is the charge of the products, whatever they are, minus the charge of the reactants, those are the charges involved.

Q: So, you have a positive charge on one site and a negative charge on the other.

M: That's correct.

Q: But which is the meaning of this positive charge on one side and negative charge on the other?

M: I don't know, I just was able to show that mathematically it was equivalent, that's all.

Q: Equivalent to what?

M: Equivalent to the final formula that I had initially, although I didn't assume that, it's just a consequence, it is equivalent to that, just a matter of mathematically equivalent.

Q: But physically . . .

M: Yes, physically, physically . . . the solvent in equilibrium with some hypothetical distribution which in, say, the 1965 paper, is something like $\rho_1 + \lambda(\rho_1 - \rho_0)$, the remainder is $\frac{1}{2}(\rho_1 - \rho_0)$, so, yes, it's related to the discussion there and if the daggered charge is in the product it would be $-$, on the other side would be $+$, but it's a hypothetical charge, that just comes out of the formalism, it is not an actual charge, no, the formalism can be expressed in terms of that hypothetical thing. It is not physical, it's hypothetical.

Q: Read your words: "Thereby, 1-0 corresponds to there being *a hypothetical charge of* $-e/2$ *on* A_1 *and* $-e/2$ *on* A_2" So now we have two different hypothetical charges

M: Let me see . . . I don't see how it could be . . . I see your question and I have to think about it. Is that different from what I said isn't it?

Q: Exactly

M: Let me think about it . . . I Think that statement there about a hypothetical charge of $-e/2$ on A_1 and $-e/2$ on A_2 is not correct.

Q: OK but in this case, you have a total physical charge of -1

M: Yes, but I think that statement isn't correct. Because $1 - 0$ is a difference of charge distributions, a difference of products and reactants charge distributions, and in one case you're going to get a $1-0$, and in the other case you're going to get a $0 - 1$, no way would you get a -1, I think that statement isn't correct.

9. Q: At this point you apparently consider *three* different ways to describe the charge distribution on TS:

(1) A -1 charge distributed over the system;

(2) A -1 *actual* charge on A_1^- *or* a -1 *actual* charge on A_2^-;

(3) A $-\frac{1}{2}$ *hypothetical* charge on A_1 and a $-\frac{1}{2}$ *hypothetical* charge on A_2.

Can you comment on the above? I am thoroughly confused...

M: I can see why you'd be confused because I think that what I said is just simply wrong. The third way is wrong, I believe it to be wrong, yes. The first two are OK, but not the third. Because $1 - 0$ is supposed to be a difference of products and reactants charge distributions. When you do that you don't get a 1/2....

Q: Hypothetically correct but from a physical point of view rather mysterious...

M: Yes, it shows that some formalism is equivalent to that, in other words the advantage is that if you want to calculate it for some system you just use those hypothetical charge distributions, you calculate the reorganization energy.

Q: But this hypothetical charge distribution doesn't keep the total charge. Because the total charge is -1 whereas the sum of the hypothetical charge distribution is zero

M: Yes, that's right, the one thing goes from −1 to 0 and the other thing goes from 1 to 0.

10. p. 4075 1st column middle: "For ΔG_r, the free energy of formation of the state (X_1, X_2, Y) from R... we write

$$\Delta G_r = D_1(X_1 - 1)^2 + D_2 X_2^2 + \lambda_o Y^2 \tag{5a}$$

where

$$X_i = \exp(-a_i x_i), \quad \lambda_0 = \lambda_0(X_1, X_2)(i = 1, 2) \tag{5b}$$

and x_i refers to the B − A_i bond displacement coordinate. In Eq. (5) there is a solvent fluctuation term $\lambda_0 Y^2$, in which a generalized fluctuation coordinate Y is introduced whose equilibrium value is *0 for the reactants' state and 1 for the products.*"

Q: (1) Is expression (5a) the expression of ΔG_r in terms of quadratic coordinates to which you were referring and that would assure the abrupt electron transfer? Why the "abruptness" should be assured by these coordinates?

(2) Note that here with "0" and "1" you refer to reactants and products.

M: (1) I don't have any resonance energy in there, so you have one electronic state, you have the other electronic state, then you go abruptly from the one to the other, if we put there a little coupling you wouldn't go abruptly, and you'd have to modify the formalism.

(2) Yes...

11. p. 4084 2nd column top: "For the present purposes, we only need λ_o in the TS region. If one wished to obtain λ_o in *the encounter complex region R, instead of the TS*, one could again use the above ideas to obtain the λ_o. For example, *in the configuration R* the '0' charge density has a negative charge on A_2 and none on A_1."

Q: Please note that here the R configuration is the configuration characterized by $Y = 0$ and by the "0" charge density in equilibrium with it.

M: Yes, that's correct.

12. Following: "In the '1' system, i.e. the $A_1^-BA_2$ system, there is a negative charge $-e$ on A_1 and none on A_2."

Q: Here the system has $Y = 1$ and apparently the charges and nuclear coordinates are in equilibrium, so are we in the "0" situation as far as charge distribution? As you see some confusion may come for the reader from using the designation "0" and "1" for two different charge distributions and for two different Y's.

M: When you have the reactants... when the solvent is in equilibrium with the reactants, there it's equilibrium with the zero charge distribution.

M305 Theory of Rates of S_N2 Reactions and Relation to Those of Outer Sphere Rupture Electron Transfers

(Continuation 4)

R. A. Marcus

1. p. 4084 2nd column bottom: "...the activation energies of the thermoneutral reactions are considerably less than the energy of the bond being broken."

Q: Does this mean that in a reaction with $\Delta H = 0$, such as the classical $H + H_2 \rightarrow H_2 + H$, the activation energy is less than the dissociation energy of the H_2 bond?

M: The answer is yes: it is less by a factor of 10. The activation energy for the reaction is of the order of ten calories and the dissociation energy is of an order of 100. As a rule of thumb, when one bond is breaking and another is forming the activation energy is fantastically less than bond energies. A factor of 10, roughly.

2. Following: "The latter result could be construed as implying an approximate constancy of 'bond order' during the reaction."

Q: Why so?

M: Because you get that factor if you assume that the bond is constant during the reaction in the BEBO formalism.

3. Following: "In the extension here to S_N2 reactions in solutions, a simple functional form ... is adopted, but now no longer involving only bond energies. Electron affinities and solvation free energies also appear, and the 'bond order' is replaced by a reaction coordinate. The three effects (i–iii) in the Introduction are again included, the solvation effect (iii) via a *partial desolvation*."

Q: What is it this "partial desolvation"?

M: Well, you have something that's solvated, then if it's an ion and it's going to form a neutral then it's going to be partially desolvated, by fluctuation. *Or, maybe you form a TS that's partially desolvated.* But what I'm saying is what would happen in forming a TS.

4. Following: "In a BEBO treatment of gas phase reactions the bond order of the rupturing bond in the reactants was denoted by n_1, that of the newly forming bond by n_2 and the energy of the A_iB bond (i = 1, 2) relative to vacuum by V_i. The V_i and n_i were related by

$$V_i = D_i n_i^{1+p_i} \tag{A1}$$

where in the exponent p_i is a quantity whose value is in the vicinity of zero, typically around 0.1. The p_i should be greater than zero, in order that V_i not to be larger in magnitude than D_i in the interval $0 \leq n_i \leq 1$. Johnston defined the reaction path by introducing a constant bond order approximation, namely

$$n_1 + n_2 = 1 \tag{A2}$$

.... The energy of the system along the reaction path *relative to that of the reactants* is then $V_1 + V_2 + D_1$, since $n_1 = 1$ and $n_2 = 0$."

Q: (1) Which is the meaning of p_i which will steadily appear in the following?

(2) Why $V_1 + V_2 + D_1$? The reaction we are considering is:

$$\mathrm{A_2 + BA_1 \rightarrow A_2}B + \mathrm{A_1}$$

M: (1) The meaning of p_i: it's just an empirical parameter, it's related in some way or another to Pauling's ideas. It is discussed in Johnston's book because there is where it all started.

(2) It should be $-D_1$ instead of $+D_1$.

5. Following: "In the case of bond energies and activation energies for neutrals, the smallness of the latter relative to the former in a thermoneutral reaction shows that the p_i in Eq. (A1) must be close to zero: With a TS having $n_1 \cong n_2 \cong 1/2$ with $D_1 \cong D_2$ (thermoneutral reaction) the energy barrier is $-2D_1(1/2)^{1+p_i} + D_1$ and so $(1-2^{-p_i})$ *must be small* and typically p_i is on the order of 0.1."

Q: Please check, I guess $(1 - 2^{-p_i})$ comes from $-2D_1(1/2)^{1+p_i} + D_1 = D_1(1 - 2(1/2)^{1+p_1} = D_1(1 - 2(1/2^{1+p_1}) = D_1(1 - 2/2 \times 2^{p_1}) = D_1(1 - 2^{-p_1})$. Why $(1 - 2^{-p_i})$ must be small?

Q: Are you using the formula with the $+D_1$ instead of that with the $-D_1$? Now I'm using the $+D_1$. What I don't understand is why you have a $-$ sign in front of $2D_1$. You should have a $+$ sign there and a $-$ sign in front of the D_1.

M: OK, good.

M305 Theory of Rates of S_N2 Reactions and Relation to Those of Outer Sphere Rupture Electron Transfers

(Continuation 5)

R. A. Marcus

1. p. 4085 1st column top: "In a BEBO treatment of gas phase reactions the bond order of the rupturing bond in the reactants was denoted by n_1, that of the newly forming bond by n_2, and the energy of the A_iB bond (i = 1,2) *relative to vacuum* by V_i. The V_i and n_i were related by

$$V_i = D_i n_i^{1+p_i} \tag{A1}$$

where in the exponent p_i is a quantity whose value is in the vicinity of zero, typically around 0.1."

Following: "The energy of the system along the reaction path relative to that of the reactants is then $V_1 + V_2 + D_1 \ldots$.

Following: "With TS having $n_1 \cong n_2 \cong 1/2$, and with $D_1 \cong D_2$ (thermoneutral reaction) the energy barrier is $-2D_1(1/2)^{1+p_i} + D_1$ and so $(1 - 2^{-p_i})$ must be small and typically p_i is on the order of 0.1."

Q: As the energy is related to vacuum, I believe that one should write the above Eq. (A1) relation as

$$V_i = -D_i n_i^{1+p_i}$$

Then $V_1+V_2+D_1$ is OK because it is $V_1+V_2-(-D_1)$ and everything falls in place.

M: I think you're right, it should be a $-$ sign.

M308 Source of Image Contrast in STM Images of Functionalized Alkanes on Graphite: A Systematic Functional Group Approach

C. L. Claypool, F. Faglioni, W. A. Goddard III, H. B. Gray, N. S. Lewis, R. A. Marcus

NOTES

1. p. 5991 2nd column bottom: "Using a McConnell superexchange formalism, the electronic coupling between the tip and sample should increase as either the HOMO energy or LUMO energy of the substituted alkane approaches the Fermi level of the graphite substrate.

Q: Is it so because the LUMO and HOMO energies are then more in resonance with the Fermi level?

M: Yes.

2. Following: "At sufficiently large electronic coupling values, the functional group image should become bright in the STM contrast even if topographic effects alone would cause the tip to approach the sample.

Q: Does "bright" correspond (as I guess) to higher tunneling current?

M: Yes.

3. Following: "A simple method for estimating the electronic coupling, and thus for predicting the image contrast of a given functional group in the STM images presented herein, would involve consideration of the ionization potentials of the various molecules of interest. When the coupling is dominated by the HOMO, functional groups with lower ionization potentials (IPs) than the corresponding alkane should appear bright in the STM image, while functional groups with higher ionization potentials should appear dark."

Q: That is, I guess, the lower IPs allow an easier exit for the electrons and so brighter images. Is it so?

M: Yes, when the IP is lower is closer to the band edge for the HOMO. When the LUMO gets into resonance with the HOMO....

4. Following: "...lower IP values generally correlate with a more diffuse orbital structure of the functional groups of concern..."

Q: Intuitively it is understandable: the easier for an electron to escape, the larger its average distance from the nucleus to which it is bound. Is it so?

M: If it is a more diffuse orbital, it means it extends out further, it is easier for the electron orbital to overlap with the ST orbital.

5. p. 5993 2nd column middle-bottom: "...in the regime of linear $I - V$ behaviour wherein the molecule experiences an *equivalent* potential drop regardless of whether the tip of the sample is *biased away from equilibrium*, the electronic coupling is the sum of the coupling matrix elements through the HOMO and LUMO, *regardless of the bias condition*. This summation over all available coupling orbitals necessarily produces a symmetry with respect to bias and provides a natural explanation for the observation of symmetry in image contrast that was observed experimentally for all systems studied."

Q: What does it mean "biased away from equilibrium"?

M: "biased away from equilibrium" I'm not sure what it means. Maybe it means the following: the forth direction vs. that reverse direction, it's like the exchange current, maybe it's just that, equilibrium means the forth and reverse being equal.

6: Following: "In this case, a large enough bias perturbation would produce a differential electronic coupling to *the LUMO with the tip* *negative of the Fermi level* and to *the HOMO with the tip biased positive relative to the Fermi level of the substrate.*"

Q: How is the relation between Fermi level of the substrate, tip positive or negative relative to it and LUMO and HOMO of the adsorbate?

M: I think the Fermi level normally is the level where the HOMO and the LUMO are equally occupied, I think so, because the occupancy in the solid state is some function of $e - e_F$, and so when e is equal to e_F . . . look at the expressions in solid state physics for the occupancy of the valence band and the holes band, it contains the energy of that band minus e_F, the Fermi level, and I think that the two orbitals are equally occupied when the energy happens to be at the Fermi level.

Everything has a Fermi level, the graphite has a Fermi level and has a HOMO and a LUMO, the adsorbate has a Fermi level and it has a HOMO and a LUMO.

Q: So, there are two Fermi levels, the Fermi level of the graphite . . .

M: When the two Fermi levels are equal there will be any net charge going from either to the adsorbate and from the adsorbate to the graphite, to the substrate.

CHAPTER 8

Dynamic Stokes Shift in Solution: Effect of Finite Pump Pulse Duration; Time-dependent Fluorescence Spectra of Large Molecules in Polar Solvents; Remarks on Dissociative Anion Potential Energy Curves for Organic Electron Transfers;Ion Pairing and Electron Transfer

Interviews on M309, M310, M311, M312, M314

M309 Dynamic Stokes Shift in Solution: Effect of Finite Pump Pulse Duration

Yuri Georgievskii, Chao-Ping Hsu, and R. A. Marcus

NOTES

1. p. 7356 1st column top: "The *loss of coherence* between the ground and excited electronic states, which is caused by solute–solvent interactions, is closely related to solvent dynamics."

M: When one excites a molecule, then of course everything should be discussed in terms of wave functions, and if one wants to look at the evolution of the wave functions one should consider that there are various types of relaxation times which involve not just the magnitudes of the wave functions but also their phases. The coherence part involves really the phases of the wave functions, the phases of the transition. I mean, often in the spectroscopic literature you'll see relaxation times divided into T_1 and T_2 and one of those refers to deactivation, loss of amplitude, the other to loss of phase, it is called decoherence. If one excites a molecule in the ground state and the molecule goes to an excited state, eventually the phase of the off-diagonal density matrix sort of disappears because of the interaction with the solvent. So, when you look at the literature and you look at excitations, then look at T_1 and T_2 relaxations, one of them is connected to the loss of coherence, loss of phase.

2. Following: "The photon echo technique, and pump-probe spectroscopy have been used to separate the *inhomogeneous (slow) broadening* of the spectral line associated with a particular electronic transition from the *homogeneous fast electronic dephasing*."

M: The essential point is this: you can have a line width and that line width can either be depending on a single Hamiltonian, you have then just a pure line, or it could be the result of absorption by many different sites, each with their own environment. The *homogeneous fast electronic dephasing* refers to a homogeneous line, to only one environment which is fluctuating, whereas that inhomogeneous broadening is a result of many different environments, each producing its own line, so there is difference between homogeneous and heterogeneous.

3. p. 7357 1st column middle: "In the harmonic oscillator approach (with finite N, which at the end is allowed to become infinite), the solvent Hamiltonian H_g in the ground electronic state can be written as

$$H_g = \sum_j \tfrac{1}{2}\left(P_j^2 + \omega_j^2 Q_j^2\right) + U_g \tag{1}$$

where ω_j, Q_j and P_j are the frequencies, coordinates and momenta of the 'normal modes,' respectively (mass weighted coordinates)."

Q: Whereas it is easy for me to imagine the different geometries associated with the normal vibrational modes of a molecule, I don't have (until now) any experience of possible collective motions of solvent molecules.

M: *This is an artificial model that is commonly used. The solvent really doesn't have normal modes, a solid does, a liquid doesn't,* so this is an artificial model that's widely used in the Physics community, and a number of chemists have taken it up. Of course there are a lot of properties this model won't do, it won't give right entropies, because it's a purely harmonic model, instead of being anharmonic, the main advantage is that you can make all sorts of calculations with it. But it's an artificial model.

Q: I realize you don't like it.

M: No, I mean I've used it but I don't like it, yes. But it has been very widely used. Well, you know, Leggett and Caldeira had back-to-back two classic papers, I think it was in Annals of Physics, some place anyway, I think it was in the 1980s, and those papers have been extremely widely circulated. I think they called it the spin-boson model. The normal modes that are used for the liquids don't have any cavities in them, no holes for ions, they don't have that sort of objects, it's a uniform medium that they use, it's really an artificial model.

4. Following: "When the resulting change of electronic state after an electronic transition leads *only to shifts of the normal modes but not to changes in their frequency*, the solvent Hamiltonian H_e in the excited electronic state can be written as

$$H_e = \sum_j \frac{1}{2}\left[P_j^2 + \omega_j^2\left(Q_j + \frac{c_j}{\omega_j^2}\right)\right] + U_e, \tag{2}$$

where the coefficients c_j uniquely characterize the shifts of equilibrium positions of the normal modes. The difference of the minima of

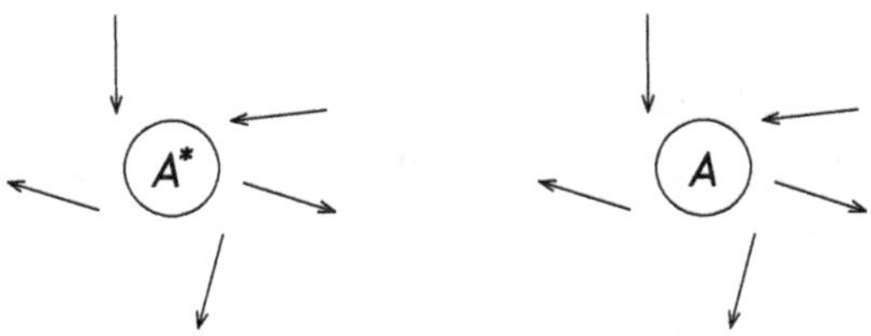

Fig. 1*.

the potential energies $\Delta U = U_e - U_g$ in the excited and ground electronic states, respectively, coincides with the *free energy difference* for this harmonic oscillator model."

Q: (1) When does a solvent become excited?

(2) Look now please at Fig. 1* where I have represented the same solvent molecules distribution around molecule A and around molecule $A*$, that is, molecule A immediately after electronic excitation. Is solvent around A in its ground state and solvent around $A*$ in an excited electronic state?

(3) There are apparently three consequences of a solvent excitation: (i) the shifts in equilibrium positions; (ii) The change in vibrational frequencies; (iii) The $\Delta U = U_e - U_g$. Can you comment and expand on that?

(4) I have supposed that an excited state of the solvent is associated with a nonequilibrium situation. Is my guess right? Which are normally the cases in which a solvent becomes excited?

(5) Why does $\Delta U = U_e - U_g$ coincide with a free energy difference?

M: (1) *Any thermal motion is an excitation of the solvent.* Independently of the solute. If you got a finite temperature, you have the solvent molecules moving, that's an excitation. And you describe that excitation in terms of instantaneous normal modes. So long as you have a finite temperature, they are excited, they are moving, in this model they are part of some vibration. I mean, if this is classical, as long as they are not stationary they are moving, there are some combinations, there is some excitation.

(2) I would say that *the solvent around A is one example of a configuration of thermal equilibrium* and the one for A^* is not in thermal equilibrium, in other words the probability of the one over A would be, say, given by some Boltzmann factor, the one upon A^* would *not* be given by a Boltzmann factor *appropriate* to A^*.

(3, i) You know, first of all, *in the liquid there're really no equilibrium positions as such, just equilibrium positions with respect to each other, or there're equilibrium positions with respect to orientation around an ion, so one has to be more specific.* Let say you are talking about positions with respect to orientations around an ion, OK? Then there is a distribution of configurations, of those little molecules around there, and what's the equilibrium are those configurations that can form with a Boltzmann factor, that's appropriate for a canonical ensemble, so there are an all set of equilibrium positions, many positions whose weight is given by the appropriate Boltzmann factor.

(3, ii) Now the vibrational frequency is in proximity of the molecule or the ion . . . that *changes the vibration frequency* of the, say, localized motion of the molecule. And you see that that's not present in that Hamiltonian you started with.

Q: The proximity with the ion . . .

M: Yes, if you think of a molecule undergoing some sort of vibration, the frequency of that vibration, or the instantaneous frequency, will depend upon its proximity to the ion, the charge of the ion, that's the change.

(3, iii) The $U_e - U_g$ involves partly the change of the electronic state of the ion, I guess, I don't remember how I defined, whether I defined the U as being only the potential energy of the solvent molecules or whether I also included the electronic energy of the ion. If I did, then the $U_e - U_g$ involves difference of electronic energy of the products compared with the reactants, and then also the change of orientations of the solvent molecules, the interaction of the solvent molecules with each other and with the ions are, and all of that is going into $U_e - U_g$.

(4) Yes, that's right. *Just fluctuations permit to become excited.*

(5) Because you consider the difference at the same nuclear configuration, so the distribution of configurations is identical, so the entropy is the same.

5. Following: "A comment is relevant here about the role of intramolecular *solute modes*.... Vibrational modes of such solutes ... must be included in any model. One can then assume that the sum in Eqs. (1) and (2) is *not only over the solvent modes but also over the intramolecular solute modes* ..."

Q: Do we now have a solvent + solute, that is, a solution Hamiltonian?

M: Yes, that's right, and you have some interaction between the two.

6. Following: "To describe the solvent dynamics related to the spectroscopy of the solute it is now customary to treat the energy difference of the excited and ground electronic states as a collective coordinate

$$X = H_e - H_g = \sum_j c_j Q_j + \Delta U + \lambda, \tag{3}$$

where the 'solvent reorganization energy' λ is given in the harmonic oscillator model by

$$\lambda = \sum_j c_j^2 / 2\omega_j^2. \tag{4}$$

The coordinate X corresponds to the optical frequency of the vertical transition *at any specified values of the nuclear coordinates of the solvent.* It can be referred to as a generalized '*solvation coordinate.*' A *similar idea* was used earlier in electron transfer theory."

Q: (1) Do you refer to the nuclear coordinates of the solvent molecules in the neighborhood of the solute molecule being excited and so interacting with this solute molecule? I mean do you refer to the solvent molecules localized in the neighborhood of the solute molecule being excited?

(2) How does the definition (4) of λ compare with your definition of $\lambda = \lambda_o + \lambda_i$? Which are the similarities and the differences? Doesn't it look more λ_i than λ_o?

M: Well, I mean you could do either that or like I did in 1960 without assuming any normal modes of solvent, OK? Then, if you do that using the normal modes of the solvent this is what you get...

(1) All of the coordinates of the system, everyone.

Q: But maybe only the ones in the neighborhood of the solute are important ...

M: No, no, everything. Those that are far away will cancel out in the difference. Everything, the entire system.

(2) Basically, since this is a gross approximation, this λ is in effect really $\lambda_i + \lambda_o$. But you see, they don't have the parallel situation, they don't have ions in there.

Q: It looks pretty much λ_i.

M: Yes, but it's really evolving out of solvent modes, it's also λ_o, *the model doesn't have an ion in there with some localized modes, it doesn't even really have a* λ_i*, but you pretend it does.*

7. p. 7358 1st column middle-bottom: "... $C_{cl}(t)$ is the classical correlation function of the solvation coordinate

$$C_{cl}(t) = \left\langle \Delta X_g(t) \Delta X_g(0) \right\rangle_{cl} \quad (15)$$

In the latter the averaging is over the equilibrium classical statistical ensemble *appropriate to the ground electronic state of the solute.*"

M: That X, that difference $X = H_e - H_g$, was the difference of the two potential energy curves at any given configuration in the 1960 paper. It's distributed according to a canonical ensemble of Boltzmann that you would have if you look at distribution when the solute is in the electronic ground state, it hasn't been excited electronically. It is a canonical ensemble for the electronic ground state.

8. p. 7359 1st column middle: "... we introduce the common assumptions that the pump pulse radiation field $E(t)$ can be described *classically* and that the dipole approximation can be used for its interaction with the solute,

$$H_{int}(t) = \frac{E(t) + E^*(t)}{2}\hat{\mu}, \quad \hat{\mu} = \mu(|e\rangle\langle g| + |g\rangle\langle e|), \quad (26)\text{"}$$

Q: (1) When should $E(t)$ be described classically and when not?

(2) What do the kets $|g\rangle$ and $|e\rangle$ mean?

M: (1) There is a quantum description of the radiation field, and I think that if you want to describe a spontaneous emission correctly, then you have to use the quantum form rather than the classical electromagnetic theory, but the main point is that if you consider just a dipole interacting with electric field, that's a classical description as far as the electromagnetic field is concerned, and if one wants to see the quantum description one would look up a book on the quantum treatment of the electromagnetic field.

(2) Oh, the kets refer to ground state and electronic state of the solute, I guess. If you put on the left side of the kets q^*'s for coordinates, you have $\langle q \mid g\rangle$ and $\langle q \mid e\rangle$, they are identical with the wave functions.

9: Following: "It is also assumed that $E(t)$ has a relatively *narrow spectrum*, so one can write it in a *quasi-harmonic form*,

$$E(t) = E_0(t)\exp(-i2\pi\nu_0 t) \tag{27}$$

where $E_0(t)$ is a function changing slowly with time."

Q: Here the field has just one single frequency and you simulate a wider—if narrow—spectrum multiplying the single frequency exponential by the $E_0(t)$ function.

M: Yes, because $E(t)$ is really a Fourier integral over a whole set of frequencies, but the light spectrum normally has some width and has not just one frequency and you approximate it sometimes with one frequency.

10. p. 7359 2nd column top: "The orientation of the solute influences only the *amplitude* of the corresponding *perturbation Hamiltonian*... as a result the *solute orientation does* not influence *the shape of the transient fluorescence spectrum* but only its *directional properties*"

Q: (1) What is it meant by "amplitude" of a Hamiltonian?

(2) Which is the relation between shape of a fluorescence spectrum and its directional properties?

M: (1) I think that what is meant is that the magnitude of that interaction depends on the angle between the orientation of the molecule and the light, so the amplitude is the magnitude of that interaction. I think that's what's intended, if it is a good use of words is another story.

(2) If you have an oriented molecule, and you shine... then depending on the angle between the orientation of the molecule and the light you get a different absorption sensitivity, the shape of the fluorescence is different at different orientations. The fluorescence behavior changes with orientation.

M309 Dynamic Stokes Shift in Solution: Effect of Finite Pump Pulse Duration

(Continuation)

Yuri Georgievskii, Chao-Ping Hsu, and R. A. Marcus

NOTES

1. (1) Last time you told that when the molecules of a liquid move, the liquid is in an excited state. If it is so, then is every normal liquid excited just for being a liquid?

(2) You also mentioned "random rotations."

M: (1) You know, the ground state for a liquid would be a state with no motion. So, it is not saying much to say that it is in an excited state, there is just motion, that's all.

(2) Yes, if you have a solution and an ion in there, the ion will *orient* the molecules around it to itself, right? So, you have *directed rotations*, they would be oriented, they won't be random, whereas in a pure liquid you get random translations and random rotations.

2. p. 7357 2nd column middle-bottom: "The average value of the solvation coordinate variation ΔX can then be obtained as a linear response to the 'applied external force,' which is a unit step function $-\theta(t)$ ($\theta(t) = 0$ if $t<0$, and 1 for $t>0$),

$$\overline{\Delta X}(t) = -\int_{-\infty}^{t} \alpha(t-t')\theta(t')dt' = -\int_{0}^{1} \alpha(\tau)d\tau. \qquad (8)$$

The generalized susceptibility $\alpha(t)$ is given in linear response theory in terms of a correlation function of the solvation coordinate,

$$\alpha(t) = -\frac{1}{i\hbar}\langle[\Delta X_g(t), \Delta X_g(0)]\rangle \qquad (9)\text{"}$$

M: Well, that is standard quantum nonequilibrium statistical mechanics in the linear regime, that's all part of a very standard formalism.

Here there's a θ function which is unit force, in practice in all books you'll see some applied force instead of a unit force, that's the only difference between this and what is in many of the books... well, there are a number of differences, that's one of them.

Q: $\Delta X = \sum_j c_j Q_j$. How's the time evolution with t? Is it just due to a dependence of ΔX on t through a dependence of $c_j = c_j(t)$ or of $Q_j = Q_j(t)$ or both?

M: That's the way ΔX depends upon time, through that combination of things, yes, that's right. In most formulations the c_j's are constant.

3. p. 7359 2nd column middle: "To find the density matrix of the solvent $\rho(t)$ with the solute in the excited electronic state, second-order time-dependent perturbation theory with H_{int} as a perturbation must be used. Under the *rotating wave approximation*, the expression for $\rho(t)$ is give

$$\rho(t) = \frac{\mu^2}{4\hbar^2} \int_{-\infty}^{+\infty} dt' dt'' E^*(t') E(t'')$$
$$\times\, e^{-i(t-t'')H_e/\hbar} e^{-it''H_g/\hbar} \rho_0 e^{it'H_g/\hbar} e^{i(t-t')H_e/\hbar} \qquad (28)\text{"}$$

Q: I shall read Mukamel's book on nonlinear spectroscopy but can you give in the meantime a short tutorial on this rotating wave approximation?

M: To learn about the rotating wave approximation you don't need to get to the complicated book of Mukamel, you'll see in many places the rotating wave approximation, in one of Feynman's books for example, I think, Feynman and Vernon and Hellwarth wrote some paper involving that, it's a very common approximation, neglecting certain terms of the time evolution.

4. p. 7360 1st column top: "Using Eq. (29) the average value of the solvent coordinate X at time t can be represented in the form:

$$\bar{X}(t) = \frac{1}{\upsilon_0} \int_{-\infty}^{+\infty} \int_{-\infty}^{+\infty} dt' dt'' f(t|t', t'') E^*(t') E(t''), \tag{33}$$

where the integral kernel $f(t|t', t'')$ is given by

$$f(t|t', t'') = Tr\left[X_e(t - t') e^{-i(t'-t'')H_e/\hbar} e^{i(t'-t'')H_g/\hbar} \rho_0\right]. \tag{34}$$

Q: Which is the meaning of the above average? Average over two times variables and t'', one for the electric field and one for its complex-conjugate? Please explain... just recall that

$$X = H_e - H_g = \sum_j c_j Q_j + \Delta U + \lambda$$

M: Well, it may be that one of the integrals just goes from $-\infty$ to t and the integrand is zero afterward. You see, often one turns on the perturbation slowly, one starts off at time equal $-\infty$ and gradually turns on the perturbation. So it's not uncommon, it depends on the actual situation, it's not uncommon to start from $-\infty$, sometimes one starts from 0, it depends on the problem, but often one approach is to gradually start from a Hamiltonian H^0 and slowly bring in the perturbation Hamiltonian. What you have there with the X is very standard in that kind of theory, I mean with that spin-boson Hamiltonian.

5. p. 7360 2nd column middle: "It is important to use a *quantum* correlation function because for a generic nonlinear system, in contrast to a harmonic one, there is no simple relation between the classical and quantum correlation functions."

M: You see in a number of books that you are going to derive certain results, classical results, *for the harmonic oscillator* and depending on what it is you are deriving, you get the same result as for the quantum results. So, for classical and quantum, for certain properties, maybe for average energies and this and that, you get similar results. Of course, time evolution and everything like that are very different, but certain properties carry over, whereas *that's not true for any anharmonic system*. It's not true for anything which is not harmonic oscillator.

6. p. 7361 1st column top: "Assuming that the main contributions to the correlation function $C(t)$ arises from *low frequency modes (classical modes)...*"

Q: Why are the low frequency modes classical?

M: Well, think of the partition function for a harmonic oscillator, when $h\nu$ is smaller than kT then that goes over to a classical partition function.

7. p. 7361 1st column middle: "This function describes a DSS which corresponds to a pulse which is much longer than the correlation time τ_c, but still shorter than *any other time scale, relevant to solvent dynamics.* "

Q: Can you say something about these other time scales relevant to solvent dynamics?

M: Yes, I know several, one is the time scale associated with Debye relaxation, another is the time scale associated with vibrational bending relaxation, and there are, maybe, some longer ones still associated with some of the hydrogen bonding arrangements, but there's certainly the main relaxation in water which at lower frequencies it is due to rotational relaxation, Debye relaxation. And I think the next one that comes up is probably the bending vibration. Those are some of the time scales. Now, the main time scale is that of Debye, but it may be that the vibrational polarization also plays some sort of a role. And each has, so to speak, characteristic time scale, ballpark.

8. p. 7362 1st column top: "for a solute with a non-Gaussian absorption spectrum"

Q: When does a solute have a Gaussian and when a non-Gaussian absorption spectrum?

M: Well, there are many inorganic studies... if you look at some inorganic complexes, you'll see that they have very definitely a non-Gaussian spectrum, you have a number of lines, typically 0 to 1, 0 to 2, 0 to 3, and so on, vibrational transitions, and you'll see them described in terms of a Huang–Rhys formula. So, many absorptions are highly non-Gaussian, there are some big organic molecules which have a log Gaussian type, a log normal type distribution, so it's a distribution... it is a Gaussian not of the ordinary functions but for log functions. Having these non-Gaussian distributions is not uncommon.

Q: Is there some reason for having a Gaussian or non-Gaussian distribution?

M: Yes, sure, the Gaussian distribution is sort of a matter of the central limit theorem. If you have a set of small systems that behave somewhat similarly, then the distribution goes over to a Gaussian. On the other hand, if you have a process where the fluctuations are very very rapid, then that usually gives rise to a Lorentzian distribution, so there are all sorts of situations where you don't get a Gaussian. But where the system is sampling many different sort of configurations, each giving rise to a spectral line, there having a Gaussian is quite common, but you also have these other distributions.

9. Following, middle: "... the low-frequency, diffusional part of the solvent's electric response... the high-frequency, inertial part"

Q: I believe the response is the one to an electromagnetic oscillating field. Why is the low frequency associated with the diffusional part of the solvent's response and what is the "inertial part"?

M: They are all related to the electromagnetic oscillating field, and the low diffusion is, you know, of molecules migrating around like rotational diffusion in the case of the Debye relaxation that occurs at relatively low frequencies, whereas strictly vibrational motions—and some of that is a very shallow part there—that is at very high frequencies.

Q: Inertial is vibrational?

M: Inertial if you have inertial term in your equations of motion …I mean, you have second derivatives of position with respect to time that describes the inertial aspect. When you are in the so-called overdamped region, then the equation which contains that inertial term is damped out and you only have a first order derivative. So, there are many aspects of the problem, Langevin equation, dynamics…related to the presence or the absence of this inertial term. When you have the inertial term, typically your correlation function for the velocity as a function of time would be a damped-out oscillation, and if you don't have the inertial term it will typically be almost an exponential decay. So, all of this is connected to nonequilibrium statistical mechanics.

10. p. 7382 2nd column middle-bottom: "A relative contribution of the ultrafast component to the correlation function $\Delta_1(t)$

$$\Delta_1(t) = \frac{Re\,[C(t)]}{C(0)} \tag{52}$$

is larger due to the 0 K fluctuations of the quantum modes, and *the oscillations with the period 10–15 fs are much stronger.*"

M: First of all, anything that says "fast" has a high frequency, so it's a quantum mode: $h\nu > kT$, and in fact maybe much greater than *kT*. So, it's a quantum mode, it's ultrafast, very fast, and now the oscillations with period 10–15 fs or stronger I don't know, much stronger that, what?

Q: It is not specified in the paper

M: Yes...I don't know...I have to look at the paper and see what was intended. But there is probably some missing word there. Probably Georgievskii did the writing on this and, you know, his Russian English may not be perfect.

11. p. 7363 2nd column top: "The probability distribution W_j of the coordinate Q_j for a single oscillator in equilibrium with the ground electronic state of the solute (the statistical state of the system before excitation) is given by

$$W_j(Q_j) = \sqrt{\frac{\omega_j \tanh(\hbar\omega_j\beta/2)}{\pi\hbar}} \exp[-\tanh(\hbar\omega_j\beta/2)\omega_j Q_j^2/\hbar]. \tag{62}$$

In the high temperature (low frequency) limit this distribution reduces to the classical one:

$$W_j(Q_j) \cong \sqrt{\frac{\omega_j^2\beta}{2\pi}} \exp(-\beta\omega_j^2 Q_j^2/2, \quad \hbar\omega_j\beta/2 \ll 1. \tag{63}$$

In the low temperature (high frequency) limit it reduces to the probability distribution corresponding to the ground state of the oscillator:

$$W_j(Q_j) \cong \sqrt{\frac{\omega_j}{\pi\hbar}} \exp(-\omega_j Q_j^2/\hbar), \quad \hbar\omega_j\beta/2 \gg 1, \tag{64}$$

which is much broader than the classical distribution when $\hbar\omega_j\beta/2 \gg 1$."

Q: Why is the high temperature associated with the high frequency and the low temperature with the high frequency?

M: Well, you should really consider the ratio $h\nu/kT$. It becomes low either if you go to high temperature or if you go to low frequency. It is really not the low temperature that produces the high frequency. So, either you have a very high frequency or you have a very low temperature, it is the ratio that counts.

12. Following: "The total distribution of all oscillators representing the solvent is given by the product of the distributions in Eq.(62)

$$W(\mathbf{Q}) = \prod_i W_j(Q_j) \tag{65}$$

The maximum of the absorption spectrum corresponds to $Q_j = 0$."

Q: $Q_j = 0$ for each oscillator I guess?

M: Yes, that's how you define your coordinate, so that the equilibrium position of each is chosen to be zero. You could choose the equilibrium position of each to have a different value for everyone, it doesn't matter, but you keep the formula as simple as possible, you don't lose anything in generality, so you just write $\omega^2 Q^2/2$ for the energy.

M310 Time-Dependent Fluorescence Spectra of Large Molecules in Polar Solvents

Chao-Ping Hsu, Yuri Georgievskii, and R. A. Marcus

NOTES

1. p. 2658 1st column top: "The dynamics of polar solvents has been studied in *charge redistribution processes* in many chemical reactions and photoinduced processes."

Q: (1) Can one say that the bimolecular electron transfer and the unimolecular electron transfer are particular cases of the more general charge redistribution processes?

(2) Does a general theory of charge redistribution processes exist of which the original ET theory is a particular case?

M: (1) A charge redistribution process can just be an electron transfer or it can be exciting a molecule that has a relatively no dipole to an excited state that has a dipole moment, or vice versa, so all of those are examples of charge redistribution processes.

Q: So, the concept is more general than ET

M: Yes, sure, like forming a polar state where you just have a dipole nonzero of an excited state of a molecule that either has no dipole or a different dipole.

Q: Is there some kind of general theory?

M: Well, you know, the theory for the actual charge transfer is different, if you are exciting to a polar excited state, unless it is a charge transfer state, but in general it isn't, so it is some different theory, you don't have that *complete transfer* of the charges, that *response of the medium*....

2. p. 2658 2nd column middle: "The short-time solvation dynamics has also been interpreted in terms of an *instantaneous normal modes* analysis of molecular dynamics simulations."

Q: (1) How can a motion to which a mass and an inertia is associated be instantaneous?

(2) How do the masses come in here?

M: Yes, well, some people said: let's treat the solvent as consisting of normal modes, but of course normal modes only really apply to a solid, not to a solvent, so they say: OK, we'll use *instantaneous* normal modes, that change with time, I don't know if that's a fruitful concept though.

(1) Well, suppose you have the liquid in one state, OK? And suppose in terms of a Fourier analysis that represents instantaneous motion in terms of the decomposition in terms of normal modes, so it's like taking any function and expanding in a complete set. Suppose only that you have chosen to use normal modes as your complete set.

Now, in a later time you then do the same and you find that you have to use different normal modes, and so the situation is quite different from the normal modes in a solid where you use the same normal modes for all time, it is just that you are trying to represent something which is not really a normal mode by normal modes, so in time you have to change normal modes and so you take *instantaneous normal modes*.

(2) Well, just think of how you do the normal modes of a molecule, the way the masses come in, the masses come in various combinations, depending on the normal mode, same here. Only here you are doing an instantaneous description, and at later time you have to use a different description, unlike the normal modes of a molecule.

M310 Time-dependent Fluorescence Spectra of Large Molecules in Polar Solvents

(Continuation)

Chao-Ping Hsu, Yuri Georgievskii, and R. A. Marcus

NOTES

1. p. 2659 1st column top: "The $p_1(\omega)$ describes the *probability distribution of the polar solvent configurations* which have a given energy difference $\hbar\omega$ between the two states *of the solute molecule, sampled from the polar solvent configurations in thermal equilibrium with the ground electronic state of the solute molecule.* Such an energy difference of the two *solute* electronic states is assumed to arise from the *polar solute–solvent* interaction."

Q: (1) If I understand it correctly, you consider the energy difference $\hbar\omega$ between ground and excited state of the polar solvent around the solute molecule in solution. In the ground state the polar solvent is in equilibrium with the ground electronic state of the solute. When the solute is suddenly excited, the solvent is not anymore in equilibrium with the solvent configuration at time $t = 0$ and the excited solute molecule is surrounded by solvent not in equilibrium with it but still in equilibrium with the ground electronic state of the solute molecule. Now please explain this story of the sampling and how you calculate $p_1(\omega)$.

(2) It is not clear if the above energy $\hbar\omega$ refers only to the difference of solute–solvent interactions or to such difference plus the electronic states difference of the solute molecule as if it were in the gas phase.

(3) I believe that the nonpolar interactions contributions to the above energies are not considered in the above discussion.

M: If you look at all the possible polar solvent configurations, there are some for which the energy difference between solute in the excited state and in the ground state equals a certain energy. A whole group of them is of equal energy, a whole group of others would involve different energy difference, a whole group of others still a different energy difference.... In that 1960 ET paper, I introduced a coordinate Γ, it was a difference of the energies of the upper state and the lower state, and if you have an n-dimensional space where n is, say, 10^{23}, then there is an $(n-1)$-dimensional space such that all points in it have the same value of that Γ, a huge space, you can calculate the free energy associated with that space, and you can then plot the free energies and that's how you obtain the parabolas. In this paper, the energy difference of the whole system and the entire surrounding in the final and the initial state is providing the same role that the Γ did in the 1960 paper.

(1) The sampling is made in the way one samples any equilibrium configurations. One samples all of the solvent configurations that are in equilibrium with the ground state of the solvent molecules and then one considers all points in the phase space samples that have the same energy difference. So, it is an ensemble of configurations

that are such that all of them have the same energy difference and so one is sampling a large number of configurations that have the same constant energy difference. The initial energy is a thermal distribution of energies, so one can calculate a free energy associated with that distribution. They just have a one additional property, that they all have that same energy difference, and then in this way one defines a new coordinate, *an energy difference coordinate*. One is sampling a space of $n - 1$ coordinates, because *one coordinate is being held fixed. It is the energy difference coordinate, you see.*

Q: Is the reference energy the energy of the ground state?

M: Well, not exactly because one is sampling the whole set of solvent configurations in equilibrium with the ground state, so it is not as though the energy of the ground state is the reference. One is sampling all energies of the systems in equilibrium with the ground state. The fact that the ensemble of all systems in which the solvent is in equilibrium with the ground state has an *average* energy just says that you're dealing with an enormous thermodynamic system which has an average energy but it also has an entropy and so on. Having an average energy is no more important than having an entropy, because in both cases you are using some statistical mechanical way of calculating the average energy or calculating the entropy.

(2) It will normally be the entire system. The difference of the entire system of the solute . . . well, we don't have the solute interacting with itself, we have dilute solutions of solute. So, it's always the solute interacting with the solvent.

Q: But you are writing there the electronic energy of the solute.

M: You have it in there, that's right, into the electronic energy of the solute you have the solute interacting with everything in its ground state and now if you go to the excited state having a constant value of the coordinate Γ with the same energy difference from the ground state, you have the electronic energy of the excited state of the solute and its interactions with all of the solvent in the configuration that you started with the solvent and you are sampling all the solvent configurations.

Q: So, from all possible configurations in equilibrium with the ground state you sample those that have an energy $\hbar\omega$.

M: The $\hbar\omega$ is the difference between the upper state and the ground state where in the upper state you calculate all of the interaction energies and in the ground state one calculates all of the interaction energies, and so the total energy in the excited state minus the total energy in the ground state, that total energy difference is the total energy above minus the total energy below for the given value of that energy difference.

Q: But the molecules at the beginning of the experiment are not in equilibrium with the solute in the excited state.

M: That's correct, they are in equilibrium with the ground state. When *you form an excited electronic state of the solute then a nonequilibrium situation arises.*

(3) When one introduces a specific model, one doesn't include the nonpolar if one focuses on the polar. One should include the nonpolar interactions, but they are normally small. They can be van der Waals interactions, that could be different in the excited state compared with the ground state, but they are normally small in comparison with the polar solvent effects, when the polar solvent effects are large. When the polar solvent effects are not large, and so when one is dealing with a lot of nonpolar solvents that are different from each other, then of course one has to include the nonpolar effects, interactions, those nonpolar interactions are different in the excited state compared with the ground state.

Q: Are they contained in the absorption spectrum?

M: The absorption spectrum would include a difference in the electronic energies of the solute and a difference of interaction energies of the solute with the solvent, whether it is polar or nonpolar.

Q: And the same for the emission spectrum.

M: The same for the emission spectrum.

2. Following: "Thus, $p_1(\omega)$ would have been the *absorption line shape in polar solutions*, if there had been *only a difference in polar interactions* in the ground- and excited-states Hamiltonians."

Following: "One might imagine that an extension of eq 2 for a phenomenological description of the time-dependent fluorescence can be written as

$$F_P(\omega, t, \omega_{ex}) = \omega^3 \omega_{ex} \int_{-\infty}^{+\infty} d\omega' \int_{-\infty}^{+\infty} d\omega'' \times g_{np}(\omega_{ex} - \omega'') f_{np}(\omega - \omega') p(\omega', t; \omega'') \quad (5)$$

where $p(\omega', t; \omega'')$ is the time evolution of a probability distribution for the energy difference of the two states of the solute that have a energy difference $\hbar\omega''$ at $t = 0$"

Q: Does this mean that the product $g_{np}(\omega)p_1(\omega)$ represents the absorption line shape which takes care of both polar and nonpolar interactions, that is, of all solute–solvent interactions?

M: You introduce a coordinate Γ which is a hyperplane, there is a way of defining a coordinate in a 10^{23}-dimensional space, and I did that in the 1960 paper, and is being done here also. Instead of calling it Γ is called ω. Now, that coordinate changes with time, because the system relaxes, so the system moves from one value of that coordinate in the 10^{23}-dimensional space to another value of that coordinate, OK? It's like moving down a parabola, that is what it really is, what it really amounts to, in the one-dimensional case, and that new value of the coordinate it's what's called ω', right? So, you're moving from one value of the generalized coordinate to a new value of the generalized coordinate, and you are integrating over all possible values, from the old value of the generalized coordinate to a new value of the generalized coordinate.

Q: And you get the fluorescence.

M: OK, fluorescence coming from the new values of the coordinate, yes, and you're integrating the fluorescence over all new values of the coordinate. Yes, because you go to all new values of the coordinate and you are fluorescing from all new values of the coordinate.

3. Following: "which then drifts to $\hbar\omega'$ at time t, if only polar solute–solvent interactions were included. This drift in the energy difference is due to the difference in charge distribution of solute in the two electronic states."

Q: This drift is evidently due to polar solvent adjusting its configuration ...

M: Precisely.

Q: To minimize the interaction free energy.

M: Well, in response from that I suppose that, in terms of a one-dimensional coordinate, it is going toward lower free energy to minimizes the free energy, typically moving on a potential energy surface, we are assuming that. The net result is for that generalized coordinate to correspond gradually in time to a lower free energy.

4. Following: "Thus $p(\omega', t; \omega'')$ can also be regarded as the time evolution of the emission spectral line shape (with emission frequency ω') when the pump frequency is ω'' for the two states solute if there were only the polar interaction with the solvent. The desired properties of $p(\omega', t; \omega'')$ are

$$p(\omega', t = 0; \omega'') = p_1(\omega'')\delta(\omega' - \omega'') \tag{6}$$

$$\lim_{t\to\infty} p(\omega', t; \omega'') = p_1(\omega'')p_2(\omega') \tag{7}$$

.... The $p_2(\omega')$ denotes the *equilibrium* probability distribution of energy difference (spectral shift), *sampled* from solvent configurations in thermal equilibrium with the *excited*-state solute charge distribution."

Q: Please explain... once more the sampling, this time from configurations in equilibrium *with the excited state.*

M: The excited state of the solute that is being photoexcited has an electric charge distribution that is different from that of the ground state of that solute and so its interactions with the surrounding solvent is different. Thereby, if one photoexcites this solute molecule by photoexcitation to some electronically excited state, that state will not be in thermal equilibrium with the surrounding solvent molecules, and some relaxation of the solvent will occur so as to be in thermal equilibrium with the new electronic distribution of the light-absorbing molecule. That is, the thermally equilibrated configurations of the solvent when the solute is in an electronically excited state differ from those when that solute is in its ground electronic state.

5. p. 2660 2nd column middle: "The solvent part can be considered *composed* of both nonpolar and polar interactions. The nonpolar interaction arises, in part, from any difference in size *or shape* of the wave functions of the two electronic states, and the polar part arises from the electrostatic interaction of the solvent polarization with the different *charge distribution* of the ground and excited states of the solute molecules."

Q: Different charge distributions are represented by wave functions of different shapes. Does this mean that they have different nonpolar interactions besides the different polar interactions?

M: "*The nonpolar interaction arises, in part, from any difference in size or shape of the wave functions of the two electronic states*" Yes, that determines the repulsion energy difference.

Q: Different charge distributions are represented by wave functions of different shapes.

M: Well, they are represented by the squares of the wave functions.

Q: Does this mean that they have different nonpolar interactions besides the different polar interactions?

M: Yes, but not due to their charge distribution, it's due to the fact that they have different wave functions, so they extend out, you know.

Q: So, they are both polar and nonpolar.

M: Yes, they are both polar and nonpolar,*the charge distribution is a first order quantity, and the polarizability is related to a sort of a second-order quantity.*

6: p. 2664 1st column top-middle: "As the techniques advanced, the dynamics in the femtosecond time regime became observable, and so a relaxation in the *intermolecular vibrations* (e.g. the *librational mode* for water.)

Q: How is it?

M: Well, that's responsible for the dielectric constant in the infrared being around 4.5 instead of about 2. About 2 would be the electronic polarizability of the water at frequencies well below the ultraviolet, well below the visible, when you get down to the infrared region, then the dielectric constant of water goes to about 4.5, and that's primarily due to that bending vibration.

Q: But which kind of motion is it this librational motion?

M: Libration is an anharmonic kind of oscillation. It is a kind of vibration but it is not a simple harmonic vibration. You know, if you have a water molecule tied to a hydrogen bond, it distorts the hydrogen bond as it moves, but the motion is not that of a harmonic vibration.

7. Following, middle-bottom: "...the assumption of fast relaxing internal modes made in the present study. The assumption itself provides a major simplification and permits the simple application of expressions for a *two-level* problem to a real system."

Q: Why are fast relaxing internal modes and two-level problems related?

M: Well, if you look at the vibrations, then you have additional vibrational states of the system. So, first you just have two electronic states, for each electronic state you have a whole group of vibrational states, and you have relaxing from each of those vibrational states. Once all those vibrational states of high frequencies relax, then you're only left with a two states system which is interacting with all of these anharmonic oscillations of the solvent. Strictly speaking the solute molecule can have a whole group of different vibrational states and each of those vibrational states can be interacting with the solvent, somewhat differently than the ground vibrational state, in that respect the problem becomes a many states problem.

8. p. 2664 2nd column top-middle: "...the *deconvolution* of an instrumental response function has been performed on the raw emission intensity data."

Q: How does one do these deconvolutions?

M: You get a certain signal, part of the signal is due to the machine itself, part of it is a legitimate signal, so you have to somehow extract the signal you want from that combined signal, basically that's the idea.

9. Following, middle-bottom: "...when the *neutral* dipolar solute was modeled as an ion..."

Following, bottom: "...when the neutral dipolar C153 solute is hypothetically modelled as an ion..."

Q: How can one model a neutral as an ion?

M: That statement "modelled as an ion" is not correct, is modeled as a *pair of ions*, an ion has a charge, a neutral doesn't, so you need a pair of ions to model it.

M311 Remarks on Dissociative Anion Potential Energy Curves for Organic Electron Transfers

R. A. Marcus

NOTES

1. p. 859 2nd column middle: "In addition to the *vertical attachment energy* there is also the *thermal dissociative attachment energy (DAE)*."

Q: In the paper you don't define DAE. What is it?

M: Well, it could be the difference between the bottom of the well, it could be the energy that you would need in order to thermally dissociate, so that could be the energy between the parallel lines just to the left to the curves in Fig. 1a, the energy to reach the crossing section, the crossing region in a thermal reaction. You see where the two curves cross in Fig. 1a. To go from the bottom to the crossing point is the energy you need in order to have a thermal detachment energy. From the bottom of the lowest curve to the intersection, if you go there, then you see that you get the attachment. You go to the final product in the lowest state there. So that's what it is. In other words: reach the crossing point, because that's what you do if you are doing it thermally.

2. Following: "For a molecule whose vibrational quantum states are *thermally distributed* at a temperature T, the DAE is typically shifted to lower electron energies *relative to the vertical attachment energy* (VAE)"

M: Typically, you're always going to the same crossing point, and if you are starting at a higher vibrational energy you need a lower energy to reach that crossing point.

3. Following: "a competition after the initial electron attachment, occurs between the re-emission of the electron and the reaching of the intersection region x_C. Systems formed by a *vertical attachment* closer to the intersection, i.e., at larger RX bond amplitudes, have a greater chance of surviving the competition of the re-emission."

M: Yes, because during the time that it spends going down that hill, going down that slide, it could be *reemitting*, if it is formed closer to that crossing point, now closer to the end, there is a fair chance of reemission.

4. Following: "Since they have a smaller value of $V^-(x) - V(x)$, they are formed at lower electron energies. Thereby, the DAE is shifted to electron energies lower than the VAE."

M: $V^-(x)$ is the highest one and $V(x)$ is the lowest one.

Q: If you go to the left of the crossing point, to the right it's the opposite.

M: The vibrational quantum states are thermally distributed: in other words they would be starting at higher than the minimum, before the vertical attachment energy, you're getting closer as you go up, as you go from one vibrational state to another when vibrational states aren't strong but think of them as parallel lines on the lowest curve, so you are going up, you're going to higher ones, you're going to have smaller vertical attachment energies, OK? Then they are also going to be closer to the crossing point, so the DAE is being shifted to lower energy, to lower distance to go to the crossing point. And now the "relative to the vertical attachment energy." I don't know why I would say relative to, I mean it doesn't have to be relative, the point is just that as you start from higher vibrational levels, you have quite less energy to reach that intersection, and of course also the vertical attachment, so I don't know why one would bring in relative to the vertical attachment energy. There maybe is some reason, but at the moment I don't see it.

It can reemit or it can reach the intersection, that's a competition. And if it reaches the intersection, it could either go over and not undergo electronic transitions it would just continue on the same curve, or it can actually undergo electronic transition to the curve that is the lowest curve there. *Anytime you get an intersection you get kind of a bifurcation, you can go one way or the other.*

Q: And now what?

M: *And now go on until the end... the DA is the thermal dissociated attachment energy* of the intersection.

M311 Remarks on Dissociative Anion Potential Energy Curves for Organic Electron Transfers

(Continuation)

R. A. Marcus

NOTES

1. p. 858 1st column: "Reaction (1) is followed by reactions of $AR^{\bullet}$.

$$A^{\bullet -} + RX \rightarrow AR^{\bullet} + X^{-} \quad (1)$$

$$A^{-} + RX \rightarrow AR + X^{-} \quad (2)$$

Competing with these reactions are the bond dissociative electron transfers ....

$$A^{\bullet -} + RX \rightarrow R^{\bullet} + X^{-} \quad (3)$$

$$A^{-} + RX \rightarrow A^{\bullet} + R^{\bullet} + X^{-} \quad (4)$$

Q: (1) I guess reaction (3) should be $A^{\bullet -} + RX \rightarrow A + R^{\bullet} + X^{-}$.

(2) Looking at the above reactions we see that there is (obviously) charge conservation. But it appears that there is also an *unpaired electron conservation*. Is it always so, I mean is this a physical law?

M: You know, it depends on the system...

Q: So it is not a general law.

M: No, no, think that you have some reactions that violate singlet triplet conservation rules, so they wouldn't have same number of paired electrons. Any reaction which opens up a bond to form two radicals clearly doesn't concern pairs, and there are a lot of those reactions.

Q: In reaction (4) you open up a bond and you have two unpaired electrons

M: Yes, right, I was talking about a simple bond fission, not with an A^{-} in there,

Q: In this case you can go from zero to two unpaired, but one unpaired to one unpaired? Like in Eq. (1).

M: I see.

Q: A is a molecule, not a radical. Of course, you can go from no unpaired to two unpaired if you break a bond, like in Eq. (4) but in Eq. (3) the number of dots is conserved.

M: I mean, certainly there would be the tendency of conservation of bond order in reactions, approximately, according to Pauling's rule, because there is a least barrier, and in fact that was behind Harold Johnstons's book on estimating activation energies in reactions, some Pauling's bond order bond length rule... So, I mean, there is probably a tendency, other conditions being equal, to conserve the number of bonds. But here you are actually breaking a bond, aren't you? Yes, in Eq. (4) you're right, you can have either zero unpaired or two unpaired electrons, but then in Eqs. (1) and (3) you conserve the unpaired electron, you exchange both charge and unpaired electron.

2. Following: "In the present paper we consider information that can be inferred about these anion potential energy curves from data on electron–molecule collisions, Eq. (5)

$$e + RX \rightarrow RX^{\bullet -} \rightarrow R^{\bullet} + X^{-} \tag{5}$$

These data include the *vertical electron attachment energy* E_{V}, the *activation energy*E_{a} for the *thermal dissociative attachment*"

Q: I believe that:

(1) The vertical electron attachment energy E_{V} is the vertical energy difference in going from the minimum of the *RX* potential energy curve to the energy value on the RX^{-} potential energy curve corresponding to the same *R–X* bond length;

(2) The activation energy E_{a} for the thermal dissociative attachment is the activation energy of the reaction

$$RX^{\bullet} \rightarrow R^{\bullet} + X^{-}$$

M: (1) Yes.

(2) Yes, that's right, at the intersection of the two curves you have a nonadiabatic transition there.

3. Following: "These data help characterize the *dissociative anion potential energy curve*"

Q: Is this curve the one, in Fig. 1, panel (a), having $R^{\bullet} + X^{-}$as asymptote or even the curve with the same asymptote in Fig. 1 panel (b) and the curve above it having $R^{-} + X^{\bullet}$ as asymptote? I think you have taken those curves from Saveant.

M: That's right.

4. p. 859 1st column top: "For purposes of the present discussion we focus on the two cases in Fig. 1. For the potential energy curve V_{RX} for RX as a function of the RX bond distance x a Morse curve will be assumed, Eq. (6), where x_{o} is the equilibrium value of x in the RX molecule. For the anion in Fig. 1a a repulsive type of Morse curve $V^{-}(x)$ will be assumed. For the moment we write it as Eq. (7), where E_X is the electron affinity of $X^{\bullet}$....

$$V(x) = D(1 - e^{-a(x-x_0)})^2 - D \tag{6}$$

$$V^{-}(x) = D^{-}e^{-2a(x-x_0)} - E_X \tag{7}$$

.... From eqns. (6) and (7) E_{v} is given Eq. (8)....

$$E_v = V^{-}(x_0) - V(x_0) = D^{-} + D - E_X \qquad (8)\text{''}$$

Q: The role of D in Eq. (6) is played by E_X in Eq. (7), and Eq. (7) *defines* D^{-}. I have tried to visualize the meaning of D^{-}in a figure. Consider Fig. 2* where I have indicated with x_0 on the x axis the minimum of the $V(x)$ curve. From Eq. (7) we can derive D^{-} as

$$D^{-} = E_v + (E_X - D)$$

which is indicated on Fig. 2*. On the other hand, we have from Eq. (7) that

$$V^{-}(x_0) = D^{-} - E_X = E_v + (E_X - D) - E_X = E_v - D.$$

The value $V^{-}(x_0)$ is also shown in Fig. 2*. Eq. (7) *defines* D^{-}.

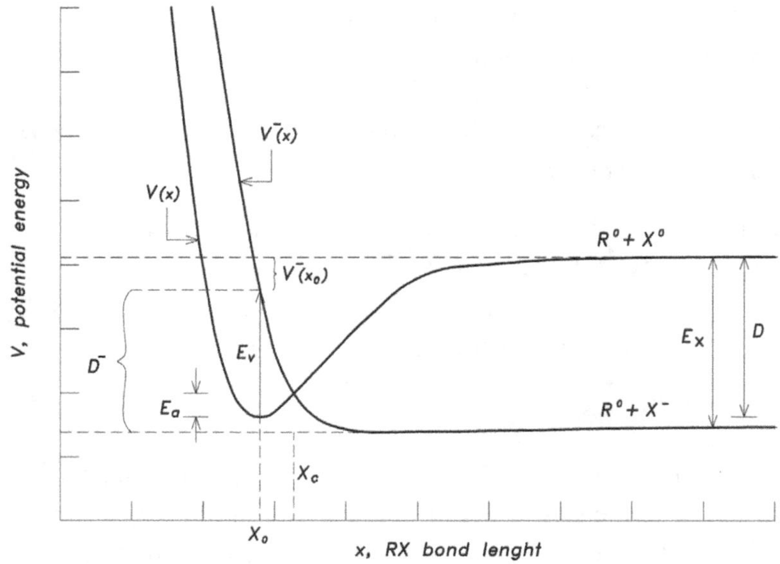

Fig. 2*.

M: I see, you have a D^- in Fig. 2* do you? Yes, that can well be. OK. Yes, that's right what you have there.

Q: The same I did for $V^-(x_0)$.

M: Yes, that's fine, that's OK what you have there, OK, all right.

5. p. 859 2nd column, bottom: "The data on activation energies for *thermal electron transfer dissociative attachment* fall, more or less, into two classes: in one class of reactions the activation energy is approximately equal to $D - E_X$. In this case an interpretation is that there are two RX^{*-} states, as in Fig. 1(b), one with the charge centered on the R (typically when R is an aromatic) and the other with the charge centered on the X. For C_6H_5Cl, for example, an initial vertical attachment to form the first state then leads, after crossing of an intersection, to the second anionic state. When both intersections are lower in energy than the asymptote of the $R^\bullet + X^-$ curve, *the rate-determining transition state is somewhere along or near the* *asymptote*. Thus, for this class, we can understand from Fig. 1(b) why their E_a values are approximately equal to $D - E_X$."

Q: (1) Apparently you are describing the reaction

$$e + RX \rightarrow RX^{\bullet -} \rightarrow R^{\bullet} + X^{-}. \tag{5}$$

Making use of the potential energy curves in Fig. 1(b), one starts at the bottom of the $V(x)$ curve representing the *RX* molecule in its ground state and at equilibrium bond length and one moves successively through two curve-crossings to the curve having the $R^{\bullet} + X^{-}$ asymptote. In order to have the dissociation of $RX^{\bullet -}$ one moves from the minimum of the curve toward the asymptote and to do so one needs to climb up the curve from the minimum to the asymptotic energy and the activation energy is just E_a shown in the figure.

(2) How does one describe the process of the electron approaching *RX* and then attaching itself to it? The PECs are only useful to describe the process (5) only after the first arrow.

(3) "The rate-determining transition state is *somewhere along or near the asymptote.*" How does one determine the TS region in this case?

M: (1) You go to an intermediate RX^{-} and that intermediate is in middle curve, so to speak, in Fig. 1b, so you're going through it, then you're passing through it, and then you're going to a crossing point.

Q: You have really three crossing points.

M: Yes, you are going to the highest crossing point, because that's the intersection between... oh, I see, if you are on RX^{-} then you go to the middle curve, so you first hop over to it and you hop over to the curve which is $R + X^{-}$. So, you actually hit two crossing points.

Q: OK, but don't you have another recrossing with $R^{\bullet} + X^{-}$?

M: You go from the bottom of the *RX* curve and then you hit the first crossing, then you continue along yet, and hit the second crossing.

Q: So, two crossings

M: Yes, two crossings, then you go by the lowest crossing there.

Q: So, I had correctly understood.

(2) First: are you talking about gas phase or solutions?

Q: Here you are in gas phase.

M: Well, first of all there is a whole theory for reactions of electrons with molecules, and the theory depends on a lot of details, you know, there is electron scattering theory, and there is electron molecule reaction theory, so, you know, one solves Schroedinger equation for all of that, so there is a lot of work going on, it's not just simply a sort of transition state theory. In that, under certain circumstances, there is some simplicity. There is a lot of study of ion–molecule reactions... a huge number of those studies beginning with Langevin in 1909 or something, with the introduction of the centrifugal potential, the charge polarization, the $1/r^4$ potential, the $1/r^2$ potential, and you calculate the rate from all that. In some of these reactions the situation would be complicated because it's not sufficient to pass over the centrifugal barrier, but there may be another barrier, that barrier may be inside, you would get first sort of cross sections and then one might be able to combine everything in a kind of transition state theory, I've never done it though. So, the main thing is that there are theories for electron–molecule collisions and reactions, you know, probably the book of Mott and Massey of 1932 might have some of those in it but, you know, there is a huge body of research done since.

Q: You quote the book of Drukarev.

(3) In that case there may be this charge polarization in which an electron is reacting with something, that's a long range interaction, that's an asymptote. In that case if you have, after this charge polarization, a potential maximum and after that the potential is sort of

going downhill, the electron gets trapped in the potential energy surface, then the first part is the rate determining step, in other words the long range cross section or whatever they call it. There would be an effective potential maximum, the centrifugal potential, if you are talking about gas phase, yes.

Q: So this curve here doesn't tell the whole story.

M: No, I didn't put the centrifugal potential in there.

6. p. 860 1st column bottom: "The value of E_{a} equals the value of V, $V(x_c)$, at the intersection, less the initial value $V(x_0)$. Obtaining x_{c} by equating $V(x_c)$ and $V^-(x_c)$, Eqs. (6) and (7), yield, after a change of variable to X, $\{X = \exp[-a(x - x_a)]\}$, the value X_{c} for X at the intersection, namely $\frac{E_X}{2D}$. Thereby, Eq. (9) holds,

$$E_a = D - E_X + \frac{E_X^2}{4D} - \Delta\left(\sum \frac{\omega\nu}{2}\right) \tag{9}$$

where we have also subtracted the total zero-point energy of RX at x_0 minus that at x_{c}. This latter difference might be of the order of 3 or so kcal mol^{-1} depending upon the system."

Q: How do you evaluate that value?

M: Probably from some experimental data on the energy barrier.

Q: It is not something that you can deduce theoretically.

M: No.

7. p. 862 1st column middle: "Apart from various constants... σ_{DA} is given... as a function of *the energy* ε *of the incident electron and of the vibrational state n of the molecule* by Eq. (13), where χ_n is the initial vibrational wave function of the molecule, x_ε *is the value of x which satisfies the Franck–Condon restriction in Eq. (14) for the attachment of an electron of energy* ε, $\frac{\Gamma(x_\varepsilon)}{\hbar}$ is the autoionization rate of the electron in the anion at $x = x_\varepsilon$, and V' is the derivative $\frac{dV^-(x)}{dx}$, evaluated at x_ε and z_ε is defined below.

$$\sigma_{DA}(n, \varepsilon) \propto \frac{\Gamma(x_\varepsilon)}{V'(x_\varepsilon)} |\chi_n(z_\varepsilon)|^2 e^{-\rho} \tag{13}$$

$$V^-(x_\varepsilon) - V(x_\varepsilon) = \varepsilon \tag{14}$$

This x_ε serves as a point from which the separation between the products of the electron dissociation attachment, $R^\bullet$ and X^-, begins along the repulsive curve $V^-(x)$.

Q: You formerly considered the vertical attachment energy of the electron to the *RX* molecule in its ground vibrational state at the coordinate $x = x_0$ of the minimum of the *RX* potential energy curve (see arrow E_v in Fig. 1 (a)). You now consider the possibility of attachment to a vibrationally excited *RX*. Please look at my Fig. 3*. There I have drawn an arrow starting from a point of coordinate

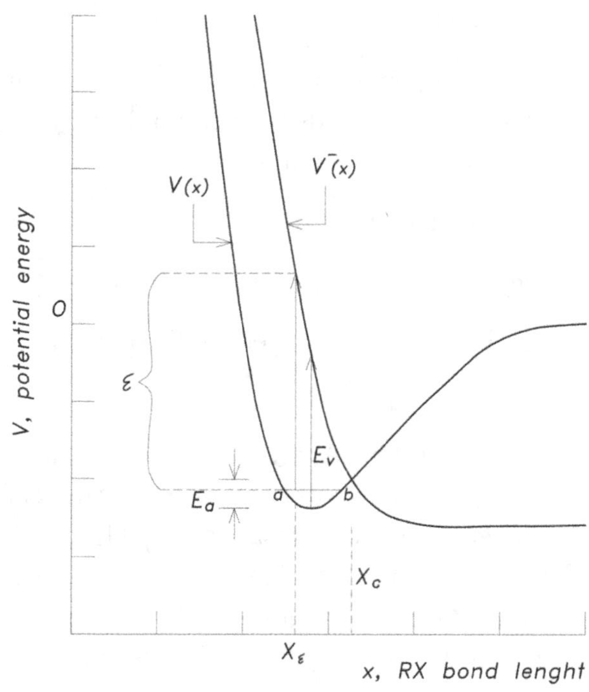

Fig. 3*.

$x = x_\varepsilon$ inside the (a, b) interval (where a and b are the turning points). The point is such that the arrow has length (energy difference) such that $V^-(x_\varepsilon) - V(x_\varepsilon) = \varepsilon$. Two questions:

Q: (1) Is it OK to draw such an arrow starting from *any point* in the (a, b) interval if just

$$V^-(x_\varepsilon) - V(x_\varepsilon) = \varepsilon?$$

(2) I believe the Franck–Condon state corresponding to the tip of the arrow is a nonquantized unbound state. Is it so? Please look at my Fig. 3*. There I have drawn an arrow starting inside the *a–b* interval...

M: Let me see, what is ε?

Q: ε is the energy of the electron.

M: You're conserving position but you should also conserve momentum as a first approximation, so instead of going to that level you should go to an equal amount of energy above. Even if you started with a higher energy you end up with a higher energy than that you have drawn. Try to conserve both momentum, classical momentum and classical position. In a vertical transition you conserve the position but you haven't conserved the classical momentum.

Q: So that arrow there is wrong?

M: Yes, you only have one part of the Franck–Condon principle in there.

Q: So how would you draw this?

M: *If you want to start that arrow on the line, I would draw to a point equally high above the upper curve, so that you have the same kinetic energy.*

Q: Equally high above the upper curve?

M: *Yes, in other words, that arrow on the line is above the minimum of the lower curve, right? All right, then extend your line to a point that's the same amount above the upper curve, then you conserve the kinetic energy.*

Q: Could you draw such a thing for me?

M: Ja, sure, let me just try to describe to you. You see where your arrow is? All right, you have your figure, now you see where your arrow is, the arrow going up vertically that goes up to the highest point that you got going up on the upper curve, OK? Extend that arrow an equal amount above the upper curve as it starts from on the lower curve, because *the arrow starts a certain amount above the minimum of the lower curve and will go an equal amount above the upper curve*. Then you conserve kinetic energy. You go an equal amount above and you get the same kinetic energy as when you started from.

Q: So the line starts where I drew it or not?

M: You can do that, start from where you drew it and then extend a little bit higher than what you got. By the way, it is equivalent to going from one potential energy curve to the other, you can see, the energy difference is equivalent, but you are starting in a different point. Now, some versions of the Franck–Condon principle that is not Franck–Condon, has the way you have drawn it, but as far as I can tell, that's not a good version. Certainly it is not in the same spirit as the Franck–Condon principle which says that you conserve, for the concerned nuclei, both the kinetic energy and the potential energy.

Q: But when you go up, the tip of the arrow then crosses somehow the V^-curve.

M: Yes, that's right, it crosses by a certain amount that's equal to where it starts from below. And in fact, if one crudely drew the wave functions, and uses a semiclassical frame, you'd see they are trying to conserve momenta, so they try to conserve that kinetic energy.

Q: You are quite right, but you don't end up anymore on top of the V^-curve

M: No, that's right, you don't, but the energy difference is the same as if you started from one potential energy curve to the other. The energy difference is the same, even if you started up a different point, you ended up at a different point, if you conserve kinetic energy of the nuclei it is the same in terms of an energy difference, as going from one curve to the other.

Q: Does the tip of the curve end up in an unbound state?

M: Yes, like to where you ended up there, especially if you ended up a little bit more, you are in an unbound state, that's right, unbound because you are above the asymptote. But of course when you come down, depending on what you do at the crossing point, you may just pass right through, if there is any probability of going to the lower curve. If you started on the $V(x)$ curve the tip has to end up on the V^- curve, if you started above then the tip has to end up above.

8. Following: "The x-dependent electronic energy of an $RX^{\cdot -}$ anion which can re-emit the electron is frequently described via a complex-valued potential $W(x)$, Eq. (5), where the $i\Gamma(x)/2$ takes into account the electron re-emission.

$$W(x) = V^-(x) + i\Gamma(x)/2 \tag{15}$$

According to the Franck–Condon principle the vertical transition to a complex-valued potential $W(x)$ as a result of a collision with an electron of energy ε, occurs at a z_ε which is a complex-valued x that satisfies Eq. (16).

$$W(z_\varepsilon) - V(z_\varepsilon) = \varepsilon \tag{16}$$

Upon expansion of $W - V$ in powers of $z_\varepsilon - x_\varepsilon$ and using Eq. (16), one sees that z_ε is given by Eq. (17), where V'_Δ is given by Eq. (18)

$$z_\varepsilon = x_\varepsilon - i\Gamma/2V'_\Delta \tag{17}$$

$$V'_\Delta = V^{-\prime}(x_\varepsilon) - V'(x_\varepsilon) \quad (18)\text{"}$$

Q: How do you define and compute $V(z_\varepsilon)$ and $V^-(z_\varepsilon)$?

M: That goes way back... Well, if you have a functional form $V(x_\varepsilon)$, then wherever you have $V(x_\varepsilon)$, if you want $V(z_\varepsilon)$, just replace the x_ε by the z_ε. In other words, the $V(x_\varepsilon)$ defines a certain functional form and now if I put a y_ε instead of an x_ε, I have the same functional form but now with y_ε, if instead of an x_ε I put a z_ε, I have the same functional form but now with z_ε. So, you write down the functional form whatever it is, exponential, polynomial, anything, whatever it is, in terms of x_ε, then you merely replace x_ε by the z_ε. That's all, it is as simple as that.

9. p. 862 2nd column top: "The exp($-\rho$) in Eq. (13) is the probability (the survival probability) that a system formed on the $V^-(x)$ curve at x_ε will survive before re-emitting the electron, and so reach the intersection x_c of the $V^-(x)$ and $V(x)$ curves, Eq. (19),

$$\rho = \int_{x_\varepsilon}^{x_c} \frac{\Gamma dx}{\hbar v_{\text{nuc}}(x)} \equiv \int_{t_\varepsilon}^{t_c} \frac{\Gamma dt}{\hbar} \tag{19}$$

where $t_c - t_\varepsilon$ is the time for separating $R^{\cdot} + X^-$ system to go from the point of formation x_ε to the crossing point x_c. In Eq. (19) $v_{\text{nuc}}(x)$ is the relative velocity of the separating nuclei at x. Beyond $x = x_c$, the $V^-(x)$ is below $V(x)$ [e.g., Fig. 1(a)] and *the competing re-emission can no longer occur.*"

Q: Why so?

M: Well, below the crossing point if you are on the $R^{\cdot} + X^-$ curve, you can't emit, you can absorb, but you can't emit. It could emit from the upper state, wherever there is an upper state, but if that upper state has become a lower state, all it can do is absorb, not emit.

Q: Ah, OK, there is a switching between absorbing ad emitting state

M: Yes. That's right.

M312 Ion Pairing and Electron Transfer

R.A. Marcus

NOTES

1. Following: "Some of these effects on k_{et} differ from those expected from a simple ion atmosphere continuum expression. Indeed, a continuum description has been found to be inadequate in some studies, such as ferrocenium–ferrocene exchange."

M: A simple continuum expression would be a Debye–Hückel type of formula and, you know, if they don't obey the Debye–Hückel, that's not surprising because Debye–Hückel is approximate, it takes only limited account of the finite size of the ions, it makes a linearization approximation, so it's never surprising when the Debye–Hückel breaks down. And especially in this case here where, you know, the contact of the ions of opposite sign coming up to react can have a major effect. So it is never surprising when you have some breakdown of continuum treatment *at finite concentrations*. It can break down for various reasons.

2. Following, bottom: "... the reorganization *parameter* λ"

Q: λ is usually designated as "parameter." Can also one speak of a ΔG^0 parameter?

M: Maybe it would be better to call it a reorganization term.

Q: But could one call also ΔG^0 a parameter?

M: Yes, absolutely.

3. p. 10071 2nd column top: "We first recall two types of ion pairs, loose and tight, which have been studied with thermodynamic measurements, optical absorption, ultrasonic absorption, NMR and other spectroscopic methods."

Following bottom: "Any equilibrium between tight and loose ion pairs is also reflected in ultrasonic absorption."

Q: How does ultrasonic absorption distinguish between loose and tight ion pairs?

M: Ultrasonic absorption deals with the propagation of the sound through the system and the propagation of sound through that depends upon whether or not the energies are calibrated, as the sound wave has more or less energy whether it goes through a peak compared with its minimum, So, the sound wave is oscillating and energy is going into and out of whatever is in the system, and if you have tight and loose ion pairs with a heat of reaction between them like a ΔH^0 for going from one to the other, then the sound wave can convert one to the other or convert part of one to part of the other, and it's a way of absorbing energy if there is a ΔH^0 associated with it. Do I give any reference for the ultrasonic absorption? That chapter at Brooklyn Poly, I forgot his name who studied ultrasonic... he came after I left...

Q: Fuoss...

M: Oh! Fuoss did that.

Q: And then also Manfred Eigen.

M: Oh! OK, Sergio Petrucci, I knew him briefly, I did not relate really...

Q: He was a guy from Rome...

M: I may have had some discussion with him...

4. p. 10071 2nd column middle: "Again, in the tight pair, when some solvent molecules are originally in the coordination shell of one of the ions, some are liberated in forming the pair, resulting in an *increase in entropy*"

M: Well, it is that but even more. The liberation of a solvent molecule corresponds to an increase in entropy, but there is also a charge coming next to a charge of opposite sign which can neutralize it, you see, and hence the whole molecular structure is less attractive for the solvent molecules, less solvated than an ion.

5. Following: "Any such decrease in ion solvation then results in an *increase in energy, an enhanced direct ion-ion Coulombic attraction notwithstanding*"

M: Well, ions solvation has a lot of energy associated with it, and I guess the net result, you know, the experimental result, is that really you're balancing degree of solvation against energy. If you lose solvation you lose a lot of energy, I mean you lose a lot of attractive energy. Now it's true that you are gaining some coulombic attraction but if you think of what you are gaining . . . well, it depends on what you gain . . . it is like having a dipole surrounded by a solvent . . . See, when you have a couple ion–ion is like having a dipole surrounded by solvent. When the ions are separated, then each of them is solvated, and that's a stronger solvation. In other words, the one charge when is near another cancels the solvation around that second charge, and that system is less solvated than the individual charges. *Increase in energy*: that means the system is less stable, is less solvated, you see.

6. Following: "Thus, an increase in temperature T actually increases the tight contribution at the expense of the loose in such a case."

Q: Isn't it counterintuitive?

M: No . . . a reaction going from the separated ions to the tight contribution, is an uphill reaction. Normally for an uphill reaction when you increase the temperature you shift the equilibrium more towards the uphill partner. Normally the equilibrium is very much stable in the lowhill part. When you increase the temperature it is going to be less stable, is shifted.

7. p. 1072 2nd column top: "**III. Fluctuational Ionic Motion and Electron Transfer.** In a third scheme for the electron transfer, the ion transfer occurs as a *reorganization*, rather than via the actual *chemical intermediates* in I and II. In the transition state there is now not only the usual reorganization of the bond lengths (and/or angles) of the reactants and of orientations of individual polar solvent molecules but also, in addition, of the position of the transferring ion."

Q: When in Section I one considers the reaction (2):

$$M^+A_1BA_2^- \xrightarrow{k_2} A_1BA_2^-M^+ \tag{2}$$

doesn't one implicitly consider a transition state in which M^+ is in a position intermediate between the initial and final one shown in Eq. (2)?

M: I see what you are saying, yes, you can regard that like as a kind of reorganization. In the transition state the ion M^+ instead of being in one or the other is some place in the middle,

Q: Exactly

M: And that's just like the reorganization, you know.

Q: So, why then is this point different from what you write in point 3 in Eq. (11)?

M: No, it's not different from that, that's an example of fluctuation in the position of M, I guess.

Q: But you distinguish it, you have three different situations.

M: Oh, I see what you are saying, yes, It depends on the way that one would treat it theoretically. Supposing the transition state is in the middle, then one would calculate the energy differently if one had a nonadiabatic treatment than if one uses an adiabatic treatment, just like with the solvation, so the details of the calculation would be somewhat different. In other words, but I don't know if that is the case here, what there would be if there would be an adiabatic... maybe this is not the answer, I'd have to look really, but if it were all purely adiabatic treatment, no reorganization, just *the usual slow transfer of charge* with the changing of charge in the transition state, that would be one way. And the other way is the sort of the nonadiabatic way, so I don't know if I mean that. *The main difference between adiabatic and nonadiabatic is really in the motion of the electrons and not in the motion of the nuclei,* and here I see distinguishing it and I don't know why. Let me just think about a moment. Now, how do I treat the ET from A_1^-B to A_2? Do I treat that non-adiabatically or adiabatically?

Q: You don't say anything about that, you just don't mention this point.

M: What was reaction (1)? Oh, ET first, all right, and you are wondering how reaction (2) differs from Eq. (1). I see now... well, it's hard to say... you see, an ET could occur without the M^+ moving, and maybe that's number Eq. (1).

Q: Yes, exactly, that is number (1).

M: Maybe ET happens before you get any ion motion. Or first you get the ion motion, then the ET. First the M^+ goes to an intermediate position and then there is ET. Apparently in Eq. (1) the ion stays where it is, and the electron transfers first. Can you see Eq. (11)? That's the third mechanism we are trying to contrast with mechanism (1). In mechanism (1) you notice that the electron is transferred before the ion is moved.

8. Following: "The following equation is derived in the Appendix for this case by solving diffusion-reaction equations using several simplifying approximations, one of which is that the reaction occurs largely near some value $x^\dagger$ of the M^+ *transfer coordinate* x, rather than over a broad interval of x."

Q: How about this rather undefined transfer coordinate? The dynamical coordinates defining the position of M^+ in space are three. Which is the meaning of this single coordinate?

M: Yes, if it is over a broad range of x, then probably one has to have a diffusional behavior... something more than just a diffusion reaction, the simplest diffusion reaction is that something diffuses to some point, then reacts at that point. Here it is reacting at all possible points, that's much more complicated.

Q: What is exactly this x?

M: The reaction occurs with M^+ at one position, it's occurring at a certain value of x, the reaction could also occur with M^+ at a different position, at a different value of x.

M312 Ion Pairing and Electron Transfer

(Continuation)

R. A. Marcus

NOTES

1. p. 10071 2nd column bottom: "Inasmuch as both types of ion pairs (tight and loose) sometimes coexist, they do not necessarily have very different dissociation constants at any given temperature."

Q: What does the coexistence of both types of ion pairs imply about the values of their dissociation constants?

M: *If they have similar equilibrium constants that means that the complexes have similar concentrations. Think of what the definition of an equilibrium constant is.* It's the complex concentration divided by the products of concentrations, so if the equilibrium constants are about the same then the complex concentrations are about the same.

2. p. 10072 1st column top: "**Electron Transfer First**. In one scheme for an ion-pair effect on k_{et}, the electron transfer occurs first, followed by an *intramolecular ion transfer*. In such a scheme the intermediate, $M^+A_1BA_2^-$, is sufficiently long-lived that the M^+ can subsequently *migrate* in reaction 2 under the influence of the charge distribution in $A_1BA_2^-$. When the ion transfer is intramolecular, we have

In such a scheme the intermediate, $M^+A_1BA_2^-$, is sufficiently long-lived that the M^+ can subsequently *migrate* in reaction 2 under the influence of the charge distribution in $A_1BA_2^-$. When the ion transfer is intramolecular, we have

$$M^+A_1^-BA_2 \underset{k_{-1}}{\overset{k_1}{\rightleftarrows}} M^+A_1BA_2^- \tag{1}$$

$$M^+A_1BA_2^- \xrightarrow{k_2} A_1BA_2^-M^+ \tag{2}$$

....

When, instead of being intramolecular, the ion transfer in reaction 2 proceeds by dissociation of the ion pair, followed by reformation of a new pair, we have

$$M^+A_1BA_2^- \underset{k_{-4}}{\overset{k^4}{\rightleftarrows}} M^+ + A_1BA_2^- \tag{4}$$

$$A_1BA_2^- + M^+ \xrightarrow{k_5} A_1BA_2^-M^+ \qquad (5)\text{''}$$

Q: (1) (i) Does the M^+ in (2) "slide" along the atoms of $A_1BA_2^-$ going through some minimum length walk?

(ii) Or are there fluctuational motions along a host of short paths?

(iii) Or is there a "migration path" to be expressed in terms of a free energy path and a free energy curve? I believe that only if the time for the process is short enough one can be sure that the M^+ on the left of 2 is the same as the M^+ to the right. Which is the order of magnitude of the time for *such* an intramolecular migration?

(2) I believe that, contrary to the above, the M^+ to the right of 4 is not in general the same as the one on the right of 5. Is it so?

M: (1) (i) The answer is yes because *it can't dissociate by definition of Eq. (2)*, so *it has to stay close* to the rest. Whether it slides . . . , just God knows what, but it has to stay close to the rest of it, so . . .

(ii) Well, it's *diffusion, diffusion is essentially a host of short paths*, you know, if you think of diffusion, the object diffusing moves a little bit one way, then it moves a little the other way, you can call it a short path. But in that case it is not really a trajectory where it slides through, that's not diffusion, *so you have to think of what the model of diffusion is, whatever the model of diffusion is that you use for diffusion.*

(iii) Well, certainly the electrostatic potential would be somewhat different at points along there, so one could write down a diffusion equation and have the free energy of the system as function of distance where it is, at the position where it is . . .

Q: So, there can be a *free energy path.*

M: *Yes, that's true of any diffusion.*

M: (2) Yes, that's true, because if the path is fantastically long it would get away of the range of attraction of the other part of the molecule, and there would be a dissociation, there would be other M^+'s too if it goes far enough away. In other words, if one tried to treat it theoretically, in some way or another, maybe by molecular dynamics simulation or God knows what, then one would put the ion close in the potential energy function, let it move, and look at the chance to reach the other end. And one would have to decide if it's done so by escaping and coming back, you have to tell when an ion gets out of range.

I don't know but perhaps one could say: all right, let's pretend for better or worse that we use a typical diffusion constant, let's draw a model of the whole system, and let's look at what the path would be like keeping the M^+ at van der Waals distance all along the path from the beginning to end, so that would give one a rough distance and the time then would be obtained by equating 2Dt to root mean square distance. So, one would use the one-dimensional Einstein's relation. The main point is: the ion has a certain distance to travel and you are giving a certain diffusion constant, you assume some value for it and you know the distance it has to travel from one end to the other and so from that you can calculate the typical time to diffuse through that distance and that would be the reaction time for that model. Keeping the M^+ at sort of van der Waals distance all along, you can sort of estimate the length of the path when you have the molecular structure in front of you and then you use Einstein's relation to get the time. You find Einstein's relation in one of his 1905 papers.[1]

3. p. 10072, 2nd column top: "**Fluctuational Ionic Motion and Electron Transfer**. In a third scheme for the electron/ion transfer, the ion transfer occurs *as a reorganization,* rather than via the *actual chemical intermediates* in I and II. *In the transition state* there is now not only the usual reorganization of the bond lengths (and/or angles) of the reactants and of orientations of individual polar solvent molecules but also, in addition, of *the position of the transferring ion.* When a large displacement of the ion is needed, it is useful to introduce the notion of *precursor and successor complexes*, indicated by the parentheses in Eq. (11).

$$\mathrm{M^+A_1^-BA_2} \rightleftarrows \begin{pmatrix} \mathrm{A_1^-BA_2} & \mathrm{A_1BA_2^-} \\ \mathrm{M^+} & \mathrm{M^+} \end{pmatrix} \longrightarrow \mathrm{A_1BA_2^-M^+} \qquad (11)$$

[1]Die von der Molekularkinetischen Theorie der Wärme geforderte Bewegung von in ruhenden Flüssigkeiten suspendierten Teilchen (Annalen der Physik, ser. 4, XVIII, pp. 549–60, 1905).

The M^+ transfer coordinate, which may be 'slow,' leads in Eq. (11) to the formation of a *precursor complex* from the reactant and to disappearance of a successor complex to form products."

Q: (1) Why $M^+A_1^-BA_2$ is an intermediate molecule while $\underset{M^+}{A_1^-BA_2}$ is not?

Q: (1) Why $M^+A_1^-BA_2$ is an intermediate molecule while $\underset{M^+}{A_1^-BA_2}$ is not?

(2) Is it may be so because $\underset{M^+}{A_1^-BA_2}$ is a form of the transition state and a transition state structure cannot be considered an intermediate?

(3) What does it mean that the M^+ transfer coordinate may be slow and apparently as a consequence we have these precursor and successor complexes?

(4) Is the difference between an intermediate and a precursor complex due to the fact that an intermediate can be physically detected but the precursor complex is a fictitious construct?

(5) Which is the geometric meaning of the $\underset{M^+}{A_1^-BA_2}$ symbol? Does it mean that the exact position of M^+ is undefined but somewhere in between the initial and final positions?

M: (1) That's because we are not using a diffusional treatment, of course the process occurs by diffusion, but we are assuming that the diffusion is relatively fast, and so the process must be like of a transition state type to reach that. In other words, if one would calculate by this mechanism a rate, the diffusion constant would never come in. In the previous mechanism it would. Just like in Eyring transition state theory, the diffusion constant never comes in. Now one is using equilibrium considerations and no diffusion constant.

Q: But why this object $\genfrac{}{}{0pt}{}{A_1^- B A_2}{M^+}$ . . .

M: Because the diffusion is fast, the slowest thing is displacing that M^+. See, diffusion can be intrinsically fast, it can be that the barrier is so high that, you know, it takes some time to get there, to form the transition state. So this would be kind of transition state formulation for the transfer of M^+ over from one side to the other.

(2) In the transition state, where you describe it in this particular model, Eq. (11), you put M^+ there in the transition state and the transition state in this case is at the intersection, and it's certainly not an intermediate, it's true, and of course in calculating the free energy of formation of that transition state, and hence a rate, no diffusion constant appears.

(3) Do you remember those diagrams that I had, Sumi and I had, where you had a fast coordinate, . . . remember that on solvent dynamics? You remember that diagram with ellipses, where you had fast and slow coordinates, if something is slow you don't necessarily go through the transition state, you may end up going to higher barrier than through the transition state.

Q: Which is the relation between the slowness of the transfer coordinate and precursor and successor complexes?

M: Well, M^+ transfer coordinate is slow, you get a two-dimensional problem now, no longer one dimensional, so you have to think in those terms.

(4) Oh, that depends . . . the way Norman Sutin does, the precursor complex is in fact an intermediate, that is an intermediate that precedes the whole reaction.

Q: And is it an intermediate at the transition state?

M: No, no, the precursor complex by definition is not at the transition state, "precursor" means *before* the transition state. Sutin has, say, two reactants coming together, to form a precursor complex, then that complex forms a transition state, then that transition state forms the precursor complex of the products, a postcursor complex, then that products' precursor product complex dissociates. If it is a precursor complex, it is not a transition state. If the expression in parenthesis were intended to be a transition state, then it's not a precursor complex. If on the other hand it was intended to be a precursor complex, then it would form the postcursor complex, which is the formula on the right, and that would form the separate products.

Q: You go from the transition state to the final product.

M: Yes, but you see, I should have had an arrow preceding the transition state then, if that's a transition state.

Q: You have a double arrow between the reactant an the transition state

M: If there is a double arrow that means the thing in parentheses is not a transition state, that shows the precursor and the postcursor. What's in parenthesis is not a transition state, what's in parentheses on the first part before the double arrow is a precursor, on the second part is a postcursor. There is no transition state shown in there. Clearly there is no transition state shown in Eq. (11), just a series of intermediates, those happen to be precursor and successor complexes.

M: (5) Well, M is roughly equidistant from A_1 and A_2.

4. p. 10073 1st column top: "**Effect of Ion Pairing on Charge Transfer Spectral Maximum,** $h\nu_{\text{max}}$. For a photoinduced charge transfer

$$\text{DA} + h\nu \rightarrow \text{D}^+\text{A}^- \tag{13}$$

where D and/or A may also bear a charge, we have

$$h\nu_{\text{max}} = \lambda + \Delta G^0 \tag{14}$$

where λ is the reorganization energy for solvent molecules and reactant's bond lengths (and/or angles) and ΔG^0 is the standard free energy change in reaction 13; i.e. ΔG^0 is the free energy of the product of the reaction minus that of the reactant prior to any subsequent reaction of the product. At a fixed position M^+, λ is expected to be approximately the same for $M^+A_1^-BA_2$ as for $A_1^-BA_2$, if one judges from the evidence mentioned earlier, where the symmetrical pair $M^{n+}A_1^{2-}BA_2^{3-}M^{n+}$ has approximately the same $h\nu_{max}$ as $A_1^{2-}BA_2^{3-}$."

Q: Whereas I can understand that $h\nu_{max}$ is the same for $A_1^{2-}BA_2^{3-}$ and for $M^{n+}A_1^{2-}BA_2^{3-}M^{n+}$ because of the symmetric positions of the M^+'s, why is it that λ is expected to be approximately the same for $M^+A_1^-BA_2$ and for $A_1^-BA_2$?

M: Yes, that means that M^{n+} doesn't contribute much to λ. And therefore, if you put an M^+ there, it's not going to change λ. In other words, the end of the sentence says that λ is approximately the same that the spectrum doesn't depend on *the λ contributed by M^+*. *Think locally*, it's true that one M^+ can roughly cancel the other, but as a first approximation think that it doesn't, so that *what's happening in the vicinity of one M^+ is not affected by what's happening in the vicinity of the other M^+*. If the two M^+'s were far enough apart it would be true, but it's just approximate, OK?

Q: So it is enough to consider just one M^+.

M: Yes, that's right, in other words, pretend for the moment that those M^+'s are far apart, then, what the spectrum is saying, is saying regardless of those M^+'s there, that the M^+ is not affecting the individual λ's.

Q: Why doesn't affect the individual λ's?

M: Because when you have the complex, let's pretend everything is far apart, then you have in the complex the spectral absorption frequency when you have the M^+'s there, but the spectral absorption frequency is about the same as when you don't have the M's there, the spectral absorption frequency is $\lambda + \Delta G^0$, so if it is symmetrical then it is not changing ΔG^0 and is not changing $\lambda + \Delta G^0$, then it is not changing λ.

Q: You start from the symmetrical structure because it assures you that ΔG^0 is the same.

M: That's right, exactly.

5. Following: "**Approximate Equations**—The various parameters in theoretical expression for k_{et} ... were obtained using Eq. (15) below, where any *high frequency modes* are treated as having a single frequency ν and contributing λ_ν to the reorganization parameter λ. The solvent (and low-frequency modes) reorganization contribution is denoted by λ_s. The rate constant k_{et} for a nonadiabatic reaction is then given by

$$k_{\mathrm{et}} = \frac{2\pi}{\hbar}\frac{|V|^2}{(4\pi\lambda_s k_{\mathrm{B}}T)^{1/2}}\sum_{\mathrm{w}=0}^{\infty}\frac{e^{-S}S^w}{w!}e^{-(\Delta G^0+\lambda_s w h\nu)^2/4\lambda_s k_{\mathrm{B}}T} \quad (15a)$$

where

$$S = \lambda_\nu / h\nu \quad (15b)$$

Q: Which is the difference between these new λ_ν and λ_s and the old λ_i and λ_o?

M: Well, in this λ_s you were including all of the classical coordinates, so this λ_s would include the old λ_ν, and what we got here is a new λ_ν, a quantum λ_ν, but those λ's are to include everything that was included before. In other words, both the previous λ's and the previous λ_ν, the classical λ_ν.

Q: So, the classical λ_ν would be λ_i.

M: Yes, the classical λ_ν would be λ_i, that's right, but of course you would remove from that this particular degree of freedom, you treat all the other vibrational λ's as classical, in this model. Classical are the λ_s, which would include now some low frequency vibrations.

Q: So, the λ_s are practically the λ_o plus the low frequency vibrations

M: Yes, that's right.

Q: So, it is part of the λ_i

M: Yes, that's right. The other part of the λ_i is now present as this new λ_ν.

6. Following: "Equation (15a) presumes that the reactant is vibrationally unexcited, *a good assumption in the inverted region* and frequently in other regions, but not when ΔG^0 is significantly positive."

M: That's right. when ΔG^0 is significantly positive and there is a big barrier, the system will make use of whatever vibrations it can, including having some excited vibrations, the high frequency vibrations. If on the other hand the barrier is low, you gain nothing from having excitation of a high frequency vibration. In the inverted region, usually a very downhill region, usually you are going from the lowest state to the ground state of the reactants for any high frequency vibration, but high frequency doesn't help you really, you have to expend energy and you gain little.

7. p. 10074 1st column bottom (and in many other points...): "... unless the odd electron in $X^{\cdot -}$ in *neat* THF is delocalized over the two naphthyls."

Q: Is here "neat" for pure? I guess so...

M: Oh, neat is pure, yes. A neat solvent means you only have the solvent. Neat for one component, yes.

8. p. 10074 2nd column top-middle: "The quantum effects are *largest in the inverted region*, and so this λ_ν is most sensitive to data in that region."

M: Well, you have to think in terms of *semiclassical theory of tunneling*. If you are in the inverted region, you are *tunneling through just a thin sliver*. Think of your two slopes and you draw a short hopping line, a tunneling line, OK? Think of two intersecting slopes, to the two curves, the region you are tunneling through is thin. Now, look *at tunneling when the curves have opposite signs, in other words in the normal region, there is a much stricter barrier*, and you can make that quantitative using semiclassical theory. But you can see it physically, just by thinking of what happens when the two curves that are intersecting have the same sign of the slopes and when they have opposite signs of the slopes.

9. p. 10074 2nd column middle-bottom: "...Eq. (16): We first replace the factor outside the sum

$$\frac{2\pi}{\hbar}\frac{|V|^2}{(4\pi\lambda_s k_B T)^{1/2}}$$

(which can be shown to be *a nonadiabatic Landau–Zener factor*"

Q: Are there Landau–Zener factors for processes other than nonadiabatic?

M: Well, it depends how you define that factor, but Zener had an expression which had an exponential which, when the exponent became small, became that factor. *Zener's expression included both adiabatic and nonadiabatic. Landau just included the nonadiabatic*, if you look at the 1932 papers and read them.

10. p. 10074 2nd column bottom: "In the models I to III for the electron/ion transfer we have considered cases where some of the transient dynamics of formation of the initial ion-paired reactant does not affect the observations. For example, in the experiments in which the two radical anions $X^{\cdot-}$ and $Y^{\cdot-}$ were studied, the system was prepared by having a solvated electron, accompanied by its cation, first react with the relevant aromatic group and thereby forming a cation–anion pair before the desired reaction occurs. In this way, it was pointed out, *any slow transient dynamics involving the formation of the subsequent* $M^{+}X^{\cdot-}$ *and* $M^{+}Y^{\cdot-}$ *ion pair was avoided.*"

M: Supposing that you start up with a system that has an ion pair, then you have no dynamics for the formation of the ion pair. If you started up with separated ions, then there would be dynamics involved in their formation. But if you already have a pair, there is no dynamics associated with this formation.

11. p. 10075 1st column middle: "The effect of M^{+} in the intermolecular case is seen to be relatively little, presumably because the counter ion no longer has to move far and, further, the reactants can orient themselves so as to avoid a large uphill ΔG^0 for any ET reaction that can occur when the transition is vertical (fixed ions)."

M: Yes, I guess it just means that the M^{+} has to move. If it moves then you don't get a large uphill ΔG^0 because if it stayed in place the M^{+} would be staying near a charge which has disappeared and hence there is a large uphill ΔG^0 due to cancelation of that favorable electrostatic interaction.

12. p. 10075 1st column bottom: "A comparison of results for pyrene and naphthalene as acceptors in the intermolecular electron transfer was revealing, and reflected the fact that *the former was a more downhill reaction and so was diffusion controlled.*"

M: *Well, the more downhill a reaction is, the faster its intrinsic sort of rate constant at contact, if its intrinsic rate constant at contact is high, then the slow step is diffusion, the reaction becomes diffusion controlled.*

13. Following: "**Appendix. Intramolecular Diffusion—Reaction of M$^+$**

We consider the *diffusion of M$^+$along a coordinate x*, where $x = 0$ and $x = a$ denote the equilibrium positions of M$^+$ in reaction 11 when the anionic charge is localized on A_1 and A_2 respectively."

Q: One normally thinks of a coordinate of a point moving along a line. How is the line in the case of a diffusion coordinate?

M: The line doesn't have to be a straight line, it may be a curved line.

M312 Ion Pairing and Electron Transfer

(Continuation 2)

R. A. Marcus

NOTES

1. p. 10075 2nd column top-middle: "**Appendix. Intramolecular Diffusion-Reaction of M$^+$**

We consider the diffusion of M$^+$ along a coordinate x, where $x = 0$ and $x = a$ denote the equilibrium positions of M$^+$ in reaction 11

$$M^+A_1^-BA_2 \rightleftarrows \begin{pmatrix} A_1^-BA_2 & A_1BA_2^- \\ M^+ & M^+ \end{pmatrix} \longrightarrow A_1BA_2^-M^+ \qquad (11)$$

when the anionic charge is localized on A_1 and A_2 respectively. We denote by $P_r(x)$ and $P_p(x)$ the probability density of finding M$^+$ at any x, for these two respective systems."

M: One way of defining it without worrying about curvature, because we are not getting into kinetic energy, one way of defining it is: draw a curve from that M^+ on the left hand side to the position where the M^+ will be after the second A, after the A_2, draw a curve, imagine a missing M^+ after the first A, all right? Draw a curve between those two M's, call that coordinate x. x is a line, not an area.

Q: Is it a precise line?

M: Yes, it is a line, you know, an approximate line, but it is certainly not an area. Now, the system may exist, does exist, in some three dimensional volume, so the strict thing to do would be connecting a volume on one side to connecting a volume on the other side, but *we are not putting in details of all sort of the Hamiltonian associated with that*, so we can be rather loose and just simply draw a line between the initial position of the M and the final position of the M, we may call that line a reaction coordinate.

Q: So it's a loose kind of model.

M: It's a loose kind of model, *we haven't written down an equation for it.* The coordinate has to be a line.

2. Following: "The *local* reaction rate constant for forming the $A_1^- \to A_2^-$ transfer at x is denoted by $k_{et}^r(x)\ldots$"

Q: Is the existence of a local rate constant an immediate consequence of the local reaction probability?

M: Well...as long as that M^+ stays in the general region of the $A_1^- BA_2$, *it doesn't dissociate*, stays in that region as it migrates, at anyone point along that line, that curve that we just discussed, you can have a certain ET rate constant, and that's defined by that $k_{et}(x)$. By the way, that means that at each x there is a certain probability of ET, so at each x there is a certain *local reaction probability* in that model.

Q: I was just asking if there being a local reaction probability, if this has as a consequence the existence of a local rate constant.

M: Yes.

3. Following: "The diffusion-reaction equations for P_r and P_p are given by Eqs. (A.1) and (A.2) where the forced diffusive fluxes are Eqs. (A.3a) and (A.3b)."

Q: Please look at the equations on p. 10075 of your paper. Wherefrom do they come from, where are they derived?

M: Well, if you think of the first equation, without the last two terms, then J is the diffusive flux, so that's standard diffusion equation theory, without the last two terms in Eq. (A1). It is standard diffusion theory to write it that way where for J you write down a flux. In fact often in writing the flux you also write down the potential, which I didn't write down there. Essentially I just derived Eq. (A1) by taking a small volume element and in that volume element you look at the change of P_1 with time and that change of P_1 is due to *something going in, something going out, and something reacting,*... The first term is something going in and out of that little volume by diffusion, the other two terms is what's disappearing by reaction.

Q: And what about the Eqs. (A.3a) and (A.3b)?

M: That's standard theory too. Whenever you have potential acting, a force acting on it, you have a diffusion term and you have a conductive term, so that part is standard diffusion theory under the influence of a potential. Anyway, it is pretty much standard diffusion theory when you have a convective term, a forced term, you can often see something like that when you have ions diffusing and moving under the influence of a potential. *The potential here is the free energy.* Electrostatic potential, diffusion of ions. *G is the equivalent of potential,* dP_r/dx is a standard diffusion contribution, that's the force contribution and that's the second term. That part is absolutely standard diffusion theory under the influence of potential energy, you'll see that everywhere.

4. Following: "The survival probability of the reactant at time t, is $\int_{-\infty}^{+\infty} P_r(x,t)dx$ and provides a description of the course of the reaction."

Q: (1) Why does x assume negative values and which is then the difference then between $P_r(x,t)$ and $P_r(-x,t)$? And Eq. (2) how are the numerical values of the real coordinate x assigned?

M: (1) Well, imagine that the ion can be lined up with A_1, is near there, but it can also be some distance away. Now assume that's it is off at infinity where it's no longer part of the ion pair. At large distance where the interaction becomes negligible, the ion pair has some strength but essentially there is zero probability of the ion being there, so it's OK to use $-\infty$. The simplest idea, crudely, is to imagine a straight line parallel with the axis of the drawing, and then the starting point is at an x right near A_1 on that straight line, the final is right near A_2 on that straight line, and, you know, it can be to the left as long as it doesn't dissociate, and *effectively in terms of probability you can call that distance infinity*.

(2) It is just a length measure whatever you want to make it...

5. Following: "By way of illustration for the present discussion we introduce two simplifying approximations: (i) the steady-state approximation, $\partial P_r/\partial t = \partial P_p/\partial t = 0$, and (ii) the dominant reaction occurs at *some x-interval* $\Delta^\dagger$ *centered at* $x^\dagger$. In this case the net reactant flow $J_r(x)$ must vanish for $x > x^\dagger$, and similarly $J_p(x)$ is zero for $x < x^\dagger$.

We can implement (ii) by setting

$$\begin{aligned} k_{et}^r(x) &= k_{et}^r(x^\dagger)\Delta^\dagger\delta(x - x^\dagger) \\ k_{et}^p(x) &= k_{et}^p(x^\dagger)\Delta^\dagger\delta(x - x^\dagger) \end{aligned} \quad \text{(A.5)}$$

where $\delta(x - x^{\dagger})$ is the Dirac delta function. Integration of Eqs. (A.3) to (A.5) yields

$$J_r\left(x < x^{\dagger}\right) = J_p\left(x > x^{\dagger}\right) \equiv J = k_{et}^{r}\left(x^{\dagger}\right)\Delta^{\dagger}P_r^{\dagger} - k_{et}^{p}\left(x^{\dagger}\right)\Delta^{\dagger}P_p^{\dagger} \tag{A.6}$$

Q: How about the units of $\Delta^{\dagger}$ and of $\delta\left(x - x^{\dagger}\right)$?

M: The unit of $\Delta^{\dagger}$ is the unit of x, and that cancels the unit of $\delta(x - x^{\dagger})$ so the product of the two, $\Delta^{\dagger}\delta(x - x^{\dagger})$, is dimensionless, *so you need the big delta*. There are different ways of treating this, the idea here is just to imagine that the reaction is occurring in a little volume Δ, where the basic transfer is largely occurring, the transition state occupies a little region $\Delta^{\dagger}$, a crude way of doing it. There are other ways of doing it, this is just a simple way of doing it.

Q: But then $\delta(x - x^{\dagger})$ you say has dimension of $1/x$.

M: *A delta function always has units of the reciprocal of the argument of it. You know, the integral of $\int \delta(x)dx$ is unity, so the unity of $\delta(x)$ have to be the unity of $1/x$.*

6. Please look at Eqs. (A.8a) and (A.8b) on p. 10076 of your paper.

Q: There I see the activation barrier $-[g_r(x^{\dagger}) - g_r(0)]$ for k_{act} in Eq. (A.8b), OK. If I look at Eq. (A.8a) for k_{diff}^{r}, I see the integral $\int_0^{x^{\dagger}} \exp\left[g_r(x) - g_r(0)\right]dx$. May I physically interpret it as the sum of all infinitesimal barriers of the type $g_r(x + dx) - g_r(x)$ when M^{+} is moving by diffusion from x to $x + dx$?

M: Well, that $[g_r(x^{\dagger}) - g_r(0)]$ I guess is the free energy change to reach the transition state, divided by kT. That whole expression, that comes about when you solve the steady state equation, becomes that k in Eq. (A.8a), it is the expression that you get for the diffusion controlled rate, and so that $g_r(x) - g_r(0)$ is the free energy change over kT, the difference between the transition state and the reactants, so that in the derivation is part of the free energy barrier to reach the transition state. What is written there is a standard expression for diffusion controlled rate constant, when you solve the diffusion equation for steady state conditions, then you find out what is the rate constant. Diffusion control is just that expression. An integral follows, involving those g's all the way from zero to the transition state, and the activation barrier doesn't involve that. Oh, certainly if you add up all of those things you would come to that sum but that doesn't say why you have $g_r(x) - g_r(0)$ rather than a $-[g_r(x^\dagger) - g_r(0)]$, it doesn't say that, you see. Just why you have that, I haven't thought through, but it comes out as part of the derivation.

Q: I believe it is so because that integral is in the denominator.

M: Yes, I know, that's right, but just why the whole expression, in other words the detailed physical insight into, just why that whole expression is in that form, I haven't thought it through, I should have done it, because I've taught it in class many times, Certainly, once you've seen that sum and that difference $g_r(x) - g_r(0)$, you can write that in terms of differences of small infinitesimal barriers, but that doesn't tell you anything. Maybe it is that, but that's always the case when you have an initial and a final, all you see is a sum. Strangely enough I haven't thought through, you know, just why the mathematics is in that form, and probably I should have thought about in physical terms why does that.

7. Please look at Eq. (A.8e) where "Δ_r and Δ_p represent the size of dominant x-intervals occupied by M^+ along x and are centered in the reactant and in the product regions in reaction 11, i.e., at $x = 0$ and $x = a$, respectively."

Q: Your "intervals" really extend—theoretically—from $-\infty$ to $+\infty$ and the different dx's making up the interval are weighted by the probabilities given by the exponentials. Is it so?

M: The ratio of forward and backward rate constants is related to the difference of the free energies, and so on one side we have essentially an equilibrium constant, a ratio of rate constants, and on the other side you have it expressed in terms of a free energy difference over kT. That is an approximation, the main point is that the correct intervals, you know, are really much more limited than that, but the probability density of those integrals are essentially zero, so then you automatically write $-\infty$ and $+\infty$.

Q: So those exponentials there give the probability to every dx.

M: Yes, that is for a very local region actually, and beyond that region the integrals become negligible.

Q: That is your definition of Δ_r and Δ_p.

Q: Yes.

8. p. 10076 1st column middle: "Δ_r represents the size of the dominant x-interval occupied by M^+ along x, and is centered in the reactant and in the product region in reaction 11, i.e. $x = 0$"

Following: "To justify the labelling of the k's and K in Eqs. (A.8a)–(A.8d) we consider the individual steps in reaction 11. We denote the four successive species in the reaction by **A, B, C** and **D**, respectively.

$$\underset{\mathbf{A}}{M^+A_1^-BA_2} \rightleftarrows \left(\underset{\mathbf{B}}{\overset{A_1^-BA_2}{M^+}} \rightleftarrows \underset{\mathbf{C}}{\overset{A_1BA_2^-}{M^+}}\right) \longrightarrow \underset{\mathbf{D}}{A_1BA_2^-M^+} \tag{11}$$

If we use a steady state treatment of a diffusion-controlled formation of **B** in reaction 11 from **A**, ... one obtains the right-hand side of Eq. (A.8a) for the rate constant, and so we have denoted it by k^r_{diff}

$$k^r_{diff} \equiv k^r_{diff}(x^\dagger) = D/\Delta_r \int_0^{x^\dagger} \exp[g_r(x) - g_r(0)]dx \quad \text{(A.8a)''}$$

Q: Can you please explain the physical reason for Δ_r to be in the denominator of Eq. (A.8a) and not in the nominator? Apparently, if the region wherefrom the reactant diffuses is larger, the diffusion velocity should be larger. Analogously, can you explain why in Eq. (A.8b) for k_{act} we have $k_{act} \propto \Delta^{\dagger}/\Delta_r$? Here I can better understand the position of $\Delta^{\dagger}$ in the numerator—if the region where the reaction happens is larger, the speed of the reaction is greater. But as for Δ_r shouldn't even here the Δ_r be in the numerator, that is, why is it not $k_{act} \propto \Delta^{\dagger} \cdot \Delta_r$?

M: In my Fig. 11 I only have one arrow . . . I am missing arrows. I'm not sure about the $x = 0$ there, the lower limit. We could do a little bit better there, you know, but it makes sure a little difference. What one would allow essentially is for M^+ to occupy regions a little bit to the left to $x = 0$, you see, so, but that's an approximate description, for our purposes it doesn't matter. Certainly, there is looseness there and I should have put a little footnote.

Well, this won't satisfy you, but *D* has units of centimeter squared per second, Δ_r has units of centimeter, the *dx* has units of centimeter, so that rate constant has units of "per second," it's the right unit for first order reaction. Remember when I was just saying that, you know, really one should go to a little bit to the left of $x = 0$, and so on, I surely had that there, in an approximate way. The rate constant will depend upon the diffusion constant, the size of the interval occupied by M^+ is some small region, typically a vibrational amplitude, so to speak. So, that's sort of fixed and that D is independent of all that, that is sort of the diffusion constant for M^+, the effective potential in fact comes in the form of the g. The M^+ is vibrating in a little region around $x = 0$.

Q: Why should Δ_r be in the denominator there?

M: Well, the D and the Δ_r are two independent magnitudes, think of M^+ vibrating around the first A, then Δ_r is a typical vibrational amplitude, has nothing to do with diffusion.

Q: And the reason for being in the denominator is that the physical dimensions are OK?

M: Well, it is more than that, I mean, that is sort of a consequence of being correct, the dimensions always have to be OK and if they are not there is a serious problem.

Q: I was asking why $k_{act} \propto \Delta^{\dagger}/\Delta_r \ldots$

M: That rate constant has units per second, so it's like the probability per unit time, so it's really the probability of being in a certain region near the transition state, which is proportional to its delta, and divided by the probability of being in the reactant state, which is proportional to its delta times some free energy barrier and times frequency.

Q: So even here the guiding idea is that of physical dimensions.

M: Yes, but also thinking of what is involved in an equilibrium constant, and thinking of how a rate constant is related to an equilibrium constant. You know, *in a crude way the rate constant is the equilibrium constant for forming the transition state times the frequency of motion through the transition state*. Now, in more sophisticated transition state theory you don't do it that way, but crudely it is so, you see.

9. p. 10076 bottom: "…the *equilibrium probability* of forming **B** from **A**,

$$(\Delta^{\dagger}/\Delta_r)\exp(-[g_r(x^{\dagger}) - g_r(0)])\text{''}$$

Following: "The *equilibrium constant* for forming **D** from **C** is

$$(\Delta_p/\Delta^{\dagger})\exp(-[g_p(a) - g_p(x^{\dagger})])\text{''}$$

Q: (1) Are "equilibrium probability" and "equilibrium constant" synonymous?

(2) Doesn't the multiplication of the standard equilibrium constant by a factor like $(\Delta_p/\Delta^{\dagger})$ imply that the Δ's must be constant?

M: (1) Let's see now... $(\Delta^\dagger/\Delta_r)\exp(-[g_r(x^\dagger) - g_r(0)])$ should really have been the equilibrium constant for forming **B** from **A** because that's an equilibrium constant,

Q: So it is wrong there.

M: Yes, it is too loose a term. I mean, strictly speaking what it is is an equilibrium constant, it's related to the probability of forming. Very loose...

(2) Yes, *they are sort of defining the typical regions due to thermal fluctuations that occupy a certain region.* Physical probability.

10. p. 10076 2nd column top: "We comment further on equilibrium constants, such as Eq. (A.8d) for the formation of **C** from **A**. The free energy for **A** is $G_r(0) - k_B T \ln q^{\mathbf{A}}_{trans}$, where $q^{\mathbf{A}}_{trans}$ is the translational partition function *of* M^+*in species* **A** and equals $(2\pi m k_B T)^{1/2}\Delta_r$, m being the mass of M^+. $G_r(0)$ refers only to the free energy of **A** when x is fixed at $x = 0$."

Q: Please explain the role of Δ_r here.

M: Well, this is a crude model for thinking of M^+ being in a region of species A, thinking that the ion is like a particle in a box, of length Δ_r and that would be its partition function. I mean, of course you can represent it instead by a little vibrational partition function, but this is crude, in either one you end up getting similar results in magnitudes. This is just a crude particle in a box model. I guess because I was using Δ for everything else, I just stayed for that.

Q: You have to acknowledge me the merit of being a careful reader of your papers...

M: You are extremely careful, you are a more careful reader than I am a careful writer.

11. p. 10076 2nd column bottom: "The formulation of k_{et} as a function of x, as in Eqs. (A.1–A.2),

$$\frac{\partial P_r}{\partial t} = -\frac{\partial J_r}{\partial x} - k_{et}^r P_r + k_{et}^p P_p \tag{A.1}$$

$$\frac{\partial P_p}{\partial t} = -\frac{\partial J_p}{\partial x} - k_{et}^p P_p + k_{et}^r P_r \tag{A.2}$$

Tacitly implies that x is at most only a minor contribution to the reaction coordinate."

Q: Please explain... x was defined (p. 10072 middle) "the M^+ transfer coordinate."

M: Yes, all right, let me think then... it's a good question, that's a good question... let me see... what I'm trying to figure out is what I really meant because it is confusing the way it is expressed. I think what I mean to say, but this is a terrible way of saying it, is that... You see, one possibility in principle is that the ET could be occurring at all values of x, but instead is just occurring at a very local region of x, and so it would have been better if I had said "*tacitly implies that when the ET occurs, it occurs only over a very small interval of x*," which is a totally different statement. Yes. Because that's basically... you know, as if I use a delta function, I think of the x where the reaction is occurring, so that's really... I think that is what I must have meant, *because x is really a reaction coordinate, you go from one point to the other*, but maybe I was thinking in terms of... you have a certain diffusion, OK? And so, x is sort of going from the starting point of the... *Oh, yes, I see now what I meant*: starting from its initial position, that M^+ eventually goes over x but *there is no reaction occurring until you get near some reorganized transition state*, where the ion is always in the middle, and at that point the reaction coordinate is really the reorganization, and not the x, and that's what I must have meant. The reaction actually occurs at a particular x, *the reaction coordinate is not that of an x motion but of a reorganizational motion.* And you see, *often if you think of a* *bimolecular reaction you think of the reactants coming together, but the reaction coordinate only at the very beginning is that of reactants coming together, but when a lot of the process is taking place and the reactants are all close to each other, the real reaction coordinate is some combination of coordinates*, and that's what I must have meant, but I should have expressed more fully. So, *x is a vehicle for going from M^+ at one end to M^+ to the transition state region. The actual reaction coordinate there is occurring at a particular x, but x is not part of that reaction coordinate, it's a region where the reaction is occurring, but the reaction coordinate is something else, the reorganization is the actual coordinate. The reaction is occurring all way over a limited region of x, you see. If it occurred over a whole set of x's then x would be very much a part of the reaction coordinate.*

12. Following: "When it *is* the reaction coordinate the transition state occurs by definition only at particular x and not over a range of x."

Q: Please explain the relation between *the* reaction coordinate and major and minor contributions to it.

M: That's right. If you define a reaction coordinate in a very generalized way, then for M^+ diffusing up to that transition state region where M^+ is in between there, that's a reaction coordinate, but is not playing a real role in the rate out of that in the diffusion part, and all of the work goes in, as far as details of reaction, at that x which is where the M^+ is right there in the middle, and there the reaction coordinate becomes a combination of other coordinates, reorganization and so on.

Well, yes, if you go back to the Sumi–Marcus paper, where you have a two-dimensional treatment, and the reaction can occur by crossing different parts of that diagonal line, then clearly the reaction coordinate involves some combination of those two coordinates.

13 Following: "In that case one replaces the pair of equations Eqs. (A.1–A.2) by a single equation, the Kramers–Klein equation for the population."

M: Well "in that case" what case is that?

Q: The case in which the reaction coordinate is really the x.

M: Oh, when the reaction coordinate is the x, then essentially that whole diffusion is the slow coordinate, and in that case then if you use Kramer's theory, if you use it in the form where he has a diffusion controlled rate, you have a diffusion constant really in front of it and that's the rate expression,

Q: But why that thing should substitute Eqs. (A.1) and (A.2)?

M: Well, because essentially, under diffusion control conditions, the concentration of the reactant is zero where the transition state is occurring, the reaction there occurs so quickly when the reactants reach there, that the effective concentration is zero, and so your rate constant is associated with the diffusion controlled rate to reach from your starting point to that point there, and what happens in the reaction is irrelevant because things are happening so quickly that they don't affect the overall rate constant.

M314 Electron Transfer Past and Future

R. A. Marcus

0. Q: Let me first ask a general question: you often refer to the ET reactions as the simplest reactions in Chemistry. But which one is the simplest between $A + B^- \rightarrow A^- + B$ and the energy transfer reaction $A + B^* \rightarrow A^* + B$?

M:*The energy transfer is not a chemical reaction, it's an energy transfer process*, there no structure or valence or change, there's no chemical change going on.

01. Q: I have a question for the top scientist in unimolecular reactions: normally on the boxes of drugs one finds an expiration date. Consider that most of them are organic compounds in the solid state. Is the expiration date due to decay due to unimolecular reactions of the molecules? If so, does this mean that most molecules used as drugs are inherently unstable in time? Or is the expiration due to reactions of the molecules with neighboring ones in the solid state?

M: I don't know the answer to your question, because the drugs contain various things besides those drugs themselves and I don't know whether they are decoupled from the drugs, or there are reactions of the drug with the surrounding material. I just don't know enough about that I wouldn't be surprised though that for reactions of molecules in the solid state there may be oxidation, with oxygen getting into it.

Q: But would you exclude unimolecular decays somehow?

M: I wouldn't exclude anything...I don't know whether oxidation is the main process or thermal decomposition, the thermal decomposition could well be unimolecular, I simply don't know.

1. p. 3. In Fig. 1 you schematize the developments in the electron transfer field after the 1940s. How and why the inverted effect is related to

(1) Chemiluminescence
(2) Solar energy conversion
(3) Photosynthesis...

M: (1) The chemiluminescence...when the ground state is unaccessible, reactions are so exothermic, an excited state may be accessible, so you get an excited state of the product form, and it can either luminesce or combine with another excited state to form a high excited state which then luminesces.

(2, 3) The solar energy conversion . . . a Nobel poster gives an example where after you excite the special pair, say, in the bacterial corpus system, the system transfers an electron to a bacterial chlorophyll next and then, rather than go back, which it could, there is a very downhill reaction that goes forward, to a pheophytin, it doesn't go back presumably because of the inverted effect, you know, it's a very downhill reaction going back.

Q: Is it today sure that it is inverted effect?

M: Nothing is sure, it's fairly sure that going back would be very downhill, that's certain, but how much of it is due to different orbitals playing a role, and so on. I don't know. But anyways, that's what the Nobel committee thought, and I'm not going to argue with it . . . that's what many other people think too. And I know in some solar cells that Harry Gray and others have work done, they say they were in the inverted region for going back, but I don't know.

2. p. 3 middle: "While fireflies have been around for quite some time, chemical 'lightsticks' were not available, and the relation of these forms of chemiluminescence to several chemical reaction steps, of which one is an ET . . . "

Q: What is a lightstick? Can you say something about ET in their case?

M: Oh, what is a lightstick? That's found in the Nobel poster, that's a stick where you bended and it gives off light, by stretching or bending you created some chemical reaction, so it's actually on the Nobel poster.

3. p. 4: "Lasers have already been applied to ETs . . . "

M: To study ET reactions, that's my guess now, I've forgotten now . . .

4. Following: "...but application of *the combination of lasers, molecular beams, and computer technology to ETs*, in clusters for example..."

M: Yes, I think that was somebody's work in clusters, I've forgotten now. In other words, if I remember it rightly, you can observe an ET in a cluster.

5. Following: "*Coherence phenomena in electron transfer, initiated in the photosynthetic reaction center* ..."

M: In this particular case when the electron goes from the special pair to the pheophytin, does it do so by superexchange, that is called the coherence phenomenon.

6. Following: "...the possibility of a material such as DNA serving as a 'molecular wire.'"

Q: Please explain in relation to ET theory.

M: There's been a lot of claims that you can get ET over long lengths of DNA. Now, clearly that can't happen as a coherent process, it's far too long, the ET dies off exponentially, it could happen if somehow you can go by hopping. Now, you can go by hopping if you can access the so to speak conduction band of DNA, the upper band of the DNA. To do that, if you attach a molecule you can excite it and then the electron can go into that conduction band.... Wasielewski, at Northwestern University, for example, has studied that process.

Q: Also your colleague there, Jacqueline Barton.

M: Yes, she studied that too, that's right, she excited something else, something inorganic, I think though that the electrochemical work she did recently has been changed, there's a matter I think of DNA bending over at the end... but I haven't discussed that with her lately.

7. p. 5: "...scanning tunneling microscopy, a form of electron transfer..."

Q: How's this form of electron transfer related to ET theory?

M: Well, it's really going from one electrode, the tip, to the other, it's a transfer of an electron, but I don't know if I should have called it an electron transfer. It is an electron transfer but it's more from one metal to another, not *through an intervening medium*.

Q: So, it has no direct relation to your ET theory.

M: No, because *there isn't a reorganization there*, and similarly for electron transfer for monolayers...that's a separate study that has been investigated.

8. Following: "...the apparent observation of STM currents in relatively thick nonconductive samples, and the similarly low-energy barriers of organic monolayers (inferred from the dependence of tunneling currents on STM tip-sample distance at a given potential difference) for STM currents remains to be explained."

Q: Why does it remain to be explained?

M: If you have a relatively thick sample, and you have a barrier of 2 eVs, then it's hard to see how you could transfer that as a coherent transfer, exponentially would be too small, it probably decays as $e^{-\beta r}$ where β is 1 per Å, so if it were thick you couldn't expect anything.

Q: So it is unknown.

M: I don't know if it occurs and if maybe there are impurities there. I've no idea.

9. p. 6: "The forward reaction involves an excited electronic orbital of the photoexcited reactant, while the back reaction involves reversion to the reactant's ground-electronic-state orbital. The latter is *probably* less extended in space than the former... Again, the back reaction is very downhill and in a region of the protein (in the photosynthetic system) that is relatively nonpolar. The latter factor results in only a small 'reorganization energy,' and so, in conjunction with the large negative ΔG^0, this back reaction *may be* in the 'inverted region,' and *hence be slow*."

Q: (1) Is it today definitely established this inverted region effect for the back reaction?

(2) How does the tunneling compare for the forward and backward reaction? Are the reactions happening through tunneling in general slow or are there also quick reactions going through tunneling?

M: (1) Let me first answer the first subquestion. *In order to establish that is definitely inverted effect, one would have to vary the free energy difference, you know, and see if it goes in the correct direction. If for example you make the free energy difference more negative does the rate decrease? That would be the key point, but there is a limit to how much you can vary that free energy, you're not dealing with the usual system that John Miller, say, did. So, I don't think you can vary that energy very much, so... rigorously tested in what would be the best way, maybe really the only way, out than pure calculation, is probably difficult to do.*

(2) The slowness of which reaction?

Q: The one in the inverted effect

M:*I don't know if the tunneling would be any different for the forward reaction compared with the back reaction. I mean, the electron tunneling would be somewhat less maybe because the orbital of the dimer, in the bacterial chlorophyll dimer, is a ground state orbital... instead the forward reaction is from excited state orbital, and an excited state orbital is extended a little bit further into space than ground state orbitals, so there's that difference but that wouldn't* *say that there is any more tunneling, you know, if the forward reaction goes by tunneling then the backward reaction is probably tunneling too. The inverted reaction is slow even in the absence of nuclear tunneling. Nuclear tunneling makes it to be a little bit less slow.* I mean, *it's the intersection of the two potential energy curves at a higher point, in the case of the inverted effect, that makes it slow. I had imagined you had the two potential energy curves and, in the case of inverted effect, the intersection of the two curves is such that both curves have slopes of the same sign at the intersection instead of opposite signs. All have the same slope sign. Now, for the reaction to happen the system has to reach that intersection, or tunnel through it, but both ways are going to be slow, one energetically slow and the other because there is tunneling, nuclear tunneling.*

10. Following: "How to utilize the charge separation in these systems so as to then harness it remains yet another major goal in application via synthesized systems."

M: If you have a charge separation, then there is maybe a driving force to bring the charges back, and maybe you can utilize that. For example, in the photosynthetic reactions... ultimately you produce photons that come out at the other hand, that due to electron transfer, and those photons can try to go back and cross the membrane, and so doing make ATP work and that is an energy producing process, in form of ATP. So, in other words, there're ions that are formed that eventually provide a driving force, in a complicated way, for other processes to occur in nature.

CHAPTER 9

Linear Response Theory of ET Reactions as an Alternative to the Molecular Harmonic Oscillator Model; Nonadiabatic Electron Transfer at Metal Surfaces; On the Theory of Electron Transfer Reactions at Semiconductors Electrode/Liquid Interfaces; On the Theory of Electron Transfer Reactions at Semiconductors Electrode/Liquid Interfaces. II. A Free Electron Model; Temperature Dependence of the Electronic Factor in the Nonadiabatic Electron Transfer at Metal and Semiconductor Electrodes

Interviews on M315, M317, M318, M322, M323

M315 Linear Response Theory of ET Reactions as an Alternative to the Molecular Harmonic Oscillator Model

Yuri Georgievskii, Chao-Ping Hsu, and R. A. Marcus

1. p. 5307 1st column top: "*Interaction with an environment* plays a crucial role in many *nonadiabatic* processes in condensed phases. Electron transfer reactions provide a major example in which *strong* electrostatic interaction of a reacting species with a polar solvent can control both *the energetics* and the dynamics of the process. The theory of *nonadiabatic transitions*, in the presence of *strong interactions*..."

Q: When I think of nonadiabatic processes in terms of your classical parabolas, I think of processes with no splitting or very small splitting, that is, of processes with very small interaction between R and P state. But here we see that these states are also characterized by strong solute–solvent interactions. Is there some relation between the *state–state interaction* and the *solvent–solute interaction*?

M: The answer to your question is no. *Is there some relation? No. The state–state interaction is really the electronic matrix element directly coupling the electronic orbitals of the reactants, that's quite independent of the solute–solvent interaction. Yes, largely independent. The interaction between the solute and the solvent are strong. The interaction energies are of the order of hundred or couple of hundred kilocalories per mole, depending on electric charges and sizes, so the interaction between the solute and the solvent is extremely strong, you know, a couple of hundreds kilocalories maybe, while the electron transfer kind of interaction, the coupling matrix element, may be less than a tenth of a calorie, it depends.*

2. p. 5307 2nd column top: "In their derivation Ovchinnikov and Ovchinnikova considered uniform systems, so dielectric image effects and any effect of the solute on the properties of the nearby solvent were not included. Image effects have been included earlier[7–9] on the assumption that *the slow and fast solvent modes allow for a clear separation.*"

Q: Separation of what?

M: This (of Ovchinnikova) is an important paper . . . they were the first ones who brought in, the best as I remember, the dielectric dispersion that, later on, was brought in by Dogonadze . . . but these were the first ones to bring it in a 1970 paper.

Slow and fast solvent modes probably refer to orientation, and so on, and vibration, vs. the electronic, and you see, what they do is really blend the whole range in theirs is a more rigorous treatment.

I assume that the electronic adjusts it immediately and the others just as slowly, and that was it, but they point out to *the whole range of frequencies*, so their derivation is more general with respect to the description of the dispersion of solvent frequencies, but less general because they don't have the dielectric image effects, they just have a homogeneous medium with no holes in it, no holes, no surface really, then there is no image effect. Image effects appear when you have surfaces, so they didn't take into account the surfaces, I don't know how big the effect is, but they didn't take that into account. We often assume that the slow and fast modes allow one to treat the quantities somewhat differently but what that is maybe saying is that I took image effects into account, then separated the slow motion, the dielectric solvent orientation polarization, for example, from the vibrations, the fast vibrations.

Q: And what is the relation with the dielectric image effect?

M: Well, the dielectric image effect is something that I wrote about in some paper... *if you have a charge near a low dielectric region, and each ion is itself a low dielectric region, then the interaction is more than just* e^2/r*, it's induced dipoles and so on, and you can treat them approximately by image charges.* They didn't consider that because they simply used a uniform medium.

3. Following: "Song, Chandler, and Marcus considered the effect of the solvent inhomogeneity due to the solute presence using the *Gaussian field model*, which could be viewed as a continuum equivalent of a molecular harmonic oscillator model."

Q: Can you please give a simple description/introduction to this model?

M: I can't because I really followed along what they did, so Song and Chandler they are really the ones to refer to... but basically I think it involved that you treat the whole system in terms of fields and you go up to quadratic in the sort of Hamiltonian for that field description, and so, if it's goes up to quadratic, that uses a Gaussian field for distribution.

Q: These guys dressed up and extended the theory but did someone derive some formulas used by experimentalists?

M: Well, the main one is Jortner who combined the work that I did with essentially work that he was familiar with in radiationless transitions, or you can say combined the work that I did and the work that Levich and Dogonadze did, to bring in quantum vibrations together with a classical description of all the other motions, so he used my formula or parts of it, for the lower frequency motions and then for the few high frequency vibrations used a quantum formula, and that gave rise to a formula which is simple when the system is really reacting from the lowest of those vibrational quantum states, and that was the case in the famous paper of Miller, Calcaterra, and Closs, and in what it produces, certainly some asymmetry in that parabolic plot of $\log k$ vs. ΔG^0.

Q: So that's basically the only practical important contribution.

M: Well, I think so, well . . . I mean you have that nice work of Ovchinnikov and Ovchinnikova, and so that gives a dielectric dispersion and then . . . in some later work of the Russians but, you know, there is a more complicated formula, you know, there is the question if it is worth the extra effort. And then some of the results that I derived, like the cross relation, are in a certain sense independent of some of these other properties, in other words there is sort of cancellation, and so the classical formula leads to roughly the same result that the quantum formula would, for some relationships.

4. p. 5308 1st column middle: ". . . a generalized coordinate *X* is introduced. (Use of the latter provides a way of avoiding the field theoretical treatment.)"

M: Well, yes, it depends on what theoretical treatment... The Ovchinnikova's theory used the field theoretical treatment due to that famous Russian...

Q: Abrikosov.

M: Yes, Abrikosov... and so they use that, so I'm not sure if that's was what I was talking about, or if I was talking about to something related to the Song, Chandler, Marcus theory. The main point is that if you have sort of a symbolic coordinate X in its quadratic domain, *that quadratic property is sort of the essential part when the free energy or energy is quadratic in X*. You know, in general it can be quadratic, cubic, can be anything, one maybe maps a three dimensional coordinate in one dimensional coordinate, often you can get the behavior of something more complicated by looking at the behavior of a simplified model, you know just like there are theorems in Hilbert space that are analogous to the same theorems for geometry, in other words certain things that are carried over with one language in one formalism and one topic may carry over to a much more complicated description in another topic, certain aspects like the quadratic nature of the free energy functional. You have a functional that is a quadratic function of some property, polarization, or molecular coordinate, or what have you, or of some more generalized coordinate such as the energy difference of the two electronic states at each configuration, the coordinate that I used in the 1960 and in 1965 papers.

5. p. 5308 2nd column top: "The Hamiltonian operator of the system reactants + solvent in either electronic state can be written as

$$H_j = h_j + H_0 + H_j', \quad j = 1, 2 \tag{1}$$

where h_j is the Hamiltonian of the 'free reactants,' H_0 is the Hamiltonian of the 'free solvent,' and H_j' is the interaction between them. The H_j' *depends on the electronic state of the reactants*. The Hamiltonian h_j includes the intramolecular vibrational modes of the reactants.... we consider structureless reactants *and so treat* h_j *as just numbers*."

Q: (1) How about the dependence of the solute–solvent interaction on the electronic state of the solvent molecules? Does one consider (does it exist) a collective solvent molecules electronic state?

(2) Why is h_j treated "as just numbers." What does it mean?

M: (2) Well, h_j contains the vibrational coordinates, but if the reactants are structureless and they really don't have any vibrational coordinates, they still have some *mean energy* but they don't have energy as a function of some coordinates. h_j refers to the reactants just moving in space, maybe the free reactants have vibrations, but if you didn't assign any structure to the reactants they couldn't have any vibrational coordinates, in other words you have a model that doesn't have any vibration, there are no vibrations and no vibrational coordinates. But you can still speak of other aspects, like the electronic energy of the molecule.

(1) Surely the solvent molecules have electronic states but there is no advantage to using those electronic states, in other words each solvent molecule can be electronically excited, if you want, by 20, 100 or whatever kilocalories of energy into it, you can produce an electronic excitation, but putting all that energy in it isn't really helping the electron transfer reaction.

Q: So, the electronic structure of the solvent only appears in determining the dipoles.

M: Yes, that's right, and also any polarizability there comes in a virtual state. *In the two stages charging process I went to a kind of virtual state, not a real state, something that produced the desired nuclear polarization.* That's maybe the virtual state, it's not a real state, it is the device to reach a nonequilibrium situation in a kind of quasi static manner.

6. Following: "The operators H_0 and H_j' act on *the nuclear wave function of the solvent.*"

M: The electronic states *of the solute* act on the nuclear wave function of the solvent, they can polarize the solvent, so to that extent they act on the electronic wave function.

Q: But the electronic wave function of the solvent appears only as dipoles and modified dipoles.

M: Yes, the main point is that of course anything which is being polarized has in its lower state some contribution from the higher states according to second order theory of polarization, of polarizability. If you have something that's polarizable, and is in a nonequilibrium situation, its energy, its free energy might be higher than what would be normal, and the system would relax to the value that gives the lower free energy, so you can imagine some nonequilibrium state that doesn't have the lowest energy but it's a hypothetical state that you bring in for some purposes and it would relax to some lower free energy.

7. p. 5308 2nd column bottom: "The electrostatic interaction between the reactants and the solvent can be written . . . as

$$H'_j = -\int \mathbf{D}_j^{(0)}(\mathbf{r}) \cdot \hat{\mathbf{P}}_0(\mathbf{r}) d^3\mathbf{r}, \quad j = 1, 2 \tag{2}$$

where $\mathbf{D}_j^{(0)}(\mathbf{r})$ is the electric field at point $\mathbf{r}$, created by the reacting species in vacuum for the jth electronic state and $\hat{\mathbf{P}}_0(\mathbf{r})$ is the operator representing the molecular solvent polarization at the same point, when the solute charge distribution $\rho(\mathbf{r})$ is set equal to zero. We use a caret notation for $\hat{\mathbf{P}}_0(\mathbf{r})$ to stress that it is an operator acting on the *nuclear wave function of the solvent.* Although we do not explicitly use the following expression for $\hat{\mathbf{P}}_0(\mathbf{r})$ in Eq. (2) it could be defined in *a long-wave approximation* as *a weighted sum* of the individual dipole moments of the solvent molecules:

$$\hat{\mathbf{P}}_0(\mathbf{r}) = \sum_n \delta(\mathbf{r} - \hat{\mathbf{r}}_n)\hat{\mathbf{d}}_n \equiv \sum_n \delta(\mathbf{r} - \hat{\mathbf{r}}_n)(\hat{\mathbf{d}}_n^0 + \hat{\mathbf{d}}'_n) \tag{3}$$

Q: (1) Where are the solvent electrons? I guess they are indirectly present determining the polar moments. Is it so?

(2) What does it mean "in a long-wave approximation"?

(3) How is the above sum "weighted"? I only see delta functions. Are they the "weighting" factors?

M: (1) They are in the polarizability and in the dipoles.

(2) You notice that's a $\hat{\mathbf{P}}_0(\mathbf{r})$ and not a $\hat{\mathbf{P}}_0(\mathbf{r}, t)$. $\hat{\mathbf{P}}_0(\mathbf{r}, t)$ brings in dependence on frequencies and so a long wavelength... that's a wavelength associated with the time behavior, because there can be a wave, you can take a Fourier transform of that $\hat{\mathbf{P}}_0(\mathbf{r})$ and you can think of a wavelength associated with the $\mathbf{r}$ description, but that's a wavelength associated with a time description, so we're really working in the static approximation for the orientational vibrational polarization, otherwise we would have a t in there.

Q: And what does this long wave refer to?

M: It refers to *low frequency motion of the solvent*. In other words, you really have a solvent dispersion and you have some very slow motions, orientational, you have some fast motion, vibrational, and you have the electronic. But we are just taking two instead of three or more categories, then the slow motion is everything but the electronic comes in in the form the refractive index.

Q: Long wave approximation is then synonymous of slow motion.

M: Yes, that's right, that's what's intended here.

(3) Well, it's weighted being centered in certain positions, you have the dipoles, I assume that's what is meant.

Q: The delta functions.

M: Yes, they are weighting the position of each dipole, that's what was intended. They're saying where the dipoles are. I imagine that is what was meant.

8. Following: "$\hat{\mathbf{r}}_n$ and $\hat{\mathbf{d}}_n$ are quantum mechanical operators which act on the wave function in the configuration space of all nuclear degrees of freedom of the solvent, after *an averaging over the total electronic wave function of the solvent has been performed*."

Q: It is the first time, I believe, that you mention the total wave function of the solvent and its averaging.

M: The wave function of the solvent would include a function of electronic coordinates of the solvent and nuclear coordinates, and we have averaged in some sense over the electronic coordinates. Now, there are more sophisticated treatments that would treat electronic and nuclear motion more or less on the same basis, on a generalized basis and not making such a sharp distinction, and I'm sure there are systems for which that is necessary but I didn't do that, the person who may have been closest to doing that for some particular problem was probably Chandler in one of his papers, the paper that he wrote with J. T. Hynes.

9. Following: "It may be stressed that *no averaging over any nuclear motion of the solvent,* such as over the intermolecular or intramolecular vibrations of the solvent molecules, is performed in Eqs. (2) and (3)."

Q: (1) You say in Question **7** that there is a "weighted sum" of the individual dipole moments of the solvent molecules. Isn't that an average?

(2) In your classical theory you consider solute intramolecular vibrations wherefrom λ_i. Can you say something about these intermolecular and intramolecular *solvent* vibrations?

M: (1) No, because we haven't multiplied by some coordinate and some distributions and integrated over them.

Q: You speak of a weighted sum but that's not an average.

M: That is not an average. By the way, you know, when one uses weighting in some other sense it could be that one has a weighting function and then if one does the integration one has an averaging but here there is really just a weighting of each point by the position of the atom and no averaging has been done yet.

(2) Well, the intramolecular vibration refers to the solvent molecules actually doing some vibrating. *Intermolecular is when it is vibrating against other solvent molecules, it's like the solvent molecule in a little cage and vibrating against other solvent molecules. That's what that would be. Or, for example, if you have some sort of binding of a solvent molecule to another solvent molecule, for example, hydrogen bonding, that would be intermolecular.*

10. p. 5309 1st column bottom: "The Hamiltonians H_j, $j = 1, 2$, describe the collective motion of the system in the initial (reactant) and final (product) electronic states of the solute. To describe the transition between the two, one must introduce the nondiagonal matrix element Δ which couples the two electronic states. *This matrix element is sometimes small and is assumed to be independent of the nuclear configurations for statistically important configurations (Condon approximation).*"

Q: Is there any relation to the Franck–Condon principle?

M: There probably is... well, in some sense... *The coupling element really depends on the nuclear coordinates, and if you neglect that dependence on the nuclear coordinates, it just depends on the electronic states, that's called the Condon approximation* and I am not sure if it is related to the Franck–Condon principle, it is often used in conjunction with the Franck–Condon principle, but I guess it's a separate thing.

11. Following: "Non-Condon effects could be of importance for solvated electrons and solutes *with weakly localized electronic clouds*, but not for tight redox couples."

M: Well, *the electronic wave functions that are used are really just the tails of wave functions so just a very small region of those electronic wave functions is really used, it's only the tails that are important*, but if you look at the intersection of those two electronic surface curves, when only hopping from one to the other in a fairly narrow range of the coordinates is involved, you have fairly sharply localized electronic wave functions and it is maybe necessary to include the electronic interaction and the matrix element does depend upon position . . . and the range of positions of the electron is extremely narrow . . . although, if the processes are really adiabatic, the electron would be distributed over a range of coordinates but then you have to modify the treatment.

12. p. 5309 2nd column top-middle: "$X = H_2 - H_1$ (7) is the *energy difference* of the system in the final and initial electronic states of the solute"

p. 5309 2nd column bottom: "In a molecular harmonic oscillator model the higher cumulants vanish exactly if *X is a linear combination of the normal modes*"

M: What's intended there is maybe some combination of the normal modes, some linear combination of the normal modes of the system. $X = H_2 - H_1$ is the energy difference that I introduced in the 1960 paper, that happens to be *a generalized coordinate*, but normally if you use *normal coordinates* then those q's aren't so simply related to difference of energies. *The use of normal coordinates is not so simply related to the generalized coordinates.* The main point is that if one refers to normal coordinates then the X would not be the energy of normal coordinates. For a normal coordinate q the energy is $kq^2/2$ *That generalized coordinate X is more general than normal coordinates, normal coordinates assume that you have a Hamiltonian* *which is harmonic, but in the actual Hamiltonian, because of the bending vibration, the breaking and forming of hydrogen bonds, only some of the vibrations are harmonic, so the generalized coordinate has nothing to do with normal coordinates, it is an energy difference. If you have a total of N coordinates where* $N = 10^{23}$*, then you can speak of an* $(N-1)$*-dimensional space such that all systems of the reactants and the products belonging to that space have the same energy difference, and so you can do statistical mechanics out of it, which is what I did in the 1960 and 1965 papers.*

13. p. 5310 1st column middle: "The correlation function $C(t)$ for the anharmonic molecular system is next expressed in terms of the dielectric properties of the solvent. To this end linear response theory is used, after introducing an effective Hamiltonian $H_{\text{eff}}(t)$:

$$H_{\text{eff}}(t) = H_1 - Xf(t). \tag{18}$$

Here, the *energy difference* X plays a role of a generalized coordinate (reaction coordinate) and $f(t)$ is a generalized dimensionless force. Thereby, $H_{\text{eff}}(t)$ can evolve from H_1 to H_2, if $f(t)$ is chosen to tend to zero as $t \to -\infty$ and to become unity as $t \to \infty$."

Q: (1) Is there a relation between X and the coordinate Γ of your 1960 paper on Discussions of the Faraday Society?

(2) Is $f(t)$ characterized *only* by the above limits?

M: Well, essentially you can start saying: take any $f(t)$ such that it has those limits, and so you start off with H_1 and you gradually move over... you know, you start off with X being the energy difference, this is saying that your $H_{\text{eff}}(t)$ goes over from H_1 to H_2, and that X in the 1960 paper really evolved over the potential energy path, really involved all of the coordinates, it was a function of 10^{23} coordinates, that X there is certainly the Γ in my 1960 paper... it is certainly the energy difference, that $X = H_2 - H_1$ is really meaningful only if the H_2 and the H_1 are functions of all of the coordinates. Now we want to apply linear response theory and so this is a standard theory, somehow you have some generalized coordinate and you apply some external force, and I can convert the electronic states from one to the other over a period of time. So, doing something like that is an application to this problem of what is standardly done in linear response theory in nonequilibrium statistical mechanics, so doing that meant that one can then apply the formalism of linear response theory, so that's what that is. So, here that $f(t)$ is, I guess, some hypothetical force that one varies and can take one from one electronic state to another electronic state. That X there, that displacement coordinate is multiplied by $f(t)$, that form is standard nonequilibrium statistical mechanics theory. X is not constant, it's a function of all of those coordinates, X is varying, it is a response, $X$is not a constant. $X = H_2 - H_1$ is just a definition of X. How X depends on all the coordinates of the system, if you specify the 10^{23} coordinates you get a certain value of H_2, a certain value of H_1, that gives you the value of X, and then, now that we have got a generalized coordinate, we can introduce the procedures of nonequilibrium statistical mechanics, and that means that you can take a generalized coordinate, multiplied by some external force and then let your system evolve, and then apply certain standard formula that is sort of a huge story itself.

Q: How about the Γ in your 1960 paper?

M: This is going far beyond that, I mean, in that paper there I didn't derive time dependent formalism and time dependent statistical mechanics, this is what is being done here, so the theory is going beyond.

Q: So, Γ would be here the X.

M: Yes, that's right. The Γ is the X, yes.

Q: And here you add the consideration of $f(t)$.

M: Yes, because here now we're introducing time dependent statistical mechanics, which we didn't do before.

14. p. 5310 2nd column middle: "$C(t)$ and $J(\omega)$ are molecular statistical mechanical properties. We turn next to their calculation in terms of the dielectric properties of the solvent, in *the long wavelength (local response) approximation* for the solvent."

M: The local response yes. The polarization is a function of $\mathbf{r}$ and t and what the Russians did was to introduce a Fourier transform both in time and in space, so they had certain integrals like the $\int \mathbf{P}(\mathbf{r}) \exp(ik\mathbf{r})d\mathbf{r}$ and if you know that dependence on k then you're really using the long wavelength approximation, in other words you are neglecting the fact that $\mathbf{P}$ is changing rapidly over short distances, and that's what that long wavelength approximation is and has nothing to do with frequency as long wave.

Q: Two different senses.

M: Yes, that's right, that was the sense, and the Russians did a lot of that work, but the formalisms aren't really new because, you know, one does know a lot about that, like that k-dependence, and *the main point is that you want a formalism that you can relate close to experiment. You want simple formalism.*

15. Following: "$\mathbf{D}^{(0)}(\mathbf{r}, t) = \mathbf{D}_{21}^{(0)}(\mathbf{r}) f(t)$ (27), is the vacuum electric field formed by the *external*, time dependent charge distribution $\rho(\mathbf{r}, t)$. The latter is a linear combination of the charge distributions of the solute in its initial and final electronic states:

$$\rho(\mathbf{r}, t) = \rho_{21}(\mathbf{r}) f(t), \quad \rho_{21}(\mathbf{r}) = \rho_2(\mathbf{r}) - \rho_1(\mathbf{r}). \tag{28}$$

Q: The time dependent charge distribution here doesn't go from the one of the reactants to that of the products but from 0 to the difference of charges between the final and the initial charge distribution.

M: That's a very interesting point Francesco, very interesting. I think it should have been stated in much more detail then. Let me come back to this stuff in nonequilibrium statistical mechanics. Normally what you do you is you apply some external field to some property and you look at how some property **D** changes with time, and this **D** you normally subtract from that the equilibrium **D**, so you have $\mathbf{D}(t) - \mathbf{D}_{\text{equilibrium}}$, OK? And that's what probably should have been done here, so that you really start from $\rho_1(\mathbf{r})$ and now you add to that some constant, say, times $\rho_2(\mathbf{r}) - \rho_1(\mathbf{r})$ during the course of the change, you see, so I think what they did here was, maybe without explicitly stating it, that they subtracted that $\mathbf{D}_{\text{equilibrium}}$, so they were looking at the change that's going on and the change that's going on depends on the density difference, the absolute value depends upon the density, so that's probably what is done there. So, all of this is closely related to the, let's say, standard treatments.

M315 Linear Response in Theory of Electron Transfer Reactions as an Alternative to the Molecular Harmonic Oscillator Model

(Continuation)

Yuri Georgievskii, Chao-Ping Hsu, and R. A. Marcus

1. p. 5310 bottom: "As a result the standard electrostatic equations for $\mathbf{E}(\mathbf{r}, t)$ and $\mathbf{D}(\mathbf{r}, t)$ can be used:

$$\nabla \cdot \mathbf{D}(\mathbf{r}, t) = 4\pi\rho(\mathbf{r}, t), \quad \nabla \times \mathbf{E}(\mathbf{r}, t) = 0. \tag{29}$$

$\mathbf{E}(\mathbf{r}, t)$ and $\mathbf{D}(\mathbf{r}, t)$ are coupled via the solvent electric susceptibility $\varepsilon(\mathbf{r}, \mathbf{r}', \tau)$ which is generally nonlocal in both the space and time domains,

$$\mathbf{D}(\mathbf{r}, t) = \int_0^\infty d\tau \int d\mathbf{r}'\varepsilon(\mathbf{r}, \mathbf{r}', \tau)\mathbf{E}(\mathbf{r}, t - \tau). \tag{30}$$

We will use a standard, *long-wave approximation* for $\varepsilon(\mathbf{r}, \mathbf{r}', \tau)$ in which

$$\epsilon(\mathbf{r}, \mathbf{r}', \tau) \cong \epsilon(\mathbf{r}, \tau)\delta^3(\mathbf{r} - \mathbf{r}') \qquad (31)\text{"}$$

Q: How is the relation between long wave approximation and the approximation in Eq. (31)?

M: I am just wondering whether we should call that long wave approximation. Probably it should be called a *local approximation*, you know, just like for the time domain. Normally in the simplest electrostatics **D** is directly proportional to **E**, $\mathbf{D} = \varepsilon\mathbf{E}$. More generally, it depends nonlocally on t, so you get that $\int d\tau$, and the $t - \tau$. In the local approximation that would be a delta function, you would have an $\varepsilon(\mathbf{r}, t)\delta(t - \tau)$, and so *that would be local in time*, you see, but many properties are not local in time but approximately so, *when it's not local in time then you get typically a frequency dependent dielectric constant*, and I guess the same is probably true for the spatial part also, and, for example, what I'm not sure of is if the most general expansion for that ε function, for that $\varepsilon(\mathbf{r}, \mathbf{r}', \tau)$, would be some sort of Fourier decomposition and *maybe the long wavelength part of that Fourier decomposition goes over to this local approximation*, but I'd really have to read up on that again and to check out the terminology, but if you look up in the most general place, whatever the most general place, it *yields a dielectric dispersion in terms of solvent electric susceptibility nonlocal in time and nonlocal in distance*, then you will probably see the correct way of phrasing it, and maybe that's the correct way of phrasing it but I just don't remember, but it corresponds to a local approximation, local with respect to **r**, that $\mathbf{D}(\mathbf{r})$ depends only on **r** and is not delocalized over some $\mathbf{r}'$, you see. So, if we take units of standard local approximation for that, then I wouldn't have any problem, but whether that corresponds to the long wavelength approximation . . . let me see . . . you see, what's confusing me is that even if we had a $\mathbf{D}(\mathbf{r}, \tau)$ you can still Fourier decompose the $\mathbf{E}(\mathbf{r})$ and so *it may be that the correct statement there might be local instead of long wave.*

Locality and nonlocality . . . maybe Ulstrup has something on that or, maybe, that book by Ulstrup and Kuznetsov, has something like that in there, I mean it will be in the literature some place and not necessarily in the Russian literature but it will be there to get the correct terminology.

2. Following: "One feature of Eqs. (29)–(31) is the inhomogeneity of the dielectric environment reflected in $\varepsilon(\mathbf{r}, \mathbf{r}', \tau)$, since the Hamiltonian H_1 includes the *nonelectrostatic interaction* between the solvent and the reacting species which defines *the structural properties of the solvent* in the vicinity of the reactants."

M: As soon as you have a certain structure property of the solvent, as soon as you put a solute in there, that is going to change the structure properties around it, it tends to reorient solvent molecules, to break up some hydrogen bond structure, might even induce some hydrogen bond structure, so the solute does affect the local structure properties of the solvent, that means that the solvent in the presence of a solute is not exactly the same as the solvent when there is no solute. The structure of the solvent depends certainly also on nonelectrostatic forces, I mean there are some electrostatic forces, but the Van der Waals is sort of nonelectrostatic. Hydrogen bonding is more complicated than electrostatic, it is not just electrostatic, so in general the structural properties depend on what are nonelectrostatic properties. I mean, locally there is some electrostatics but the situation is more complicated than that.

3. p. 5312 top: "The frequency ω_{op} corresponds to the *transparency region* which separates the frequency region of the solvent nuclear motion and the one of the solvent electronic motion."

M: That region *is transparent in the sense that the imaginary part of the dielectric constant is zero there. The absorption is related to the imaginary part of the dielectric constant, the simple response is related to the real part, so there is a real part, it is such that is not absorbing but there is a real part.* In other words, the ratio of the local electric field to the local dielectric displacement will not be unity. On the other hand, there won't be any absorption there at that frequency, but there will be a response of the medium. In other words, the dielectric constant has a real and an imaginary part, there are the response functions, and the imaginary response is related to an absorption. *Here only the real part is nonzero*. In that region . . . the frequency is too high for the nuclear motion to respond to it, so only the electronic motion is responding. *In that transparency region the frequency is too high for the nuclear motion to respond either by absorption or by being polarized, the electronic motion is not absorbing but is being polarized.*

4. Following: "Calculations based on $\mathcal{F}_\omega$ can be used for realistic solute charge distribution, i.e., those not containing point charges, Eq. (36) would yield an infinity which cancels when $\mathcal{F}_\omega - \mathcal{F}_{\rm op}$ is calculated."

M: It may be that if you look at Eq. (36) and to the limit of the charge distribution in a very tiny region, atomic, those integrals will blow up, the integrals may explode, because of ee/r^2 and so the whole system blows up when $r \to 0$.

Q: A situation similar to that of the self-energy of the electron?

M: Yes, that's right.

5. Following: "To avoid this infinity, it is useful to introduce the energy $\mathcal{F}'_\omega$ of the 'dielectric' in the external electric field $\mathbf{D}_{21}^{(0)}(\mathbf{r})$:

$$\mathcal{F}'_\omega = \mathcal{F}_\omega - \frac{1}{8\pi}\int [\mathbf{D}_{21}^{(0)}(\mathbf{r})]^2 d^3\mathbf{r} = -\frac{1}{2}\int \mathbf{P}_\omega(\mathbf{r}) \cdot \mathbf{D}_{21}^{(0)}(\mathbf{r}) d^3\mathbf{r}. \tag{40}$$

That is, $\mathcal{F}'_\omega$ differs from $\mathcal{F}_\omega$ by the frequency-independent *self-interaction term* in Eq. (40)."

M: $\mathcal{F}_\omega$ is the second term in the first equality, so $\mathcal{F}'_\omega$ differs from $\mathcal{F}_\omega$ by the second term in the first equality, the second term is the self-interaction term, *there is no polarization, no frequency dependence.*

6. Following: "When the solute can be modeled with point charges q_j, the representation of $\mathcal{F}'_\omega$ in terms of an electrostatic potential $\phi(\mathbf{r})$ is *convenient*, Eq. (40) corresponds to

$$\mathcal{F}'_\omega = \frac{1}{2}\sum_j q_j \phi'_\omega(\mathbf{r}_j), \tag{41}$$

where $\phi'_\omega(\mathbf{r})$ is *the part of electrostatic potential created by the solvent polarization with the dielectric function* $\varepsilon_\omega(\mathbf{r})$."

M: It is simpler to work with scalars than with vectors. In Eq. (41) you have scalars, in Eq. (40) there are vectors and, especially in the case where you have point charges, everything becomes very simple with scalars.

7. Following: "In the case when the solute is modeled by a point dipole $\mathbf{d}_0$, it is more *convenient* to express $\mathcal{F}'_\omega$ in terms of an electric field,

$$\mathcal{F}'_\omega = -\frac{1}{2}\mathbf{d}_0\mathbf{R}_\omega, \tag{42}$$

where $\mathbf{R}$ is the *reaction field*, equal to $-\nabla\phi'_\omega(\mathbf{r})$, created by the solvent at the site of the dipole."

Q: Why does one use a potential in the case of charges and a force (a force field) in the case of dipoles?

M: That's the simplest expression that is sort of simpler than doing an integral.

8. Following: "Any spatial dependence of ε_ω reflects the structural inhomogeneity of the solvent in the vicinity of the reacting species.

In the simplest approximation, the polarizability of the solute is neglected and it is so assumed that $\varepsilon_\omega(\mathbf{r}) = 1$ in the region occupied by the reactants. An improved approximation for the *electronic polarizability of the solute would be $\epsilon_c \cong 2$ inside the cavity, corresponding approximately to an electronic polarizability of the solvent.*"

Q: Please explain... why $\epsilon_c \cong 2$?

M: The polarizability of the solute and the solvent are probably in the same ballpark. 1.77 is the square of the refractive index and $1.77 \approx 2$.

9. p. 5312 2nd column top: "The dielectric properties of the solvent enter into Eq. (43)

$$\alpha_\omega = 2q^2(\varepsilon_{op}^{-1} - \varepsilon_\omega^{-1})(a^{-1} - r_0^{-1}) \tag{43}$$

in the form of a factor $1/\varepsilon_\omega$."

Q: Why do you mention ε_ω only?

M: The dielectric properties of the solvent are represented also by $\varepsilon_{\mathrm{op}}$ but ε_ω *varies a lot,* $\varepsilon_{\mathrm{op}}$ *doesn't vary a whole lot from solvent to solvent, no comparison, but sure the dielectric properties of the solvent hinge on both, absolutely*, but in fact ε_ω depends more than $\varepsilon_{\mathrm{op}}$ on the solvent, $\varepsilon_{\mathrm{op}}$ surely depends on the solvent but ε_ω varies immensely but, in terms of $1/\varepsilon_\omega$ and $1/\varepsilon_{\mathrm{op}}$ since $\varepsilon_{\mathrm{op}}$ is so small, it may be that in going from solvent to solvent the $1/\varepsilon_{\mathrm{op}}$ varies more than the $1/\varepsilon_\omega$, because ε_ω is rather large, so $1/\varepsilon_\omega$ is relatively small. So, there is no reason to exclude $1/\varepsilon_{\mathrm{op}}$, not to mention it. ε_ω varies a lot but, of course, if one considers the reciprocals then $1/\varepsilon_{\mathrm{op}}$ does vary a lot and dominates over $1/\varepsilon_\omega$ usually. One should have mentioned both, yes.

10. Following: "Another example is the Onsager model... In this model the solute has a spherical shape but the charge distribution is approximated by a dipole d_0 in the center of a polarizable sphere of radius a and the electronic dielectric constant ε_c inside the sphere... Using Eq. (42) one then obtains that the α_ω in Eq. (35) in this case is

$$\alpha_\omega = \frac{(\varepsilon_c + 2)^2}{3} \frac{d^2}{a^3} \left[\frac{1}{2\varepsilon_{\rm op} + \varepsilon_c} - \frac{1}{2\varepsilon_\omega + \varepsilon_c} \right] \tag{46}$$

Comparing Eqs. (46) and (43) one can see that the factor $1/\varepsilon_\omega$ for the homogeneous medium is changed to the factor $1/(2\varepsilon_\omega + \varepsilon_c)$ in the Onsager model."

Q: Even here the factor $1/\varepsilon_{\rm op}$ which has been analogously modified is not mentioned. Is there some reason for that?

M: Yes, one should have mentioned it, except that we have focused on ε_ω because it varies a lot, but $\varepsilon_{\rm op}$, even though it doesn't vary a lot, is a very important term, so one should have mentioned that.

11. p. 5313 1st column top: "The spectral density *of the harmonic system* is temperature independent. For real *nonlinear systems* the spectral density was found to depend strongly on the temperature for low frequency component of $J(\omega)$ ascribed to diffusive motion of the solvent molecules."

M: Well, if you have a truly harmonic potential, then certain properties are independent of temperature. *If you go up in energy for an anharmonic oscillator the frequency decreases, eventually becomes zero, for a harmonic oscillator your frequencies are independent of energy, so they are independent of temperature. For an anharmonic oscillator typically you are going to have some different frequencies, some lower frequencies at higher energy, because you are dealing with an anharmonic potential.*

12. Following: "It is worthwhile noting also that while *for a harmonic oscillator solvent* the spectral density is, of course, the same in classical and quantum mechanics, classical and quantum mechanics will give different results for $J(\omega)$ of a real nonlinear solvent."

M: *If you look up in quantum mechanical books, you find that for harmonic systems certain properties are the same in classical and in quantum mechanics, they are the same for certain responses. Harmonic systems are special with respect to certain properties. Not for everything of course, the classical harmonic oscillator for instance doesn't have zero-point energy.*

13. Following: "The spectral density which enters into Eq. (16) can be readily rewritten as a Fourier transform of the imaginary part of the correlation function $C(t)$:

$$J(\omega) = 2\int_0^\infty dt\, \mathrm{Im}[C(t)] \sin \omega t \tag{47}$$

Using this equation as *a definition* of a spectral density of *effective* harmonic oscillators *a real anharmonic system could be mapped, if one wished, onto the harmonic oscillator model.*"

M: Supposing that $C(t)$ is obtained from the nonlinear system, then that $J(\omega)$ gives you the spectral density, then the question is: how would you go from the spectral density to a description of some properties of a collection of harmonic oscillators. In other words, once you have a $J(\omega)$, the upper term of that $J(\omega)$ comes from harmonic oscillators, there is a connection between $J(\omega)$ and the distribution of harmonic oscillators at different frequencies and so one might be able to invert, you see, because you start out with anharmonic oscillators, and that gives you the $C(t)$, and then you calculate the $J(\omega)$. Look up some book which discusses harmonic oscillators, some statistical mechanical books, time dependent statistical mechanics, and look up to see how one can invert $J(\omega)$ to get a distribution of harmonic oscillators.

14. Following: "In the present paper we considered electron transfer reactions. Our considerations, however, are applicable to other nonadiabatic processes in a polar environment in which the change of a charge distribution occurs *in the process of a nonadiabatic transition.*"

Q: Can you give an example of nonadiabatic process not involving electron transfer?

M: If you change excitation energy. Energy transfer.

15. p. 5313 1st column bottom: "By using Eq. (32)

$$\nabla \cdot \mathbf{D}_\omega(\mathbf{r}) = 4\pi\rho_{21}(\mathbf{r}), \quad \nabla \times \mathbf{E}_\omega(\mathbf{r}) = 0, \tag{32}$$
$$\mathbf{D}_\omega(\mathbf{r}) = \varepsilon_\omega(\mathbf{r})E_\omega(\mathbf{r}),$$

it is assumed that the dielectric response which appears in this equation in terms of the electric susceptibility ε_ω is *local* in space. This assumption may not be accurate for a field which varies considerably on a molecular scale and one can use a spatially *nonlocal* dielectric response function to calculate the electric field in the vicinity of the solute. At present, however, only nonlocal spatial dependence (or, equivalently, a k-dependence in the Fourier space) of the *static* electric susceptibility has been estimated, namely from neutron diffraction measurements."

Q: How would a *dynamical* electric susceptibility be in relation to this local vs. nonlocal problem?

M: If you have an $\varepsilon(\mathbf{r}, \mathbf{r}')$, *where* $\mathbf{r}$ *refers to a polarization and* $\mathbf{r}'$ *refers to a field* that's called a *nonlocal response*, in other words if the polarization at a point is directly related to the field at that point and ε is a function only of $\mathbf{r}$, then you don't have this $\mathbf{r}$ and $\mathbf{r}'$ formula, so that's called a nonlocal expression, and it may be that one of the Russians, I wouldn't be surprised, treated that possibility and maybe others have too, the formulas become more complicated.

M315 Linear Response in Theory of Electron Transfer Reactions as an Alternative to the Molecular Harmonic Oscillator Model

(Continuation 2)

Yuri Georgievskii, Chao-Ping Hsu, and R. A. Marcus

1. Following: "At any specified *solvent nuclear configuration* the solvent polarization operator $\hat{\mathbf{P}}_0(\mathbf{r})$, which refers to the hypothetical neutral state of the solute, can be viewed as *a real function of* $\mathbf{r}$ *and not only as an operator.*"

M: It is not only an operator. You know, some operators are just that, for example, if you consider the Schrödinger equation, the kinetic energy operator and the potential energy operator, the potential energy operator is a real function of the coordinates. So that's example of another system in which the operator is a function of **r**, it is not only an operator but also a real function.

Q: The electron transfer, for instance $A^+ + B^- \to A + B$ is a *local* phenomenon. Why does one consider the nuclear configuration of the *whole* solvent instead of that in the neighborhood of the reacting ions or molecules, something that intuitively seems more reasonable?

M: The whole solvent is considered just for generality, because *where do you stop considering the nuclear configurations*? So, you consider everything, you see.

2. Following: "The electronic part of the Hamiltonian operator H^{el}, which acts on the electronic wave function Ψ of the solvent and which describes *the solvent interacting with itself and with the charge distribution* $\rho(\mathbf{r})$, can be written as

$$H^{(\text{el})} = H_0^{(\text{el})} + \int \rho(\mathbf{r})\overline{\phi}(\mathbf{r})d^3\mathbf{r}, \tag{A1}$$

where $\overline{\phi}(\mathbf{r})$ *is the electrostatic potential.* The $\overline{\phi}(\mathbf{r})$ in Eq. (A1) must be considered as a quantum mechanical operator (actually, as an operator function parametrically dependent on **r** acting on the electronic wave function Ψ"

Q: (1) Is the electronic wave function of the solvent the wave function of the polarized solvent in the presence of solute, that is, is it a functional of the form $\Psi = \Psi[\rho(\mathbf{r})]$?

(2) Is the electrostatic potential $\overline{\phi}(\mathbf{r})$ due both to the charged solute particles and the polarized solvent at a fixed nuclear solvent configuration or only to the solute?

M: (1) If that $\rho(\mathbf{r})$ includes not only the solute charge at any point but the polarization, if it does include that, and I'm not sure if it does, then the functional form is of that sort, if $\rho(\mathbf{r})$ didn't include the polarization, you'd have to put in also the polarization with $\rho(\mathbf{r})$.

(2) I think it would be for everything, yes, I think so.

3. Following: ".... In Eq. (A1) H_0^{el} is an operator which refers to the solvent and includes the interaction of the electrons with the nuclei and with themselves and the kinetic energy operator of the electrons of the solvent. *The part of the solute–solvent interaction which is not solute–state specific is also included in* H_0^{el}."

Q: Which ones are the interactions which are not solute state-specific?

M: I don't know, but let me guess... I'm trying to think of what might be not solute state specific... For example, if you have in the solute a whole group of electrons, then you have, you know, *the electron being gained or lost*. Now, the different states of the solute correspond to that extra electron whether it is excited or not, or whether it is lost, whereas all of the other material is more or less present to the same extent regardless of whether that extra electron is there or not. So, maybe what was intended was that quite a bit of the electronic energy of the solute doesn't enter in the job, sort of cancels, is the same before and after the electron transfer, that's probably what was meant, so you don't have to consider the thousands of volts, hundreds of volts of electronic energy due to all the inner electrons, that sort of cancels throughout, you see. In that way, by removing quite a bit of that which is not solvent state specific, you are focusing really on the changes that are going on, that was probably the reason for the preceding statement.

4. Following: "It is usually assumed that the response in $\phi'_\rho(\mathbf{r})$ *depends negligibly on the nuclear configuration of the solvent.*"

M: If you move the nuclei a little bit, the electrons associated with them move a little bit, and that may have changed the response of the electrons of the solute for something, but just in a small way, you know, *I mean the point is that the electronic energy is so high that the small energy moving the nuclei a little bit here or there isn't going to change sort of the electronic polarization of the molecule much. You see, the electronic polarization, the polarizability, depends on a gap of a couple of electron volts, and moving a solute or a solvent molecule around a little bit of 0.025 eV and so on isn't going to change that gap that is responsible for the polarizability.*

5. Following: "If one wished, one could readily estimate W' in terms of the *electronic* dielectric response function $\varepsilon_{\text{op}}(\mathbf{r})$, obtaining as a result,

$$W' = \frac{1}{8\pi}\int \varepsilon_{\text{op}}(\mathbf{r})[\mathbf{E}_{j,\text{op}}(\mathbf{r})]^2 d^3\mathbf{r} - \frac{1}{8\pi}\int [\mathbf{D}_j^{(0)}(\mathbf{r})]^2 d^3\mathbf{r} \quad \text{(A7)}$$

where $\mathbf{E}_{j,\text{op}}(\mathbf{r})$ is the electric field in *the environment* with the dielectric response function $\varepsilon_{\text{op}}(\mathbf{r})$. $\mathbf{E}_{j,\text{op}}(\mathbf{r})$ satisfies Eq. (C3) below

$$\nabla \cdot \mathbf{D}_{\text{op}}(\mathbf{r}, t) = 4\pi\rho(\mathbf{r}, \mathbf{t}), \quad \nabla \times \mathbf{E}_{\text{op}}(\mathbf{r}, t) = 0,$$
$$\mathbf{D}_{\text{op}}(\mathbf{r}, t) = \varepsilon_{\text{op}}(\mathbf{r})\mathbf{E}_{\text{op}}(\mathbf{r}, t), \quad \text{(C3)}$$

with $\rho_j(\mathbf{r})$ replacing $\rho(\mathbf{r}, t)$ in the right-hand side of the first equation in Eq. (C3)."

Q: (1) I believe this is the part of the charging work where the solvent electrons are in equilibrium with the solute field but the nuclei are not.

(2) Please explain the physical meaning of the two terms on the right-hand side of Eq. (A7).

M: (1, 2) Right, the (A7) equation is sort of basic electrostatics and how one defines something. Now, for example, that W' is supposed to be sort of a charging term, right? W' would be equal to zero if that $\varepsilon_{\text{op}}(\mathbf{r})$ were unit, $\mathbf{D}_{\text{op}}(\mathbf{r}, t) = \varepsilon_{\text{op}}(\mathbf{r})\mathbf{E}_{\text{op}}(\mathbf{r}, t)$, so this simply reduces to zero when there is no electronic polarizability, so the purpose of that second term in Eq. (A7) is to insure that. Technically the way you do it is that you charge something up with the electronic polarizability, you subtract from that the charging something up in the vacuum, the difference is the interaction, you see.

M315 Linear Response in Theory of Electron Transfer Reactions as an Alternative to the Molecular Harmonic Oscillator Model

(Continuation 3)

Yuri Georgievskii, Chao-Ping Hsu, and R. A. Marcus

1. p. 5315 1st column top: "We next introduce an electric field operator $\hat{\mathbf{E}}(\mathbf{r}, t)$ as

$$\hat{\mathbf{E}}(\mathbf{r}, t) = -\nabla \int \hat{\mathbf{P}}_0(\mathbf{r}) \cdot \nabla \frac{1}{|\mathbf{r} - \mathbf{r}'|} d^3\mathbf{r}' + \mathbf{E}_{op}(\mathbf{r}, t) \qquad \text{(C2)}$$

The first term in Eq. (C2) is an electronic field generated by the solvent polarization $\hat{\mathbf{P}}_0(\mathbf{r})$. The second term $\mathbf{E}_{\text{op}}(\mathbf{r}, t)$ is an additional electric field which is due to the external charge and to the instantaneous electronic response of the solvent to the external charge distribution $\rho(\mathbf{r}, t)$.... The electric field $\mathbf{E}_{op}(\mathbf{r}, t)$, *which is assumed to be independent of the solvent nuclear configuration*..."

Q: Why this independence? Do you mean that the electronic response of the solvent molecules is about the same whatever the positions of their nuclei relative to the external polarizing field?

M: *If you turn the solvent molecules around, reorient them, that hardly changes their electronic response, it changes their nuclear response, you know, their response at low frequencies, but their electronic response is hardly changed, so the electronic response is largely independent of just how the nuclei are arranged.*

2. Following: "We note that the operator $\hat{\mathbf{P}}(\mathbf{r}, t)$ of the total solvent polarization can be written as a sum of two terms

$$\hat{\mathbf{P}}(\mathbf{r}, t) = \hat{\mathbf{P}}_0(\mathbf{r}) + \mathbf{P}_{op}(\mathbf{r}, t). \tag{C4}$$

The first term $\hat{\mathbf{P}}_0(\mathbf{r})$ is the operator of the molecular solvent polarization $\hat{\mathbf{P}}_0(\mathbf{r})$, Eq. (3),

$$\hat{\mathbf{P}}_0(\mathbf{r}) = \sum_n \delta(\mathbf{r} - \hat{\mathbf{r}}_n)\hat{\mathbf{d}}_n \equiv \sum_n \delta(\mathbf{r} - \hat{\mathbf{r}}_n)(\hat{\mathbf{d}}_n^0 + \hat{\mathbf{d}}_n') \tag{3}$$

which corresponds to a hypothetical neutral state of the solute with $\rho(\mathbf{r}) = 0$. $\hat{\mathbf{P}}_0(\mathbf{r})$ itself consists of two parts, one part arising from all unperturbed dipole moments of the solvent molecules, and the second part being a collective electronic response to the first part [cf. the discussion after Eq. (3)]."

From p. 5309 1st column top: "In this equation the coordinates of the center of mass of the nth solvent molecule are denoted by $\hat{\mathbf{r}}_n$ and its net dipole moment by $\hat{\mathbf{d}}_n$. This $\hat{\mathbf{d}}_n$ is not only a function of the internal nuclear configuration of the nth solvent molecule but also of the other solvent molecules, since they induce an electronic polarization in it. As a result, the solvent polarization operator $\hat{\mathbf{P}}_0(\mathbf{r})$ itself consists of two parts, one part arising from all the unperturbed dipole operators $\hat{\mathbf{d}}_n^0$ of the solvent molecules, and the second part related to $\hat{\mathbf{d}}_n'$ being a collective electronic response to the first part."

Q: How about the above use or nonuse of carets? If instead of the above equation we would have:

$$\hat{\mathbf{P}}_0(\mathbf{r}) = \sum_n \delta(\mathbf{r} - \mathbf{r}_n)\hat{\mathbf{d}}_n \equiv \sum_n \delta(\mathbf{r} - \mathbf{r}_n)(\hat{\mathbf{d}}_n^0 + \hat{\mathbf{d}}_n')$$

without carets on the $\mathbf{r}_n$, why would this last formula be not correct?

M: It would be correct but then those quantities would be classical and not quantum. If you want to get the polarization of the system, then you take that operator and you operate on whatever the state function is, but if those quantities would be classical then you can just drop the carets.

Q: But why the carets are just on $\mathbf{r}_n$ and not on $\mathbf{r}$?

M: Oh, well, $\mathbf{r}$ is a variable, you can call it whatever you want. But yes. . . that's a good question. . . actually I suppose one could drop the caret on the $\mathbf{r}_n$, because it's a simple distance operator, the same as a distance.

Q: So, one could write the way I wrote.

M: Yes, I think so, that's my impression. You want to have an operator, that's the main point, if you drop the caret on the $\mathbf{r}_n$ you still have an operator on the $\hat{\mathbf{d}}_n$, OK? So that can operate. . . let me see. . . since x as an operator has the same symbol as x as a variable, I think that would be all right, that's my impression, I believe that is OK.

3. p. 5315 2nd column bottom:

$$\text{“}\mathbf{E}(\mathbf{r}) = \mathbf{D}^{(0)}(\mathbf{r}) + \int_{\text{out}} d^3\mathbf{r}'\mathbf{T}(\mathbf{r}-\mathbf{r}')\mathbf{P}(\mathbf{r}') \qquad \text{(D1)}$$

.... $\mathbf{D}^{(0)}(\mathbf{r})$ is the external electric field, $\mathbf{E}(\mathbf{r})$ is the local electric field at the point $\mathbf{r}$, and $\mathbf{T}(\mathbf{r}-\mathbf{r}')\mathbf{P}(\mathbf{r}')$is given by Eq. (D2), $\mathbf{T}$ being a tensor:

$$\mathbf{T}(\mathbf{r}-\mathbf{r}')\mathbf{P}(\mathbf{r}') = \frac{3\mathbf{P}(\mathbf{r}')\cdot(\mathbf{r}-\mathbf{r}')(\mathbf{r}-\mathbf{r}')}{|\mathbf{r}-\mathbf{r}'|^5} - \frac{\mathbf{P}(\mathbf{r}')}{|\mathbf{r}-\mathbf{r}'|^3} \qquad \text{(D2)}$$

Equation (D2) gives the *electric field* produced at the point $\mathbf{r}$ by a dipole $\mathbf{P}(\mathbf{r}')$ at the point $\mathbf{r}'$.”

M: I have taken Eq. (D2) itself from earlier classical theory that I have been doing in the early 1950s on treating dielectric constants from a molecular point of view, I think that's where that equation originally came from. You may perhaps find it, I'm not sure, maybe in the book of Böttcher on the theory of electric polarization and I don't know if it has that, but I know that there are some statistical mechanical people from Belgium who treated the statistical mechanics of dielectric media, and had expressions that look something like that, only they would write in terms of $\cos\theta$, and so on, dot products.

Q: Why that 3?

M: I don't remember...I'm trying to remember if it's related to orthogonal polynomials, $3P(\cos\theta)$ or something or another, in other words that 3 occurs in some other formula too, if you look up the orthogonal polynomials there may be something there, so this may be part of a decomposition of a big operator in those terms, more than that I don't remember.

M317 Nonadiabatic Electron Transfer at Metal Surfaces

Shachi Gosavi and R. A. Marcus

1. p. 2068 1st column "When a continuum of donor or acceptor levels is involved in the electron transfer, as is the case in a metal electrode, the right hand side of Eq. (1)

$$k_{\mathrm{ET}} = \frac{2\pi}{\hbar} FC|H_{\mathrm{DA}}|^2 \qquad (1)$$

is integrated appropriately over these levels. The rate constant for electron transfer can then be written as[15]

$$k_{\mathrm{ET}} = \frac{2\pi}{\hbar} \int d\epsilon \frac{e^{-(\lambda - e\eta + \epsilon)^2/4\lambda k_B T}}{(4\pi \lambda k_B T)^{1/2}} |V(\epsilon)|^2 f(\epsilon) \qquad (2)\text{''}$$

Note 15: "This expression for the rate constant includes *the assumption that the solvent's dielectric polarization is not 'sluggish*.'"

Q1: Please explain Note 15...

M: In other words, you can use equilibrium statistical mechanics to calculate probabilities, instead nonequilibrium.

2. Following: ".... For the Franck–Condon factor we have substituted a classical value (*the final ratio of* k_{ET}*'s for different metals will be insensitive to this approximation*)."

Q: Why so?

M: I suspect because...*for different metals the quantum corrections for the nuclear motion will probably roughly cancel out, you know, if you use different metals that's only changing the position of the Fermi level, the density of states, it's not changing the structural reorganization, and so under those conditions I think that the* *quantum Franck–Condon factors, their ratio, would largely be the same because it's the same compound, so there is the same sort of quantum corrections roughly.* One could test that by putting in instead of that classical expression a quantum expression and then look at the ratios, at the rates, and so on. I remember that *when we looked at the quantum corrections on the cross relation, there wasn't much effect, if you're in the normal region, because there is a lot of this cancellation.*

3. p. 2068 1st column bottom: "For a single band case, $\eta = 0$ and $\lambda \gg \epsilon$ *(as is usually the case)* Eq. (2) simplifies..."

Q: Why $\lambda \gg \epsilon$?

M: Because usually this is energy relative to the Fermi level, so you just go over a *kT* or so of the Fermi level.

Q: So, it is relative to the Fermi level.

M: I believe so, yes. $\varepsilon = 0$ is the Fermi level.

4. p. 2068 2nd column bottom: "To make the calculations of Hsu one needs the difference in energies of a virtual superexchange state and the donor/acceptor state at the transition state. This difference can be calculated using a formula[27] ...."

Note 27: "Using the work functions of the metals (5.7 eV for Pt and 5.31 eV for Au) and the formal potential of $pyRu(NH_3)_5^{2+}$/ $pyRu(NH_3)_5^{3+}$ (which is −0.08 eV above SCE) we find that the difference in energy of the two states is

$$\Delta E = E_{\text{metal}} - \Psi_B - \Delta\varepsilon_i$$

where E_{metal} is −4.97 eV for Au and −5.16 eV for Pt. Ψ_B is the ionization potential of the bridge and $\Delta\varepsilon_i$ is the difference in energy of *the electronic state of the electron in the bridge* and of *the HOMO of the bridge*."

Q: (1) Is the electronic state of the electron in the bridge just the electronic state of the bridge to which the transferring electron belongs? I mean by electronic state of a molecule doesn't one intend the state of all the electrons of the molecule, the bridge in this case?

(2) If I consider the MO to which the transferring electron belongs, does not one intuitively thinks that MO is the HOMO?

M: Yes, I think that it is just that to which the transferring electron belongs, and one is presuming that the energy of all the other states sort of cancels out . . . but one could phrase it a bit more generally than that, basically it is the difference in energy when the electron is in there and when is not in there. And if the electron is coming from the metal then certainly energetically it would be best if it came from the HOMO.

5. p. 2070 2nd column middle: "**M** is a $J \times J$ matrix (J is the number of orbitals per atom . . . . Using the tight-binding (TB) approximation, we assume that any given plane interacts within itself,

$$\mathbf{M}(n\hat{\mathbf{R}}_p, n\hat{\mathbf{R}}_p) = \mathbf{A} \tag{A8}$$

. . . . **A**, being the self-interaction of the plane, is Hermitian."

Q: Why is it Hermitian?

M: Self-interaction of the plane?

Q: You have planes in a crystal.

M: In the usual Quantum Mechanics you want every operator to be Hermitian, so you choose something which is Hermitian.

M318 On the Theory of Electron Transfer Reactions at Semiconductors Electrode/Liquid Interfaces

Yi Qin Gao, Yuri Georgievskii, and R. A. Marcus

1. p. 3358 2nd column top: "In this study the authors chose a series of viologen ions with *very similar molecular structures* and *thus presumably similar reorganization energies* but with *very different free energy changes for electron transfer reactions.*"

Q: Why from the fact of very similar molecular structures follows that the reorganization energies will be very similar but this will not be true for the free energy changes which, on the contrary, will be very different?

M: Well, *the point is that supposing you had a substituent to an aromatic molecule, that's one of the reactants, then that substituent, whether it is electrophilic or nucleophilic, can change the free energy enormously. You know, practically everything else is unchanged, the solvation is largely unchanged, most of the vibrations are not really that much changed, and so on.* This putting substituents on assuming that the reorganization was constant was done for many systems, by Sutin for example, in the days when he was testing the free energy relationship. Now, it is not a perfect assumption and, for example, there is some work in a totally different kind of problem, the work of Lewis at Rice University, he used the theory to use the cross relation to predict and confirm his methyl transfers, in arenesulfonates, and there the molecule was largely varied with the substituents, you see, there it was presumed that λ was hardly affected, but the ΔG^0 was affected quite a bit. Edward Lewis, E. S. Lewis, the son of G. N. Lewis, did this work on the transfer of methyl cations between arenesulfonates, and he got marvelous agreement, because he was able to do not only the cross reactions but all other reactions, he got the best agreement that I have seen, even though the theory wasn't designed for that kind of reactions. But you know, a number of people have taken the theory and applied or derived relations related to it, for reactions that aren't ET: atom transfers, proton transfers, in this case methyl cation or maybe something else too. A chap by the name of S. Wolfe, I believe Saul Wolfe, he was at the University of Toronto, among others has done that and J. R. Murdoch, he had been at UCLA but left. Wolfe wrote some papers on applying the equations to reactions that weren't electron transfers.

M318 On the Theory of Electron Transfer Reactions at Semiconductors Electrode/Liquid Interfaces

(Continuation)

Yi Qin Gao, Yuri Georgievskii, and R. A. Marcus

1. p. 3358 2nd column middle: "By applying a treatment for liquid/liquid interfacial electron transfer reactions to semiconductor/liquid interfaces and by assuming an '*electron ball*' *model* for the electron in the semiconductor, Lewis.... In the model of Lewis the electron in the semiconductor is represented by a spherical donor with radius around 10 Å, and any reorganization around the electron in the semiconductor was neglected."

Q: Can you please briefly describe this "electron ball" model?

M: Yes, the electron ball model was something that Nathan S. Lewis at Caltech derived, I think it was just a ball that had a charge in it and used the interaction of that ball but I don't remember any of the details, it is his model, not mine.

2. p. 3359 1st column bottom: "In the present paper, the electron transfer reactions at semiconductor/liquid interfaces are treated nonadiabatically."

M: *The main criterion in an approximate way is that you calculate not only a coupling strength but you calculate a Landau–Zener factor,* *and if it turns out to be small the process is nonadiabatic, if it turns out to be near unity it is adiabatic.*

Q: (1) If I remember well, I have the impression that when you are confronted with a new system where electron transfer reactions happen, you begin to tackle the problem nonadiabatically. Is it so? Is there some reason for that?

(2) Is it so that nonadiabatic processes are more common that adiabatic ones? Or maybe more general?

M: (1) *That is certainly true for systems with potential energy curves with very small splitting and the potential energy curves are largely, usually, diabatic ones. Then we pretend that the splitting at the intersection is relatively small, so from an energetic point of view that is certainly true. Then, whether we take the transmission coefficient to be unity or use the Landau–Zener factor probably depends on the circumstances. For a long-distance ET we would use the appropriate coupling factor, decreasing exponentially with distance. If it's a short distance ET, we may well assume that from an electronic nuclear point of view the ET is essentially adiabatic, in other words that the Landau–Zener factor is close to unity.*

(2) *Well, yes, if the reactants are far apart then of course the reaction is nonadiabatic* and *the rate constant depends upon the nonadiabaticity but also depends on the energetics, and the rate constants can vary by units orders of magnitude and a lot of that you can capture by the free energy term, and so, you know, the* $\lambda + \Delta G^0$ *term. The long-range ET is certainly nonadiabatic, if the reactants can come into contact then you assume the process to be adiabatic.*

Q: So, the distance is giving the direction for the first guess, somehow.

M: Yes, *what we have done is in the paper with Sutin and probably before that... if the reactants are essentially in contact we pretended that the reactions are adiabatic*, and there are some people who in more recent years, in the past 54 years, they have tried to calculate some of these nonadiabatic factors, Marshall Newton, for example, Sutin, but others too have done that sort of calculations since then. I have not followed that literature, but there is some literature up there where people have tried to calculate essentially nonadiabatic factors.

3. p. 3359 2nd column middle: "In addition, *surface states* are included for the current vs applied potential behaviour observed at InP/$Me_2Fe^{+/0}$, $PV^{2+/+}$ interfaces, as *one way of accounting for the nonideal behaviour* observed for this system. The *maximum rate* mentioned earlier, which is the principal focus of our attention, is calculated under conditions where *surface states are unimportant.*"

Q: Can you please explain the relation between surface states, ideal behavior and maximum rate?

M: I should have to look at Lewis' papers to find what ideal behavior is. The *ideal behavior*, in my guess, is what happens with simple electron transfer and that, if you get deviations from that, because of some electron transfer by the surface states, it won't fit into that simple formula and so you get nonideal behavior, that's all my guess, what Lewis called ideal behavior probably is behavior in the absence of surface states.

Typo: Misprint in Note 20: "Y. D. Demkov...". It is really Y. N. Demkov for Yury Nikolayevich Demkov.

4. p. 3360 1st column bottom: "Considering first the acceptor as being at position **r** in the solution, the rate constant is

$$k_f^t(\mathbf{r}) = \sum_{\mathbf{k}} k_f^s(\mathbf{k}, \mathbf{r}) \tag{10}$$

Here, $k_f^t(\mathbf{r})$ is the rate constant for the total current from the semiconductor to the molecule, expressed as a sum of the currents from all the electronic states of the semiconductor, **k** denotes the electronic state of the semiconductor with the wave vector **k**, and $k^t_f(\mathbf{r})$ varies with the position **r** of the acceptor relative to the electrode."

Q: I believe you have already studied somewhere such position dependent rate constants.

M: Yes, I think in the paper with Paul Siders, we had a position dependent rate constant.

5. Following: "It can be further written as

$$k^t_f(\mathbf{r}) = \frac{2\pi}{\hbar} \int_{\epsilon} \mathrm{FC}(\epsilon) f(\epsilon) \sum_{\mathbf{k}} |V(\mathbf{k}, \mathbf{r})|^2 2\pi \delta(\epsilon_{\mathbf{k}} - \epsilon) d\epsilon \qquad (11)$$

.... After denoting by $\overline{V}(\varepsilon, \mathbf{r})$ the averaged coupling of all the states with energy ϵ, Eq. (11) can be written as

$$k^t_f(\mathbf{r}) = \frac{2\pi}{\hbar} \int_{\epsilon} f(\epsilon)\rho(\epsilon)\mathrm{FC}|\overline{V}(\epsilon, \mathbf{r})|^2 d\epsilon \qquad (13)$$

where $\rho(\varepsilon)$ is the density of states, i.e., $\sum_{\mathbf{k}} 2\pi\delta(\varepsilon_{\mathbf{k}} - \varepsilon)$"

Q: Why $\rho(\varepsilon) = \sum_{\mathbf{k}} 2\pi\delta(\varepsilon_{\mathbf{k}} - \varepsilon)$?

M: *That's because the matrix element can depend upon k, if you just use the density of states then either you're neglecting the effect of the particular state on the matrix element or you are defining an effective density of states which is effectively averaged over the state dependent matrix element. Often, when applying some sort of nonadiabatic theory, I am just writing down the density of states, but that isn't correct because the particular state coupling can depend on the state, is not the same for all states. Yes, that's a mistake sometimes I have made. That* $2\pi\delta$ *is a density of states, by the way, that includes the energy* ε*, that's right. If we just did some averaging, that would be a density of states, yes, but when you weight by that* $\delta(\mathbf{k})$ *you no longer have just simply a density of states. In other words, what some people* *have done is that somehow they forget about the state dependence of V, and bring it outside of the sum and then you have some strange V, but what you see in the Eq. (11) is a more correct expression.*

6. Following: "and $|\overline{V}(\varepsilon, \mathbf{r})|^2$ denotes $\frac{\sum_{\mathbf{k}} |V(\mathbf{k},\mathbf{r})|^2 \delta(\varepsilon_{\mathbf{k}}-\varepsilon)}{\sum_{\mathbf{k}} \delta(\varepsilon_{\mathbf{k}}-\varepsilon)}$."

Q: The weighting term $\frac{\delta(\varepsilon_{\mathbf{k}}-\varepsilon)}{\sum_{\mathbf{k}} \delta(\varepsilon_{\mathbf{k}}-\varepsilon)}$ has the form of a normal weighting term but made up of delta functions.

M: That's a way of defining an effective density of states, it is really a definition, that's all.... What it is, is that if you have the $\sum_{\mathbf{k}} \delta(\varepsilon_{\mathbf{k}} - \varepsilon)$, that's a density of states, and so what we did there is multiply and divide by the same thing, and call one thing then an effective V, coupling, it is just a matter of if you want to introduce an effective coupling, this is the way of doing it....

Q: Yes, exactly, the expression in the denominator, except for the 2π, is the same...

M: Yes, that's right, the 2π cancels in both I guess. I'd have to look at the 2π question, but... I don't know why 2π should be there, that 2π may have something to do with normalization but, apart from that, I'm not sure why that 2π is there. Is it written some place in the article that the density of states is that thing with the 2π?

Q: Yes, it is in the paper.

M: I am not sure why that 2π is there... it is very strange to me.... Why that 2π, what is the origin of it.

Q: It appears in the formula $\rho(\varepsilon) = \sum_{\mathbf{k}} 2\pi\delta(\varepsilon_{\mathbf{k}} - \varepsilon)$, reference 22.

M: I guess one would have to look up to see what the density of states is, and the thing may be different whether it is a surface density of states, or a volume density of states. You see, I don't remember whether we are talking here of a surface density of state or a volume density of states. It may have something to do with going from **k** to E, you know, from **k** to energy, I don't see why, but you are summing over **k** states, then sometimes 2π's come... you have something like $2\pi\mathbf{k}d\mathbf{k}$ for the number of states in the range **k** to $\mathbf{k} + d\mathbf{k}$, but how it's related, that I don't know, I would have to look at their paper to know why.

7. p. 3363 1st column middle: "The density of states is obtained by randomly choosing a certain number of wave vectors in the calculation which show *a statistical number of the energy eigenvalues.*"

p. 3363 2nd column top: "...the density of states as a function of energy is calculated using *a statistical number of the eigenvalues.*"

M: Well, suppose that every time you have a state you draw a vertical line, you plot a vertical line indicating that you got a state. All right, now if you are considering a range, *you have to consider an energy range of states broad enough that you have in it quite a few vertical lines.*

8. Following: "Except for the *intrinsic surface states*, which lie in the band gap..."

Q: Are the intrinsic surface states the ones not due to impurities?

M: I suppose so, I'd have too look back and see what we meant by that. Maybe that would imply surface states simply due to the fact that the bulk in the crystal structure comes to an end, and if the structure comes to an end then without impurity you now have, in a tight binding approximation, a different Coulombic situation for the atoms on the surface compared with the other atoms and that can be enough to produce a surface state. With tight binding equations in that case you find the wave function for that particular state to be localized near the surface instead of extending throughout the crystal. Just breaking crystal symmetry can be enough to produce a surface state. Now, surface states don't always exist, that you see, for example, in some powder emission spectrum that you plot vs. energy.

9. p. 3364 1st column middle: "To be consistent with the experiments where *the only rate-limiting step* is the electron transfer process the acceptors are considered homogeneously distributed in the solution."

Q: I imagine you tacitly do that in order to avoid the limiting step do to diffusion. Is it so or are there other rate limiting steps avoided thanks to the homogeneous distribution of acceptors?

M: Yes, that's right, that the acceptors are homogeneously distributed when diffusion is not a factor. Now, of course, as you know, in electrochemistry at high driving potentials then eventually diffusion becomes rate controlling.

10. p. 3364 1st column middle-bottom: "Under the assumption that the coupling decays exponentially with distance with a decay exponent β_s, the range of contributing distances is $\frac{1}{\beta_s}$."

Q: The formula for exponential decay is, I believe, $e^{-\beta_s x}$. For a distance $x = \frac{1}{\beta_s}$ the coupling becomes $e^{-1} \approx 0.37$, that is, the coupling becomes about $1/3$ of the maximum at $x = 0$. Can we then say that the range of coupling is conventionally measured by the distance at which the coupling becomes $1/e$ of its maximum value?

M: Yes, as a ballpark statement, very ballpark. You don't need to introduce that, you can just simply integrate $e^{-\beta_s x}$. If you want to have a very rough interpretation you can say that, but you certainly you don't need it. When I integrate $e^{-\beta_s x}$, I get $\frac{1}{\beta_s}$, so I'm going to call $\frac{1}{\beta_s}$ range, typical distances that are within $\frac{1}{\beta_s}$ are the typical distances contributing for electron transfer, that doesn't affect any quantitative description.

11. p. 3364 2nd column middle: "For a semiconductor/electrolyte interface... the change of electrostatic potential across a semiconductor/liquid interface exists mainly within the semiconductor, because of the low concentration of the charge carriers in the semiconductor."

M: Yes, when you have a high concentration of *metal* electrons the potential inside the *metal* is essentially zero, the change of potential is essentially zero.

12. Following: "In this case, the change of applied potential changes only the concentration of carriers at the interface and does not change the free energy ΔG^0 of the electron transfer reaction."

M: In the semiconductor case, when you can think of the electron acceptors and donors drifting toward the surface, so you have diffusion toward the surface, then really when you change the potential you change the concentration of them at the surface and so *the potential is just affecting the concentration of carriers at the surface. In other words, the carrier is the same but the number of carriers at the surface depends upon the potential.*

13. p. 3365 1st column top: "... we shall also assume that the electron transfer between a surface state and the bulk state is much faster than *the electron transfer between the surface state and the molecular acceptor.* In the preceding treatment we neglected the interactions which might lead to such a coupling, e.g. radiationless transitions."

M: *The transition between a surface state and a bulk state... an analogous thing is a radiationless transition, it's a transition between two different kinds of states in the molecule, here the molecule is the whole semiconductor*, just think of a radiationless transition between one electronic state of the molecule and another, here you think of an electronic transition between a surface state and a bulk state. The analogy with a radiationless transition is that you have two different states, in this case surface states and bulk states, two different classes of states, while in radiationless transitions you have two different electronic states of a molecule, but the idea that you go from one to the other is the same in both cases.

14. Following: "The resulting expression for the current density corresponding to the electron transfer from the surface states of the semiconductor to the acceptor is

$$J_f^{ss} = e[A]\frac{2\pi}{\hbar}\frac{1}{\sqrt{4\pi\lambda k_B T}}\frac{1}{\beta_s} \times \int_\epsilon |V_{ss}(\epsilon)|^2 e^{-[(\lambda+\Delta G)^2/4\lambda k_B T]} f(\epsilon)\rho_{ss} d\epsilon \qquad (45)$$

where ρ_{ss} is the density of surface states, and

$$\Delta G = \Delta G^0 - \varepsilon \qquad (46)$$

is the driving force for the electron transfer from *a surface state with energy* ε to the acceptor."

Q: If $\varepsilon = 0 \Rightarrow \Delta G = \Delta G^0$. The energy of the surface state is ε with respect to what?

M: ε represents the energy difference of the state compared with the conduction band. I'd have to look how we defined the ΔG^0. ε is the energy with respect to the conduction band.

Q: Is the energy ε of a surface state <0 or >0?

M: *It would depend on whether it is electron transfer from the surface state or to the surface state*, you can argue that way, physically whatever it makes sense. So, it depends upon whether you are talking into or out of the surface state, whatever physically makes sense, if it is more difficult to transfer either out of the surface state or it is easier to transfer in the surface state. *The surface state is always below the band edge.*

15. p. 3366 2nd column bottom: "For a metal/liquid interface, the electron transfer reaction happens largely at the Fermi energy, and the coupling strength between the metal electrode and the metal acceptor can be characterized by

$$\Delta(\varepsilon) = 2\pi \sum_{\mathbf{k}} |V_{\mathbf{k}}|^2 \delta(\varepsilon_{\mathbf{k}} - \varepsilon) \tag{55}$$

where $\Delta(\varepsilon)$ is evaluated at the Fermi energy. $\Delta(\varepsilon_f)/\hbar\omega_{\max}$ *has been then taken as the criterion for distinguishing nonadiabatic from adiabatic reactions* for a metal/liquid interface, where $\omega_{\max}$ *is the 'fastest phonon mode' contributing to the electron transfer reaction.* For an $\omega_{\max}/2\pi$ of the order of $10^{13}\,\mathrm{s}^{-1}$ this $\hbar\omega_{\max}$ is about 0.03 eV"

M: See, there we are getting away from Landau–Zener description to a... what's the reference?

Q: Demkov, Schmickler...

M: Yes, I guess to some quantum treatment of the electron transfer, and I haven't done that, so it is based on the quantum treatment of electron transfer by the coauthors.

16. p. 3367 1st column bottom: "The difference between the calculated results for the Si/viologen$^{2+/+}$ and InP/$Me_2Fe^{+/0}$ system may be due to *several effects.* One factor is the *difference in size of the acceptor state*, since the LUMO of a viologen$^{2+/+}$ ion is calculated in the present work to be *delocalized* over the whole ring system, and the LUMO of a $Me_2Fe^{+/0}$ ion is *localized* mainly on the Fe^{3+} ion... and so *distant from the periphery of the molecule*... using calculations we have compared the calculated maximum rate constant for the Si/viologen$^{2+/+}$ system, $1.6 \times 10^{-16}\,\mathrm{cm}^4\,\mathrm{s}^{-1}$.... with a maximum rate constant... for the Si/$Me_2Fe^{+/0}$ system, $0.17 \times 10^{-16}\,\mathrm{cm}^4\,\mathrm{s}^{-1}$. Thus, it is seen that the viologen$^{2+/+}$ pair has a closer *effective contact* with the Si interface than does the $Me_2Fe^{+/0}$ pair."

Q: You introduce here a new concept of "effective contact" which apparently defines the sum of two effects, that is, the size of the LUMO delocalization and the size of its distance from the closest contact point of the acceptor with the electrode surface.

M: Well, yes, there are several effects that come in. Suppose that you have an orbital distributed over the periphery of the large end of the molecule, OK? You have a molecule, supposing an orbital be distributed that way, suppose it for the sake of argument. Then of course here is a large molecule, and *it is always the part of the orbital that is close to the other reactant that's going to really couple.* So, when you introduce the wave function of the whole reactant that is distributed over that large region you're going to have very small coupling contribution of the region away from the other reactant.

Q: So it depends on how oriented is this orbital.

M: How oriented but also on what the size is. If the orbital is distributed over a very large volume, a very large area, then remember that the coupling matrix element varies exponentially more or less with the distance.... Remember that you integrate the wave function times the other wave function and the other wave function decreases exponentially, so the whole of that orbital doesn't really contributes as much as the part that is near the other reactant.

Q: But this effective contact is not a rigorous kind of concept

M: Yes, no, that rigorous no. The rigorous thing would be to actually calculate the electronic matrix elements for both systems.

17. Following: "...we attribute the higher experimental rate for InP to the presence of *surface states* in the InP..."

p. 3367 2nd column bottom: "While the inclusion of the surface states helps to explain the current vs. applied potential behaviour obtained in the experiments, the *nonideal behaviour* being explained by the Fermi–Dirac occupancy of the surface states, the nonideal behaviour can also result from many other mechanisms, such as the recombination of charge carriers in the solid bulk."

Q: Apparently the ideal behavior is that in which the rate constants are determined only by the size of the interaction strength. Is it so?

M: Yes, I guess so. If you have anything else than the effective wave function, if you have anything besides the simple expression, then that is what gives rise to what is called the nonideal behavior.

Q: So, you can have parasitic effects.

M: Yes, I suppose so, but what they are I don't know. The semiconductors you know, are curious systems, you can get different kinds of defects, you can have a disordered surface, you know, God knows what.

M322 On the Theory of Electron Transfer Reactions at Semiconductors Electrode/Liquid Interfaces. II. A Free Electron Model

Yi Qin Gao and R. A. Marcus

0. Let me begin with a general question. Imagine we have your classical couple of *R* and *P* potential energy curves with their splitting.

(1) Do to such two potential energy curves always correspond two free energy curves?
(2) Do such free energy curves present a splitting in free energy?
(3) Which is the relation of such a splitting to that of the potential energy curves?
(4) How does one compute the splitting?

M: (1) *The potential energy surfaces have* 10^{23} *dimensions. It's the free energy curves that are one-dimensional.*

(2, 3) *If you define a global reaction coordinate, as I did, which is a difference of the two potential energy curves, a difference that serves as a global reaction coordinate that can define an $N - 1$ dimensional surface, on that surface I can take a canonical distribution, and that gives the free energy. The splitting is really in the potential energy, and it's only on the average that you could calculate some sort of a weighted, sort of potential energy splitting. There isn't really a splitting of the free energy, there is really some average value over the $N - 1$ dimensional surface.*

Q: So, you don't see any splitting of the free energy curves.

M: *Well, the splitting would come from the splittings of the potential energy curves. If you follow the potential energy curves in the case of the diabatic, then of course there's no splitting, so you look at where the splitting is and say: well, now this splitting depends just upon on where I am on the potential energy curves, at the intersection of those curves, then what you end up with is some sort of Boltzmann average at the intersections, where you do average over the energy differences, the splittings.*

(4) *If you have the unperturbed curve and you have the Hamiltonian, you know what the perturbation is, it is usually some Coulombic interaction, you can calculate the splitting. You would get an average splitting canonically distributed and so on, you have some average and if you want you can consider it as the splitting of the free energy curves.*

1. p. 6351 2nd column bottom: "... the current density J_f due to electron transfer from the semiconductor to the molecule is proportional to *both* the concentration [A] of the molecules in the liquid and *the density of electrons n_s at the semiconductor surface,*

$$J_f = ek_{et}n_s[A] \tag{1}$$

Following:

$$\text{“}J_f = ek_f[A] \tag{4}$$

where k_f . . . *pseudo-first-order* rate constant . . . and from Eq. (1),

$$k_f = n_s k_{et} \tag{6)”}$$

Q: A pseudo-first-order rate constant appears to be the product of a second order rate constant and of a constant concentration. Is it so?

M: Yes. In semiconductors you speak of concentrations of holes and concentration of electron carriers, you know, so it is part of the lore of the semiconductor terminology.

2. p. 6353 1st column middle: “. . . the coupling matrix element denoted by $V_{\mathbf{k}}$ between the semiconductor state . . . and the molecule, can be written as

$$\begin{aligned} V_{\mathbf{k}} &= \frac{V_1\langle\psi \mid \Psi_{\mathbf{k}}\rangle_1 - V_1\langle\psi \mid \psi\rangle_1\langle\psi \mid \Psi_{\mathbf{k}}\rangle}{1 - |\langle\psi \mid \Psi_{\mathbf{k}}\rangle|^2} \\ &\cong V_1\langle\psi \mid \Psi_{\mathbf{k}}\rangle_1 - V_1\langle\psi \mid \psi\rangle_1\langle\psi \mid \Psi_{\mathbf{k}}\rangle, \end{aligned} \tag{15}$$

where $\langle\cdots\rangle_1$ means integration over the space occupied by the semiconductor.”

p. 6354 1st column top: “Since $\langle\psi \mid \psi\rangle_1\langle\psi \mid \Psi_{\mathbf{k}}\rangle$ is typically small as compared to $\langle\psi \mid \Psi_{\mathbf{k}}\rangle_1$, the $V_{\mathbf{k}}$ in Eq. (15) can be approximated by $V_1\langle\psi \mid \Psi_{\mathbf{k}}\rangle_1$.”

Q: I believe that it so because $\langle\psi \mid \psi\rangle_1$ is very small.

M: Let me see, what is $\langle\psi \mid \psi\rangle_1$?

Q: That’s the probability density inside of the semiconductor.

M: And ψ that's the wave function of the particle, right? Yes, so that's something about the contribution of the probability contribution of the particle that is actually inside the semiconductor, all right, so that is normally probably small. You know, it is a weak overlap, so that $\langle \psi \mid \psi \rangle_1$ is small, so in that case it looks that the interaction is very low, it is a second order interaction, yes, right.

3. Following: "Following Bardeen, this quantity can be written as an integral over the space occupied by one of the reactants, here the semiconductor S, we note that

$$\langle \psi | T + V_1 | \Psi_{\mathbf{k}} \rangle_1 = E_{\mathbf{k}} \langle \psi \mid \Psi_{\mathbf{k}} \rangle_1 \tag{16}$$

.... But we also have $T|\psi\rangle = E_2|\psi\rangle$ in the region outside the molecule, where E_2 is the eigenvalue for the molecule. Thereby,

$$\langle \Psi_{\mathbf{k}}^* | T | \psi^* \rangle_1 = E_2 \langle \Psi_{\mathbf{k}}^* \mid \psi^* \rangle_1 = E_2 \langle \psi \mid \Psi_{\mathbf{k}} \rangle_1. \tag{17}$$

We have from Eqs. (16) and (17),

$$-\frac{\hbar^2}{2m} \int_1 (\psi^* \nabla^2 \Psi_{\mathrm{k}} - \Psi_k \nabla^2 \psi^*) d^3\mathbf{r} = (E_{\mathrm{k}} - E_2 - V_1) \langle \psi \mid \Psi_{\mathrm{k}} \rangle_1.$$

(18)"

Q: **Erratum**: I believe the asterisk on the first ψ should be shifted to the second $\Psi_{\mathbf{k}}$, that is, Eq. (18) should be

$$-\frac{\hbar^2}{2m} \int_1 (\psi \nabla^2 \Psi_{\mathrm{k}} - \Psi_{\mathrm{k}}^* \nabla^2 \psi^*) d^3\mathbf{r} = (E_{\mathrm{k}} - E_2 - V_1) \langle \psi \mid \Psi_{\mathrm{k}} \rangle_1.$$

M: Yes, I believe what you have is correct.

4. Following: "When $\langle D|H|D \rangle$ and $\langle A|H|A \rangle$ are set equal in the transition state, $E_{\mathbf{k}}$ and E_2 are not quite equal, but the difference between them is neglected. We thus obtain, on neglecting the terms mentioned earlier,

$$V_{\rm k} = -\frac{\hbar^2}{2m}\int \mathbf{n}\cdot(\psi^*\nabla\Psi_{\rm k}) - \Psi_{\rm k}\nabla\psi^*)ds, \qquad (19)\text{''}$$

where **n** is a unit vector normal to the surface of well S and pointing outward from S, that is, in the direction of negative z, and ds is the area element of the surface of well S.

Q: **Erratum**: Correcting as above, we should have, I believe,

$$V_{\rm k} = -\frac{\hbar^2}{2m}\int \mathbf{n}\cdot(\psi\nabla\Psi_{\rm k}) - \Psi_{\rm k}^*\nabla\psi^*)ds$$

M: OK.

5. Following: "Setting $z = 0$ at the semiconductor surface, Eq. (19) then becomes,

$$V_{\rm k} = -\frac{\hbar^2}{2m}\int_{z=0}\{\psi(\partial\Psi_{\rm k}^*/\partial z) - \Psi_{\rm k}^*(\partial\psi/\partial z)\}ds. \qquad (20)\text{''}$$

Q: **Erratum**: I believe here it should be

$$V_{\rm k} = -\frac{\hbar^2}{2m}\int_{z=0}\{\psi(\partial\Psi_{\rm k}/\partial z)\Psi_{\rm k}^*(\partial\psi^*/\partial z)\}ds.$$

M: OK. I don't know how we got so careless then. I wonder how we got that careless. I mean, it certainly looks that way. The equations you wrote are correct. I'm surprised, but that's the way it is.

6. p. 6354 2nd column top: "The normalization of $\Psi_{\mathbf{k}}$ will be made with respect to a large volume υ, most of which encompasses the semiconductor surface.... A normalization to a delta function could have been introduced instead if we had introduced a z-dependent electric field inside the semiconductor similar to the actual field. However, the present procedure is simpler and should suffice for our purpose."

M: Yes, in one of the papers I used the z-transform. What was often done in semiconductors and in metals is to use a slab method, because they use fairly elaborate electronic structure methods there, the slab is a finite object, but you could also treat the semiconductor as semi-infinite. If it is semi-infinite then your normalization is usually to a delta function, and that is what I did in a paper that I wrote on semiconductors' z-transforms. If you look at z-transform under my name in the publications list you will see a paper where I focused on just that, and in fact I made some use of it in papers with that Chinese fellow... Hui Ou-Yang and that Swedish fellow... Källebring.

7. p. 6355 1st column middle: "For a low-doped semiconductor... the occupation of its conduction band at the surface is low enough that the occupancy probability, $f(\varepsilon_{\mathbf{k}})$, of the state **k**, the kinetic energy of which is $\hbar^2 k^2/2m$, can be treated as obeying Boltzmann statistics. The sum in Eq. (9)

$$k_f^{\mathrm{t}} = \frac{2\pi}{\hbar} \sum_{\mathrm{k}} \mathrm{FC}(\epsilon_{\mathrm{k}}) f(\epsilon_{\mathrm{k}}) |V_{\mathrm{k}}(\mathbf{r})|^2 \tag{9}$$

can be written as an integral over **k**-states, when properly normalized. The number of electrons in the semiconductor conduction band in the volume υ is $n_s\upsilon$, and the probability of finding one of these electrons in $dk_x dk_y dk_z$ is the Boltzmann factor

$$\exp(-\varepsilon_{\mathbf{k}}/k_B T) dk_x dk_y dk_z \bigg/ \int_{-\infty}^{+\infty} \int_{-\infty}^{+\infty} \times \int_{-\infty}^{+\infty} \exp(-\varepsilon_{\mathbf{k}}/k_B T) dk_x dk_y dk_z.$$

When multiplied by $n_s \upsilon$ it becomes the probability that a state is occupied. The sum in Eq. (9) thus becomes

$$\sum_{\mathbf{k}} \mathrm{FC}(\varepsilon_{\mathbf{k}})|V_{\mathbf{k}}(\mathbf{r})|^2 f(\varepsilon_{\mathbf{k}}) = \frac{n_s \upsilon \iiint \mathrm{FC}(\varepsilon_{\mathbf{k}})|V_{\mathbf{k}}(\mathbf{r})|^2 e^{-\varepsilon_{\mathbf{k}}/k_B T} dk_x dk_y dk_z}{\iiint e^{-\varepsilon_{\mathbf{k}}/k_B T} dk_x dk_y dk_z} \quad (29)$$

Since $|V_{\mathbf{k}}|^2$ is inversely proportional to υ, the υ cancels. Equation (9) yields

$$k_f^{\mathrm{t}} = n_s \upsilon \frac{2\pi}{\hbar} \frac{\iiint \mathrm{FC}(\epsilon_{\mathrm{k}})|V_{\mathrm{k}}(\mathbf{r})|^2 e^{-\epsilon_{\mathrm{k}}/k_B T} dk_x dk_y dk_z}{(2\pi m k_B T)^{3/2}/\hbar^3} \quad (30)\text{''}$$

Q: How do you get the denominator of Eq. (30) from that of Eq. (29)?

M: You know, k is proportional to p, the momentum, you put in momentum terms, you integrate and now you have the classical partition function.

8. p. 6356 2nd column middle: "Another factor responsible for the larger calculated maximum rate constant at the Si/viologen$^{2+/+}$ interface is the character of the acceptor orbital. The use of a p_z-like orbital yields more efficient coupling between the semiconductor and the acceptor for the Si/viologen$^{2+/+}$ interface than does the d_{z^2}-like orbital that is used for the InP/$Me_2Fe^{+/0}$ interface, both for the tight-binding and for the free electron calculations. *The relative inefficiency of the d-electron in electron transfer was described in an earlier paper*[40]

In the earlier paper: Shachi Gosavi and R. A. Marcus: "Nonadiabatic Electron Transfer at Metal Surfaces" on p. 2069 2nd column middle you wrote: "The fact that the d orbitals are localized and not strongly coupled to the environment is well-known from field-emission experiments. Thus, even though the d electrons are present at the Fermi level, they interact very weakly with external fields or ions and these electrons tunnel out from the metal much less than s electrons at the same energy. This fact has been observed in field emission as well as in ion neutralization experiments. It is perhaps not surprising that these d electron localization also manifests itself in a reduced contribution to the rate constant in electron transfer experiments."

Q: Why do *d* electrons behave this way?

M: All right, let's see... *the d electrons I think are buried inside the atom, and so you can add d electrons and not change the properties a lot filling in transition metal orbitals... so I think that d electrons, although they have high angular momentum, somehow for whatever reason seem to be buried inside the atom and, if so, they overlap less*, but I'm just vaguely remembering, and I don't remember that well enough to be absolutely sure. You know, for example, if you take the f electrons in the rare earths maybe or the actinides, one or the other, and you start filling them up, you don't change the properties a lot, the elements have a lot in common, it is a family, the electrons would be buried. If you look up some place at the overlap integrals involving those wave functions, I think you find them small. I think maybe one should look at the references, if we mention some reference, experiments and interpretations on those experiments, you see.

9. NOTE 32: "The minima of $E(\mathbf{k})$ for Si shows *ellipticity* and two different effective masses can be distinguished parallel and orthogonal to the axis (100)."

M: Yes, *you know in some systems the* $E(\mathbf{k})$, *when plotted vs. k, in two dimensions would be a circle, there would be a set of circles, if they are the same in both x and y directions. But now, if because of the crystal structure at the surface, the lattice is different in the two directions x and y, then you no longer have a symmetrical plot of curves of constant k, the plot is different in the x and y directions, so those curves of constant energy in* k_x *and* k_y *space are ellipses.*

M323 Temperature Dependence of the Electronic Factor in the Nonadiabatic Electron Transfer at Metal and Semiconductor Electrodes

Shachi Gosavi, Yi Qin Gao, and R.A. Marcus

1. p. 72 2nd column top: "When $\eta = 0$, Eq. (5) becomes k^0, the standard rate constant

$$k^0 = \frac{2\pi}{\hbar}\frac{e^{-\lambda/4k_\mathrm{B}T}}{(4\pi\lambda k_\mathrm{B}T)^{1/2}}\int_{-\infty}^{\infty} e^{-\epsilon^2/4\lambda k_\mathrm{B}T}(\epsilon)|V(\epsilon)|^2 d\epsilon \qquad (8)$$

Typo: Inside the integral after the exponential one should have $g(\varepsilon)$ instead of (ε).

We first consider, for comparison, the simplest case: both the dependence of $V(\varepsilon)$ on ε and the $\frac{\varepsilon^2}{4\lambda k_B T}$ term Eq. (8) are neglected. In that case the integral in Eq. (8) is a standard integral [20][1] yielding

$$k^0 \cong \frac{2\pi}{\hbar}\frac{e^{-\lambda/4k_\mathrm{B}T}}{(4\pi\lambda k_\mathrm{B}T)^{1/2}}\pi k_\mathrm{B}T|V(0)|^2 \qquad (9)$$

....

We next consider the case where the dependence of $|V(\varepsilon)|^2$ is neglected, as before, but where the $\frac{\varepsilon^2}{4k_\mathrm{B}T}$ in Eq. (8) is included.

Typo: it should be $\frac{\varepsilon^2}{4\lambda k_\mathrm{B}T}$.

In this case we have

$$k^0 \cong \frac{2\pi}{\hbar}\frac{e^{-\lambda/4k_\mathrm{B}T}}{(4\pi\lambda k_\mathrm{B}T)^{1/2}}\pi k_\mathrm{B}T|V(0)|^2\langle e^{-\epsilon^2/4\lambda k_\mathrm{B}T}\rangle \qquad (11)$$

where we have used [20] $\pi k_\mathrm{B}T = \int_{-\infty}^{+\infty} g(\varepsilon)d\varepsilon$ and where $\langle\,\rangle$ denotes an average over the distribution function, $\frac{g(\varepsilon)d\varepsilon}{\int_{-\infty}^{\infty} g(\varepsilon)d\varepsilon}$."

Note [20][1] (Integrals from the Gradshteyn–Ryzhik's tables of integrals):

$$\int_{-\infty}^{\infty} g\left(\frac{\varepsilon}{2}\right)^2 g(d\varepsilon) = \frac{(\pi k_B T)^3}{4}, \quad \int_{-\infty}^{\infty} g\left(\frac{\varepsilon}{2}\right)^4 g(d\varepsilon) = \frac{(\pi k_B T)^5}{\left(\frac{5}{16}\right)},$$

$$\int_{-\infty}^{\infty} g\left(\frac{\varepsilon}{2}\right)^6 g(d\varepsilon) = \frac{(\pi k_B T)^6}{\left(\frac{61}{64}\right)}$$

Q: Which is the meaning of the symbol $g(d\varepsilon)$ in the above integrals? Is it maybe a **misprint** for just $d\varepsilon$? How do you get $\pi k_B T = \int_{-\infty}^{+\infty} g(\varepsilon)d\varepsilon$ from the above integrals?

M: I think there is a **misprint**, there shouldn't be a $g(d\varepsilon)$, it doesn't make sense, it's probably just $d\varepsilon$, absolutely, yes. That $\pi k_B T = \int_{-\infty}^{+\infty} g(\varepsilon)d\varepsilon$, that's a standard integral when you put in what $g(\varepsilon)$ is. I don't remember the details now. But if you put in the $g(\varepsilon)$, then what you get is a standard integral.

2. p. 74 1st column bottom: "...for an averaged value of $\varepsilon \approx \lambda/4$ in the sampling of εs *for the exothermic direction*"

Q: Why this "exothermic direction"? Is it maybe that the electron transfer at $\varepsilon = 0$ happens with $\Delta H = 0$ and it happens with $\Delta H < 0$ when from energies are above the Fermi level?

M: ε is the energy in excess of the Fermi level, right? And so when $\varepsilon \approx \lambda/4$... I guess that, when you are looking at that, you are looking at reactants going in the forward direction, you are looking at reactants going over a certain barrier, and I think that is the reason why you are sampling in the forward direction, that is sort of reaching that intersection going in the forward direction, the main point is that you are going in the exothermic direction, you are overcoming a barrier, so it is probably an exothermic reaction that is just overcoming that, I imagine, yes.

3. p. 75 1st column top: "Although the extent of ε is nonzero is $\sim\pm 0.3$ eV . . . the validity of the $\lambda \ll \varepsilon$ approximation . . ."

Q: **Typo**: I guess it should be $\lambda \gg \varepsilon$.

M: Oh, yes, sure, ja, that's definitely a **misprint.**

CHAPTER 10

Theoretical Investigation of the Directional Electron Transfer in 4-Aminonaphtalimide Compounds; Variable-range Hopping Electron Transfer through Disordered Bridge States: Application to DNA; A Model for Charge Transfer Inverse Photoemission; Mechanisms of Fluorescence Blinking in Semiconductor Nanocrystals Quantum Dots; Diffusion-controlled Electron Transfer Processes and Power-law Statistics of Fluorescence Intermittency of Nanoparticles

Interviews on M331, M337, M339, M341, M342

M331 Theoretical Investigation of the Directional Electron Transfer in 4-Aminonaphtalimide Compounds

Yi Qin Gao and R. A. Marcus

1. p. 1957 2nd column, legend to **Figure 3**: "Schematic diagram of the free energy *vs* nuclear coordinate for the ground and excited states for the fluorophore. |**g**⟩ is the ground state, |**e**⟩ is the fluorescence excited state, |⟩$\mathbf{e}_{\mathrm{CT1}}$⟩ is the charge separated, nonfluorescent state due to the electron transfer from the distal nitrogen."

p. 1957 2nd column middle: "**III. The Electronic States**

We consider a mechanism for the florescence quenching *and for its recovery* for the 4-aminonaphtalimide compounds in which there are the ground state, $|\mathbf{g}\rangle$, and at least two excited states, one of which, $|\mathbf{e}\rangle$, is less charge-separated and is fluorescent. The other labeled $|\mathbf{e}_{\mathrm{CT1}}\rangle$ in Figure 3 is more charge—separated and relatively nonfluorescent. The state $|\mathbf{e}\rangle$ is created by the excitation of a HOMO of the fluorophore and contains one electron in the HOMO and one in the LUMO of the fluorophore (Figure 4)."

p. 1958 1st column, legend to **Figure 4**: "The various diabatic states and the relevant orbitals:

$$|\mathbf{g}\rangle,\ (\mathrm{HOMO})^2(\mathrm{LUMO})^0(\mathrm{N4})^2(\mathrm{ND})^2;$$

$$|\mathbf{e}\rangle,\ (\mathrm{HOMO})^1(\mathrm{LUMO})^1(\mathrm{N4})^2(\mathrm{ND})^2;$$

$$|\mathbf{e}_{\mathrm{CT1}}\rangle,\ (\mathrm{HOMO})^2(\mathrm{LUMO})^1(\mathrm{N4})^2(\mathrm{ND})^1,$$

$$|\mathbf{e}_{\mathrm{CT2}}\rangle,\ (\mathrm{HOMO})^2(\mathrm{LUMO})^1(\mathrm{N4})^1(\mathrm{ND})^2$$

Q: (1) Why is the $|\mathbf{e}\rangle$ state the excited *fluorescent* state and the $|\mathbf{e}_{\mathrm{CT1}}\rangle$ state the charge-separated *nonfluorescent* state and not vice versa?

(2) Is the nuclear coordinate of the $|\mathbf{e}_{\mathrm{CT1}}\rangle$ state more distant from the nuclear coordinate of the $|\mathbf{g}\rangle$ state than that of the $|\mathbf{e}\rangle$ state?

(3) Why is the energy of the minimum of the $|\mathbf{e}_{\mathrm{CT1}}\rangle$ lower than the energy of the minimum of the $|\mathbf{e}\rangle$ state?

M: Frequently when you excite one of these molecules, you excite it to a state which is intrinsically fluorescent, unless it is a charge-transfer state. If the state is a charge-transfer state then normally, when you excite it, there is weak, very weak absorption, so usually you excite to *a locally excited state,* and *then* you get an electron transfer, an internal one, to a charge-transfer state. Now, when you have a charge-transfer state, there is very little overlap of those wave functions with those of the ground state, so the state doesn't fluoresce with much of an intensity.

(1) That is for the reason I just said, and now the $|\mathbf{e}\rangle$ state is not necessarily a charge-transfer state, the $|\mathbf{e}\rangle$ state can be a locally excited state.

(2) No, the stable environment that is required for the charge-transfer state, if it is a polar system, is quite different from that for this uncharged state, so there is a big shift to that. The $|\mathbf{e}\rangle$ state *doesn't have much of the actual charge-transfer state*. "Locally excited state" is the standard terminology that people use in that field, in other words, people that do those experiments use that terminology. What they mean is that supposing you have two species, *A* and *B*, and what you are going to have is a charge-transfer spectrum, you excite *A* and it first goes to another state in *A*, that is called a locally excited state, and then, in this mechanism, an electron goes over to the other species and then the locally excited state becomes a charge-transfer state, this is standard terminology. In a locally excited state normally, often, there isn't much of a charge-transfer, it is just that is locally excited to a higher state which is not of the charge-transfer type, then later it can often go to a charge-transfer state. You can excite directly to a charge-transfer state but this indirect excitation is probably the more common.

Q: How is wave function of the CT state?

M: It is one where the electron instead of being on the molecule you started with, it is over on the other molecule, in a locally excited state the electron is still on the same molecule you started with, but in the charge-transfer state the electron is moved to the other molecule. In other words in this case you have, say, two molecules *A* and *B*, and you excite *A*, this state is called locally excited, if it transfers an electron to *B*, that's a charge-transfer state. *The CT state is a state made up of both A and B*, one has lost an electron and the other has gained an electron. Usually in electronic structure theory for these aromatics you excite to another state in the aromatic, with light, and then the electron may jump over to a partner there, if there is a partner available, and form a charge-transfer state. Occasionally you can excite directly to the CT state, it depends on the system, on how much overlap of the wave function, of the CT function.

(3) That would depend upon the solvent. If your solvent were non polar, then it wouldn't be true.

2. p. 1958 1st column: "**IV. Crude Theoretical Estimate of the Electronic Coupling Matrix Elements**

We first consider a very approximate estimate of the difference in coupling of the fluorophore to the two distal nitrogens by noting that the distal N on the 4-substituent is *three bonds removed* from the 4-N, which is found in the calculations given later to make a significant contribution to the HOMO of the system and is part of the initial acceptor state $|\mathbf{e}\rangle$ for the electron-transfer reactions (Figure 4). With the use of a decrease of the square of the coupling matrix element V of a factor 3 per bond...

If the *contact value* is of the order of 2 eV...

On the other hand, the distal nitrogen in the 9-substituent is three single bonds and six double bonds away from the 4-N...

Q: Is the value of 2 eV a typical contact value?

M: If you have two atoms that are bonded to each other, that would be the contact value, and that would be the value of β, you know, the exchange integral, and that is usually a couple of electron volts.

3. p. 1958 2nd column top-middle: "It seems highly unlikely that the $|\mathbf{e}\rangle$ to $|\mathbf{e}_{\mathrm{CT1}}\rangle$ transfer in Figure 3 would be in the deep inverted region."

Q: Why?

M: That would have to be based on the energetics of transferring an electron from one state over to the other, on what the effect of the solvation would be on lowering one state with respect to the other.

Usually when you are in the inverted region you have an excited state much higher than the ground state and here you have two excited states, so, you know, here the chance that one would be much lower than the other is less than if one of them is the ground state and the other in an excited state.

4. Following: "Further, if the system were in the deep inverted region, a larger V would actually retard electron transfer, unlike its effect in the normal region."

M: Yes, the larger V there separates the two curves more, so I think that's probably it that the electron actually has a bigger tunneling to go through. If V becomes huge, those curves become so separated that the electron can't really tunnel from one to the other.

Q: It is unbelievable how many subtle points are connected to those two parabolas.

M: Yes, that's right. You must realize . . . when I first drew the parabolas as such, I think I did it in the 1960 paper, you know, it took me 4 years to do that. Actually around that time, around 1957 I think, Hush wrote a paper, and I mean this shows how different the situation was, and in that paper I think he had something that you can't use ordinary potential energy curves to describe electron transfer.

5. p. 1958 2nd column middle-bottom: "The two distal amino groups of **9** were then both chosen to be an NH_2 group. In the calculation, *the acceptor level was chosen to be the HOMO of* the molecule and the donor states were identified as the levels having electrons localized at the distal amino groups."

Q: I guess when going from $|\mathbf{e}\rangle$, $(HOMO)^1$ $(LUMO)^1$ $(N4)^2$ $(ND)^2$ to $|\mathbf{e}_{CT1}\rangle$, $(HOMO)^2$ $(LUMO)^1$ $(N4)^2$ $(ND)^1$.

M: Yes, I am referring to processes occurring in the excited states.

6. p. 1959 1st column bottom: “This radiationless transition rate is enhanced when the $(\text{HOMO})_1(\text{LUMO})_1$ and $(\text{HOMO})_2(\text{LUMO})_0$ surfaces are more displaced in coordinate space from each other.”

Q: Is it so because when they are more displaced the crossing point and the activation barrier lowers?

M: What figure should I look at?

Q: Figure 3.

M: If one state is sitting right over on top of the other, but quite above it, of course the usual transition rate would be very small, there would be practically no overlap, right? So that’s when the state above is not shifted, you just look at the state above the ground state. But if you shift the excited state now, then you get less tunneling distance to go from one to the other.

7. p. 1959 2nd column middle: “The pK_a of the 4-N is -0.5, a very low value, which indicates that the lone pair electrons on the 4-N are bound so tightly that an added H^+ binds only weakly to it. The pK_a of the distal N group is between 6 and 8, depending on its nature, indicating a much weaker binding of the lone pair to the N. The difference in pK_a corresponds to an energy difference of 0.39 eV in proton binding affinities.”

Q: Is the proton binding affinity the ΔG^0 of $\Delta G^0 = -RT \ln K_a$?

M: Yes, you can use that, I would think so.

8. Following: “Thus, *the creation of a hole* is expected to be energetically much more favourable at the distal N than at the 4-N, and the $|\mathbf{e}_{\text{CT1}}\rangle$ state is expected to be in a higher energy than $|\mathbf{e}_{\text{CT1}}\rangle$.”

M: You see, you might create a hole when there is no H^+ attached or you might create a hole also when there is an H^+ attached, and maybe we were comparing those two cases. Not that the H^+ is creating a hole, but one is maybe comparing what the effect of an H^+ is on creating a hole. Supposing you have a charge transfer of an electron on the nitrogen to some other place, what is the effect of a proton on the energetics of that charge transfer? When you transfer an electron from that end you're creating a hole, what is the effect of an H^+ on the energetics of creating a hole?

9. Following: "While the state $|\mathbf{e}_{\text{CT1}}\rangle$ may therefore not be directly involved in the present electron transfer, its presence could only enhance the pathway to the 4-distal nitrogen via a superexchange mechanism."

M: If you have a state that is sort of electronically related to a state on either side of it, then when you calculate the matrix elements, they could be enhanced by a superexchange.

10. Following: "The low yield of the fluorescence for **4** and **5** in Table 1, ~ 0.05, corresponds to an intramolecular electron-transfer rate on the order of 10^{10} s^{-1}, if the natural fluorescence lifetime is on the order of 5 ns."

Q: How do you deduce the electron-transfer rate from the fluorescence lifetime?

M: Well, you are using the natural fluorescence lifetime and the quantum yield of the fluorescence, you combine those magnitudes. You see, the fluorescence yield would be the fluorescence rate constant divided by fluorescence rate constant plus the charge-transfer rate constant, so putting the fluorescence yield to 5% or whatever it is, and now putting for the fluorescence rate constant k_f, the value that corresponds to the fluorescence lifetime of 5 ns, that is, you know, 1/5 ns, then solve for the other k, the charge-transfer rate constant, and you get 10^{10} s^{-1}.

M337 Variable-range Hopping Electron Transfer through Disordered Bridge States: Application to DNA

Thomas Renger and R. A. Marcus

1. p. 8404 1st column top: "Similarly small β values were reported by Schuster et al. and were discussed in terms of a *phonon-assisted polaron hopping model.*"

Q: (1) Please explain the "hopping" vs. "transfer" and vs. "tunneling." For instance: How would a "phonon-assisted polaron hopping model" differ from a "phonon-assisted polaron transfer" or from a "phonon-assisted polaron tunneling"? Are "hopping" or "tunneling" just different mechanisms for transfer? If so, how do they differ?

(2) For long time this "hopping" concept didn't appear in your papers. Why and when does it become important?

M: (1) Yes, well, typically if you got some distance to go for the electron, you can do it either by hopping or by tunneling. If it is tunneling, it is a coherent sort of wave function that describes the whole process, the electron doesn't spend any significant time in between the two sites. If it's hopping, it hops from site to site. The description is very different.

Q: In the sense of wave functions?

M: Yes, in the case of hopping, you can use a wave function for the first hop and then a different wave function for the next hop, because you are then in different positions, one different from the next one, whereas in the tunneling you use the same wave function.

Q: Do you use hopping when instead of going just between two sites you go through a series of different sites?

M: Yes, you travel between, you sort of pass several sites. And in that case there's a very little distance dependence whereas if you tunnel there is an exponential decrease with distance. So that's the difference between the tunneling and the hopping. And the phonon assisted, the polaron assisted, whatever it is . . . *Phonon assisted* means that *vibrations are involved in the polarization*, I suppose, and the polaron is where the electron is distributed either over a local site or a somewhat delocalized site, and the main point is that you go from one site which may involve one region or some polarized region around it, and you jump to another one.

(2) It becomes important where you have several sites in between or, for example, if you have just one site in between, as in the photosynthesis, then maybe that site in between is used for a superexchange, but I don't remember what we finally concluded, I think we finally concluded that in the intermediate site there was an actual intermediate, it was some years ago so I've forgotten.

2. Following: ". . . depending on the energetics of the system studied, either *the superexchange mechanism or the hopping mechanism* dominate the observed electron/hole transfer leading to the *strong or weak distance dependence of the rate constant* respectively."

Q: Can you please explain the relation between the two mechanisms *and* the dependence on distance of the electron/hole transfer rate constant?

M: Yes, if it is superexchange, that is the tunneling, tunneling that whole region, the rate constant exponentially decreases with distance, whereas hopping is different. The actual tunneling mechanism is one of superexchange, if one wants to be technical.

3. Following: "Numerical calculations applying a Redfield relaxation model were performed on a *model donor–bridge–acceptor* system. The latter was *coupled to one effective high-frequency mode,* of which *the potential energy minimum was shifted by the reaction*"

M: One possibility is the following: when you do a somewhat elaborate treatment of these problems, then there is a transfer coefficient, there is also a shift due to the interaction, and so the energy gap, so to speak, is changed by the interaction. In other words, when one reduces the number of coordinates, one works in a reduced dimensional space, then typically one has both a transfer and an energy shift, and I don't know if that's what's meant there or not. But if you look at Feshbach, for example, for treatment of a big matrix condensed down to a small matrix, you see that there is a *transfer part* and an *energy shift part* that comes in.

4. Following: "The same result was obtained in theoretical studies performed using a *Liouville pathway correlation-function approach.* The latter included *the coupling of the electron to a manifold of vibrational modes* and contained *the reorganization of a whole set of nuclear coordinates involved in electron transfer.*"

M: I imagine that probably that refers to his using a reduced Liouville equation, and sort of working with that formalism. The person who is using that formalism a lot has been Mukamel. In a Liouville sort of approach there are all sorts of manipulations that you do and that is probably referring to that.

5. p. 8404 2nd column middle-bottom: "Bixon and Jortner explained an experiment on hole transfer between guanine triplets in terms of *thermally induced hopping.*"

p. 8405 2nd column middle: "The hole . . . thermally activated . . . may hop through the bridge . . ."

Q: (1) Is there a nonthermally induced hopping?

(2) Please consider your classical pair of parabolas: can we say that

(i) In order to have electron transfer the reactants must be thermally activated by a suitable fluctuation to reach the crossing region and that

(ii) The electron transfer is the hopping of an electron from the R to the P parabola?

M: (1) Yes, there is a nonthermally induced hopping. For example, if you photoexcite a molecule, and then the electron hops to the nearest site, the system is already up there in energy, it doesn't need the extra thermal energy to get up there, you are already at the place where you can have electron transfer, you may photoexcite a molecule so that it has already enough energy to get up there, it doesn't have to be thermally activated then, it could be *photochemically activated.*

(2i) Yes, within the framework of those parabolas that's absolutely right.

(2ii) No, you see, normally when you refer to hopping normally you are speaking of a succession of events, not just one, so normally one wouldn't call it hopping then, it's when you are going from site to site that would you call it hopping, that's the usual wording.

6. p. 8405 1st column middle: "Berlin et al. gave an explanation of the Giese experiment in terms of a *phenomenological model* that used a description of bridge in terms of a tight-binding band."

Q: When would you consider a model to be phenomenological?

M: Maybe when they don't put in the molecular details, when they apply a formalism that could be applied to anything without putting in enough molecular details like what are the compounds involved. You know, tight binding is a kind of phenomenology, and then . . . it depends on what one means by phenomenology . . . often one means that one sets up a formalism without thinking too much about the details.

Q: Some kind of rough formalism?

M: Well, for example, in *dye sensitization*, something on which I am working right now, you write down a diffusion equation, write down a recombination, and write down a light absorption, without going into the specifics of what the recombination is, that's a phenomenology. Dye sensitization. . . that's where you excite a dye, the electron goes from dye to a neighbor, so you can write down formulas without going into the details, you can write down that you excite, you transfer, you recombine, you can write as rate equations, that's phenomenology, it doesn't tell you the details of steps. . . that's theory. The idea is that in a dye sensitized process you excite dye, electron can go from dye over to the other species, a process can happen there including to recombine with dye the large particle, you can write down a series of equations for that, that's called phenomenology, it doesn't tell you the details.

7. p. 8405 1st column bottom: "In the experiments of Giese et al. a single guanine is oxidized following a continuous *UV irradiation*. The reaction scheme is described in Figure 1, in which the rate constant for *hole injection into a guanine* is denoted by P.

Q: Is the hole injection simply the extraction of an electron from guanine by UV irradiation so the guanine becomes a guanine cation? I mean is the "hole injection" nothing but the process $G + h\nu \rightarrow G^{+} + e^{-}$?

M: I guess so, I don't know if there is an electron acceptor some place nearby, but that's right, it's a G^{+}, it's the hole yes, that's right, I just don't remember where the e^{-} is.

8. p. 8408 2nd column middle:

$$\text{"}H_{\text{vib}}(i) = \sum_{\xi} \frac{\hbar\omega_{\xi}}{4}((Q_{\xi} + 2g_{\xi}(i))^{2} + P_{\xi}^{2}) \qquad (43)$$

The *equilibrium statistical operator* of the vibrations in the ith PES is W_{eq} (i)

$$W_{\text{eq}}(i) = \mathrm{e}^{-H_{\text{vib}}(i)/(kT)} \qquad (44)$$

Q: What is the meaning of W_{eq} (i)?

M: W_{eq} (i) is used to weight various quantities that you calculate in some ensemble. You use that exponential divided by a sum over all possibilities.

9. p. 8408 2nd column bottom: "The well-known *classical limit of the rate constant* in Eq. (45) is

$$k_{i\to j} = \frac{2\pi}{\hbar}\frac{|V_{i,j}|^2}{\sqrt{4\pi\lambda kT}}e^{-(\lambda-\hbar\omega_{ij})^2/(4\lambda kT)} \qquad (49)\text{"}$$

Q: Under which condition are we allowed to use the above classical limit expression?

M: Probably when the spacing of the various states is small compared with *kT*.

10. p. 8409 1st column middle-bottom: "...the high-frequency normal modes are typically more localized than the low-frequency modes..."

Q: Is it may be so because a more localized mode implies a lower mass moving?

M: Or really a difference of the moving mass compared with all the other masses. If you want to have it localized then there should be either a big mass difference compared with the other masses or a big force constant difference. In other words, there has to be a big frequency difference. If you have different motions, and a very different frequency, for example, if you take H_2O where you excite a high overtone of one of the O-H's, then instead of using delocalized wave functions for the H_2O you use the localized bond wave functions.

11. p. 8411 1st column middle: "The transition between the tunnelling and the hopping regime is somewhat smoother ($N = 4,\ldots,8$) in the absence of disorder, while in the presence of disorder a sharp transition is obtained at $N = 3\ldots$ one sees that *disorder in site energies decreases the localized states hopping rate and enhances the delocalized states hopping*..."

Q: (1) What is meant by "disorder in site energies"?

(2) Which is the relation between disorder and localized and delocalized states hopping rates?

M: (1) Well, if you have a whole set of different sites, then you have a hopping from site to site or you have a tight binding approximation. Disorder means that those sites aren't of the same energy, you have differences in energy, you have a disorder. It depends on what one is talking about, if one is talking about hopping, then you go from site to site and you use a little bit of the site... and of course those sites may have different energies, that corresponds to a disorder. When you are going from site to site, if that's the mechanism, then that is one way of transferring. The other mechanism is if you have instead tunneling through everything and then the going from one end to the other is a coherent motion whereas the site to site hopping was an incoherent motion. Now, if the whole process is coherent, which sometimes is, then you can use some electronic structure theory, for example, tight binding or any one of a number of other approaches to calculate the entire tunneling rate. So, those are the two limiting situations, one site to site hopping, which is incoherent transfer, or the other, coherent transfer, where you have to treat the whole process all at once in a wave mechanical way. The α's, in tight binding, may be a little bit different, it depends on whether the atoms are the same or not, it depends upon on what is next to the atoms, so when you consider those sort of α's, tight binding α's, there is sort of local energy that you put into tight binding approximation, that particular quantity α may differ from site to site, in that case you have disorder, or it may be the same, in which case you have order. Tight binding is the solid state equivalent to molecular orbitals, and you recall that in molecular orbitals you have α's and β's, and in the tight binding, the disorder refers to the α's, primarily, so all the same or not all the same. There the αs appear diagonal in the $\alpha - E$'s, appear diagonal in the tight binding or molecular orbitals approximations, the β's appear off-diagonal. What we are using there is really some information drawn from molecular orbital theory.

Q: Can you have even intermediate kinds of things?

M: Well, anything can be intermediate, yes.

(2) If you think about hopping from one site to another and there is a big discrepancy in energies, a big difference in energies, that kind of hopping would be localized. If on the other hand two energies are the same, you can have kind of a resonance and then you can easily hop from site to site. So, the closer the sites are in energy to each other, then the more you have sort of a delocalized wave function.

Q: Are the delocalized states more in resonance?

M: Well, the delocalized states are typically more in resonance, yes. If you think of two states, if you think of the two states Hamiltonian, and you look at any particular state combined by those two states, it's more equally divided if the states are of equal energy.

12. p. 8411 2nd column top-middle: "The more general theory, more general in the sense of being more microscopic, that was used to obtain a rate constant . . . for hole transfer between *partly delocalized states*"

M: Something can be delocalized over the whole system or something can be just delocalized over approximately three or four parts of the system.

13. Following: "...the part of the earlier theory that describes the transfer between different electronic states was treated in *the Markov approximation*..."

M: In general, the Markov approximation is that when you are looking at a sequence of events, the probability of one of them just depends on the initial state there, it doesn't depend on any event of the past history.

14. Following: "...as expected, electronic delocalization leads to smaller reorganization energy of charge transfer."

Q: Is it so because when the charge is more delocalized its density is lower and so its polarizing strength is also minor?

M: Yes.

15: Following: "Disorder in energies tends to localize the electronic states."

M: Yes, right, that's related to Anderson localization that people have treated to describe what happens if you have a disordered material vs. an ordered material, and to what the wave functions look like there. There is a lot of sophisticated theory done by Anderson at Princeton in the 1950s, I think, on that problem, and so one speaks of Anderson localization. That is, under certain conditions, the electron instead of being extended throughout the medium, gets localized and the details of that is related to the disorder question.

16. Following: "Here, localization due to static disorder was taken into account and dynamic localization, the so-called *self-trapping* was neglected."

M: *Self-trapping is when the system digs its own hole by polarization of the medium around it.* That's what self-trapping is.

M337 Variable-range Hopping Electron Transfer through Disordered Bridge States: Application to DNA (Continuation)

Thomas Renger and R. A. Marcus

1. p. 8411 2nd column bottom: "In the present formulation, we have not included a possible dependence of the reorganization energy λ on distance between hopping sites. We expect that a distance dependence will mainly have an influence on the yield for short bridge lengths where tunnelling dominates."

M: Yes, there certainly should be dependence on the distance, I guess, we just didn't bother with that. Or, other, the guess is that it's so low that that term becomes small.

2. Following: "However, our focus here is on *the thermally activated hopping observed for longer bridges.* In the case of *nearest-neighbor hopping*, there is just a single λ, and in the case of *variable-range hopping*, it can be expected that an effective λ can be introduced that takes into account an average over different distances."

What the *variable-range hopping* is, is explained in the following page 8413:

p. 8413 1st column top: "An interesting effect is observed in the presence of static disorder. The latter barely changes the transfer efficiency in the first limit of strong electronic coupling, but it strongly decreases the transfer efficiency via localized states hopping. This decrease is a result of *local barriers created by the disorder in the bridge*. Because the *localized states hopping* is a nearest-neighbour hopping, such barriers in the bridge will be critical bottlenecks for the overall transfer rate, and thus the efficiency goes down with increasing disorder. In the case of *partly delocalized states*, the hole (electron) can tunnel through such *local barriers in the bridge,* a phenomenon that is termed *variable-range* hopping."

Q: Which is the nature of the local barriers and how are they created by the disorder in the bridge?

M: If there is no disorder in the bridge, a fair number of the bridge sites have the same orbital energy, then there would be hopping between sites where $\Delta G^0 = 0$. Now, if you had disorder in the bridge, then of course some of the sites would be up, some would be down, so some steps would be accelerated, some would be decelerated, but, you know, if the orbital energy of a site is largest, the corresponding step becomes the slowest and it is rate determining, so that you're probably losing more, you may lose more, by having disorder, because some of the disorder would correspond to an endothermic process, you see. So, you gain something but you lose something, and it may well be that if there is enough disorder then where there is the biggest barrier, you know, the biggest contribution to ΔG^0, you have a bottleneck.

3. p. 8412 2nd column middle: "...a delocalization of electronic states leads effectively to a smaller horizontal displacement of the free energy surfaces of the different states ..."

M: Well, if you have a delocalization, in other words if the electron spreads among several sites there, the electron got there from another site and now, as spreads among several sites, you have effectively a larger radius of the region that the electron is occupying, and a smaller λ. A smaller λ corresponds to a smaller horizontal displacement, I mean you can either change the force constant or you change the horizontal displacement, either one will produce a smaller λ.

4. p. 8412 2nd column bottom: "In the limit where the electronic coupling is comparable to or larger than the reorganization energy of a local hole state, the splitting between electronic eigenstates of the bridge will determine the gap between donor and bridge."

Q: Using your classical couple of parabolas, even drawing them very close to each other so that λ is very small, it seems to me geometrically impossible that the splitting could be greater than λ...

M: That happens when the splitting becomes very large. Suppose you have two parabolas, OK?, and now split them, split them so much that one would be a distorted parabola, it can have a bump in it, and the other upper surface goes up. Two parabolas can become one parabola if there is enough splitting. Just imagine, draw two parabolas. And now split them and keep on lowering down the lower parabola, and you are correspondingly raising the other. Now take the lower cusp and lower it.

Q: Until?

M: Until you have got that cusp lower than the two lowest points of the two original parabolas.

Q: Then the lower surface becomes one single parabola.

M: That's right, and that happens.

Q: It goes so much down that the two original parabolas merge, somehow.

M: That's right.

5. p. 8413 1st column middle: "...*variable-range hopping*... Because of this phenomenon, the transfer efficiency between extended states does not depend as critically as the one for the localized states on the disorder. Hence, in a situation where the *localized states channel* dominates the *thermal activated transfer*, an increase in disorder in bridge energies will change the branching ratio of the *two channels* in favour of the *delocalized states channel*."

M: Variable range hopping would be that you don't hop between localized adjacent states, you can hop between non adjacent states, that would be variable range I guess... I think the main point is that the hopping is not between adjacent sites but between sites that are somewhat removed from each other, because the electron is sort of spread out a bit more, and with the hopping then it goes from one group of sites to another group of sites, I guess, and in between within a group is not hopping, so the variable range part may refer to the fact that you have a variable distance between hops, as now you are hopping between group and group. I'm not sure, Renger introduced that term, or somebody else did, I guess, and I'm not sure how it is defined... when you have delocalization over group of states and then delocalization further along over groups of states, you have then a variable range, because you can have a group of one size and a group of a different size.

6. Following: "Finally, we note that the present formulation in terms of *two channels* is an approximation of the of the real situation where just *one channel* exists that contains partly delocalized states but takes into account a *dynamic localization of the states by so-called self-trapping.*"

M: Well, it's known in spin resonance, that you can have two sites and if there is a rapid exchange between them then you get a single spectral line instead of two. Narrowing then is what one would have from the two sites together, and that's called motional narrowing, and the theory of that is given in Kubo's book, there is a book of about three or four authors on nonequilibrium statistical mechanics, and there you see something related to this dynamical process, so one limit is in the environment changing slowly, you have then what is called the static limit, and in the other limit you have a dynamic limit. Kubo gives an expression that reduces to either limit as you proceed to a limit. So, to fully understand this sort of process, I think one has to understand the Kubo's formalism, but the main point is that you can have two separate some sort of Hamiltonians, and so you have a broad situation, or you can have change going so fast that effectively you just have a single Hamiltonian, and then you have a narrow motion. Another way of seeing it is probably related to the fact that if you have sort of a two sites system, then maybe you have absorption at the two sites. But now, if you have a big splitting, that is a big dynamical effect, then you have a single well, and you have a narrow spectrum. So there are probably various ways of looking at it, but the most fundamental is that of Kubo, and that gives you the effect of the difference between dynamic disorder, or dynamic effects, and static, sort of static effects.

7. Following: "A more exact but more complicated formulation (*involving the solution of a non-linear Schrödinger equation*) in terms of *solitary electronic states* can be found in a series of papers by Fischer and co-workers."

M: Well, if they have something that is sort of polarizing itself, the wave function of the electron depends in part on the square of the wave function of the electron and, you know, the charge distribution, so you end up having a nonlinear Schrödinger equation. I think if you look at the equation for a polaron, you will see that the potential depends upon the absolute value of ψ squared, and that appears in the equation for ψ, so that's clearly a nonlinear equation.

8. p. 8413 1st column bottom: "In the comparison between theory and experiment in Figures 4 and 5, for short ($N < 3$) bridges, the *superexchange mechanism* dominates, so an exponential distance dependence of the yield results, whereas for $N > 3$, the hole transfer involves *thermally populated bridge states*."

Q: (1) I believe the intermediate superexchange state is not thermally populated.

(2) Why and how are superexchange and tunneling related?

M: (1) No, it isn't, that's right, you tunnel through.

(2) Well, because in tunneling the system essentially spends no time in the intervenient *region*, in superexchange the system spends essentially no time in that intervening *third state*.

Q: Is there any place where both superexchange and tunneling are treated together?

Not that I know of. You see, the point is that *superexchange* is kind of *one of the mechanisms of tunneling*, then there would be other cases where you get tunnel but you wouldn't call it superexchange. You know, if you just had a square barrier, then there is no superexchange there, you just have a tunneling.

Q: And that is what you find in standard Quantum Mechanics textbooks.

M: That's right, but if you have a three states system instead something as a square barrier then you could have the equivalent of that but it is different. I don't think that you will find, my guess, something which compares the two.

9. p. 8413 2nd column top: "The flat distance dependence results from the efficient transfer between the *partly delocalized states due to a small reorganization energy* as discussed above and because *the splitting between the bridge states is larger for longer bridge lengths* thus decreasing the energy gap between donor and bridge even further and so promoting thermal activation."

Q: Is it also so that once a delocalized state is made up by a linear combination of states of sites, say, 3, 4, 5, and 6, the hole is present at the same time at site 3 and at site 6 so that, once this delocalized state is implied in the transfer, the distance between site 3 and site 6 becomes automatically equal to zero?

M: Yes, that's right, because the wave function is delocalized over that region, so you don't speak of its hopping from one site of the region to the other. If you have a wave function spread over something, you never speak of the system as hopping from one part of the wave function to another.

Q: I have put things in a very simple way for myself...

M: No, no, fine.

Q: But this for nonspecialist future readers of the book may be useful.

M: Yes, absolutely.

Q: I'm thinking of the book by Pauling, "The Nature of the Chemical Bond."

M: That was an amazingly popular book.

10. p. 8418 2nd column middle: "The latter λ was decomposed into a contribution $\lambda_s = 0.23$ eV from low-frequency *solvent* modes and a contribution $\lambda_i = 0.99$ eV from a high-frequency ($1500\,\text{cm}^{-1}$) quantum mode."

Q: Is $\lambda_s = \lambda_o$?

M: Yes, I think it is just that... You wonder that I used to start with the λ_o, that was solvent reorganization energy, and now λ_s is used. We are referring in terms of modes here, in other words we got into a molecular description, whereas with the other λ we didn't break it up into dynamical modes.

Q: So, it is the same thing but seen from two different points of view.

M: Yes, that's right, one is sort of maybe thinking a little bit more molecularly about it.

M339 A Model for Charge Transfer Inverse Photoemission

Shachi Gosavi, R. A. Marcus

1. p. 4, 1st column middle: "In the solution experiments, in contrast, both the initial and the final energy levels are typically *bound levels*... Since the energy difference between the two levels has a maximum of about 3.5 eV ...due to *solution stability constraints*"

Q: What do you mean by "solution stability constraints"?

M: You might be amused by one thing...I was just reviewing a paper...previously a person had written a paper saying that the transfer coefficient wasn't that explained by the Marcus's theory, because he got deviations from 0.5...so in the 1965 paper I had the corrections when there is an asymmetry of force constants, and now I have a paper by the same person in which he has used the equations of that 1965 paper, so you see how things continue even though it's now 46 years later, he is using the equations that are in that paper, but anyway I thought it was interesting that some of these things continue to live. It never occurred to me that one would actually test those equations, or anything like that, or use those equations because they are correction for asymmetry of force constants, but now with the techniques in cyclic voltammetry apparently they can get the transfer coefficients so accurately that they can test the equations.

Q: But on many occasions you have written equations that have been used later.

M: Well, sometimes anyway, yes.

Q: For the RRKM theory

M: Oh, absolutely, but that was only 10 years later, this is 46 years later!

Q: And what about the experimental discovery of the inverted free energy effect, how many years later?

M: That was about 25 years. Although Al Bard says, and it's true, that he had the first indirect evidence of it because he worked with some very exothermic reactions and for a couple of reactions he had a 100% formation of the excited states, and of course because that is a reflection of the inverted effect, so in that case that was just a matter of about 10 years.

Q: That is already extremely exceptional, I mean not many theoreticians have discovered things in advance.

M: Some yes...

Q: I believe in particular among the theoretical Chemists, because I mean I don't want to be flattering you but I mean to discover things in advance.... you can count people on the fingers of one hand, I mean... like Dirac or... not very many.

M: Well, that for those who have two hands, they can use another hand.

Q: At most with two hands you count all of them.

M: Yes...

M: "*Bound levels*"... Yes, the electron is fixed, you know, within the molecule, so the electron is bound, whereas if you have an electron in a metal is sort of unbound, right? Or, if you have an electron that is moving in a solvent, then until it gets localized it is essentially unbound.

"*Solution stability constraints*"... Probably it has something to do with ionization. You know, if you have too large an energy difference, that probably allows to ionize something, I suppose, in other words, for most reactions you wouldn't find free energy differences greater than that amount, maybe you would in some electronically excited states, but even there... I mean that's a very large free energy difference, 3.5 eVs. When you start having maybe 5 or 6 you ionize the molecules. In solution ... you can get an excited state in solution, but it is not usual to have that much in excess... In gas phase you get up to a state of maybe 10 eV, yes, but that wouldn't be stable in liquid, you probably go over to ionization, so I think that is what it is, is that typically for solution reactions you are rather limited, but it is not such a big limit for the range of reaction free energy to study.

2. p. 6 1st column middle: "Step one of the model is the electron injection via an electron transfer to the metal. *For simplicity*, a nonadiabatic expression (weak interaction) is used, but *the main features* would apply to the adiabatic case also."

Q: What do you mean by "for simplicity" and which ones are main features and which ones are nonmain features?

M: Well if it's a nonadiabatic treatment you write down a golden rule or something like that, it is so simple, in an electrode you have all of those acceptor levels, if you could assume a nonadiabaticity then that is a relatively simple problem. If it's adiabatic then you treat a whole set of levels interacting with the electron.

Q: Which are the main features?

M: Yes, you would say: all right, you're starting off at one level, you're going to go into one of many possible levels, in the Fermi level, and you treat them all as additive, so you treat the problem as a two states problem, although it is really a many states problem, and in the nonadiabatic limit that's OK. In the adiabatic limit, you know, or in the mixed regime too, if the electron doesn't go in a one level maybe it will go into the other and go into the other, you've sort of a *multiple curve crossing problem*, a huge problem.

Q: Can you have a multiple nonadiabatic?

M: Yes, you can have that too, it is just that you can treat each of processes as separate events. Yes, it is like a one-step event, you know. And indeed the common expressions that you have going into a many levels problem treat them as separate events.

3. p. 6 2nd column middle: "... $e\eta$ is the overpotential, $E_F - E_0, \ldots$"

Q: Does this mean that the potential applied to a metal electrode shifts the Fermi level, I mean that $e\eta = E_F - E_0$?

M: It does. Yes, the Fermi level changes linearly, with a slope of unity, with the electrode potential.

Q: Can you write $e\eta = E_F - E_0$?

M: Yes... and now remember that E_F depends on the applied potential.

4. p. 6 2nd column bottom: "The emitted light intensity is now given by

$$I(\eta, \omega) \propto \omega^3 \frac{2\pi}{\hbar} \int d\epsilon \frac{e^{-(\lambda - e\eta + \epsilon)^2/4\lambda k_\mathrm{B} T}}{(4\pi\lambda k_B T)^{1/2}} |V(\epsilon)|^2 f(\epsilon - \eta) \quad (7)$$

where the mean energy level of the metal *above the Fermi level* into which the electron is injected E_inj *is the same as the overpotential* $e\eta$. The ω is in units of energy."

Q: (1) Apparently from the above we have then $E_\mathrm{inj} = e\eta$. How does it compare with the above $e\eta = E_F - E_0$? Do we have $e\eta = E_F - E_0 = E_\mathrm{inj}$?

(2) What is ω? And why is it to the third power?

M: (1) Well... I guess E_inj occurs at a particular E, doesn't it?

Q: Yes.

M: All right, I believe the answer to that is yes, because if you are injecting... yes, I think the answer for that is yes, that is the injection energy. Normally, if you're ejecting from the metal, sorry, from the anion, you're injecting at an energy which is E_0, and the excess energy that can go into light is a difference between that and the Fermi level of the metal. So, what we called injection energy is really the excess energy that will come off as light.

Q: The difference $E_F - E_0$ is then the difference above the Fermi level.

M: That's right, yes, and that can come off as light.

(2) ω is a frequency, frequency of the light. If you talk about emission, emission depends on the fourth power of the frequency, under certain conditions, for this particular problem it depends on the third power.

5. p. 7 1st column middle-bottom: "Tight-binding wave functions of metals are of the form

$$\Psi_k(\mathbf{r}) = \sum_j \sum_n u_n \phi_n(\mathbf{r} - \mathbf{R}_j) \exp(i\mathbf{k} \cdot \mathbf{R}_j) \tag{10}$$

where... $\phi_n(\mathbf{r} - \mathbf{R}_j)$ are the individual wave functions of each orbital... $\phi_n(\mathbf{r} - \mathbf{R}_j)$ is assumed to be $\phi_n(\mathbf{R}_j - \mathbf{R}_j) \exp(i\mathbf{k}(\mathbf{r} - \mathbf{R}_j))$ in the Wigner–Seitz cell of the atom and zero everywhere else."

Q: I believe there is a Typo in the last formula.

M: Let me see... there's a **Typo**, a misprint there, there it shouldn't be $\mathbf{R}_j - \mathbf{R}_j$, that's for sure, that first $\mathbf{R}_j$ should be **r**, that's like in the exponential.

6. p. 8 1st column bottom: "In the case of a metal with a surface, only a wavevector conservation of $\mathbf{k}_{||}$, the component of the wavevector parallel to the surface, exists... When the optical transition arises from a wave function that is situated fairly deep in the metal, there is *also* an *approximate* conservation of the $\mathbf{k}_\perp$component."

M: The question is: what is the optical contribute to the momentum? Because a photon has momentum and so there is conservation of some sort, and, first of all, let's say that the photon comes in at right angles, so it has no parallel component, but why $\mathbf{k}_\perp$ isn't changed a little bit, that I don't know. Why $\mathbf{k}_\perp$ isn't changed by the equivalent of one photon momentum unit I don't know and I don't know what the typical magnitudes of **k** are, and maybe the typical magnitude of **k** is so large that one little difference doesn't matter. The point is that when you are at the band edge, then you don't have a **k** just in one direction. When you are in the metal far from the surface then, you know, for over some correlation length you can speak of a **k** in the direction perpendicular to the metal, but when you are right at the band edge or when you are right at the surface, then you don't have a pure **k** or anything like this. So, things are sharper when you sort of have essentially that free motion. For parallel motion you always have it, except for impurities or imperfections.

7. Following: "In the present tight-binding treatment, we have not found it necessary to invoke *indirect transitions* because of the accessibility of *direct transitions* ..."

p. 18 2nd column middle-bottom: "If no direct transitions can occur, then the electrons radiate only through indirect transitions..."

M: Yes, one would have to become familiar with transitions in semiconductors and metals, when you can go vertically in a transition that is a direct transition. If the vertical transition is not available at the wavelength, then you sort of violate the Franck–Condon, you go at an angle, and that is, at best as I remember, an indirect transition. But one would have to look up optical treatment in metals and what a direct and an indirect transition is. I think that if there is an available level at that particular **k**, at that energy, then it is possible to have a direct transition, but when you go someplace else and maybe you're violating the **k** magnitude, and you do something, I don't remember what it was, that's an indirect transition. It is something to look up in books on absorption of light by metals.

8. p. 9 1st column top-middle: "In the metals considered in this article...no *image states*..."

Q: What is meant by "image state"?

M: If you bring a charge up to a metal, then you'll induce an image.

Q: But it looks like that they belong to the metal, don't they?

M: No, they are images.

9. p. 9 2nd column top: "The surface states . . . have a *dispersion* of the form $\hbar^2(k_x^2 + k_y^2)/2m^* - E_{\text{surf}}(\text{eV})$"

M: The surface states have an effective free motion on the surface, so that m^* is an effective mass, you notice that the form gives the kinetic energy for motion in a plane, on the surface. So, what that is, is that you have a surface band plus the sort of the effective kinetic energy part, that's what you have in three dimensions when you have three k's. There, that form with two k's is the way of writing in an effective mass form the energy of a surface state.

10. p. 11 1st column middle: "The detection of *polarized light emission* provides information about the *optical matrix element* of the radiative transition."

M: That's right, that means that how the system reacts toward light emission is given by the detailed form of the optical matrix element.

11. p. 15 1st column middle-bottom: "The decay of the integrated intensity at high E_{inj}s occurs because the upper band in the tight-binding model has a relatively *flat energy dispersion*, and the density of states reduces with energy on going to larger energies."

M: I don't remember why the upper band in the tight binding model has a relatively flat energy dispersion.

Q: What does it mean, by the way?

M: The effective mass has to be so large that there is no k dependence that would be flat.

Q: "and the density of states reduces with energy"?

M: Well, that would be related. If you have a curvature of a band, you can probably get a density of states from that, from the curvature, and maybe if that is flat it is conceivable that the density of states goes to zero, I'd have to think about that...

12. p. 16 2nd column middle: "We assume that the emitted photons arise from $\mathbf{k}_{||}$-conserving direct transitions. In the direction perpendicular to the surface a Lorentzian *broadening* is introduced *to include the effect of the surface*."

M: Well, you don't have a sharp line, when something isn't conserved you don't have a sharp line.

13. Following: "The model also assumes no $\mathbf{k}_{||}$-conservation and so no possibility of *direct* transitions. All the spectral intensity arises from *indirect* transitions and so the band structure information is introduced into the problem only via the *density of states* in the metal at a given energy. Thus, any information about *band gaps* is omitted."

M: Well, if you sort of go at an angle instead of going vertically, that is an indirect transition. If you look at the surface texture of the energy levels as a function of $\mathbf{k}_{||}$ for a metal, or if you look at the full volume as a function of $\mathbf{k}$ for a metal, if for that particular $\mathbf{k}$ there is a band gap, and overall there is no band gap in the metal, but at a particular $\mathbf{k}$ there is one, and if you are relaxing that prescription on having a vertical transition, namely because you are not having transitions in the interior where $\mathbf{k}$ would be preserved, then it is the density of states that comes in. The band gap question doesn't come in because you're not requiring that the transitions be direct.

14. p. 18 2nd column middle: "For the free-electron model, we first calculate the contribution to $H_{k_1 k_2}$ ... from direct $\mathbf{k}_{||}$-conserving transitions. We use a two-band approximation for the band structure of the metal ... yielding

$$|H_{k_1k_2}|^2 = \left| T \int_{-\infty}^{\infty} \int_{-\infty}^{\infty} \int_{-\infty}^{\infty} e^{-i(k_{1x}x+k_{1y}y)} \sin(k_{1z}z) \right.$$
$$\left. \times\, e^{-i(k_{2x}x+k_{2y}y)} \sin(k_{2z}z) e^{-z/l} dx dy dz \right|^2 \qquad \text{(A.2)}$$

.... We calculate next the contribution of indirect transitions to $H_{k_1k_2}$ ... A broadening term similar to the mean free path broadening but present in all directions is introduced. This term, like the l of the direct transitions decorrelates the $\mathbf{k}_1$ from the $\mathbf{k}_2$. *A large broadening allows a calculation of* $|H_{k_1k_2}|^2$, *by replacing the integral by a constant multiplied by T. This constant, which we shall call* T_{ind}, *serves to reduce the entire spectral intensity of the indirect transitions relative to the direct ones.*

Upon introducing this approximation into the expression for the coupling we have, for $|H_{k_1k_2}|^2$

$$|H_{k_1k_2}|^2 = T_{\text{ind}}^2 T^2. \qquad \text{(A.3)}$$

M: What I see there, by the way, are the envelope wave functions, in the metal, and, yes, basically that's what we've done and then the atomic part is sort of incorporating into that T there.

Q: How about this broadening term?

M: Well, I suppose, you know, that if you have a sharp line, then you would only have direct transition, but if you allow putting in something which allows k to change, I guess in k space, you don't have a delta function with respect to k, so you're allowing indirect transitions. Yes, if you put that in, then indirect transitions become sort of more likely.

M341 Mechanisms of Fluorescence Blinking in Semiconductor Nanocrystals Quantum Dots

Jau Tang and R. A. Marcus

1. p. 1, 1st column top: "The developments in low-dimensional materials have led to the study of quasi-zero-dimensional systems, known as quantum dots . . ."

Q: What is exactly meant by "quasi-zero-dimensional"? Even if the quantum dot is a nanocrystal isn't it a three-dimensional object?

M: A line is a one-dimensional system, a plane is a two-dimensional system, a dot is a zero-dimensional system.

2. Following: The QDs have unusual optical and electrical properties such as narrower transition linewidth, larger oscillator strength . . ."

Q: "narrower", "larger" . . . with respect to what? With respect to molecules, I guess?

M: The cross section of the dot, you know, is 20 Å, 30 Å, so it's a lot to absorb light, a high cross section, and a molecule is usually quite a bit smaller than that. The dot has a larger oscillator strength, its absorption is broad but when you look at the experimental fluorescence, the fluorescence is narrow, so it has a narrow fluorescence.

3. p. 1 2nd column middle: "A possible connection of the power-law statistics to the *first passage theory* in real space or in *configuration (energy)* space . . . In the first passage theory, one considers *one-dimensional (1D) free-space diffusion* with an *absorbing end point*. If the *absorption rate* at the boundary point is much greater than the *in-chain hopping rate*, the rate of overall population change follows $t^{-3/2}$ power law."

M: What is involved is an electronic transition, like in ET reactions: two states, those two states have an energy difference. In an N-dimensional system you can collect everything that has the same energy difference that forms an ensemble of systems of one dimension less than the total number of dimensions, 10^{23}, so that *the energy difference serves as a reaction coordinate*. The system moves, it can move from one value of the energy difference to another value of the energy difference, the same thing as in electron transfer, and it is in the 1960 paper in *Discussions of the Faraday Society* that I introduced that coordinate, the energy difference of the two states that are undergoing electron transfer, I did the same thing here.

The two states are here the light state and the dark state. First passage theory is that supposing you are starting from some point, then you have some other point down the line, and you have diffusion, and you look at what is the rate of the first passage, the first time you reach that point, because if you're doing a general diffusion problem you can diffuse at that point, then diffuse back and so on. But what is the rate for first reaching that point? You can solve the diffusion equation and you put the boundary condition there as $c = 0$, so if it's zero it can't diffuse back, and then you get a t^{-3} law. You have fluctuations, and if you reach the intersection of two potential energy curves you have the possibility of a transition there. And you reduce the whole thing to one coordinate, an energy coordinate, and the potential energy curves become free energy curves.

Don't forget that it is just a theory, I would stake my life on the electron-transfer theory, I won't stake my life on the application to quantum dots. It may be OK, but, you know, I hope and I hope to live for quite a bit longer, so... all right, OK, now what's happening is that the configuration of the system is undergoing slight changes, the positions of the nuclei, maybe there is a charge that is jumping around, there are structural changes. In effect, in QDs, when you look at the spectrum of them, you see what is called spectral diffusion, the energy difference actually jumps around, it diffuses, and that's because of structural changes going on all the time, very small changes.

Q: In the structure

M: Yes, that's right, extremely small, maybe there are some ions around there that move, I don't know, maybe it's a defect that's moving around.

Q: The entire system.

M: Yes, that's right, the entire system.

Q: (1) What is "absorbed"?

(2) What is it meant by "in-chain hopping rate"?

M: (1) It's a system that's transitioning from one electronic state to the other, it is a transition, and so absorption just means that the transition is occurring. Absorption from the initial state over to the final state.

(2) If you think of a one-dimensional problem and if you discretize it, then it is hopping from site to site and back, and forth and so on, that's a model of diffusion, it is not essential, but it happens to be a model that is around.

Q: Always of the entire system.

M: It's the entire system. Along the energy coordinate it is the energy of the entire system that we are talking about.

4. p. 2 1st column middle: "This model is based on the reaction-diffusion (the 'stochastic Liouville' equation for a *one-dimensional diffusion process in energy space*... slow *structural diffusion of the QD*..."

Q: (1) Does "one-dimensional" mean that the process happens along a parabolic free energy surface with a one-dimensional, that is, a single abscissa coordinate—the reaction coordinate q—and a one-dimensional single ordinate G?

(2) Is "the" reaction coordinate for a reaction always unidimensional or is it possible to have multiple reaction coordinates, say $q_1, q_2, q_3 \ldots$? In the case in which in an elementary reaction one gets from the reactants only to a precise set of products is the reaction coordinate single even if it may have many components contributing to it?

M: (1) That is what it does in the theory, yes.

(2) You remember that in the Sumi–Marcus papers we had a two-dimensional sort of reaction coordinate, and depending on the viscosity you could cross the intersection line in different ways.

Q: So, you may have components.

M: Yes, that's right, and if it's very viscous you won't go very far along the viscous coordinate before you go along with the vibrational coordinate to cross the surface, if it is not viscous you may go all the way to where that transition state line intersects the x-axis.

5. p. 2 1st column bottom: "QDs are known to have a very narrow fluorescence spectrum, but a broad absorption spectrum. As illustrated in Fig. 1(a), this fact indicates that a *photoinduced electron-hole pair* relaxes rapidly to a band-edge state $|L^*\rangle \ldots$"

From the legend of Fig. (1a): "The radiationless decay from the higher *excited electronic states* to $|L^*\rangle \ldots$"

p. 3 1st column middle: "The model consists of two neutral states and two charge-separated states."

Q: Do we have higher excited electronic states or a photoinduced electron-hole pair? Apparently from the last line above at least the $|L^*\rangle$ state should be neutral.

M: Yes, in one model you excite to a broad range of states and where you excite to depends on the wavelength of your exciting light. If it

is very much near the red, you will excite only to the lowest state in the upper state and normally it is assumed that then it rapidly relaxes and now you have the system in the lowest upper electronic state and you will either fluoresce from there, a narrow fluorescence, or, by diffusion, one parabola may cross to the other and you may get a transition. In that case you go from light to dark. OK? Now, in actual fact things are a bit more complicated than that, and the details still have to be worked out, in fact experiments still have to be worked out, so it is an ongoing research, because when you excite at too high energy you get a quick relaxation but there is a certain probability that at high energy you may get a transition to a dark state, you see. I am not sure if this answers your question or not, but let me just say the following: when you excite a molecule, if you measure the absorption spectrum often it is pretty broad, then the fluorescence spectrum sometimes is a mere image, in certain systems there is time for everything to cascade down to the low standing level and then the spectrum becomes a narrow fluorescence spectrum. So, there a lot of activity, a lot of work is done. I mean, for example, on quantum dots the absorption spectrum is very broad, the fluorescence spectrum is very narrow, on the other hand in certain aromatics you have that the fluorescence spectra are mere images of the absorption spectra. In each case one can understand both those situations from the theory.

Q: You write: "a *photoinduced electron-hole pair* relaxes rapidly to a band-edge state $|L^*\rangle \ldots$" And the evidence for that is a narrow fluorescence, if it didn't relax rapidly you'd have fluorescence from a whole range of states, and you'd have a broad fluorescence. This excited electronic state you call "photoinduced electron-hole pair"

M: Yes, that's what they are, the upper state. You have excited an electron, you have left a hole, that is what excited states are in QDs. In fact, that is what excited states are in any molecule.

Q: So, excited states in molecules are always, in a sense, photoinduced electron-hole pairs.

M: Yes, because the electron has gone up, upstairs to some state, and of course you left behind a vacancy that is a hole.

Q: In this case yes, but in many cases when you excite a molecule electronically you have a different structure of the nuclei . . .

M: Yes.

Q: You have a different distribution of the electrons around a different distribution of the nuclei.

M: Yes, that's what you have here too, it's the same as in an ordinary molecule, in any excitation you've excited an electron to some upper state, so the electron is in some state that it wasn't in before, and it may be the only electron in that state, because it is up to higher energy, and what you have left behind is a vacancy. It is so for every molecule.

6: p. 2 1st column bottom: "In Fig. 1(b) parabolic potentials with q as the reaction coordinate are assumed, where q_0 is the horizontal displacement between the free-energy parabolic potential well for $|L^*\rangle$, $U_L(q) = \kappa q^2/2$, and that for $|G\rangle$, $U_G(q) = \kappa(q+q_0)^2/2 - E_g$, where E_g is the free-energy gap between the minima of these two potential wells."

Q: Please look at your Fig. 1(b).

(1) There the parabola with minimum at $q = 0$ is the one for $|G\rangle$ and not the one for $|L^*\rangle$.

(2) Imagine now that $E_g = 0$, so that for both parabolas the minima are at $U_L(q) = U_G(q) = 0$. If we now consider the parabola describing the parabola $U_L(q)$, we see that the formula describing it should be $U_L(q) = \frac{\kappa(q-q_0)^2}{2}$ because in this way $U_L(q) = 0$ for

$q = q_0$. The formula $U_L(q) = \frac{\kappa(q+q_0)^2}{2}$ is then wrong. But maybe you tacitly mean $q = \Delta q$, where Δq's are increments or decrements from $q = 0$ or from $q = q_0$? If so the formulas in the paper are OK but one should probably mention the incremental meaning of q.

(3) In the parabolic free energy curves does the "force constant" κ have some physical meaning?

M: (1, 2) The way you have written is the correct way of doing it. It may well be that a mistake was made.

(3) In the general case any physical meaning isn't immediately transparent. Remember that the force constant is a projection from a 10^{23} dimensional space onto a one-dimensional space, its physical meaning may be a bit obscure, because it involves many coordinates, I don't know if I can say more than that because, you know, if one had just one class of coordinates one could probably give it a physical meaning, but with everything thrown in I'm not sure that there is a physical meaning... in other words, it's not just a matter of being a vibration, in which case it becomes a vibrational force constant, the energy coordinate is dealing with all sort of change of anharmonic motions, 10^{23} of them, and so on, so I'm not sure that it has a simple physical interpretation, but its value may well be, you know, in the low-frequency range, because there are so many low-frequency coordinates that come into consideration ...

7. p. 2 2nd column middle: "The differential equation for classical diffusion on a harmonic potential is well known and can be applied here. The rate equation for the *probability $f(q,t)$ of finding a QD at q in the state $|L^*\rangle$* is given by

$$\frac{\partial}{\partial t} f(q,t) = D_1 \left(\frac{\partial}{\partial q^2} f(q,t) + \frac{1}{k\Delta_1^2} \frac{\partial}{\partial q} \left(f(q,t) \frac{\partial}{\partial q} U_1(q) \right) \right) \tag{1a}$$

where D_1 is the diffusion constant related to τ_1, the diffusion correlation time constant by $\tau_1 = \Delta_1^2/D_1$, and

$$\kappa\Delta_1^2(\hbar\Omega/2)\coth(\hbar\Omega/2k_BT) \approx k_BT \tag{1b}$$

where... Ω is *the structural vibration frequency* of the QD."

Q: (1) Are we considering here an ensemble of many QDs or a single QD?

(2) Which is the meaning of the constant Δ_1^2? Algebraically it is the geometrical mean of τ_1 and D_1. But does it have a physical meaning?

(3) What is it the "structural vibration frequency of the QD"?

M: (1) We are considering one quantum dot, and describing the motion in that quantum dot.

(2) Roughly some mean square displacement for a hop. When you see something like that, then think of the Einstein relation, $(\Delta x)^2 = 2Dt$ for a one dimensional process, where Δx is the mean diffusion jump, so this Δ_1^2 is not exactly the mean diffusion jump, but it's roughly so. Probably if you look on top of the page where you see $\tau_1 = \Delta_1^2/D_1$, you can see that it behaves in an Einstein's like relation, so I guess that probably it is, really. I imagine, probably largely a mean square jump, a diffusion jump, *in the sort of structural changes in the quantum dot*. It is not the diffusion of a particle, the relation is being applied to spectral diffusion, the diffusion is here different from ordinary diffusion, it is where the energy levels fluctuate because of vibrational changes. The charge is moving around, so it is here related to a diffusional motion of whatever is causing *those fluctuations in the spectrum, called spectral diffusion.*

(3) Yes, it is along the reaction coordinate, so to speak, everything is reduced to one coordinate.

8. Following: "There are higher electronic states, but there is relaxation to $|L^*\rangle$, the excess energy being released to the lattice *also assists in diffusion.*"

Q: How does it?

M: Well, first of all I'm not sure that it does. But if you have some excess energy, it is being released because you are going down in the electronic level, you know, it is becoming vibrations. Now, whether those vibrations catalyze the diffusion or not I don't know, there is evidence down at low temperatures that they definitely do, that when you go at higher intensity there is higher diffusion rate, higher absorbed intensity. Whether that happens at room temperature is another question. But that is certainly one of the key questions, that is: does a diffusion constant depend upon excitation energy, on the light, the wavelength that you are putting in, does it depend on the excess energy? The answer to that I think at the moment isn't really known, although if the experimentalists are doing the right experiments they might be able to find out. Part of the problem is that at low temperature there is something to slow enough, like diffusion, the spectral diffusion, that you can see whether the spectral diffusion depends upon the energy or something related to it. At room temperature I don't know that those spectral diffusion experiments have been done, the spectral diffusion may be too fast to measure, and the spectral diffusion means that you are having these fluctuations of the energy levels, of the upper and the lower state, and if it is occurring too quickly you can't resolve it, but down at 4 K they follow that spectral diffusion, they look at how that line fluctuated and now gradually reaches the saturation, so there are experiments we refer to in the papers. There is much still to be known in these QDs, it is not a closed subject. There are a huge numbers of papers now on dots .... Bawendi is the preminent person there, there are books now...it is just a big field.

Q: Is it like ET in the case of photosynthesis?

M: No, I think photosynthesis is a much cleaner system because here in the quantum dots we have a whole surface with all sorts of defects, and so there are all special things you don't know. So, even though the photosynthetic system is a bigger system, in a sense it's cleaner cut, you know the components. Here when the QD is in the dark you may have a charge localized on the surface, it can be any place, and what are those surface states like, most of that is not yet known.

9. Following: "For an initial *population* at any given point q of the reaction coordinate, the *probability density* $G(q, q'; t)$ of finding *this* QD that was initially at q to be at another point q' at time t is given by

$$G(q, q'; t) = \frac{1}{\sqrt{2\pi\Delta_1^2(1 - \exp(-2t/\tau_1)}} \times \exp\left[-\frac{(q - q'\exp(-t/\tau_1))^2}{2\Delta_1^2(1 - \exp(-2t/\tau_1))}\right] \quad (2a)$$

Q: Here the term "population" refers clearly to a system of many QDs but the "this" instead of a "a" seems to refer to a single QD.

M: It should be *population probability*.

Q: In the sense of a wave function?

M: Yes, in the broad sense where you are on that one-dimensional coordinate.

Q: For a single QD.

M: Yes, for a single QD.

Q: Because normally when one speaks of population one thinks of many . . .

M: That's correct, if I had to rewrite it I would call it population probability. Occupation probability.

10. p. 3 1st column top: "... $\Lambda = \kappa q_0^2/2$ is the 'excitation reorganization energy,' 2Λ is the Stokes shift..."

Q: Why this relation between reorganization energy and Stokes shift?

M: Well, the Stokes shift is the difference between going from the bottom of the lower parabola straight up and going from the bottom of the upper parabola straight down, that difference is a Stokes shift, and you will find that it is just what it says, it is 2Λ, I mean just make two parabolas, look at that difference and you'll see it's twice the reorganization.

11. Following: "Unlike the *static origin* assumed in a *distributed rate model*..."

M: Suppose that you excite the electron and you eject it outside the quantum dot. The distributed rate model is one model for trying to understand the occasional case that you get a dark state. It's not our model, it's somebody else's model. Suppose you have that, so the electron is ejected out and now you have a dark state. You can get Auger's processes of what's left. It's discussed in later papers. Now, when the electron is ejected out, it should come back with various probabilities. That coming back, that is called distributed kinetics, there is not a single wave function for coming back. By the way, that is a very common model, I mean: that's not the model we use, we use a very different model, but that is a very common model and a lot of people believe it, distributed kinetics due to different injection distances.

Q: Your model is very elegant, from an aesthetic point of view...

M: Yes, that's right.

12. p. 3 1st column middle: "$|D\rangle$ describes a charge-separated state with a charge in the core and a countercharge assumed to be trapped in surface states just below the edge of the *quasiconduction* band or just above the edge of the *quasivalence* band."

Q: What does "quasiconduction" and "quasivalence" mean?

M: If the quantum dot were infinitely large it would be like a solid and you would have a continuous set of energy levels, but instead you have a particle in a box like system, confined, so that means you have discrete levels.

13. Following: "Figure 2(a) schematically represents a total energy for each state, where the transition between $|L^*\rangle$ and $|G\rangle$ is the bottleneck process responsible for intermittency."

Q: There is also a transition between $|L^*\rangle$ and $|D\rangle$.

M: Actually, all three states are involved. You know, if you don't get to $|L^*\rangle$ you don't get to $|D\rangle$ and I am not sure that bottleneck is a good way of describing the process, neither is a good way to have two transitions, one that brings you up, and the other to go over to $|D\rangle$. Now, you might say that you want to go from $|G\rangle$ to $|D\rangle$, but you can't go from $|G\rangle$ to $|D\rangle$ unless you can go from $|G\rangle$ to $|L^*\rangle$, so that is a bottleneck. I don't regard it as a good way of phrasing the process. You know, in a number of times when I am involved in joint papers something steps through that I don't bother because it doesn't matter that much, but probably I should. I think I would have phrased the thing in a different way if I would have written the paper. A dark state is a somewhat different electronic state, a state where fluorescence can't compete successfully with some competing recombination process, typically an Auger process. And in a dark state frequently one particle from a previous excitation has been put some place in a surface trap. You now have three particles, you have one left over that matches the one that's kept aside in some surface trap, and you have the two new ones, electron and hole, so you got to have two electrons and one hole or two holes and one electron. Once you have that sort of a situation, then, before it fluoresces, a very likely process is one in which, because of the Coulombic interaction, one electron will go up and the other will go from the conduction band down in the valence band and there has been no fluorescence, that is called an Auger process.

"but to get to the dark state you need to go over ..." It's an electronic transition that somehow casts an electron or hole into some trap state.

Q: So, both processes are really the bottleneck.

M: Yes, I mean somehow Tang said that if you want go from $|L^*\rangle$ to $|D\rangle$, you first have to go from $|G\rangle$ to $|L^*\rangle$, so he's calling that a bottleneck.

14. Following: "Photoemission from $|D^*\rangle$ to $|D\rangle$ for QDs was observed on a rough gold surface... and this fact indicates that the *surface-enhanced radiative rate* has become comparable with the rate of the Auger nonradiative process previously responsible for the QD being dark."

Q: How and why does the surface enhance the radiative rate?

M: *The same phenomenon occurs in surface-enhanced Raman spectroscopy (SERS), same thing*. In other words in the surface phenomena you have something bumping on metal surfaces, and if you are looking at Raman, you have the *surface-enhanced Raman spectra*, you know, that was someone involved in cold fusion...

Q: Martin Fleischmann...

M: And later someone at Northwestern... You remember this surface-enhanced Raman effect? This is the same thing, only it is not Raman, whatever produces that produces this. The main point is that if you want understand that part, try to understand the surface-enhanced Raman, and then, when you understand that, then you see that, when you have a rough metal surface, it can really enhance the response, produce much stronger electromagnetic field for the optical process, so that it speeds up the optical transition, it really reduces the fluorescence time, so that the time is of the order of picoseconds instead of nanoseconds by enhancing the local electromagnetic fields involved in that process. This is the famous SERS, and you know, Martin Fleischmann was a pioneer and some fellow at Northwestern, Van Dyne, was a pioneer, so it is a very commonly used method, a very sensitive method for detecting things, in other words the unevenness of the surface creates local huge electric fields that are maybe a thousand times the ordinary field, and so that enhances the response.

Q: Because somehow small metal spikes form on the surface?

M: Well, I'm trying to remember now, I've never done the electromagnetic theory, you know, quite a few people have, but when you have that uneven surface and then you look at the local magnetics... yes... it is probably like due to small spikes. When you have a small spike static electricity comes off, you got a tremendous concentration, tremendous local field and so on...

15. p. 3 2nd column top: "As illustrated in Fig. 1(c), at birth of the *on state* a QD jumps from the *off state* at the energy level crossing at $Q = Q_c$. This one-dimensional coordinate Q describes an ensemble of points in a hyperplane in N-dimensional space. For each *event of a light or dark period*, Q starts and ends at Q_c, the crossing point of the parabolas for $|L^*\rangle$ and $|D\rangle$."

p. 4 1st column middle: "The blinking statistics $P_{\text{on}}(t)$ for *the on events of the neutral QDs* (or $P_{\text{off}}(t)$ for *the off events of the dark QDs*) are defined as the *waiting-time* distribution function for a QD that is initially in the neutral 'light' states $|G\rangle$ and $|L^*\rangle$ (or charge-separated states $|D^*\rangle$ and $|D\rangle$) and is turned into a dark state (or light state) between t and $t + dt$ per unit dt."

Q: It seems to me that here we have two different concepts: *(1) Light or dark states, that is, on or off states; (2) On or off events.*

(1) Is a *light state* or *on state* a state of the QD in which the QD is in one of the two neutral light states $|G\rangle$ or $|L^*\rangle$?

(2) What is the on-event? Intuitively I would think that it is the switching on of the on-state or light state. But from Fig. 2(b), because of the directions of the arrows there, it seems that the on-event is just the crossing from a light to a dark state, whereas this crossing refers only to the end point of the on-event.

(3) What exactly is the "waiting time"?

(4) Is the blinking of a single QD due to a continuous oscillation between light and dark states?

M: (1) $|G\rangle$ is a ground state, isn't it? Yes, I mean when the QD is in the light state it is sort of going between the $|G\rangle$ and $|L^*\rangle$ states back and forth during the course of excitation before the emission.

(2) Yes, that's right, that's what it says there, and...

Q: I think one should change the directions of the arrows there.

M: Yes, that could be... well, an on-event is whatever it's defined, you could define it anyway you could define an on-event as going from an on to an off, or you could define as going as forming an on from an off, and so you may choose to define it one of those two ways, but you could define it either way, you know, you could always define the way you want. The on-state is certainly what is like, yes, but you can say that an on event is forming an on-state or you can say that an on-event is disappearing an on state, you can define an on-event either way, whichever way you want. But it is important then that if you use the word on-event again, then you stay with that definition. I mean, with definitions the beauty of them is that you can define something any way you want, but once you have defined that way you have to be consistent.

(3) It is a distribution of rates. The waiting time distribution is a waiting time of breaks, and so phrasing that last way "events between t and $t + dt$ per unit dt" means that you are talking about the rate of the process.

(4) The blinking is due to... Well, part of the time the QD is in the on-state, and in that case you get fluorescence. If it jumps to what is called the off state, then because of Auger process the energy is dissipated not by fluorescence but by internal conversion, and so that's a dark state, so you have intermittent light states and dark states. I wouldn't call it an oscillation because oscillation implies a regular frequency.

16. p. 3 2nd column bottom: "The initial condition for motion on the light or dark state is a *delta-function population* at the crossing."

p. 4 1st column middle: "Similarly, for a *single dark QD* on $U_2(Q)$, the population distribution $\rho_{22}(Q,t)$ satisfies

$$\frac{\partial}{\partial t}\rho_{22}(Q,t) = \frac{1}{\tau_2}\left(\Delta_2^2\frac{\partial^2}{\partial Q^2} + \left(1 + Q\frac{\partial}{\partial Q}\right)\right)\rho_{22}(Q,t) - \frac{2\pi|V_{12}|^2}{\hbar}\delta(U_1(Q) - U_2(Q))\rho_{22}(Q,t)) \qquad \text{(4b)}$$

Q: What is meant by "delta-function population"? Aren't you dealing with a single QD? Can one speak of population even *for a single QD*?

M: This is common in the treatment of diffusion in statistical mechanics. A common problem is called the Green function for that diffusion, you start $x = 0$ and you look at probability distribution at other x's at times t, that's the Green's function for diffusion. So, you start saying you are definitely at zero, and you have your delta function, and your initial conditions for solving the diffusion equation.

Q: And you speak of population even for a single object.

M: Yes, sure, that's population probability, you know, its definition, one can call it whatever one wishes as long as one's internally consistent. Now, usually when speaking of population one is referring to ensembles, but if you think of population as being probability distribution, it's OK to use it here.

Q: In the same sense as the square of a wave function, somehow.

M: Yes, that's right.

17. p. 3 2nd column bottom: "As illustrated in Fig. 2(b), the potentials are defined as $U_1(Q) = \kappa_E(Q + Q_1)^2/2$ and $U_2(Q) = \Delta G^0 + \kappa_E Q^2/2\ldots$"

Q: Contrary to the representation in Fig. 1(b), where the minimum of the parabola for $|L^*\rangle$ state was at $q_0 > 0$, now, in Fig. 2(b), the minimum of the $|L^*\rangle$ state is at $-Q_1 < 0$. I believe then the above formula for $U_1(Q)$ should be $U_1(Q) = \kappa_E(-Q-Q_1)^2/2$. But maybe see question 5.

M: The way it is written there one of them has a minimum at $Q = -Q_1$ and the other has a minimum at $Q = 0$.

Q: But they are inverted this time because in Fig. 1(b) the light state is written to the right.

M: Yes, that's unfortunate, he shouldn't do that way. Yes ...who knows whether the dark state is to the right or to the left...

18 p. 2 2nd column middle "... D_1 is the diffusion constant related to τ_1, the diffusion correlation time constant..."

p. 4 1st column bottom: "...the critical time constant $t_{c,k}$..."

p. 4 2nd column top: "For time t comparable to the saturation time $1/\Gamma_k$, but shorter than the effective diffusion time constant τ_k, $P_k(t)$ for the blinking statistics....

$$P_k(t) \approx \frac{\sqrt{t_{c,k}}}{2\sqrt{\pi}} t^{-3/2} \exp(-\Gamma_k t), \quad k = 1 \text{ (on) or } 2 \text{ (off)}. \quad \text{(6a)"}$$

p. 10 1st column bottom: "The distribution of the blinking statistics $P(t)$..."

p. 11 1st column top:

$$\text{“}\Gamma_k \equiv \frac{(Q_c - x_{0,k})^2}{4\tau_k \Delta_k^2} \equiv \frac{E_{a,k}}{2\tau \kappa_E \Delta_k^2}, \quad E_{a,k} \equiv \frac{(\lambda \pm \Delta G^0)^2}{4\lambda}. \quad \text{(A9b)”}$$

Q: Why is the time $1/\Gamma_k$ (at which $\exp(-\Gamma_k t) = \frac{1}{e}$) the "saturation time"?

M: $1/\Gamma_k$ is a time when the exponential starts taking over, but I don't know why he called it the saturation time, I just call the time $1/\Gamma_k$. I don't need the word saturation, and also we have a simple looking expression for $t_{c,k}$, this expression that he has there is OK but it looks complicated, we have something that brings out what is happening at equilibrium . . . This doesn't violate anything that he has done, but I wouldn't call it saturation time. He may have meant saturation in the sense that that's where the power law saturation stops and you go over to an exponential. Maybe he uses because he is from Taiwan and doesn't have the precise English, it could be, but I just don't know, that's something you can ask him, just out of curiosity, why he called the saturation time, but I don't see that there is saturation.

Q: Tang is the only one of your coworkers I happened to know in 2006. He looks a very clever guy.

M: He is, he is smart and his computations abilities are very high, very very high. So, if you have any computational problem, he may well be able to handle it. He got some award for interpreting some NMR data, he is very skillful there.

Q: I shame myself that when I came to your place I was a pure dilettante . . .

M: Well, as long as you enjoyed yourself . . .

19. p. 4 2nd column middle: "In the present model the role of light absorption in spectral diffusion is *to catalyze* the rate of attainment of a thermal distribution of q (*i.e., change τ_k rather than create an athermal distribution*)."

M: Well, I think that, . . . it is not clear if this is true at room temperature also, but at low temperature the light apparently makes the spectral diffusion go faster, in other words you reach the thermal distribution by shining light on it, faster when the more intense the light, and you see that in work of Bawendi, and you could rescale everything so that you can plot sort of the approach of the spectral distribution to an equilibrium internal distribution. So there at low temperature light catalyzes this diffusion, makes it go faster. The same may or may not be true at room temperature, because you can get diffusion by thermal excitation more easily, and the light may not have any role and so we don't know if light also catalyzes spectral diffusion at room temperature. It certainly does do that at low temperature and maybe by the excitation, the local heating, you know, relaxation, God knows what . . . does it. I think though that that's related to effective light intensity, on the width of the fluorescence line, and if you go to higher intensities, if I remember rightly, your line has a bigger width because of more spectral diffusion. I think it was Bawendi who first studied that, initially down at low temperatures but probably ultimately at room temperature. Well, anyways, it is the light that causes charged particles there and maybe also deforms the crystal a little bit, and causes a broadening of the absorption band, and that is called spectral diffusion, and there have been many studies of spectral diffusion, by Bawendi and I guess two others too, and that spectral diffusion increases as you increase the temperature and also increases as you increase the light intensity, so both things come in and you would normally expect something, you know, where things are moving around. Now why it depended on light intensity? Maybe the higher the light intensity, the more you are shaking up the crystal, so the more the diffusion. In one case, I remember, and this was on one of the papers with Tang, that if you plotted the amount of spectral diffusion, versus intensity then at room temperature, somehow I think you could rescale it, and all the points from the different systems would fall on the same line, and when you plot the different curves width versus time, for different intensities, you can make them fall on the same line in a simple way . . . physically it is this way but I forgot the details.

20. Following: "That is, ultimately, the light absorption leads to an attainment of a thermal equilibrium, *whose position* is independent of the *light intensity*."

Q: That is, of the photons concentration? Does this have to do with the fact that catalysts can act in small quantities?

M: I think that should do with the fact that if you have two states and you're diffusing and the light catalyzes it, it can make you to approach the desired distribution between those two states more quickly, I think. Like, for example . . . yes, I think I can give you an example. If you look, at an early paper with Jau, the spectral diffusion and the effective light intensity, you will see that you get several lines which approach saturation, and I think when we varied the light intensity we could superimpose all the curves by plotting the spectral diffusion vs. the product of the intensity times the time, and it reaches the same final limiting value, all the curves overlapped but the final limiting value was the same independent of the light intensity. So, you could look at that particular paper where we plotted something related to the spectral diffusion at low temperature at different light intensities. And I think we may have matched on to the same product of the light intensity and the time and they all the curves fall on top of each other, so that meant that the final equilibrium value is independent of intensity. All have the same value.

Q: This independence on the intensity, that is, the act that only a small amount of photons is enough to catalyze isn't somehow related to fact that in catalysis you somehow need small amounts of catalysts?

M: Well, I don't know if you need small amounts, here you may need large amounts. So, I mean, it is different from ordinary catalysis, you know, the mechanism is different, yes. The main point is that probably the light works by accelerating the diffusion, whereas in ordinary catalysis you can have reactions without focusing on diffusion, so it's a different phenomenon.

M341 Mechanisms of Fluorescence Blinking in Semiconductor Nanocrystals Quantum Dots (Continuation)

Jau Tang and R. A. Marcus

1. p. 4 2nd column middle: "Equation (6a)

$$P_k(t) \approx \frac{\sqrt{t_{c,k}}}{2\sqrt{\pi}} t^{-3/2} \exp(-\Gamma_k t), \quad k = 1 \text{ (on) or } 2 \text{ (off)} \tag{6a}$$

can be rewritten as

$$\frac{2\pi}{\hbar}|V_k|^2 t_{c,k} = \sqrt{4D_k t_{c,k} \times \sqrt{2\kappa_E \lambda}} \qquad \text{(7a)''}$$

Q: It is not Eq. (6a) that can be rewritten as Eq. (7a) but it is Eq. (5c)

$$t_{c,k} \equiv 2\kappa_E\lambda \left(\frac{\hbar\Delta_k}{\lceil V_k\rceil^2 \pi\sqrt{\tau}_k}\right)^2 \tag{5c}$$

that can be recast as Eq. (7a) in the following way:

Considering that $\Delta_k^2 = D_k\tau_k$ we can rewrite Eq. (5c) as $t_{c,k} = 2\kappa_E\lambda\frac{1}{|V_k|^4}\left(\frac{\hbar\sqrt{D_k\tau_k}}{\pi\sqrt{\tau}_k}\right)^2$ $\Rightarrow$ taking the square root of both terms and shifting places of $|V_k|^2$ and $\sqrt{t}_{c,k}$ $\Rightarrow$ $|V_k|^2 = \sqrt{2\kappa_E\lambda}\frac{1}{\sqrt{t}_{c,k}}$ $\left(\frac{\hbar\sqrt{D_k}}{\pi}\right)$ $\Rightarrow$ multiplying both terms by $t_{c,k}$ $\Rightarrow$ $|V_k|^2 t_{c,k} =$ $\sqrt{2\kappa_E\lambda}\sqrt{t}_{c,k}\left(\frac{\hbar\sqrt{D_k}}{\pi}\right)$. Multiplying now both members by $\frac{2\pi}{\hbar}$ we get $\frac{2\pi}{\hbar}|V_k|^2 t_{c,k} = \frac{2\pi}{\hbar}\sqrt{2\kappa_E\lambda}\sqrt{t}_{c,k}\left(\frac{\hbar\sqrt{D_k}}{\pi}\right) \Rightarrow$ simplifying one gets Eq. (7a).

M: OK.

2. Following: "The right-hand side of Eq. (7a) is approximately the *energy spread in q space* at time $t_{c,k}$ due to diffusion ... *The reciprocal of the right side serves as the state's density*... The density of states near Q_c at time t is approximately $\sqrt{2\kappa_E\lambda}/\sqrt{4D_k t}$"

Q: **Typo**: Misprint: Shouldn't it be $1/\sqrt{2\kappa_E\lambda} \times \sqrt{4D_k t}$?

M: OK.

3. p. 5 1st column top: "The mean-square displacement from the crossing point at equilibrium ($t = \infty$), $\langle(Q(\infty) - Q_c)^2\rangle$, is related to $\Gamma_k\tau_k$

$$\Gamma_k\tau_k = \frac{\langle(Q(\infty) - Q_c)^2\rangle}{4\Delta_k^2} - \frac{1}{4} = \frac{E_{a,\lambda}}{2k_B T} \tag{7b}$$

where $E_{a,k}$ defined in Eq. (A9b)

$$E_{a,k} \equiv \frac{(\lambda + \Delta G^0)^2}{4\lambda} \tag{A9b}$$

is the energy difference between the crossing point and the bottom of the potential well $U_k(Q)$...

If the potential $U_k(Q)$ were flat the right-hand side of Eq. (7b) would vanish..."

Q: What is it meant by this "flat" potential? The right-hand side Eq. (7b) and of Eq. (A9b) become zero for $\lambda = -\Delta G^0$. Do you maybe mean by "flat" the potential whose curve is crossed at its bottom?

M: OK

4. p. 2, legend to Fig. 1(a): "... W is the *photoexcitation rate*."

p. 5 2nd column top: "With an *excitation intensity* of 400 W/cm^2 the time interval between absorption of photons is $1/W \sim 1\,\mu s \ldots$"

Q: The symbol W suggests power, that is, the dimension [energy]/[time] but if its reciprocal has the dimension of [time] do you maybe mean by W the [number of absorbed photons]/[time]?

M: With the symbol W we have indicated two different magnitudes, something to be avoided. $1/W$ refers to the inverse of W as photoexcitation rate.

5. p. 6 2nd column middle: "In the fitting of Figs. (4a) and (4b) to *the quantum form of Eq. (1b),*

$$\kappa\Delta_1^2 = (\hbar\Omega/2)\coth(\hbar\Omega/2k_BT) \approx k_BT \qquad \text{(1b)"}$$

Q. You mean that the quantum form is $\kappa\Delta_1^2 = (\hbar\Omega/2)\coth(\hbar\Omega/2k_BT) \approx k_BT$ and the classical limit form is $\kappa\Delta_1^2 \approx k_BT$.

M: OK

6. Following: "Empedocles and Bawendi observed that the reduction of the photon energy by 250 meV reduced the spectral diffusion by 25%. Further experimental study can determine whether this increase occurs in the transient part of $\sigma(t)$ or in the saturated part $\sigma^2(\infty)$, or both. If is observed only in the transient and not in the saturated value, then *the role of the photon absorption is 'catalytic,' namely to change the τ_1 and not the $\kappa\Delta_1^2$* present in the expression for $\sigma^2(\infty)$."

Recall that, from Eq. (3), $\sigma^2(t) = 2\kappa\Lambda\Delta_1^2[(1-\exp(-t/\tau_1)]$ and that $\tau_1 = \Delta_1^2/D_1$ is the diffusion correlation time constant, and that $\kappa\Delta_1^2 = (\hbar\Omega/2)\coth(\hbar\Omega/2k_BT) \approx k_BT$.

M: All right: They have spectral diffusion, eventually you fill up the whole absorption line, part of the absorption line, then another, and you get another, and another, another, you fill a whole part of the absorption line. Now, the saturated value is related to $\kappa\Delta_1^2$, and the dynamic value is related to τ_1. You see that e^{-t/τ_1}, and the saturated part is the factor that is in front of the (1-exponential)...so that is what happens to long times, so you can distinguish what happens at long times ...the photon energy isn't affecting the saturation, if you can tell whether the saturation is catalyzing changing the τ_1 or changing the final value is independent of the exciting light, then is not changing that final value of the saturation value of the spread in energy. You see, they measure the square...The main point is that in the square you have a saturated value which is that factor (1-exponential) and you have τ_1 and by looking the data you can tell when you're changing energy, you are changing one or you are changing the other.

7. Following: "Equations (1b) and (3)

$$\kappa\Delta_1^2 = (\hbar\Omega/2)\coth(\hbar\Omega/2k_BT) \approx k_BT \tag{1b}$$

$$\begin{aligned}\sigma^2(t) &= \langle(\Delta U(q) - \langle\Delta U(q)\rangle)^2\rangle = \int_{-\infty}^{\infty} dq(\kappa q q_0)^2 G(q, q; t) \\ &= 2\kappa\Lambda\Delta_1^2[1 - \exp(-t/\tau_1)] \end{aligned}\tag{3}$$

imply that in the present model the excess energy acts by affecting τ_k and not the other terms."

Q: Why so?

M: I don't know, I suspect that when they looked at different energies they found the saturation value. In fact, I think they were able to scale things and put everything on the same plot, there they have the same saturation value, the scaling there was affecting just the excess energy, and so I think what they ended up doing was to plot the energy times τ. In other words, you can scale things and that means that the value is the same to the equilibrium value, it isn't changed.

8. Following: "If, instead, it has an effect on $\sigma^2(\infty) = 2\kappa\Lambda\Delta_1^2$, the model would need to be modified to include an additional contribution to $2\kappa\Delta_1^2$."

M: Well, the way it stands it shouldn't depend on the intensity, so if $\sigma^2(\infty) = 2\kappa\Lambda\Delta_1^2$ does depend on intensity you have to add something else in, you have to modify the theory.

9. p. 8 1st column middle: "In Eq. (6a) and (6b)

$$P_k(t) = \frac{\sqrt{t_{c,k}}}{2\sqrt{\pi}} t^{-3/2} \exp(-\Gamma_k t), \quad k = 1(\text{on}) \text{ or } 2(\text{off}) \tag{6a}$$

$$\Gamma_{\text{on}} = \frac{(\lambda + \Delta G^0)^2}{8\tau_1 \kappa_E \lambda \Delta_1^2}, \quad \Gamma_{\text{off}} = \frac{(\lambda - \Delta G^0)^2}{8\tau_2 \kappa_E \lambda \Delta_2^2} \tag{6b}$$

with $P(t) \sim t^{-1.5} \exp(-\Gamma_k t)$, Γ_k is proportional to the light intensity, and one can use energy density $(I \times t)$ as a *universal* variable to describe the bending."

Q: What do you mean by "universal"? Just "general"? As far as I know the term "universal" is normally used for constants such as Planck's h, the velocity of light, and so on.

M: Universal in the sense of general...

10. p. 9 1st column top: "When the crossing point occurs at the bottom of the potential-energy well, diffusion to the sink is not needed *in the center for the wave packet to reach the crossing point*, and the largely exponential decay ensues."

M: The wave packet needs not to diffuse because it is already there at the crossing point.

Misprint: p. 11 bottom: "**APPENDIX B: FURTHER REMARKS ON *Q* AND *Q***"

The first ***Q*** is a ***q***.

11. p. 11 2nd column middle: "During the sampling period of 0.1 s, the instrument integration time (*bin time*) immediately after the birth of the light state..."

Q: What does "bin time" mean?

M: Bin in the sense of time interval during which one measures the number of blinkings.

M342 Diffusion-controlled Electron Transfer Processes and Power-law Statistics of Fluorescence Intermittency of Nanoparticles

Jau Tang and R. A. Marcus

1. p. 1 1st column top: "Recent advances in nanoscience and nanotechnology and their potential applications have generated wide interest: The development of techniques in probing single molecules has provided a tool to study *its* intrinsic properties and *its* interaction with the surroundings."

Q: **Typo**: Shouldn't the its be their?

M: Let's see... it should be their, absolutely.

2. p. 1 1st column middle: "There exists in the literature two approaches, partial ordering prescription (POP) with a time-dependent but *nonretarded* diffusion coefficient, and a chronological ordering prescription (COP) with convolution of a time-retarded *diffusion kernel*"

M: See Mukamel. There can be diffusion coefficients of the form $D(t)$ and there can be convolutions of $D(t)$s.

3. p. 1 1st column bottom: "We consider a POP type 1D non-Markovian equation, with a population sink at the potential energy crossing ($Q = Q_c$) between $U_1(Q)$ for the 'light' state $|1\rangle$ and $U_2(Q)$ for the 'dark' state $|2\rangle$. One has

$$\frac{\partial}{\partial t}\rho_k(Q,t) = D_k(t)\frac{\partial}{\partial Q}\left(\frac{\partial}{\partial Q} + \frac{1}{k_B T}\frac{\partial}{\partial Q}U_k(Q)\right)\rho_k(Q,t) - \frac{2\pi|V_k|^2}{\hbar}\delta(U_1(Q) - U_2(Q))\rho_k(Q,t) \qquad \text{(1a)''}$$

Q: Please compare the above equation with the following ones from **M341**:

$$\frac{\partial}{\partial t}\rho_{11}(Q,t) = \frac{1}{\tau_1}\left(\Delta_1^2\frac{\partial^2}{\partial Q^2} + \left(1 + (Q+Q_1)\frac{\partial}{\partial Q}\right)\right)\rho_{11}(Q,t) - \frac{2\pi|V_{12}|^2}{\hbar}\delta(U_1(Q) - U_2(Q))\rho_{11}(Q,t) \qquad \text{(4a)}$$

$$\frac{\partial}{\partial t}\rho_{11}(Q,t) = L_{11}\rho_{11}(Q,t) - \frac{2\pi|V_{12}|^2}{\hbar}\delta(U_1(Q) - U_2(Q)) \times (\rho_{11}(Q,t) - \rho_{22}(Q,t)) \qquad \text{(A1)}$$

where $L_{11} = D_1\left(\frac{\partial^2}{\partial Q^2} + \frac{1}{\Delta_1^2}\left(1 + (Q+Q_1)\frac{\partial}{\partial Q}\right)\right)$, $D_1 = \frac{\Delta_1^2}{\tau_1}$.

M: Equation (A1) is the simple form which would have a time dependence, either in the form of diffusion constant, or in the form of an integral. So, Eq. (A1) would be the counterpart of Eq. (4a).

Q: Except that there are two densities, you see, in Eq. (4a) you only have ρ_{11} and in Eq. (A1) you have ρ_{11} and ρ_{22}.

M: I think that Tang uses Eq. (A1) when he is treating an ensemble and Eq. (4a) when he is treating a single molecule. I think.

Q: Now, why do you use two indexes in Eqs. (4a) and (A1) and only one in Eq. (1a)?

M: Yes, that's probably because in the general case you would also have a ρ_{12}, a phase. The diagonal elements are $\rho_{11} - \rho_{22}$, but in the other thing, just a ρ_k, and so on, you don't talk about off-diagonal elements, so the ρ_{11} is probably carried over from a formulation that had off-diagonal elements, the ρ_{11}'s are probably the same as the ρ_1's. Because normally you would write ρ_{11} or ρ_{22} when you're also considering ρ_{12}, an off-diagonal matrix element, dephasing, my guess for Eq. (1a) is that he assumed that everything was dephased, so I don't know if you have anything there, in that paper, for ρ_{12}.

Q: For ρ_{12} no.

M: Then he assumed that everything was dephased, that's ρ_{11} and ρ_1 are the same.

Q: Is V_k the same as V_{12}?

M: That V_k is the same as V_{12}, yes.

4. p. 1 2nd column middle: "For a Debye medium...one has $\tau_{L,k} \equiv \tau_{D,k}\varepsilon_\infty/\varepsilon_0$...and a time-independent diffusion constant $D(t) = \Delta_k^2/\tau_{L,k}$..."

Q: Even here I do not understand why you use double indexes for τ. According to the above, *for the light state* $|1\rangle$ *the index k should have the value of 1*. If so, which is the meaning of $\tau_{L,1}$?

M: I don't know what that k stands for.

Q: From **M341** $k = 1$(on), $k = 2$(off)

M: Oh, I see what you're saying.

Q: If 1 means L, then you have a repetition.

M: Good question, good question, I think you may be right, you don't need that, Jau might have had something in mind. On the surface you don't need the two symbols, but there may be some subtlety.

Q: Even worse, if $k = 2$, we have $\tau_{\text{light, dark}}$.

M: Yes . . . Does he have τ_L any place other than that?

Q: He has also only τ_L and only τ_D.

M: You know, I'm vaguely remembering . . . there is something about different relaxation times depending on when you are doing at constant dielectric displacement or constant electric field, and that's probably discussed maybe in Froehlich's book, I don't remember where, that you get different relaxation times depending on where you're relaxing from a constant E or a constant D . . . and so when I see that E_0 . . . then maybe that is a relationship between the difference between relaxing in D vs. relaxing in E, so there may be subtleties there that I don't know about.

CHAPTER 11

Single Particle versus Ensemble Average: From Power-law Intermittency of a Single Quantum Dot to Quasistretched Exponential Fluorescence Decay of an Ensemble; Explanation of Quantum Dot Blinking without the Long-lived Trap Hypothesis; Photoinduced Spectral Diffusion and Diffusion-controlled Electron Transfer Reactions in Fluorescence Intermittency of Quantum Dots; Chain Dynamics and Power-law Distance Fluctuations of Single-molecule Systems; Determination of Energetics and Kinetics from Single-particle Intermittency and Ensemble-averaged Fluorescence Intensity Decay of Quantum Dots

Interviews on M344, M345, M350, M351, M352

M344 Single Particle vs. Ensemble Average: From Power-law Intermittency of a Single Quantum Dot to Quasistretched Exponential Fluorescence Decay of an Ensemble

Jau Tang and R. A. Marcus

M: You illuminate the QD and the QD blinks, all the papers with Jau are on QDs.

1. p. 1, 2nd column top: "In the present study we use this simplified two-state model with *shallow energy traps* (<50 meV) which is *distributed over the surface rather than being localized.* We do not consider here *deep trap states* (several hundred meV and 1 eV), which could *cause intermittency via an Auger-assisted process.*"

M: Well, *if you get an electron transfer between an ion on the surface, maybe a Selenium ion, say, and the valence band, that would be a jump, maybe an electron jumping*. When you excite a semiconductor, a hole is created, a hole is being created by some excitation of an electron to an excited state. Now, if there is a dangling Selenium ion on the surface of the QD, an electron from that dangling Selenium ion, which is not stabilized, might hop into the hole in the valence band. It doesn't happen all the time, but it can happen. The hole is in the Selenium ion on the core... Now, *it can't easily happen that the process may go like that, jump like that, because that means converting electronic energy into vibrational energy*, so you need to have another change going on and Pavel and I discussed that, it can become in the upper excited state from a 1S to a 1P, it goes up, it goes down, let's call it an Auger process. Those traps are deep traps, they are not just slightly off the valence band, and I discussed the question above a little bit more in a paper in the Transactions of the Royal Society. At that time, when Jau and I wrote that, we hadn't really been thinking deeply about what this does other than just going off the conduction band, nor we, both Pavel and I, talked about the point, we wanted just off the conduction band.... So, anyway, we used that, that was before we thought the problems through and even now that we thought through more, there is still considerable lack of knowledge and uncertainty. We are dealing with systems where there are definitely more questions asked than have been answered, it is quite different from many systems in electron transfer. There are surface states around, the system is largely surface, no doubt, and so there are all sorts of uncertainties. So, that is a very murky situation right now.

Q: A working process.

M: Yes, very much so and I think I could do a better work probably by rejecting these QDs, not beautiful simple systems, with a variety of different kinds of surface states, and probably various time scales, so it is kind of a messy system. One should do better than that.

Q: The Selenium ion an anion.

M: Selenium is like the sulfur. And Cadmium in a crystal is a Cadmium cation.

2. p. 2 2nd column bottom: "In this study of ensemble-averaged fluorescence decay, Chung and Bawendi measured the time evolution of fluorescence intensity *with QDs initially in darkness prior to light illumination.*"

Q: Can you say something about how are these experiments performed? Apparently one shines light on an ensemble of QDs initially in the dark states, then the light is turned off and one observes the fluorescence decay of the blinking QDs. Is it so?

M: Well, typically one is shining light and the QD is blinking. But there are experiments where one can wait until the system is in the dark state and then turn the light off, don't turn it off when it is in the light state, because it's fluorescing back and forth between light and dark. Should one turn it off when it is in the dark state, what you find then is that it takes about an hour to recover, *which implies that in the dark state there's a charge distribution which is quite different from when it's in the light state*. Light catalyzes the going back and forth between light and dark, because when you turn the light off, it just stays there for a long time, whereas when you have the light going on and off you normally you don't have those long living dark states.

Q: If I understand well, if you keep the light on, then you have a quick change between light and dark state

M: That's right.

Q: Whereas when you turn the light off...

M: While it's dark, yes...

Q: Then the switching takes a long time.

M: That's right, about an hour. If you look at the fluorescence of an ensemble of dots in the Chung and Bawendi sytem, it takes about an hour to establish to come to the steady state... Fluorescence takes time.

Q: So, at the same time you illuminate the thing and the thing blinks.

M: Yes, it's blinking, eventually it reaches a steady state, where you have an ensemble of course, when you have a whole collection of QDs doing that. It turns out that establishing a steady state is almost a separate issue, because by and large it's not the short trajectories that you're looking at. So, ordinarily, you can't see, in a single molecule experiments, increments of fluorescence due to some of these very long trajectories, it takes so long, there is very little signal then. The basic idea is this, that it has been shown that with a QD it takes about an hour to reach a steady state, but in single molecule trajectories you just go for at most 10 seconds, so you never reach the steady state. Now what that means then is that doing steady state experiments will complement those from single molecule experiments, because they will provide information not available to single molecule experiments.

3. p. 6 1st column middle: "Existing experimental results involve deep traps which are short lived. For internal consistency an explanation is needed as to why a deep trap state is short-lived when a conduction band edge state is not. This deep trap state in CdSe is likely *a hole in the surface Se dangling bond.*"

M: Well, what I don't know is if there are any traps shallower than deep trap states, it may be that problems that Jau and I were treating could be treated with a few trap states, that remains to be seen, and it may be that we need another set of states. Maybe we don't need anything other than deep trap states, and we need to modify the type of treatment that Jau and I had. You see, we wanted states with sort of shallow curves right near each other, the initial and the final state.

Q: Why did you have shallow instead of deep traps?

M: Well, we are thinking in terms of parabolas, near each other, so they are all just at a small energy difference, but that may be questionable. So, there are subtleties that we hadn't thought about then.

4. Following: "The energy fluctuation at the surface state can be caused by stochastic environmental changes or *migration* of the trap state to others at a slightly different energy..."

Q: Migration or diffusion?

M: Diffusion is what's meant, it's not moving under the influence of a field or anything like that.

M345 Explanation of Quantum Dot Blinking without the Long-lived Trap Hypothesis

Pavel A. Frantzusov and R. A. Marcus

1. p. 1 1st column middle: "This power law is unchanged for the different temperature... and *atmospheric conditions (from air to vacuum)*."

M: I think they have QDs under different conditions, just with a vacuum above. QDs could be deposited on a plate and you can evacuate everything. Now, maybe there are some of them that are actually observed in solution, I should know that and I don't. There are experiments by Bawendi in which instead of using a glass or silicon plates or something they use a metal underneath, and that actually accentuated the local electric field and maybe some dots that would have been dark actually are a little bit lighter... so, just like in surface enhanced Raman, there you can see the dark states. You get enhancement when you have a rugged surface, a metallic surface, as you do in Surface Enhanced Raman... and I think Silver has an uneven surface, then you get local very high electric fields, and the high Raman intensities, you get a factor of a million enhancement of the signal, and so that happens with other objects too. The fluorescence of the object that was producing the fluorescence was enhanced a lot, by a similar mechanism.

2. p. 1 2nd column middle-bottom: "...qualitative quenching of the luminescence is found after the addition of *an electron or hole quencher...*"

M: I think Pavel referred to a cresol crystal on the QD, you quench the fluorescence with a Cresol crystal. That's work by Guyot-Sionnest. You can put that crystal on a QD and that acts as a hole trap. So, there are experiments where in fact this chap put that there and that created a hole in those QDs. He put something, an aromatic, that served as a hole trap, on the surface of the QD. In other words, he could chemically cause a chemical reaction in which a hole was created in the added surface material and the electron has to be someplace else, but that means it has to be in the conduction band, it's the only place where it could be and still be negative. You need electro-neutrality. Cresol is an electron quencher, the main point is that a hole was created in the cresol and that left an extra electron in the QD in an excited state.

3. p. 2 1st column middle: "According to the model of Shimizu *et al.* the electron transfer event switching the luminescence intensity happens only when the excited state and the trap state are in resonance. The model provided a *promising concept of a slow diffusive coordinate*... *The weak point in this model* is an assumption of an existence of a single electron trap state placed in the narrow region of energies and located *in the surrounding* near each given QD. In order to resolve *this problem* Tang and Marcus assumed that the trap is the crystal-induced surface state."

M: Well, a crystal induced surface state is always a shallow state, and it could be right near the conduction band edge or right near the valence band edge, and has a small but definite energy. I think the Shimizu's model and the model I was using are very similar... it's a model. Pavel had a different idea on how things go, so he made that statement, this was his paper, so I just let him go with it. I never really quite understood what Pavel was doing there, but he had a model where things weren't actually trapped, there were rapid resonances, rapid nonradiative transition to the ground state, but I never really followed that, he may be right, but I never followed that. That is one thing one of these days I shall look into but I don't know the details now.

Q: What is the advantage of your model?

M: Well... I think what we were thinking at the time is that it would be accidental that you meet all the energies, you can get a narrow state which is very close in energy, almost in resonance with it, but the point is that we were thinking in terms of Auger processes where you can get something jumping to one state while another has a continuum almost to jump into... you have the transition, you are thinking just in terms of single state to single state transition, but if the transition involves something from single state going into a continuum, then that can match a single state to a single state, you see.

Q: And Shimizu with this problem of the surroundings?

M: I don't think they gave any specification, I don't think they went into any detail.

It is a messy problem, very messy. On the other hand, there are some interesting phenomena that are to explain and one should really redo some experiments, one should do them more thoroughly, some people are trying to do something, but it's a messy problem because of all sorts of surface states, you see, but nevertheless it is instructive and could lead to something from it, applying to transfers between organic molecules and surfaces, so there could be a benefit from it, but it has been a painful experience.

M345 Explanation of Quantum Dot Blinking without the Long-lived Trap Hypothesis

(Continuation)

Pavel A. Frantzusov and R. A. Marcus

0. You have often suggested experiments to be done. Are there cases in which your suggestions have been particularly important?

M: I only know one, because I don't know that the experimentalists have tried to carry out any other. And that was a collaboration with Pelton, we wrote a PNAS paper on it, at Argonne, and that was testing the change in slope of the log plot predicted and he got the change in slope, but whether that is verification of the theory or whether it's a consequence that would happen for other theories I don't know, I haven't looked into that. Around 2006, I'm just guessing. There are other obvious experiments that one can suggest, you know, designed to find out in this "on" state when there is bending, how does it scale with time when you grow the intensity, how does it scale with the intensity, so there are experiments to do but so far nobody has tried to do anything.

1. p. 1 1st column top-middle: "It is known from many single-molecule experiments that under *continuous excitation* the luminescence emission switches 'on' and 'off' by sudden stochastic jumps... *the distribution of the 'on' or 'off' times is usually exponential or near exponential.*"

M: Calculate the number of times it switches from on to off in a time and vice versa. The way they usually define it is the survival probability is dP/dt, that is what they count.

2. p. 1 2nd column middle: "A now commonly accepted idea on the nature of the trapping state... was proposed by Efros and Rosen (a *long lived trap* hypothesis). They suggested that the luminescence is quenched if one of the carriers is trapped *in the surrounding matrix... such a charged dot...*"

Q: (1) Apparently what happens is that either the electron or the hole exits from the QD leaving it electrically charged and goes in the "surrounding matrix." Is it so?

(2) Can you explain what this surrounding matrix is made of?

(3) Why is such a trap long lived?

M: (1) Well, I'm not quite sure if that is actually what they meant, but I suspect that that is what they meant, but another view, not necessarily theirs, is that the carrier goes to a trap on the surface of the QD, yes.

Q: That is your idea.

(2) I forgot if they have some of their dots sitting in solution, on a slide or what, or if they are dry... it may be that there is always something added to the outside of the dot to protect it from oxygen... so it could be imbedded in there and I forgot whether in some cases there is actually a solution or something like that, it is a good question, I don't know.

(3) Well, the idea is that, if it is dark, the trap has existed that way for a long time. Now, why it is long-lived depends in part on what is needed to detrap it, to detrap you have to conserve energy in a detrapping, and so that means... well, that something else has to go up while something goes down to a trap or vice versa, so, you know, it takes a joint effort to do that, a joint probability, and that may be slow, so the trap exists there for a long time.

3. p. 2 1st column top-middle: "It is known from the experiment that *the energy of the excited state* displays a *long-correlated stochastic motion* with *a characteristic time* of hundreds of seconds."

M: Well, the absorption is broad so goes to many states and quickly cascades down in a very narrow fluorescence. The fluorescence... that energy quickly goes down probably in picoseconds, the excited state goes in some picoseconds to the ground state. So I don't know if the use of the word energy of the excited state... I'm not sure that it is good phrased because the fluorescence is very narrow.

Q: Is the characteristic time a mean time period after which the moving energy level gets back to a value it formerly had?

M: Well, the energy level in this case would be the lowest thing in the excited state, right? In other words, you quickly cascade down to that and when you have a power low, there isn't really a mean time period. By the way, you can't define it, normally, the survival probability goes as $\frac{1}{\sqrt{t}}$, the derivative goes as $\frac{1}{t^{\frac{3}{2}}}$, so the survival time, the mean one, would be $\int_0^\infty t\frac{1}{\sqrt{t}}dt$ and that doesn't converge.

4. Following: "At the cryogenic temperatures the diffusion is light induced, and *shift in the luminescence line* depends directly on the number of absorbed photons."

Q: Do you refer to a shift in frequency or in intensity of luminescence of the line?

M: Well, I don't know about that, I mean surely that diffusion is light induced, I mean that what there Pavel presumably means is that the frequency of the line is changing, I don't remember that, I don't even know if it's true, but that diffusion is light induced at low temperatures that is clearly true because you can scale the results at different intensities, they lay on the same line if you assume that diffusion is proportional to light intensity. But that shift in luminescence line, that I have to look at, I don't know what is based on.

5. p. 2 1st column middle: "... Tang and Marcus assumed that the trap is the *crystal-induced* surface state"

M: That state is an image state. You know, in a metal you have image states... it is an image state at the edge of either the conduction band gap or the valence band gap. So, we assume that it is in the conduction band gap, but I am not sure that is the best assumption.

6. Following: "The third model proposed by Margolin *et al.* gives us an explanation in which the blinking by three-dimensional hopping diffusion of the photoejected electron is *in the surrounding media.*"

Q: From the above it is not clear if the blinking happens in the surrounding media or if the blinking is explained by what happens to the electron while it is in the surrounding media.

M: The latter.

7. Following: "The positively charged QD stays "off" until the electron returns back.

Q: Apparently a condition for the "on" event is that the electron will be back inside the QD after its wandering in the surrounding media.

M: Absolutely, that's right. While it is outside you have an extra particle, so to speak, in the QD, that causes Auger processes, prevents the fluorescence of further excited electrons, instead of fluorescing will be killed by an Auger process.

8. Following: "We explore here an alternative mechanism for the luminescence intermittency in which it does not assume any long-lived trap state. The QD always returns back to the ground (neutral) state after photoexcitation (directly or via a surface state). The 'on'-'off' switching of the QD luminescence intensity is caused by large variations of the nonradiative relaxation rate of the excited electronic state to the ground state via surface hole trap states. The hole trapping is assumed to be *induced by an Auger assisted mechanism.*"

Q: Is this a new theory with respect to the original theory of yours and of Tang.

M: Very different. It doesn't actually have long-lived traps, it just has recycling traps.

M345 Explanation of Quantum Dot Blinking without the Long-lived Trap Hypothesis

(Continuation 2)

Pavel A. Frantzusov and R. A. Marcus

1. p. 2 2nd column middle: "The size of the colloidal QDs used is smaller than a size of the Bohr exciton in the bulk [∼112 Å for CdSe...]. So they are in the so called *strong confinement regime*, where the electrons and holes could be considered as independent particles, taking into account the Coulomb interaction as a perturbation."

Q: What is this strong confinement regime? What if the QDs weren't in the strong confinement regime?

M: Let me just say something at the beginning then. Pavel's idea is that, instead of having traps, what those sort of extra states do in the band-gap is to provide a way of going from the excited state to the ground state but just like not a traffic stop but just like a brief route, OK? So that when the energy levels sort of match those levels, then that route is open. When the energy levels don't match, then that route is closed, and that explains light and dark. But there is no trap there. Now, I mean, that is his view and fine, OK, you know, we presented that in that paper, but my own view that I prefer although, you know, I have no idea if it is right, is that there are actually traps, and it is not just a matter of providing a vehicle, the idea is that some an electron, or a hole, actually gets stuck, actually gets stuck in a trap, and then you have an odd number of electrons, but those are two different points of view.

Q: But he speaks of trapping and detrapping

M: He speaks of trapping and detrapping, I speak of trapping and detrapping, and many others speak of trapping, but I'm not sure that he speaks of trapping, he tries to avoid traps. I thought of this as a preamble to give an overview.

Now about the strong confinement regime... Well, first of all, supposing that you are in the infinite solid, and now you had an exciton, you had a charge... you had a hole there and an electron excited around it, and suppose you regard that as a hydrogen-like system. And you use a dielectric constant in the hydrogen atom-like treatment, so it differs from the ordinary hydrogen atom-like treatment because that has a dielectric constant of 1. This would have a dielectric constant of whatever it is... 10... I forgot what... Now you calculate the Bohr radius for that, which you can do, old quantum theory, and you calculate the radius for that and you find that it is 120 Å. That means that is what it would be if you had that pair moving around freely in the solid. But if you have a QD, the QD is maybe 20, 30, 40 Å diameter, then of course that means that this exciton can't be treated that way, it is confined, that is a strong confinement regime.

In other words you can't use a simple hydrogen atom-like treatment for it, you have to put in a boundary, so that is what they do.

Q: By the way, I am reading "The Structure of Physical Chemistry" by Hinshelwood...

M: When I was a student we used his Chemical Kinetics book, and after some time after that he wrote a book on the cell, he wanted to do some kinetics, I guess, about biology in the cell, but I never looked at that.

Q: It is written in a wonderful English, people don't write that way anymore.

M: In the British school system, I know when our boys were briefly in it, we were on sabbatical at Oxford, they wrote a lot of essays, lot of essays, yes.

Q: A joy to read.

M: OK, yes, I never did look at that one but his earlier book was part of a course that I have taken, that I read intensively.

2. Following: "The structure of the delocalized electronic electronic levels in the QD could be described using simple particle-in-the-box model (see Fig. 1). The levels in the conduction band correspond to the single electron states in the spherical well. For the electron states with the *envelope angular momentum* 0, 1, 2... we use notations $1S_e$, $1P_e$, $1D_e$, etc. with a subscript *e*. The valence band levels correspond to the single hole states ($1S_{3/2}$, $1P_{3/2}$, $2S_{3/2}$, etc.), the subscript denoted the total angular momentum which is the sum of the *envelope angular momentum* and the band-edge Bloch function angular momentum."

Q: What is it this envelope angular momentum?

M: Think of a spherical box. Well, actually you really should use a more detailed theory, like, say, tight binding, or something better

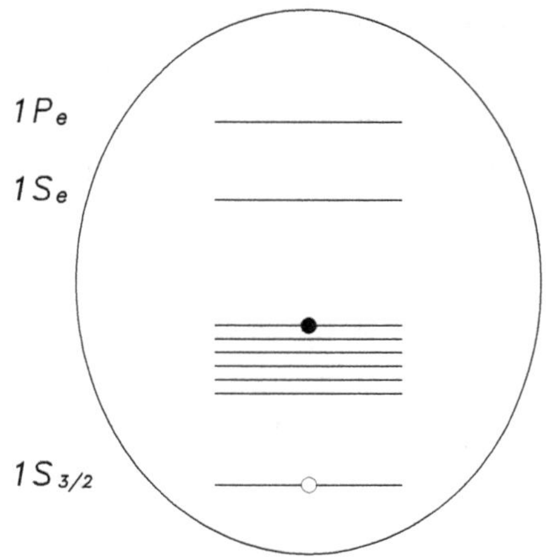

Fig. 1*.

than that, and now if you have only atomic orbitals, then if you look at their amplitude as a function of space, that amplitude is called the envelope, and that amplitude looks like the spherical box particle in the box equation...It gives you an idea of the coefficients of the atomic orbitals as a function of space. It's the same thing right when people treated aromatic systems, and they treated, say, by linear combinations of atomic orbitals, but also for quite a while after that there is a free electron treatment, the free electron treatment is the envelope treatment.

3. p. 3 1st column top: "The emission spectrum of the CdSe QD *photoluminescence* has a narrow peak at the energy E_{ex}. There is *also* a broad and less intense *luminescence* maximum at 300–700 meV lower than the excitation energy, presumably due to the *electron or hole* transition *into the surface states* in the QD... We can conclude that the surface states are placed lower than the Fermi level and 300–700 meV higher than the $1S_{3/2}$ state."

Q: Please look at your Fig. 2 where the states are shown. Consider also that we are now considering emissions. If the emission is "into" the surface states, the emitting electron could be due to the electron leaving the excited state and ending down in a surface state, a situation not considered in Fig. 2 where only holes are considered in the surface states. I have shown the final state in my Fig. 1*. If it is the hole that makes a transition into the surface states, then this hole should leave a state *higher* than the surface states, but even this case is not depicted in the Figure nor is considered in the text. Or do you consider maybe an emitting transition due to a hole going from the surface states into the $1S_{3/2}$ state? But the hole was originally formed in the $1S_{3/2}$ state. Why should it climb to the surface states to go then back to the $1S_{3/2}$ state?

M: Well, most of the emission is of course directly to the discrete levels below. Only just part of the emission is to the surface states. Only a small part of it.

Q: If the emission is into the surface states, that part, the emission could be due to the electron leaving the excited states an ending down in a surface state.

M: That's right.

Q: A situation that is not considered in Fig. 2 where only holes are considered in surface states, I have shown the final state in my Fig. 1*.

M: Oh yes, you know, that is an interesting question because I think when the surface states are close to the valence band they are supposed to be hole states, and when they are close to the conduction band they are supposed to be electron states. But it is more sophisticated than that. If you look at a later paper that Pavel and I wrote, what we have is that *those surface states being states that are potentials for holes, an electron can't go into those. An electron can go to surface states that are near the conduction band*, so those are hole states and, some place in that paper, you will see where we make use of some work of Guyot-Sionnest where he has shown what a $1S_e$ going to $1P_e$ is, and that just more or less matches an electron going from one of those surface states down to the vacant $1S_{3/2}$ state. It is from Guyot-Sionnest that Pavel and I got the idea of having that one 1S to 1P transition.

Q: If it is the hole that makes the transition into the surface states then this hole should leave a state higher than the surface states but even this case is not depicted in the Figure, nor is considered in the text.

M: We have shown there that the surface state can take a hole, there would be also surface states near the conduction band that we didn't show.

Q: But if the hole goes into the surface states shouldn't the hole be higher in energy than the surface states?

M: Well, it is higher in energy by being down in the valence band, a hole state in the valence band is higher in energy than a hole state... things are reverse for holes.

Q: When it is in the $1S_{3/2}$ is then in a higher energy state.

M: Yes, that's right, because you have a hole there.

Q: So, for the holes when you go higher up the energy goes down.

M: You go to lower energy, just the reverse of the electron. When your hole goes up an electron goes down. So when a hole goes up you go to lower energy. It's an electron that's coming down.

4. Following: "A broad *surface state emission peak* with the energy 400 meV lower than E_{ex} was found..."

Q: Is this is an emission due to the electron in the surface states as in my Fig. 1* recombining with the hole in the ground state?

M: Well, I'm not sure now... I have to look at that broad peak and see it again, to look at it in context to see just what is involved there.

I have completely forgotten that and I know that there is some work that when an electron from the surface state can go down into a hole, the $1S_e$ can go up to the $1P_e$, in the excited state, the electron, and that is when, I think, Guyot-Sionnest observed that, and he prepares his hole, I think, by putting orthocresol on the surface of the QD, so that it helps create a hole there, but I have forgotten what the spectrum looks like, so I have to look at the spectrum to be sure of what is going on, I have completely forgotten, I have to see what that broad spectrum near the surface states is. Does Pavel give a reference to that?

Q: I'm not sure.

M: That's what I would have to look. I'm glad you raised because I will look at.

5. p. 3 1st column middle: "The 15 Å radius CdSe QD consists of about 1000 atoms, more than 400 of them existing on the surface. So, there are *at least 400 dangling bonds* in the QD. It was demonstrated that only Cd atoms on the surface (but not all of them) are *connected to the TOPO ligands*. As a result, we have *at least 200 surface states* in the band-gap corresponding to the Se dangling bonds."

Q: What are the TOPO ligands? If I understand correctly, one considers first 200 dangling bonds belonging to Cd atoms and 200 belonging to Se atoms. The 200 belonging to the Cd atoms are "connected to the TOPO ligands" and do not take part in the formation of the surface states, which are formed from the remaining 200 dangling bonds of the Se atoms.

M: I think so.

6. p. 3 1st column middle-bottom: "Absorption spectroscopy of *the negatively charged QDs* shows a peak in the *infrared* region corresponding to $1S_e \rightarrow 1P_e$ transition of the excess electron in the conduction band.... The *infrared* absorption spectroscopy of the QD ensemble *after* the *visible excitation* also shows a peak centered at the same energy with a similar width... The absorption line rises in a few picoseconds *after* the *visible excitation* and *decays very slowly*, at times $\tau_n \sim 1\,\mu s$. It is the clear evidence that *fast hole trapping* occurs in a large fraction of the QDs in the ensemble."

M: They either use the visible or they use the infrared, I think. In the common experiments they just use the visible, but there are some experiments where they use, I think, the infrared, where Guyot-Sionnest uses infrared, but maybe where he has sort of cresol attached to it, I am not sure, because I think that when he puts orthocresol then he has some systems, which are stable for a while, where you have an extra electron on the $1S$ and there is a hole in the orthocresol, and then with that extra electron you can absorb infrared radiation, so one would have to look at Guyot-Sionnest experiments to see what he did. You of course do the excitation in the visible because otherwise you never had a hole some place and then you would use an infrared laser to see that other absorption. The absorption going from the 1S to 1P in the conduction band, that absorption is not in the visible, I think that absorption is in the infrared, but I'm not sure, you should look it up, ... 3 eV, let's see, what is that, that's $.3 \times 8{,}000/3$ wave numbers, yes, that's in the infrared.

Q: You excite in the visible.

M: That's right, and now supposing you have got an electron sitting for a while in the $1S_e$, you now study that excited system with an infrared laser.

Q: So you first work with the visible and then with the infrared.

M: That's my guess but I'd have to look up the paper of Guyot-Sionnest and see what he did.

Q: (1) How can it be that the absorption decays very slowly while the hole trapping is fast?

(2) You are discussing here the negatively charged QDs and so, I believe, their electronic states are not the ones shown in Fig. 2 which refers to neutral QDs.

M: (1) Well, it all depends on what fast and slow is. The time scale of the usual observations of the blinking QDs is, say, in milliseconds to seconds, so anything that's shorter time than milliseconds is fast, microseconds is fast, compared with the observation times blinking in QDs. Now, the question is: what absorption line is that? The absorption line I suppose he is talking about is $1S_e \rightarrow 1P_e$, that's in infrared, he's looking an ensemble too, that's a big difference by the way, there is a huge difference between an ensemble and the single molecule. Things that you see in ensemble that are important maybe unimportant in the single molecule, very, very different, so let me say a little bit more about that as there's been later work which brings us out too. In the single molecule case, essentially all of the emphasis is up to a 10 seconds, 100 seconds. If you look at an ensemble, how much it fluoresces, change that has been in that time is negligible, and only after maybe an hour, the fluorescence in ensemble decreases to a steady state value, about 25% of this other value, in other words processes that are important for an ensemble are not seen in the single molecule. And the whole business in the single molecule is on something that is negligible from the point of view of total fluorescence in the ensemble.

Q: I think this is one of your last discoveries.

M: Well, it is something that I have emphasized, yes. I think I may have mentioned that in one of the articles that I wrote, I certainly mentioned it in the talks. So, what you see in ensemble can be misleading, unless you really interpret it carefully, when comparing it with what actually exists in the single molecule. I am glad you called my attention to this because I shall talk a little bit about QDs in the meeting that is coming out, and I certainly will look at this again because I have completely forgotten this and this is one more example of the fact that when you have an ensemble you have to be careful if you are trying to interpret things that are going on in the single molecule. Because if you have an ensemble, that doesn't necessarily tell you what is happening at the single molecule level. In fact, it tells you perhaps very little of what is happening at a single molecule level. I could give an example of something that takes an hour in the ensemble level, like that case we just described, then that's telling you something quite different from what the single molecule is doing.

(2) Well, they would have the same electronic states, but the energy levels would be modified because you got an extra charge there.

7. Following: "While the hole is trapped, the electron is located in the $1S_e$ state in the conduction band and can be excited to the $1P_e$ state by IR irradiation."

Q: (1) Please look at Fig. 2: Is the initial state what is indicated in the figure as "hole trap state" and is the final state the "intermediate state" there?

(2) Why is the hole represented *on top of the band* in the hole trap state and *on the bottom of the band* in the intermediate state?

M: (1) When it starts off... that is the initial state after some excitation, it is not the initial state before you start to irradiate it, I don't think so.

Q: I don't think so either.

M: The difference between the intermediate state and the hole trap state is that the electron that was up in the $1P_e$ has gone down, has relaxed to the $1S_e$, then it has got the hole raised to the top of the surface band. The hole trap state is when the electron is in the lowest state it could be, which happens to be the $1S_e$ state, and the hole is in the lowest hole state it could be which is shown there on the top of that surface band.

Q: It can be excited to the $1P_e$ state by IR irradiation.

M: Yes, that is what was done, it was absorbed that way, by Guyot-Sionnest.

Q: So you go from the hole trap state to the intermediate state.

M: Yes, if you were now shining infrared radiation on it, but in that diagram infrared radiation isn't being shone on it. In the intermediate state if you were doing that excitation, that hole there would not be at the bottom of that surface band, it would be at the top of the surface band.

Q: So this situation is not shown there.

M: That's right, no, it is not shown because you have already got the relaxed sort of system, when you are doing the shining of the infrared radiation.

Q: So when you have shined the infrared radiation you have something like the intermediate state but with the hole on the topmost of the surface states.

M: Yes, that's where the hole would be if it is relaxed.

Q: At this point you have already answered the next subquestion.

M: Because it relaxes in the hole trap state.

8. Following: "The slow decay of the absorption corresponds to a nonradiative phonon-assisted recombination of the trapped hole and the electron through the large energy gap."

M: I think that relaxation is supposed to be pretty fast, so it means that the $1P_e \rightarrow 1S_e$ is fast, I think that is picoseconds, and the slow step is going from the surface state to the ground state.

9. Following: "That wide spectrum of the $1P_e \to 1S_e$ energy difference

$$\varepsilon = E_{1P_e} - E_{1S_e} \tag{2.3}$$

of about 200 meV is difficult to explain by the size distribution of the QDs in the ensemble. *The energy difference ε scales as R^{-2}* with the radius of the dot R . . . we suggest that the energy difference between $1S_e$ and $1P_e$ electron states ε *is a subject of a light-induced diffusion*, analogous to the diffusion of the luminescence line energy E_{ex}.

Recall p. 2 2nd column bottom: "The low excitation energy of the QD is calculated as

$$E_{ex} = E_{1Se} - E_{1S3/2} + U(1S_e, 1S_{3/2}) \qquad (2.1)\text{"}$$

Q: (1) Why does the energy scale as R^{-2}?

(2) I believe one should write: "is subjected to a light-induced diffusion" because the subject here is the light and the object is the electronic energy difference ε which is subjected (i.e., the object of an action on the part of the subject).

(3) Please check if I correctly understand: in the above one considers *two luminescence lines*, one related to the energy difference $E_{ex} = E_{1Se} - E_{1S3/2} + U(1S_e, 1S_{3/2})$, *in the neutral* QD, in which the excited state represented in Fig. 2 goes to the ground state represented in the same figure and a line related to the energy difference $\varepsilon = E_{1P_e} - E_{1S_e}$ *in the charged* QD in which the excited state goes to the ground state by recombination with the *hole outside the QD*.

M: (1) I imagine that is the particle spherical box formula, those are energies of the body, I mean just like a particle in a box scales as 1 over the square of the length of the box, I imagine this is the same origin.

(2) OK.

(3) Well, I don't know if a hole is actually outside the QD or on some surface of the QD. When Pavel is speaking of a charged quantum dot probably he is having a hole on the surface and an extra electron then in the body of it.

10. p. 3 2nd column middle: "As is seen from experiment, the *deep trap emission* rises in a few picoseconds after the light excitation demonstrating *surprisingly high trapping rate for such a large energy gap*. The overlap integral between the delocalized $1S_{3/2}$ state and the *localized surface state* must be very small in comparison with the overlap of the delocalized states. So the phonon-assisted hole trapping process is assumed to be very slow in this case."

M: I think he is referring to surface states.

Q: (1) To the ones represented in the figure? Are we dealing with the charged QD or with the neutral one?

M: (1) Yes, which are sort of part of the QD. And the charge... of course you can actually have a charge to it, I mean, people have done that electrochemically, but I think he is referring here to the charge being where the other charge, the hole, is on the surface. Say, you can regard that, if you want, as a charged dot or not.

Q: (2) Suppose we are dealing with the neutral one whose states are shown in Fig. 2.

(i) Which states are the ones separated by the "large energy gap"?
(ii) Is the *deep trap emission* mentioned above due to an electron in the surface states combining with the hole in the $1S_{3/2}$ state?

M: (2i) Well, considering the $1S_e$ and the ground state, the band gap of those states are separated by a large energy gap, going from the excited state in the conduction band to the surface states.

(2ii) No, there're surely no electrons in the surface states, it would be question of a hole.

Q: What is then this trap emission?

M: It may be that you have a hole in the surface state and an electron in the $1S_e$ state. Now, if an electron came down from where it is to the surface state, in principle it could do that, whether it does it or not I don't know, then in principle that could be an emission. Well, if, for example, suppose you had that surface traps developed, that means that a transition becomes possible, is from the $1S_e$ to that hole in that surface state, I don't know if that is what Pavel is talking about or not, that becomes possible, one would have to look at that surface mentioned and what wavelength to set, one would have to look at the data that he is referring to.

11. Following: "We suggest that the trapping is an Auger-type *resonant* process (see Fig. 2): *the excess hole energy provides to the electron excitation.*"

Q: Please check my understanding of the story following the panels in Fig. 2: By illuminating the system one goes first from the ground state to the excited state. The trapping process is the one in which in Fig. 2 one goes from the excited state to the intermediate state. Here I see that both the electron and the hole are excited. The intermediate state looks like a further excitation. From the intermediate state one goes to the hole trap state. Here the lowering in energy of the electron is compensated by the raising in energy of the hole. Which one of the above processes is the Auger-type process and which one the non-Auger process?

M: The hole going up is really going down in energy, yes. The intermediate state is an excitation of the relaxed state, yes. Anytime you have two electrons changing their state simultaneously, one going up, the other going down, that's an Auger process.

Q: Just a compensation in energy.

M: Yes, that's right.

12. p. 4 1st column middle: "The characteristic laser intensity $I = 1\,\mathrm{kW/cm^2}$ corresponds to a *excitation rate* $k_I \sim 10^7\,\mathrm{s}^{-1}\ldots$"

M: The excitation rate would be the number of excited atoms produced per second in that system, and you can get that from the intensity and the absorption cross section.

13. Following: "The *radiative lifetime* of the QD excitation is about 150 ns at 10 K. It decreases with increasing temperature and reaches the value ~20 ns at room temperature. Such a dependence is attributed to *the splitting in the fine structure of the excited state* (*existence of a dark exciton*)."

Q: (1) Is the radiative lifetime the time during which the QD radiates?

(2) Please explain its dependence on temperature.

M: (1) Yes, that is like the fluorescence lifetime of a molecule, usually that is nanoseconds, you are at low temperatures, it is 150 ns, the same as the fluorescence lifetime of a molecule.

(2) Yes, right, let me say something there, although my knowledge is very limited. If the system is in a lowest phonon level, and that is the way it is, then there are certain selection rules that prevented from relaxing, in other words the fluorescence may be forbidden, I forget the details, though only when you have some *excitation by temperature* and you are in a different phonon state is that allowed, that fluorescence is allowed, and so I think that is what is happening, that has been studied in detail by somebody at Mac Gill, and probably others, I haven't followed that but I think that's the essence of that. The splitting is probably due to phonon interaction.

14. Following: "The population probability of the excited state is always much less than unity since $k_I \ll k_r$. So the effect of a *double excitation* and of an induced emission can be neglected."

Q: What is a double excitation?

M: Well, you may actually have that when you excite two electrons in very intense light sources, one excitation by one photon, and later another by another photon, and before the first one can relax you have such a high excitation.

Q: So, you can add energies.

M: In principle yes. I don't know if that has been done or not, but there has been a lot of work done at high energies, I think that has been done but I'm not sure.

15. Following: "The rate of the phonon-assisted $1P_e \rightarrow 1S_e$ transition *without the hole excitation* (non-Auger process) $k_e \ldots$"

Q: Please look at your Fig. 2. There, there is a trapped hole excitation. Is this process nonradiative?

M: Yes, that is probably nonradiative, it is probably just relaxation, that's probably the process Pavel is talking about.

16. Following: "The nonradiative relaxation of the trapped state (trapped hole + $1S_e$ electron) to the ground state of the QD (the electron-hole recombination) $k_n \ldots$"

Q: Which is the mechanism of this last nonradiative process? Does it differ from the preceding one?

M: If you excite aromatics you get some decay which is nonradiative, you maybe tunnel from one excited state to another excited state, you have a vibrational Condon overlap, it is the same idea as what you have in nonradiative transitions, nonradiative decays of excited states, you know, excited states of molecules can decay nonradiatively, same thing.

17. Following: "The rate of the Auger assisted trapping as a function of the reaction coordinate ε defined by Eq. (2.3), $\varepsilon = E_{1P_e} - E_{1S_e}$, is given by the Fermi Golden rule formula,

$$k_t(\varepsilon) = \frac{2\pi}{\hbar} V_{eh}^2 n[\varepsilon + E_{1S_{3/2}} + U(1)P_e, TS - U(1S_e, 1S_{3/2})] \tag{2.7}$$

Where V_{eh} is the matrix element of the electron–hole interaction, $n(E)$ is the density of the surface states as a function of energy and $U(1P_e, TS)$ is the averaged Coulomb interaction between the electron in $1P_e$ state and the trapped hole."

Q: Let me first rewrite Eq. (2.7) above inserting the value of $\varepsilon = E_{1P_e} - E_{1S_e}$:

$$\begin{aligned} k_t(\varepsilon) = \frac{2\pi}{\hbar} V_{eh}^2 n[&E_{1P_e} - E_{1S_e} + E_{1S_{3/2}} \\ &+ U(1)P_e, TS - U(1S_e, 1S_{3/2})] \end{aligned} \tag{2.7}$$

Now, the lowest excitation energy of the QD is $E_{ex} = E_{1Se} - E_{1S3/2} + U(1S_e, 1S_{3/2})$ (2.1) so the energy inside the square parentheses in Eq. (2.7) can be rewritten as $E(1P_e) + U(1P_e, TS) - E_{ex}$, which looks like the excitation energy of the intermediate state minus the excitation energy of the excited state *except that* in the intermediate state a hole appears in the intermediate state (see Fig. 2) and the energy of this hole doesn't appear in the above formulas.

M: Well, usually when you are applying Fermi golden rule you have a transition going to some state, then that state can relax, so that intermediate state is what is happening before it relaxes. It is like all of that is done on nonradiative transitions in the case of aromatics, that's what it is, an excited state going to the ground state, then that ground state relaxes. The reason is just a matter of conservation of energy. The point is this: you have an excited state that *at the same energy* goes over to the ground state, that means that the ground state has a lot of energy, lot of vibrational energy now, you go from the excited state at the same energy, it's a radiationless transition, you have to conserve energy, so now the ground state has a huge amount of vibrational energy, and now it relaxes.

18. Following: "The fast $1P_e \to 1S_e$ relaxation generates a finite width $\Gamma = \hbar k_e \sim 200\,\mu$ eV for the $1P_e$ state energy."

M: Anytime you've got a state that's relaxing, by the uncertainty principle that gives a spread in energy. $\Gamma = \hbar k_e$ because it is $\Gamma = \frac{\hbar}{\tau}$.

19. Following: "According to our previous suggestion the *density of surface states* $n(E)$ consists of *many peaks* at the energies $E_1, E_2, E_3, E_4, \ldots$ with spacing of about 1 meV. The small lifetime of the *final state* $1P_e$ broadens *each peak in Eq. (2.7)* up to the width Γ."

Q: (1) You write: "the *density of surface states* $n(E)$ consists of many peaks at the energies $E_1, E_2, E_3, E_4, \ldots$ with spacing of about 1 meV." Is $n(E)$ a *density* of states or a *group* of states?

(2) Where are the peaks in Eq. (2.7) to be seen? Maybe inside n?

(3) Pavel calls $1P_e$ "final state" evidently referring to the *excited state* $\xrightarrow{k_t}$ *intermediate state* process (see Fig. 2). Because of its small lifetime the state $1P_e$ is broadened. But why does it broaden other states, the peaks in Eq. (2.7)?

M: (1) He may be talking about a whole band of surface states that are narrowly separated by 1 meV, you have slightly different nuclear configurations, they give you a group of surface states which are separated by 1 meV, in his way of thinking. In other words, think for the moment of one surface state being, say, a hole in the Selenium, OK? Now, supposing that you have an electron missing in one of the surface Selenium ions, it can be either in that Selenium ion, it can be in another or another, or another, and not all of those Selenium ions are exactly the same, you know, the environment is very uneven, so you have really a distribution of energies, rather than Selenium ion just being in a single energy, because it can be in quite different *surface states*, and can migrate between them, that energy is somewhat different, there are slightly different energies, so you have a whole set of possible surface states maybe if the electron can be in, say, N positions, and not all of them have the same identical energy because the nuclear configurations aren't quite the same, then instead of a *degeneracy* of N states you have *splitting*, of 1 meV say, of N states, that is probably what he is thinking of. By $n(E)$ *Pavel clearly means a density of states, he uses that in the Golden Rule*. I think that probably what he means is that the density of states consists of a whole group of peaks *closely* separated, so that if you lump them together you have a density of states.

M: (2) That's usually what you do when you speak of a density of states in a finite system, it really is a set of discrete states. The density is a coarse grain way of doing it. I mean you speak of a density of states in an aromatic, final states in radiationless transitions, you know, you really have separate states, right? But you lump them together and you divide by the energy of the lump and that is called the density of states.

Q: *The peaks are difficult to see.*

M: That's right, the same way in aromatics, you have discrete states but you lump them all together and you call the density of states.

(3) Because it is not only the $1P_e$ that has a small lifetime but if that electron is gone into one of those trap states, that trap state has the same small lifetime.

M345 Explanation of Quantum Dot Blinking without the Long-lived Trap Hypothesis

(Continuation 3)

Pavel A. Frantzusov and R. A. Marcus

1. p. 1 1st column top: "...the statistics of the single QD luminescence switching events (blinkings). It is known from many single-molecule experiments that under continuous excitation the luminescence emission switches 'on' and 'off' by sudden stochastic jumps. Such a behavior was observed in dye molecules, light harvesting complexes and polymers. In these non-QD systems the *distribution of the 'on' and 'off' times is usually exponential* or near exponential."

Q: Does the exponential distribution of "on" and "off" times refer to

(i) The *number of switches* per time bin which decreases exponentially per unit time?
(ii) The *number of "on" times* per time bin which decreases exponentially? (If so, doesn't the number of "off" times increase exponentially?)
(iii) Or do the *"on" and "off" times* refer to the time duration of the "on" or "off" time intervals which decreases (or increases?) exponentially?
(iv) Or is the *overall time in the "on" state in a time bin*, sum of the duration of the single "on" times that decreases exponentially? And in this case does the time in the "off" state increase exponentially?

M: (iii) Basically the latter, (iii), and really what it is in each case is the survival probability, actually what they really plot is what's called the survival probability density, what they really plot is dP/dt.

(iv) No, they don't sum all the times.

2. p. 2 1st column top: "... the energy of the excited state displays *a long-correlated stochastic motion* with a characteristic time of hundreds of seconds."

M: Well, probably there are two correlation functions: the survival probability is measured for by that derivative, measured for hundreds of seconds, one correlation function is something that decays exponentially, and another decays more slowly than that, it follows typically a power law, so that has a correlation time which is not exponential.

3. p. 3 1st column top: "The emission spectrum of the CdSe QD photoluminescence has a *narrow peak* at the energy E_{ex}. There is also a *broad and less intense luminescence maximum* at 300–700 meV lower than the excitation energy, presumably due to the electron or *hole transitions into the surface states* in the QD."

Q: Recall that $E_{ex} = E_{1Se} - E_{1S3/2} + U(1S_e, 1S_{3/2})$. Please look now at Fig. 2. The narrow peak is due to passage from excited state to ground state. Is the broad and less intense luminescence due to a direct passage from the excited state to the hole trap state, a direct transition not shown as such in the figure, or does he refer to a luminescence emitted when passing from the hole trap state to the ground state?

M: *Normally electrons cascade down to the lowest excited state, essentially, and fluoresce from there.* It depends where it is. If it were 0.3 V, before were that energy, that'd be from the trap state, if it were the band-gap −0.3 V, then it would probably be from the conduction band to the trap state. Something like that, yes.

That distance is about 0.3 V, I'm not sure they see that luminescence, *I know they see the one from* $1P_e$ *to* $1S_e$ *luminescence*, I'm not sure if they see that particular luminescence though. But *they see the Auger process that accompanies it.* I know what he says, but that was an old paper, so I don't know if it's up to date.

The fluorescence that I was familiar with... some work that was done by Guyot-Sionnest where he created holes by putting something on the QD, I think it was some aromatic compound, and in a way he created a hole there, and the extra electron went in a $1S_e$ state, so one could see the absorption there, $1S_e$ to $1P_e$, and that's in the infrared, so what I don't know is whether fluorescence is from the trap state to the valence band, and if so it'd be about 0.3 eVs, because that fluorescence is in resonance with the $1S_e$ to $1P_e$ transition, roughly.

4. p. 4 2nd column bottom: "The standard theory predicts well-known Breit–Wigner shape of the decaying level. It was demonstrated however that the shape of the single level connected to the quantum chaotic system transforms into the Gaussian form in the so-called strong coupling limit. In this case the *trapping rate* could be cast as

$$k_t(\varepsilon) = \sum_i A_i \exp\left(-\frac{(\varepsilon - \varepsilon_i)^2}{\Gamma^2}\right) \tag{2.8}$$

where ε_i is expressed via the energy E_i of the ith surface state in the band,

$$\varepsilon_i = E_i - E_{1S_{3/2}} - U(1P_e, TS) + U(1S_e, 1S_{3/2})."$$

Q: (1) Which is the meaning of ε_i? Until now the following two energy differences had been defined:

$$E_{ex} = E_{1Se} - E_{1S3/2} + U(1S_e, 1S_{3/2}) \tag{2.1}$$

$$\text{and} \quad \varepsilon = E_{1P_e} - E_{1S_e} \tag{2.3}$$

and by the way, why do we have the U interaction term in Eq. (2.1) and no interaction term in Eq. (2.3)? I believe because in Eq. (2.1) one considers the total energy of the system, while in Eq. (2.3) one considers an energy difference between two levels. For symbolic consistence, the ε_i should be similar to ε and not have the U terms, but really we see that there they are. Moreover, whereas in Eq. (2.1) the states $1S_e$ and $1S_{3/2}$ are considered with their interaction, in $\varepsilon_i = E_i - E_{1S_{3/2}} - U(1P_e, TS) + U(1S_e, 1S_{3/2})$ there is not a $U(TS, 1S_{3/2})$, that is, an interaction between the trap state of energy E_i and the hole of energy $1S_{3/2}$, but the term $-U(1P_e, TS) + U(1S_e, 1S_{3/2})$ appears.

(2) What about the meaning of $\varepsilon - \varepsilon_i$? Let me write down in full the last difference:

$$\varepsilon - \varepsilon_i = E_{1P_e} - E_{1S_e} - E_i + E_{1S_{3/2}} + U(1P_e, TS) - U(1S_e, 1S_{3/2})$$

Apparently it should be the energy difference between the energy of the intermediate state—where the electron is in the $1P_e$ state, the trapped hole in the E_i state and they interact with interaction energy $U(1P_e, TS)$—and the excited state, where the electron is in the $1S_e$ state, the hole in the $1S_{3/2}$ state and their interaction energy is $U(1S_e, 1S_{3/2})$. If now I write down this energy difference I get:

$$\varepsilon - \varepsilon_i = E_{1P_e} + E_i + U(1P_e, TS) - E_{1S_e} - E_{1S_{3/2}} - U(1S_e, 1S_{3/2}).$$

As you see the terms are the same in both expressions but the signs of the terms E_i and $E_{1S_{3/2}}$ are opposite in my expression. Note that the terms E_i and $E_{1S_{3/2}}$ refer to holes. Is it so because the holes energy decreases when it climbs the energy?

(3) Each component of the sum in Eq. (2.8) resembles an Arrhenius rate constant but there are squares and the kT term in the denominator is missing.

M: (1) It may be that he's assuming that those terms fluctuate less than the other terms. Usually when you make an assumption . . . you make some assumption but you try to minimize them, and we know very little about fluctuations in those other terms, we know very little about the fluctuations, period. So he's probably making a minimal assumption. *TS* is a trap state. My guess is that he was largely looking at fluctuations in the main terms rather than in the interaction terms, that's my guess. I think that E_1 is composed of several terms, one of which is that E term and that $E_{1S_{3/2}}$ is composed of several terms, one of which is that U term, and he doesn't want to consider fluctuations in the E terms, so he's considering fluctuations in whatever is left over. *You have fluctuations of diagonal terms, and fluctuations of interactions.* By subtracting the interactions, he is just looking into fluctuations of the diagonal terms, the energy levels themselves, the unperturbed energy levels.

Q: What do those interactions there?

M: Well, it may be that those interactions fluctuate less, when the structure fluctuates. It may be that the assumption is that the diagonal term, the most important term, the interaction of the electron with the surroundings, with the lattice, that fluctuation is much bigger than the fluctuation of the interaction. From each of them subtract an interaction term, from each of them . . . just the main part of it. I can see why he has two U's in the Eq. (2.1), I can see why he has two U's there. Because he wants to subtract from E_i something, in order to have what's left as the main fluctuating term. He thinks that the U part of the E_i doesn't fluctuate much, and the U part of the other E doesn't fluctuate much, but the other parts of those E's fluctuate, that's what he wants.

Q: He subtracts part of the energy

M: That's right. So, he gets the part of the E that does fluctuate and the part of the $E_{1S_{3/2}}$ that does fluctuate.

(2) "*What about the meaning of $\varepsilon - \varepsilon_i$?*" He doesn't want to have the U's in there as part of the ε's. He wants to get rid of those. I would have thought that he doesn't want the U's as part of the ε's, so I would think that he wouldn't want those U's in there, and if he's ended up with having the U's in there, then I think he would be unhappy with that, in other words I think that the reason for defining that ε_i that way is to work with a quantity where the U's were not doing the fluctuating. If that's the case, then you wouldn't have any U's in the $\varepsilon - \varepsilon_i$, why would be there I don't know.

Q: Not only that: I have a problem with a couple of algebraic signs

M: He has corrected the E_i for the U, but why he hasn't corrected the $1P_e$ for a U I don't know. But the E_i is corrected for a U and the $E_{1S_{3/2}}$ is corrected for a U, because he didn't want those parts in there. But why he didn't have the E_{1P_e} corrected for a UI don't know.

(3) There is no intention to be related to Arrhenius in any way. There is no temperature there, just... it may be that one has a *distribution of rate constants* that's Gaussian, I don't know, I suppose, they speak of *rate constants that are distributed and not all to the same energy*. What he says is that you have a distribution of rate constants above an energy, in other words not a single rate constant.

Q: And the distribution is given by that function there.

M: That's the assumption. It's not uncommon. In fact another common one has assumed that the rate constants are distributed at uniform density over a certain narrow region of energy.

5. Following p. 5 1st column: "The behaviour of the trapping rate (2.8) is illustrated in Fig. 3, when the rate increases many orders of magnitude in the vicinity of ε_1."

p. 5 2nd column middle: "As seen from Fig. 3 there is some value ε^* very close to ε_1 separating two regions with the small and the large trapping rates

$$k_t(\varepsilon) \ll k_t^* \quad \text{for } \varepsilon < \varepsilon^* - \delta$$
$$k_t(\varepsilon) \gg k_t^* \quad \text{for } \varepsilon > \varepsilon^* + \delta \qquad (2.12)$$

... As a result the luminescence intensity ... is a nearly step function of ε."

p. 5 1st column bottom: "In the proposed model of the QD blinking *the* $1P_e \rightarrow 1S_e$ *energy gap plays a role of the slow variable.* It performs the light-induced stochastic motion generating huge variations of the trapping rate."

Fig. 4 A schematic picture of the mechanism of the QD blinking. The $1P_e \to 1S_e$ energy gap $\varepsilon(t)$ is the stochastic Markovian process. The thin curve shows the stationary distribution of ε. The dependence of the QD luminescence intensity on ε has a sharp threshold at ε^*. So, the luminescence is "off" when $\varepsilon > \varepsilon^*$ (hatched area) and "on" otherwise.

p. 5 2nd column middle: "Figure 4 gives an illustration of the model. When ε which was initially smaller than ε^* crosses the border and becomes larger than ε^*, the luminescence switches from "on" to "off." It switches back to "on" when *the trap energy* becomes smaller than ε^* again."

Q: (1) From the definition of $\varepsilon_i = E_i - E_{1S_{3/2}} - U(1P_e, TS) + U(1S_e, 1S_{3/2})$ we have $\varepsilon_1 = E_1 - E_{1S_{3/2}} - U(1P_e, TS) + U(1S_e, 1S_{3/2})$. Why is ε_1 apparently particularly important? By the way: please look at Fig. 2: I believe the E_1 state is the one occupied there by the small circle on the lowest line of the surface states.

(2) From all of the above can you please explain the blinking mechanism?

(3) Where in particular do you see or read the trap energy in the above figures?

M: (1) Let me tell you what the main physical idea is: I won't go into the details, for the details it would probably be best to consult Pavel. But let me tell you of *the main physical ideas. You know that* $1P_e - 1S_e$ *energy difference. That difference is fluctuating. Regardless of whether the states are occupied or not, is fluctuating. We also have a series... of trap states which compare with the top of the valence band. That energy difference of the trap states is approximately the same as the* $1P_e - 1S_e$ *difference. So that when those two energy differences are in resonance you can transfer... you can make the*

$1S_e$ go up to $1P_e$, and the other to come down to the ground state from the trap state, that's is the main physical idea, and so these energy differences can drift in and out of resonance with each other. That's the main physical idea.

Q: This paper of Pavel is not the most clear I have ever read....

M: No, and you see, I don't completely feel comfortable with it anyway. But, you know, I don't know, he may be right, and so on, but so much is not known about these systems. The one point that bothers me the most, and I don't know if he ever really addressed, is that he doesn't want the QD to lay in a trap state but on the other hand I don't think he's adequately explained why when you would turn off the light, the QD in existence would dark for so long if it's not in a trap state. So... but maybe his idea said: well, as long as you have the light on it... it maybe goes in into a trap state when light is off but the main point is that when the light is on, it spends very little time in that state. Anyway, that's one point that I never really discussed with him, but he's a very smart guy and thinks of all the things and so on, so one can't dismiss anything, but is best to talk with him on some of the details. *There is a physical idea, that's the most important point, that the QD doesn't spend long in those trap states* and maybe it doesn't spend long because the light is on all the time, but when the light goes off then of course it does spend time in there, but maybe he regards that as irrelevant because he may say: well, when the light is on it's never staying there, I don't know.

Q: There is apparently a difference between trap states and deep trap states.

M: *You know, there can be all sorts of trap states, there can be trap states off the conduction band, trap states off the valence band, and one doesn't know what one has... those states have not been studied correctly.* There is something like when you have metals that you do some photoelectric emission, or you do some inverse photoelectric emission, but you can really get a detailed picture of the sort of surface states, you can know a lot about the people who have done that, decades ago, but this is now a highly disordered sort of system, on the surface, and is more complicated, you know, so... and one doesn't have the direct experiments.

Q: When you have a very intense light the energies of the photons can add up...

M: You can have a nonlinear phenomenon.

Q: Einstein's ideas about photoemission were accepted because the electrons would not be emitted from metals even if one would use very intense infrared or red light, only with ultraviolet light.

M: *Maybe you could do, that would be a multiphoton process. I don't know if you can do with metals, you can do it with molecules. You know, with molecules you can have something which is proportional to the first power of intensity, something like that. Now you can get to such high intensities that even though one quantum alone wouldn't do it, a group of quanta together in a coherent way would do it.*

M345 Explanation of Quantum Dot Blinking without the Long-lived Trap Hypothesis

(Continuation 4)

Pavel A. Frantzusov and R. A. Marcus

1. Let me first recall the following definitions:

p. 3 1st column bottom: "Absorption spectroscopy of the negatively charged QDs shows a peak in the infrared region corresponding to the $1S_e \rightarrow 1P_e$ transition of the excess electron in the conduction band. *The center of the peak* ε_0 corresponds to *the energy difference between* $1S_e$ *and* $1P_e$ *states predicted theoretically* ($\sim$300 meV...)"

p. 3 2nd column top: "...wide spectrum of $1P_e - 1S_e$ *energy difference*

$$\varepsilon = E_{1P_e} - E_{1S_e} \tag{2.3}$$

of about 200 meV..."

p. 4 2nd column middle: "...the density of surface states consists of many peaks at the energies of $E_1, E_2, E_3, E_4, \ldots$"

p. 5 1st column bottom: "...*the stochastic variable* $\varepsilon(t)$..."

Fig. 4: "A schematic picture of the mechanism of the QD blinking. The $1P_e - 1S_e$ energy gap $\varepsilon(t)$ is the stochastic Markovian process. The thin curve shows the stationary distribution of ε. The dependence of the QD luminescence intensity on ε has a sharp threshold at ε^*. So, the luminescence is 'off' when $\varepsilon > \varepsilon^*$ (hatched area) and 'on' otherwise."

p. 5 2nd column top: "As seen from Fig. 3 there is *some value* ε^* *very close to* ε_1 separating two regions with the small and the large trapping rates

$$k_t(\varepsilon) \ll k_t^* \quad \text{for } \varepsilon < \varepsilon^* - \delta \quad k_t(\varepsilon) \gg k_t^* \quad \text{for } \varepsilon > \varepsilon^* + \delta \tag{2.12}$$

Q: Please look at Figs. 2 and 3: (1) I believe the E_1 level there is the one occupied by the little circle in the "intermediate state" figure because that's the highest among the energies of the hole surface states.

(2) The matching between electron energies and hole energies is possible only until the *increase in electronic energy* ε is balanced (Auger process) by a *decrease of the hole energy, that is, by the hole climbing up the set of surface states.*

(3) Please look at my Fig. 2*. where I have drawn (from left to right) first the excited state of your Fig. 2, then the intermediate state of Fig. 2 and then I have added another intermediate state with a higher

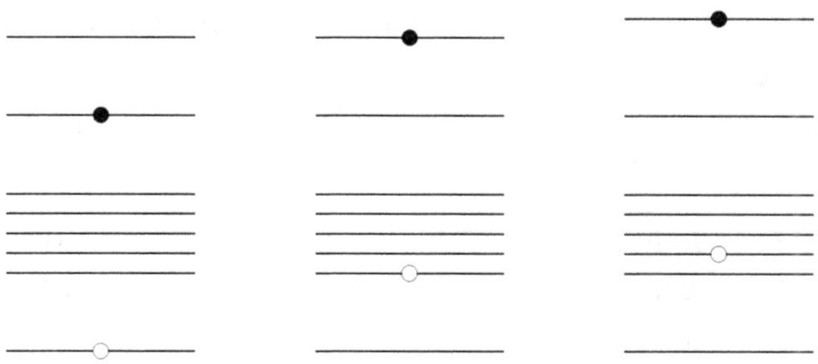

Fig. 2*.

value of $\varepsilon = E_{1P_e} - E_{1S_e}$ to which corresponds a lower hole energy. Is my interpretation of the story correct? By the way: if it is so, the energy of the hole depicted in the intermediate state in Fig. 2 is the $\varepsilon^* \sim \varepsilon_1$ energy, I believe.

M: (1) Yes, that's right.

(2) Yes, *when the hole climbs up an electron climbs down.*

(3) Yes, OK. You have a hole in the valence band and an electron in the $1S_e$ band, right? Correct.

2. Please look at Fig. 3 where $k_t(\varepsilon)$ shows a periodic behavior as a function of $\varepsilon - \varepsilon_1$. How do you explain this periodicity from equation

$$k_t(\varepsilon) = \sum_i A_i \exp\left(-\frac{(\varepsilon - \varepsilon_i)^2}{\Gamma^2}\right) \qquad (2.8)?$$

Consider that $\varepsilon_i = \varepsilon_1 + (i - 1) \times 1$ meV.

M: You see, that's a sum over ε_i, so, when ε gets close to one of the ε_i's you're going to get a small k, when it's in between you have a large k. Sorry, when ε is close to anyone of the ε_i, then that exponential factor becomes unity, and so if the A's are roughly equally valued or something, you get a local maximum. That's the image of the sum, the sum creates the periodicity. If the sum is periodic, you know, at special levels, then you get a periodicity. It depends on the spacings of the i's there. The spacing depends on what the ε_i's are, the A_i is just a magnitude. Here Pavel is clearly taking the A_i's to be equal.

3. p. 5 2nd bottom and p. 6 1st column top: "The probability distribution function of the coordinate ε at the time t satisfies the following diffusion equation

$$\partial_t \rho(\varepsilon, t) = D\partial_\varepsilon(\partial_\varepsilon + (\varepsilon - \varepsilon_0)/\Delta^2)\rho(\varepsilon, t) \quad (3.1)$$

where $\Delta \sim 50$ meV is a root mean square deviation of the (Gaussian) steady state distribution of ε, and $\varepsilon_0 \approx 300$ meV is the center of the distribution... We introduce the difference $x = \varepsilon - \varepsilon_0$ as a new reaction coordinate. This x is a function of the nuclear coordinates of the system mentioned earlier. The master equation for the distribution function $\rho(x, t)$ follows from Eq. (3.1):

$$\partial_x \rho(x, t) = D\partial_x(\partial_x + x/\Delta^2)\rho(x, t) \quad (3.2)$$

The threshold coordinate value between 'on' and 'off' regions is $x^* = \varepsilon^* - \varepsilon_0$. When the edge energy $\varepsilon^* \sim 300$ meV is very close to ε_0, the following condition is fulfilled:

$$|\varepsilon^* - \varepsilon| \ll \Delta.$$

Under this condition one can set $x^* = 0$ without loss of generality."

Q: Why $\varepsilon_0 = \varepsilon^*$? Shouldn't ε^* be just the threshold energy between light and dark states?

M: Yes, probably, I would think so and, you know, I was always queasy about this.

4. p. 4 1st column bottom: "The rate of the phonon-assisted $1P_e \rightarrow 1S_e$ transition without the hole excitation (non-Auger process) k_e .... *The nonradiative relaxation of the trapped state* (*trapped hole* + $1S_e$ *electron*) to the ground state of the QD (the electron–hole recombination) k_n ..."

Q: (1) I believe the two above processes are nonradiative processes leading to the ground state. How do the mechanisms of the two differ?

(2) How does one follow experimentally the velocity of processes happening in the dark?

M: (1) One is making use of conservation more or less of electronic energy and the other is converting electronic energy to phonon energy.

(2) Well, you can look at a position of a line before the dark and after the dark, and see that has moved quite a bit. In other words, you can look at the spectral diffusion there. Then in some cases you are not measuring rate constants, rate constants are being inferred as a result of some mechanism, but they are very uncertain.

5. p. 5 2nd column bottom: "The probability distribution function of the coordinate ε at the time $t \ldots \rho(\varepsilon, t) \ldots$ We introduce the difference $x = \varepsilon - \varepsilon_0$ as a new reaction coordinate... The master equation for the distribution function $\rho(x, t) \ldots$"

p. 6 1st column top-middle: "The *distribution of the 'off' times* $p_{\text{off}}(t)$ is a derivative of the *survival probability of the 'off' state*,

$$p_{\text{off}}(t) = -\frac{d}{dt} P_{\text{off}}(t) \tag{3.3}$$

The survival probability is equal to unity at zero time and goes to 0 at infinite time. It could be expressed as an integral of the *survival probability distribution function* $\rho_{\text{off}}(x, t)$,

$$P_{\text{off}}(t) = \int_0^\infty \rho_{\text{off}}(x, t) dx \qquad (3.4)\text{"}$$

Q: Here we have three magnitudes: $\rho_{\text{off}}(x, t)$, $P_{\text{off}}(t)$, and $p_{\text{off}}(t)$. Please check my understanding.

(1) With the integral (3.4) we consider that an "off" state may be due the reaction coordinate x belonging to any point of the interval causing the "off" state;

(2) I am uncertain about the meaning of $P_{\text{off}}(t)$, the survival probability of the "off" state. Imagine the QD crosses the appropriate threshold energy and starts being in the "off" state. Is the survival probability measured by the time in which it remains in the "off" state or, given a time bin interval Δt, is the survival probability given by the time sum of the times in which the QD is in an "off" state in the time bin interval?

M: (1) Maybe OK for this, but if you have an intersecting parabolas model then you can have the same value of x but you can be on two different potential curves. And I don't know if there is any subtlety like that in here.

(2) "*Imagine the QD crosses the appropriate threshold energy and starts being in the 'off' state.*" At that instant the survival probability of the "off" state is unity.

Q: At the beginning?

M: Yes. *As soon as the "on" state starts, or the "off" state starts, whatever it is, the survival probability of that state starts off at unity.* By definition.

Q: What is the survival probability function?

M: *If you think of a first order decay, and you plot C/C_0, it will start with 1 and will decay exponentially, that exponential is the survival probability function.*

Q: Like in radiation.

M: Yes, that's right, nothing more.

6. Following: "$\rho_{\text{off}}(x, t)$ satisfies Eq. (3.2),

$$\partial_x \rho_{\text{off}}(x, t) = D\partial_x(\partial_x + x/\Delta^2)\rho_{\text{off}}(x, t) \tag{3.5}$$

with an *absorbing boundary condition at the border* (*the first passage problem*)

$$\rho_{\text{off}}(x,t)|_{x=0} = 0. \tag{3.6''}$$

M: Well, if you have, say, diffusion to a point, and if you say at that point you're going to make the concentration be zero, then you are getting... then you're saying that there is an infinite sink at that point, and that's the first time that it will reach that point, because if it had reached before it would have disappeared, it would be zero. *Imagine something diffusing, towards a main hole, OK? Now, as soon as it reaches that main hole, it's going to drop down, that's the first passage.* And if you say that you're gonna put a boundary condition, it's going to be zero there, that means you are getting it disappeared at the first passage.

7. Following: "The condition (3.6) is not exact because of the finite size of the transition region δ near $x = 0$. However, it is valid for times much larger than

$$\tau_0 = \frac{\delta^2}{D}. \tag{3.7}$$

The *initial value of the coordinate x at zero time is taken equal to* δ,

$$\rho_{\text{off}}(x, t = 0) = \delta(x - \delta) \tag{3.8''}$$

Q: Apparently there is probability distribution function equal to a delta—a peak-like function—at $x = \delta$. How does this relate with the stationary distribution function represented on the ordinate of Fig. 4?

M: Well, *if you have something diffusing in time, I think it is not a stationary distribution is it? It depends on what time it started, a stationary distribution is when something is independent of the time you started.*

Q: So you start with a Dirac delta function and then things spread out somehow.

M: Yes, that's right, something like that.

Q: Things spread out to reach a Gaussian distribution.

M: Yes.

8. Following: "The solution of Eqs. (3.5), (3.6) and (3.8) is well known

$$\rho_{\text{off}}(x,t) = \rho_0(x,t) - \rho_0(-x,t) \tag{3.9}$$

where $\rho_0(x,t)$ is a solution of Eq. (3.5) with the initial condition (3.8) but without the boundary condition (3.6)."

M: You see, that equation there... *symmetry means that is automatically going to be zero next to $x = 0$. It's the analogue of the method of images in electricity.*

9. p. 5 2nd column bottom: "The probability distribution function of the coordinate ε at the time t satisfies the following diffusion equation

$$\partial_t \rho(\varepsilon,t) = D\partial_\varepsilon(\partial_\varepsilon + (\varepsilon - \varepsilon_0)/\Delta^2)\rho(\varepsilon,t) \tag{3.1}$$

where $\Delta \sim 50$ meV is a root mean square deviation of the (Gaussian) steady state distribution of ε, and $\varepsilon_0 \approx 300$ meV is the center of the distribution."

Following: "We introduce the difference $x = \varepsilon - \varepsilon_0$ as a new reaction coordinate."

p. 6 2nd column middle-bottom: "The distribution $\rho_{\text{off}}(x,t)$ at small times is much narrower than Δ. So, *an unbiased diffusion approximation could be used* which gives the power law result

$$p_{\text{off}}(t) = \frac{\delta}{2\Delta}\sqrt{\frac{\tau}{\pi}}t^{-3/2} \qquad \text{(3.14a)''}$$

M: You have written down the solution of a nonbiased diffusion equation. Do you have something that derived that solution? I know that Pavel sent me an e-mail, one time, deriving that solution, he derived by a method of images. So he didn't actually solve the full equation formally, he derived by method of images and this is the solution that occurs when you use that, but that method of images is only valid after a certain time interval, it is not valid in very short time, it is valid after that, but why it is called unbiased diffusion... I think a nonbiased diffusion equation is one that doesn't have that $(\varepsilon - \varepsilon_0)/\Delta^2$ term, because that is a forcing term, that biases it, so I am almost certain that that is what that means. In a work that we wrote, we solved a biased diffusion equation, where the slope of the parabola affected the diffusion rate. Anyway, that's what I think that is. In the method of images you have some point on the left of the origin and then on the right you have its image.

10. Following: "The exponential dependence at large times in Eq. (3.14b)

$$p_{\text{off}}(t) = \frac{\delta}{\tau\Delta}\sqrt{\frac{2}{\pi}}\exp(-t/\tau) \tag{3.14b}$$

corresponds to the decay of the *quasistationary* distribution $\rho_{\text{off}}(x, t)$ as a whole. Specifically, there is a *reactive flux* proportional to the survival probability $P_{\text{off}}(t)$, generating the first order steady-state decay rate $1/\tau$."

M: Well, *anytime something is disappearing at a sink that means that there is a reactive flux, the reaction is going into the sink.*

11. p. 7 2nd column middle: "Early spectroscopic experiments showed that the emission spectrum of the colloidal QD *ensemble* is quite wide (about 80 meV). It was explained by the inhomogeneous broadening due to the QD size distribution... Unexpectedly, it was found that the homogeneous luminescence spectral line measured in the *single* QD experiments is also much broader than expected width $\hbar/\tau_r$, where τ_r is the radiative lifetime. Posterior investigations showed that *the energy of the emitted photon E_{ex} displays a long correlated stochastic motion* with a characteristic time of hundreds of seconds."

M: Well, *maybe that is the spectral diffusion of the line.* If you look at reference in one of Tang's papers, to work of Bawendi on spectral diffusion, you look at the time scale that Bawendi uses by looking at Bawendi's paper. Then you will see this order of time.

12. Following: "The distribution function $\rho_E(E_{ex}, t)$ of the energies thus satisfies a diffusion equation

$$\partial_t \rho_E(E_{ex}, t) = D_E \partial_E [\partial_E + \varphi'(E_{ex})] \rho_E(E_{ex}, t) \qquad (4.2)$$

where the '*diffusion coefficient*' $D_I = \alpha_E^2/\tau_I$ is proportional to the excitation light intensity I and where a unitless '*potential*' $\varphi(E_{ex})$ corresponds to the steady-state distribution of the emitting energies $\rho_{st}(E_{ex})$,

$$\rho_{st}(E_{ex}) = \exp[-\varphi(E_{ex}). \qquad (4.3)$$

If the steady state distribution has a Gaussian form the function $\varphi(E_{ex})$ is parabolic,

$$\varphi(E_{ex}) = \frac{(E_{ex} - E_0)^2}{2\Delta_E^2} \qquad (4.4)$$

where E_0 is a center of the steady-state distribution $\rho_{st}(E_{ex})$."

M: *Every time you got a distribution of something, like in that case of trap states, then you can say that that distribution can be regarded as derivable from a potential. If you have a diffusion equation, you can arrange it so that when you go to equilibrium you can actually extract that distribution out of that equation. So, it is a lot of work where you put in the diffusion equation, and put in $U(t)$, or gradient of that, and then, if the things have reached steady state or equilibrium, then the derivative of what is inside there is zero and eventually you find* *out that* $\rho \propto e^{U_k/kT}$. *There is a whole big formalism about that,* it's done all the time in problems like this and Pavel is just applying that, as far as I can tell. Anytime you have a distribution…then you can derive it from a potential. If you solve the differential equation you get a distribution function, it is a lot of work, but it can be done and it is done.

13. p. 8 1st column bottom: "The observed time dependences of the luminescence line with the CdSe QD at the temperatures 10–40 K are seen in Fig. 5 to be fitted by the function

$$\mathrm{FWHM} = \Delta\sqrt{8\ln 2[1 - \exp(-2t/\tau_E)]} \tag{4.7}$$

As also seen there, the parameters for the spectral diffusion depend weakly on temperature in the given range. For the excitation intensity $I = 85\ W/\mathrm{cm}^2$ the averaged time between excitations is $\tau_I \sim 1\,\mu s$. The results of the fitting allow the determination of an *averaged length* a_E *of the single jump*

$$a_E = \Delta_E\sqrt{\tau_I/\tau_E}.$$

Its value varies from $0.1\,\mu$eV to $0.3\,\mu$eV for the data presented in Fig. 5, when the temperature varies from 10 K to 40 K.

We suggest above that the energy difference between $1P_e$ and $1S_e$ states in ε, which is the function of the nuclear coordinates of the system, is also a subject of a light-induced diffusion with a mechanism analogous to the mechanism of the diffusion of the emission energy E_{ex}. The single jump size

$$a = \Delta\sqrt{\tau_I/\tau} \sim 1.5\,\mu\mathrm{eV}$$

is much larger than a_E (0.1 to $0.3\,\mu$eV). *The probability density of the delocalized electron near the surface is larger for* $1P_e$ *than for* $1S_e$ *state.*

Q: Why so?

M: "*The observed time dependences of the luminescence line with the CdSe QD at the temperatures 10–40 K are seen in Fig. 5 to be fitted by the function*" Yes, *that is a standard function in diffusion problems.*

Well, *the* $1P_e$ *sticks out more than the* $1S_e$. *You know, it has angular momentum, which means it sticks out quite a bit.*

14. Following: "Thus the sensitivity of the $1P_e$ energy to the surface modifications could be much larger than the $1S_e$ energy. *It gives a possible explanation of the difference between a and* a_E."

Q: How and why?

M: I'd have to look back at what a and a_E are, but *if a is related to an internal spectral diffusion and* a_E *is related to diffusion that involves a surface state, then that would follow.*

M345 Explanation of Quantum Dot Blinking without the Long-lived Trap Hypothesis

(Continuation 5)

Pavel A. Frantzusov and R. A. Marcus

1. p. 5 1st column bottom: "In the proposed model of the QD blinking the $1P_e - 1S_e$ energy gap ε plays a role of *the slow variable.* It performs the *light-induced stochastic motion generating huge variations of the trapping rate.*"

M: Things have changed a fair amount since that paper, some of the essential ideas I think are OK, but some of the details are not. I mean, for example, it always bothered me that there is an asymmetry between "on" and "off." In a Tang paper we attributed the curvature to the derivative of the potential energy function, that is a pretty much standard in diffusion theory and that is what we attributed to. We no longer did so now, because that didn't explain the asymmetry, why it is that only the dark state showed curvature and the bright state didn't. So that was always the failing with that, and then in the Joy and Zhu paper we changed the origin of the exponential. In Tang's it followed because of the slope of the potential energy surfaces at the intersection, the forced diffusion caused an exponential. Now we give it a different origin, it's due to expulsion of an electron when you have an Auger process where one electron now has an energy one band gap above the end gap edge, conduction band gap edge. So, the origin now is Fermi's golden rule, and in the paper with Joy Anne we explained why that doesn't happen with the dark state.

2. Following: "The *motion of the stochastic variable* is *much slower than the excitation–relaxation transitions* of the QD, one can utilize a *local steady state approximation*. Accordingly, the luminescence intensity at any given time t is calculated using *a stationary solution*

$$w = \frac{k_I}{A + k_t/k_t^*} \tag{2.11}$$

where k_t *is treated as a constant*, $k_t = k_t[\varepsilon(t)]$.

As seen from Fig. 3 there is some value ε^* very close to ε_1 separating two regions with the small and the large trapping rates

$$k_t(\varepsilon) \ll k_t^* \quad \text{for } \varepsilon < \varepsilon^* - \delta, \quad k_t(\varepsilon) \gg k_t^* \quad \text{for } \varepsilon > \varepsilon^* + \delta \tag{2.12}$$

...As a result, *the luminescence intensity* (*2.11*) is a nearly step function of ε. It causes *the telegraphlike behaviour of luminescence intensity* as a function of time...Fig. 4."

Q: (1) Here two motions are mentioned as being one much slower than the other. Why is the motion of the stochastic variable much slower than the excitation–relaxation transitions of the QD?

(2) Apparently because of the slow motion of $\varepsilon(t)$ one can use a steady state approximation for k_t which allows one to treat it as a constant during the time in which the system is "on" or "off."

(3) Here the telegraph-like behavior in Fig. 4 is attributed to the luminescence intensity (2.11). But on the ordinate ε is shown . . .

(4) From Fig. 4 one has the impression of very short "on" and "off" times because there is no indication of time units on the abscissa. I guess—but please correct me if I'm wrong—that to every short time interval ("point," say, on the abscissa) corresponds a quick excitation–relaxation process and that an "on" or an "off" state corresponds to a whole set of excitation–relaxation processes all of them belonging to the same region above or below the threshold energy ε^*. Do the telegraph-like oscillations referring to the "on" state show a telegraph-like changing intensity of light emission? I mean: is there a blinking due to the on-off switching and a fluorescence intensity blinking due to sudden successive changes of light intensity during the "on" period?

M: (1) *Because that slow variable probably involves fluctuations in positions of the ions, and that's a slow process.*

(2) Well, no, k_t actually, you see, varies with energy, so it is not a constant.

Q: But Pavel says that it is a constant.

M: Well, maybe in one local region where some coordinate is hardly fluctuating. It is a constant but locally it is just a function of energy that is not a constant. You see, you can see that as a function of the energy which is a slow variable in time. In other words, maybe what Pavel is saying is that k_t is actually fluctuating very wildly but you can sort of locally average or something like that, but ε is the slow moving variable changing with time. You can have several processes that are going on, that are very fast, maybe establish a steady state at that point in time, that steady state may then change slowly with time. This is quite common, to have some transfer processes which are much faster than some overall process, so you can sort of speak of them as occurring and say what has happened at that time as a result of their occurring, but the net behavior will depend upon a parameter which is very slowly varying with time. I think that that is what is involved here. What is w now?

Q: w is the luminescence intensity.

M: Yes, all right, then he is saying that when you have a particular value of ε, you have a particular value of luminescence intensity, it depends on how big k_t is compared with the fluorescence with time, OK? All right, so at any particular value of ε you have a particular value of k_t and that gives you a particular value of the luminescence at that time t, that is sort of a *local* steady state for that given value of ε, but ε fluctuates, so the steady state fluctuates. Often you have some very rapid processes occurring on a fast time scale. You treat those separately. Those processes may depend parametrically on some variable that varies slowly. When you do that, it is like a two time scales problem. That's very common. In other words... let me give you an example: when you write down in a chain reaction the rate of concentration of free radicals with time, normally there is some overall change in reaction, reactants disappearing, the products are formed, that is one time scale. But the time scale for the rate of change of the concentration of a particular free radical forming and dissociating is very fast, so you use a steady state for them but that doesn't mean a steady state for the whole process that changes with time. It is the same thing. A steady state in some process that gives rise to that Eq. (2.11) but that whole process is slowly fluctuating with time.

(3) Meaning that going on changing intensity, slowly changing intensity with time, telegraph, on off on off...

Q: But on the ordinate he shows ε and not w.

M: Yes, that is the slow variable. Think of the energy as being the energy difference of two states that are undergoing spectral diffusion, so the energy difference of those two states is fluctuating. Actually, what it is, is that the energy difference of $1P_e - 1S_e$ is compared with the energy difference associated with the trap motion, that surface state motion, *the difference of two energy differences.* w depends on a particular value of ε because for any particular value of ε you have a particular k_t.

Q: But one has the impression when one looks at this telegraph-like behavior there are sets of oscillations.

M: No, that tells you of k_t, that doesn't tell you about the telegraph of going on and off, when k_t is high of course then it would be off, when k_t is low it would be on, but that tells you only indirectly about the telegraph-like behavior.

Q: When a set of oscillations in not in the dark region.

M: Not the oscillations that you see there, in k_t, they are going to a set of dark and light regions, separately. When the energy difference is near the top of one of those maxima then you have a rapid decay that is probably maybe a rapid dark region, a rapid light region...

Q: Some oscillations are in the dark region.

M: Figure 3 is just describing different regions of the ε coordinate, where you can have light or dark, and at any one time you are in one of those regions, and you're either near the top in which case there is very little light coming out or you are near one of the bottom ones with k_ts where there's a lot of light coming out, that's all that is telling you, it is not telling you how you are moving in that space, that's a separate stochastic equation for how ε depends on time. ε fluctuates as a function of time and, depending on what ε is, you are either in a light or dark period, this is just telling you nothing about the fluctuations of ε with time, nothing about that stochastic process, but is telling you: if you happen to be in a region with a certain range of ε, then you have a dark period. If you are in another region you have a light period that is what is telling you.

Let me just say something more. In Figure 3 you can see that to the left of the curve the k_t becomes so small that that you are always going to be in a certain state. *The biggest change is being from the left of the vertical line in Figure 3, that's the biggest change*, so when he has things in one side, he soon has a pretty high value of k_t and that's probably then when you have rapid decay all the time without turning off light, whereas when you are outside of that region, to the left of the vertical line, you are largely giving off light.

Q: So these very quick oscillations have nothing to do with dark and light.

M: No, that's right, *everything to the right is roughly dark, and everything to the left of the vertical line is roughly light.*

Q: But if you look t these oscillations somehow you can group them.

M: Yes, but that's just small changes compared with the big change.

Q: So you can have a light state in which there is a changing light intensity.

M: *The light state is going to be on the left, it's going to be out of resonance.*

Q: But I mean: when you are in the light state are you going to have a steady intensity of the light or not, or it oscillates a little bit?

M: *Your incident light intensity is steady but your fluorescence comes out in separate bursts, separate emissions, so it doesn't emit fluorescence intensity, it emits one, then another, then another, then another*, like a stochastic process. In other words, in a continuous train... an emission comes out, another emission comes out, then another emission... you know, here is some kind of stochastic process there.

Q: During the light time.

M: *That's right, you shine then it emits, later then emits another burst, and so on during that whole light process.* And then you measure a light... *So you see what Pavel has here is that the thing could be dark without actually being trapped, just in a region of the energy diagram where k_t is pretty large and so that dominates things, that is his key idea.* I personally don't agree with it but that's his key idea.

3. p. 9 1st column top: "The present model explains the important properties of the QD blinking, but some features were left unexplained.

One of them is *a large difference between the cutoff times for the 'on' times and 'off' times found* in *single* QD experiments and *ensemble* luminescence experiments. The cutoff time *for the 'off' distribution* is about 40 times larger than *for the 'on' distribution* at room temperature."

Q: (1) Which is the meaning of the cutoff time for a single QD vs. that for a distribution?

(2) Why is there a large difference?

M: (1) You see, one has exponential decay and the other doesn't. That's what he means by cutoff times, but I think that now with the modern developments we understand it, but we didn't then and I didn't even a year ago. See, that's one thing about Pavel, he is very frank, he is very honest, what he knows he states, when he doesn't know, he also states. He is a real scholar.

Q: Well, from this point of view I am a scholar too...

M: Ah, ah, ah... I really didn't imply that. I was just trying to say that even though I don't agree with Pavel now on mechanisms I have a tremendous respect for him.

(2) What was not understood at the time and I don't think I thought I understood until a couple of months ago. And I understood really mainly on the basis of ideas that were introduced by others, the Debye exciton.

4. Following: "This fact could be explained if *the mean jump length* depends on the $1P_e - 1S_e$ energy gap ε. *It is equivalent* to the introduction of the ε dependence of the diffusion coefficient in Eq. (3.1)

$$\partial_t \rho(\varepsilon, t) = D\partial_\varepsilon(\partial_\varepsilon + (\varepsilon - \varepsilon_0)/\Delta^2)\rho(\varepsilon, t). \tag{3.1}$$

Such dependence could be the result of the *large* difference *in the length of the jumps made with or without photon irradiation* at room temperature."

Q: Why does the mean jump length dependence on ε explain the different cutoff times of above?

M: You mean is this a cutoff business?

Q: Yes.

M: I don't understand that, my interpretation of the difference of cutoffs of light and dark periods is totally different from that.

5. Following: "While the *diffusion coefficient* is reasonably larger in the 'on' region in comparison with the 'off' one, the *cutoff time* on the 'on' time distribution . . . would need to be much smaller than the corresponding 'off' cutoff time . . ."

Q: In yours and Tang's model I believe the inverse proportionality above can be explained because the greater the diffusion coefficient the shorter is the time for reaching the crossing of the parabolas. Is this explanation valid even for Pavel's model?

M: I don't even know that. That's questionable. I'm not sure if we know it today, I think. I was at a meeting in San Diego. Bawendi describes in very sophisticated experiments where he may be learning about diffusion coefficients . . . dark state, but that's all very recent, at that time he didn't know about what the difference of the diffusion coefficients in light and dark states does, no idea. Yes, no idea.

I think it is now thought by leading people that the reason why the on region has a cutoff and the other doesn't are totally different. There's no evidence that I know of, that the diffusion coefficient is different, it may be different in the light and dark states and certainly at the time and even today one may not know.

6. Following: "Another possible explanation of the cutoff times difference is the influence of the *competing trapping process*, namely Auger-type excitation of the $1D_e$ electron state. It is natural to assume that the $1D_e - 1S_e$ energy gap is also the subject of the light induced diffusion. Its motion should be correlated with the motion of the $1P_e - 1S_e$ gap. So, when the $1P_e - 1S_e$ energy gap ε becomes small enough, the $1D_e - 1S_e$ could reach the resonance with the upper edge of the trap energies band and *quench the QD emission.* It can be shown that this cutoff mechanism could apply for the 'on' times, *but because of the positioning of the states could not apply to the 'off' times*."

Q: It is evident that an "on" time can be quenched by this mechanism because the $1D_e$ substitutes the $1P_e$ in the quenching job. Can you explain why it doesn't work for the "off" times?

M: No, I think that's outdated now. Maybe it's true that one might from a historical point of view try to go into things that may have been wrong, but I'm not sure how much one learns from that. You see, *these are details that we have no information about,* like the difference of the diffusion constants and so I'm not sure how much one would learn from that. You know, for example, old quantum theory has certain aspects that are right . . . and certain that are wrong, and it is instructive in that case to compare old quantum theory with modern quantum mechanics, so there are definitely, you know, strikingly interesting things but it shows the limitations of the time and how theories were overcome, that's of great historical interest, but here the theory rests on quite uncertain assumptions, you see.

7. p. 9 2nd column middle: "An important component of the present mechanism is the suggestion that radiationless relaxation goes via the *hole trap state.* This suggestion could be directly tested by the single QD near-infrared spectroscopy. According to the present model each single QD should demonstrate *deep trap emission.* The luminescence intensity in the spectral region should be switched by sudden jumps, correlated with the blinks in the visible spectrum. *The population of the trap states is larger during 'off' periods*, so the deep trap emission should be more intense that time."

Q: Please look at Fig. 2. As I understand it, during the "off" periods the holes go in the surface states and from there the above "deep trap emission" should come. How would that happen? By which mechanism do the holes in the surface states become emissive?

M: Yes, *that idea of Pavel is very good*, the idea that you would have the $1P_e - 1S_e$ resonance as energy difference in resonance with a trap to valence band state, I think that was an excellent idea of him, that provided a vehicle for various things to happen.

"*The deep trap emission, from there the above 'deep trap emission' should come*" Yes, if it does, maybe more a radiationless transition rather than an emission, maybe the change there is radiationless, you won't see any emission. We don't know that those trap states emit. That's a speculation, emission from trap states I don't think has ever been observed, if it has I'm not aware of it.

8. Following: "The time-resolved measurement of the relaxation rate of the excited state in the single QD could be done by the *time-correlated photon counting procedure together with the maximal likelihood estimation method.*"

M: With the maximal likelihood method you're trying to fit a whole set of data. You assume some functional form for the data, whether the data may follow a power law or power law times an exponential, you assume that, you usually take the log of that, and you maximize things so as to choose a parameter to fit the data. In other words, it is taking some assumed law, and then trying to fit the data with that law. And you do with the form of the log because that converts a multiplication to a sum, you see, that's easier to work with. One of my students has just recently written a paper on that, that we're going to submit shortly, I think she was very good... The maximal likelihood method in science goes back more than a hundred years, but its application to QDs is very recent. It was applied in the literature and then, when she applied it, the best function didn't go through the points... I asked her to look into it, and it turns out that for the particular case of dyes on surfaces the power law is minus one and there the maximal likelihood method as done in the conventional way breaks down, and she found a way of doing the method without its breaking down, so... she has done a nice piece of work there, we have written down a draft... she has written down a draft for the paper and that is a little bit more about maximal likelihood method, you assume some functional form, you have the way of trying to fit it to the data, you evaluate the parameters from that fitting.

M350 Photoinduced Spectral Diffusion and Diffusion-controlled Electron Transfer Reactions in Fluorescence Intermittency of Quantum Dots

Jau Tang and R. A. Marcus

M: In this paper, Jau Tang is trying to summarize everything that was done, that we did in the previous papers, I'm not sure if there's anything at all new in that, but it was a convenient summary. It summarizes things.

1. p. 2 1st column middle-bottom: "Efros and Rosen considered a rate equation among three states with 0, 1 and 2 electron–hole pairs..."

Q: Is each state made up by the couple of an electron state and a hole state?

M: Yes, that's right. *In this mechanism you need to have at least two excitons to be able to form a trap state. If you just have one exciton, you just get fluorescence. And if you have no exciton, you get nothing.*

2. p. 2 2nd column top: "Klafter and Szabo explored two-state single-molecule trajectories arising from a multi-substrate kinetics."

Q: What is it?

M: *I can tell you what the usual alternative theory is, and that is probably what they are talking about. You can eject an electron to different distances and tunnel back, this is the alternative, tunnel back from each of these distances, each of those possibilities has a different rate constant, its tunneling depends on distance, and so that means that if you look at the whole rate possibilities, this is going to have a whole set of possible rate constants for return.* In that case, if you have a distributed set of rate constants, then, under certain conditions, even if each rate constant were referring to a unimolecular decay, if you have a whole set of them, you can get a power law out of that as a net result, in other words when you have many rate constants that appear, in other words when you have *a distribution of rate constants, then if you put all of that together in an appropriate way then for the time behavior you would come out not as a single exponential, not as a double exponential, but actually as a power law.* But there has to be the right distribution of all of those exponentials.

3. p. 3 2nd column top: "...to describe $\langle I(t) \rangle$ for an ensemble one needs to use *coupled rate equations including both forward and reverse reactions*, whereas to describe $P_{\text{on}}(t)$ or $P_{\text{off}}(t)$ of a single QD *only a forward or reverse rate equations* needs to be used."

Q: I guess because in the ensemble there are QDs in the "on" state and others in the "off" one.

M: Yes.

4. Following: "Therefore, the *nonergodic nature of QDs* needs to be understood from the *context of the fact* that in $P(t)$ of single-particle studies a QD is distinguishable whereas in ensemble measurements of $\langle I(t)\rangle$ QDs are indistinguishable."

Q: What do you mean by "context of the fact"?

M: Well, I don't know that specific point, but *the essence of that is that when you are looking at an ensemble you are looking at the behavior of a whole number of dots each of which started their fluorescence at different times, each of which ended their fluorescence at different times, that's what you do with the whole ensemble. When you're looking at an individual one, you're looking at when it started, when it stopped, when it started, when it stopped, you're looking at those individual times, so you're looking at far more detail than you look at an ensemble, in an ensemble you just mention the intensity from all of those states as a function of time, you've no idea of when one started and the other left off and so on.*

Q: And this is the nonergodic nature?

M: Well... *the nonergodic behavior is that when you're looking at the individual trajectory, you're just exploring a limited part of phase space, not covering the whole phase space, you have not time to do that.*

5. p. 4 2nd column bottom: "Initial observations of $t^{-3/2}$ power-law intermittency in QDs prompted the hypothesis of 1-D diffusion of the electron/hole among discrete localized surface traps. However, such a discrete hopping mechanism is faulty. To account for 7 to 8 decades for the observed $P(t)$, the number of the localized traps has to exceed 500 . . . ; in addition, these localized traps have to be aligned in a 1-D fashion for this 1-D mechanism to work. A QD of 2 to 4 nm in radius would not contain so many surface traps that can line up in a 1-D array"

Q: Why should the hopping be only along a single direction? Aren't we dealing with 3-D objects whose surface is 2-D? Should the traps in 1-D be aligned along a radius of the QD or along some line inside or on the surface of the QD?

M: I don't know, I never really looked closely at that argument, I mean *the one dimensional, in the way we are using it, is the same as in the ET theory, where one has really a* 10^{23}*—well, not quite—dimensional problem, and we are looking at an energy difference and using that as a coordinate, calculate the free energy as a function of that coordinate, that's how we get to 1-D description.*

Q: But that's a quite different kind of story.

M: Yes, that's right, but he was trying to make the point, somehow, that if you are trying to just do it, considering the going in and out of actual traps, instead of considering this energy coordinate fluctuations, the traps maybe would have to be much more varied in distance in order to account for so many decades, by sort of a jumping from a trap to the main bulk, I think he thought that you couldn't possibly get such a big difference of rates . . . . I don't know if that is true or not, I haven't thought about that, I probably should have . . . but I let it go.

6. Following: "Therefore, the 1-D diffusion is likely to operate in *a configuration space, such as an energy space . . .*"

Q: By the above do you mean a reaction coordinate given by the system's energy depending on the coordinates of all the QD's atoms?

M: Yes, but *it is the energy difference of the two electronic states that is the coordinate. And basically you convert it to a free energy difference.*

7. Following: "To be more general and to accommodate a power law with an exponent that could differ from the ideal −3/2, we considered a *non-Markovian anomalous diffusion* in the diffusion-controlled electron transfer model between a *dark charged state* and a *light neutral state*, and the *ordinary Markovian diffusion* is thus treated as a special case."

Q: (1) Why is the dark state charged and the light state is neutral?

(2) What is meant by non-Markovian anomalous diffusion and by ordinary Markovian diffusion?

M: (1) One view is this: if the dark state is charged, then when you excite the next electron you now have two electrons say in the upper state because you have a charged QD, and you have a hole. Now, if you didn't have that second electron in the upper state one electron would just come down to the hole and fluoresce, but when you have two electrons in the upper state you have the choice, the possibility that one goes down while the other goes up and it's coming down without fluorescing because is just balancing the energy of the other that is going up, so this competing process is radiationless, is balancing energy by having two electrons move instead of just one. That's called an Auger process, if you look Auger processes in atomics physics you will see that they involve several electrons, one going someplace and the other going someplace else.

If a *hole* goes into a −2 then it makes the charge −1, because a hole is positive. A hole is a lack of a negative charge, so it is like a positive charge. When you excite an electron from the valence band to the conduction band by absorption of light you have a negative charge up there and you have a positive charge left behind. The valence band has lost one electron so one now has a positive charge in it. The hole is a vacancy in a whole band. The loss of an electron is equivalent in the solid state to the gaining of a hole.

(2) That deals really with the statistical mechanics ways of treating problems, and to understand that one would have to know what the Langevin equation is and what the generalized Langevin equation is. And the generalized Langevin equation has a memory function in it which makes it more complicated. Rather different ways of treating something, and there is a reason why a something has a memory function and the other thing doesn't. So, to get at that, one would have to know the whole formalism. And when a process doesn't have a memory function it's called Markovian, because something for the next step is determined only on what that step is, not in the past history, not in the past memory. If you have a sequence of events, it only depends in the last event and not in the whole prehistory. Just in the last one, just depends on that information, not on anything that happened before. And the difference, in some phenomenon, if you are plotting a magnitude vs. time and the phenomenon has no memory function the function is sort of decreasing exponentially, if it has a memory function it actually goes through an oscillation, becomes negative and then positive. Like for example, if you have a molecule in a liquid and you look at its velocity as a function of time, if there is no memory and you start off at the final velocity, it will just decay, but if there is a memory and it's a molecule in a liquid, it will bounce and now turn back, its velocity becomes negative, so your correlation function now has an oscillation in it.

Q: And the charge in the outside is negative and in the core is positive?

M: Yes, when you have a simple Debye relaxation you have a simple exponential decay and then everything is simple with one relaxation time that corresponds to a diffusion constant that is a constant and related to the relaxation time, but if you have the Cole–Cole plot, then they have a whole distribution of relaxation times if I remember rightly, and so you don't get a simple type physical theory there, they have a fractional thing which I think some people have attributed to a whole distribution of relaxation times. In other words, if you look up the Debye formula and look up the Cole–Cole formula for real and imaginary parts of dielectric constant you will see that they differ in that one has some sort of a fraction in it, that the other doesn't, and that fraction is one way of sort of... well, it amounts to one form of having a whole distribution of relaxation times, not just one. I think that somebody may have given a model that could lead to the Cole–Cole plot or the Davidson–Cole plot, but I don't remember if they actually did or not, I know there is something by Watts, some Watts relaxation, but I don't remember that at all.

8. p. 5 2nd column middle: "After the initial transient, a *quasi*-equilibrium is established between $|0\rangle$ and $|1\rangle$"

Q: I believe the equilibrium is the one between emitted and absorbed photons. Why "quasi"?

M: Yes, that's right, in other words *you turn on something, everything does start instantaneously, that you have some going on and off of going to the excited state and coming down and you are building up a kind of quasi equilibrium there. Quasi? Well, it is a steady state, not an equilibrium, so one may call it a quasi-equilibrium.* Jau wrote that, and I'm not sure that that is what he had in mind, but I assume that that is what he had in mind.

9. p. 6 1st column top: "... for single QDs one should use a decoupled rate equation."

Q: Decoupled from what?

M: Well, *when you are considering an individual QD, you start with it being in an on period and you are looking at its probability of going off, so it is uncoupled from the equation for the QD that starts from off and going on. When you're doing an ensemble, you're not solving* *one differential equation, you're solving a pair of coupled differential equations. OK. When you're doing an individual QD you're just looking at one of those equations to get a distribution of times . . .*

10. p. 7 2nd column middle: "Because diffusion is light-induced, as light intensity increases τ_1 is shortened, and according to Eq. (16)

$$\Gamma_k \tau_k \equiv \frac{E_{A,k}}{2k_B T} \tag{16}$$

Γ_1 increases . . ."

M: Yes, incidentally that is only proven so far for QDs around 10 K, it is not clear if it is true if a QD is at room temperature. But recently Bawendi may have some measurements on spectral diffusion at room temperature, so then one will know better for room temperature. But that intensity dependence I think it is only established for low temperature QDs, no longer a prime source of interest. I mean not in QD theory. They're probably of interest if anybody is trying to use for some sort of quantum computation.

Q: In Pavel's paper, the τ was defined as "cutoff time" and as "diffusion relaxation time." How to compare the two definitions?

M: *They're totally different, the cutoff time here means that you have* $e^{-t/\tau}$ *to represent a tail of the exponential, the cutoff. The* τ*, in the case of diffusion, means that if you have a mean square jump length that is called a* Δ^2*, that* Δ^2 *divided by* τ*,* Δ^2/τ*, is the diffusion constant, the two* τ*'s have a total different meaning.*

Q: Tang uses the same symbol.

M: Well, he shouldn't then.

11. p. 8 1st column middle: "In the above *single-QD* intermittency study only *the short time power-law* behaviour was of major interest.

To calculate longer time behaviour for a single QD or the ensemble fluorescence intensity over the entire time span..."

Q: Why short times for the single QD and long times for the ensemble?

M: Well, *it's clear from some experiments of Bawendi and Chung that when you look an ensemble to establish a steady state, it took about an hour or so, whereas a single quantum dot doesn't cover anywhere near that time, maybe they go up in bursts of hundred seconds or ten seconds, so the single quantum dots experiments don't cover the time necessary to reach the steady state. The single is too weak at that time. But that just says that the most important trajectories for establishing a steady state are not those that are studied by quantum dots.*

12. p. 8 1st column bottom: "The coupled rate equation for electron transfer in the COP scheme is given by

$$\frac{\partial}{\partial t}\rho_1(Q,t) = \int_0^t d\tau L_1(t-\tau)\rho_1(Q,\tau) - \frac{2\pi|V_1|^2}{\hbar}\delta(U_{12}(Q))(\rho_1(Q,t)-\rho_2(Q,t)) \quad (19)$$

......

where $L_1(\tau) \equiv \Delta_1^2\varphi_1(\tau)\frac{\partial}{\partial Q}\left(\frac{\partial}{\partial Q} + \frac{\partial}{\partial Q}(U_1(Q)/k_BT))\right)$, and $\overline{\varphi}_1(s) = (s\tau_{D,1}+1)^{1-\beta_{CD}}/\tau_{L,1}$."

Q: Is $\varphi_1(\tau)$ a transform?

M: See, there you have a nonlocal L, that's when you look at things that go from the usual diffusion equation to more elegant, no, to more general equations. There are various ways of doing it, and one is by having an integral formulation like that instead of a differential formulation, and the other is by having a time dependent diffusion constant. He did a Laplace transform of that $\varphi_1(\tau)$. $\varphi_1(\tau)$ itself is the untransformed variable, that $\overline{\varphi}_1(s)$ is a transformed variable. You have to take the inverse of that $\overline{\varphi}_1(s)$ in order to find out what $\varphi_1(\tau)$ is. $\varphi_1(\tau)$ is just the ordinary function and he hasn't given the ordinary function, but he has given the transform. The way he got that transform from, I don't know. Probably one of the COP, POP alternatives has a diffusion constant that varies with time and the other uses an integral formulation like he has here.

Q: Which is the meaning of that $\varphi_1(\tau)$?

M: It looks as though it's effectively related to a diffusion constant. Because you see, if that L were a delta function of $(t - \tau)$, $\delta(t - \tau)$, you would have an ordinary diffusion operator, so this is where you take the diffusion operator and you make it more complicated and that is one of the standard of the two alternative treatments, one of which is to use ordinary time dependent diffusion operator and not having an integral, and the other is having an integral of it.

13. Following: "Eq. (19) *becomes Markovian* if $\beta_{CD} = 1$ where $\varphi_1(t) = \delta(t)/\tau_{L,1}$."

p. 6 1st column bottom: "For a Debye dielectric medium $\overline{\chi}(s) = 1/(1 - s\tau_D)$"

p. 6 2nd column medium "For a Cole–Davidson dielectric medium $\overline{\chi}(s) = 1/(1 - s\tau_D)^{\beta_{CD}}$"

Q: (1) Which is the meaning of the β_{CD} exponent?

(2) Why in the Debye case does the equation become Markovian and what does that mean?

M: (1) Well, you know, when you plot the dielectric constant, real, and imaginary parts, then you get this Cole–Davidson plot, as best as I remember, or where you plot the real part vs. the frequency or the imaginary part vs. frequency, frequency of dielectric, you get these Cole–Davidson plots, I forgot which is more advanced, Cole–Davidson or Cole–Cole. Anyways, books are written on that, and so he's relating this sort of diffusional property to some of the language that occurs when you don't have simple Debye relaxation. When you have simple Debye relaxation you have a simple exponential decay, that's all straightforward, then you would have diffusion constant independent of frequency, and so on. So this is associated with that Cole theoretical work and experimental work which has gone on really with the real and imaginary parts of dielectric constant and one has got a simple relationship to a Debye type theory and when it doesn't.

(2) You know, essentially, if you have $dn/dt = n/\tau$, then n is a function of time... indirectly from that equation, and *that is kind of a Markovian process, how n changes depends upon the n just at any particular time, but when it depends on the whole time sequence and so has many more parameters to that, that is non-Markovian*, it is Markovian if it depends just on the previous time that you're talking about, that's more common, it means that if there is a whole set of events, each of them depends on just the previous event, not on the whole past history. When the process depends on the past history it is a process with a memory function. The rate of change of n depends only at n at that time and not on some past history of n, so this is an example of a Markovian system. If you had something more complicated instead of n you would have all the past ns.

Q: People's life is essentially non-Markovian then.

M: *Yes, that's right, ... absolutely ... You may forget the recent things but remember things as a child, that would be anti-Markovian.*

14. Following: "If an ensemble of QDs is initially at Q_c, the energy level crossing, with $\rho_1(Q, 0) = C\delta(Q - Q_c)$ and $\rho_2(Q, 0) = (1 - C)\delta(Q - Q_c)$..."

"If a QD ensemble is initially in $|1\rangle$ as illustrated in Fig. 10 *with a Boltzmann distribution* $\rho_t(Q,0) = (1/\sqrt{2\pi k_B T/\kappa})\exp(-\kappa(Q - Q_{0,1})^2/2k_B T)$"

Q: (1) Is the C a molar fraction?

(2) Why do you have a Boltzmann distribution at $|1\rangle$ and you don't have it at the energy levels crossing Q_c?

M: (1) Yes, in other words *when you're dealing with an ensemble, it is not initially at the intersection, is initially at the bottom of the well, then you start exciting*... oh, I see, a mole fraction, where is C?

Q: C is in front of ρ_1 and ρ_2. A quantum dots fraction I guess.

M: Yes, when an ensemble... what's wrong in it a little bit is that the ensemble is never initially at that one point, what he is saying there is that supposing you have a hypothetical ensemble that began at that point, which you never do, OK?, How can you do? Then some of the QDs would be on one electronic state, a certain fraction C, and then the other fraction 1-C would be on the other electronic state, but you never do that though. Normally you start at the bottom of the well, so he may have said that... just introduced the fact that you no longer as an individual trajectory start at the intersection, instead you start at the bottom.

Q: Why at the bottom you have a Boltzmann distribution and you don't have a Boltzmann distribution at the intersection?

M: Well, at the intersection... remember that the system at the intersection has a free energy which represents an ensemble in a 10^{23} dimensional space, and in that subspace you may have a Boltzmann distribution, but this is a projection... well, not a projection, this is the free energy as a function of that... well, *you don't start at the intersection when you're starting with an ensemble of dots, that would be too magical that you would be starting in the point where the two electronic energies are the same, that only happens in single dots,* *when you switch on light off after light, that by definition you start at the intersection*, but typically there is a thermal distribution of states at the bottom of well.

Q: So this is an ensemble of QDs distributed in energy with a Boltzmann distribution.

M: That's right, and the other thing is a purely hypothetical possibility which is never realized. I think there Jau is just trying to show the contrast between some very hypothetical thing that doesn't exist and the actual thing which does.

Q: I understand that there are in the past history some points that are extremely interesting, but then there is a lot of scientific noise, so to say.

M: Yes, that's right, exactly, and one occupies too much with the noise than with the more interesting points.

15. p. 9 1st column bottom: "Eqs. (25–27) have been derived previously by Weiss... and were applied by others. These equations were previously expressed in terms of $\overline{P}_k(s)$ and $\langle \tau_k \rangle$. Here, we show $\langle \tau_k \rangle^{-1} = \gamma_k$, and is simply the non-adiabatic electron transfer rate. More importantly, via $\overline{g}_k(s)$ Eqs. (21–23) provide formulas calculable for the entire time span (*for anomalous normal diffusion model*), and link $\langle I(t) \rangle$ and $P_k(t)$ to measurable molecular-based quantities."

Q: (1) Have you been the first to interpret the QDs' blinking phenomenon in terms of electron transfer and electron transfer rates?

(2) What is an "anomalous normal" diffusion model?

M: (1) It is quite possible, it's quite possible, I think the first to interpret it in detail probably, yes... Now, of course, the Pavel interpretation is a kind of electron transfer, but the mechanism, the details are different, there wasn't really an actual electron transfer, it is sort of an electron resonance kind.

(2) Anything that causes a deviation from a Debye model for a dielectric.

16. p. 10 2nd column top: "To obtain desired accuracy, we have included 25 terms in Taylor's series expansion of $\overline{g}_k(s)$ in either regime of a small and large s"

Q: Is there a limit to the theoretical number of terms one considers or is it just a matter of computer time?

M: Just a matter of computer time. Also, at some point, I don't know about this specific problem, you have to worry about whether the series is truly converging, sort of power series or not, because there are some series that converge up to a certain number of terms and then diverge, and they are good as their last term of its convergent part but after that the series would diverge, such series are called asymptotic series, so in this particular problem I don't know.

17. Q: Can you please explain and summarize the physical ideas about single particle vs. ensemble behavior?

M: I think so. *When you are looking an ensemble and its kind of phenomena, you turn on the light, say, and then you get some emission, as a function of time, from all of these QDs. Some of the QDs would be emitting, are in the light state, some are in the off state, you have no idea because you're not looking at that level of detail, you just know that you have a certain total intensity vs. time. And those experimental results by Chung and Bawendi show that you have to wait about an hour in the experiments before you finally reach the steady state. OK? That is the ensemble measurement. In a single molecule measurement instead you look at the individual QD, you look to see from its fluorescence whether it is on or off, you continue looking at it , then you see at some point if it goes on and at some point if it goes off, you continue looking on it, you see when it comes on again, you continue looking, you get a whole trajectory of on off periods, which* *have different time lengths, then you plot the probabilities of some of those things, the log of the probabilities vs. the log of time, and you get a power law, so there you're looking at individual start-off behavior, like counting a whole set of results and putting into little boxes and you say: Ok, this will last between such and such a time, that will last for so many years, last between such and such a longer time, some specific time Δt, so you compare the log of the distribution vs. the log of time it will last, that is the individual single molecule information, so you get totally different information, along just total intensity vs. time.* One can define the survival probability after being in a dark state, the survival probability is a function of time, and one can define the survival probability as a function of time being in bright state. If you take the derivative of that survival probability, the log of that is plotted vs. the log of the time that is the actual thing that is being plotted there.

M351 Chain Dynamics and Power-law Distance Fluctuations of Single-molecule Systems

Jau Tang and R. A. Marcus

1. p. 1 2nd column: "... the Langevin equation of overdamped oscillators in an N-unit chain can be expressed by

$$\zeta \frac{d}{dt}\mathbf{Q}(t) + \omega^2 \mathbf{R} Q(t) = F(t)/m \qquad (1a)$$

and the generalized Langevin equation is given by

$$\int_0^t d\tau \zeta(t-\tau) \frac{d}{d\tau}\mathbf{Q}(\tau) + \omega^2 \mathbf{R} Q(t) = F(t)/m \qquad (1b)\text{"}$$

Q: (1) The Langevin equation is $m\frac{d^2\mathbf{x}}{dt^2} = -\lambda\frac{d\mathbf{x}}{dt} + \eta(t)$. In Eq. (1a) there is no acceleration term. Why so? Is it so because there is no center of mass movement?

(2) The Rouse coupling matrix $\mathbf{R}$ is defined as $\mathbf{R}_{ij}^{\mu\nu} = \delta_{\mu\nu}(2\delta_{ij} - \delta_{i,j+1} - \delta_{i+1,j})$. Is it really a whole set of matrices each for a fixed couple of indices?

(3) In Eq. (1b) is the term $\frac{d}{d\tau}\mathbf{Q}(\tau)$ inside the integral sine? (I believe so...).

M: (1) Yes, *usually one sees a Langevin equation for velocity but one also sees, and I think it is sometimes called, the Langevin equation in the overdamped regime, and that's what you have here. In the overdamped regime you neglect the acceleration, OK? It is changing all the time, fluctuating rapidly, is overdamped, so there's no acceleration, no oscillations, and so you get this equation there.*

M: (2) I'm just wondering what the μ and ν refer to...

Q: μ and ν *refer to x, y, and z*

M: (2) The Rouse coupling. I guess that it couples the i's and j's, it would be a diagonal thing when $i = j$, and then, when $i \neq j$, it links, I suppose, the nearest neighbors states, I suppose, so it looks though it is a matrix that has diagonal terms and has links to its nearest neighbors, so the diagonal terms I suppose may be stretching force constants for, you know, each link, and the other is a connection of each link, the force on each link, on neighbors.

(3) That is inside the integral, yes, the generalized Langevin equation has a time dependent friction constant, so that's what you got there, you see, same thing but inside the equation in order to be time dependent.

2. Following: "$\mathbf{F}(t)$ in Eq. (1b) represents Gaussian noise and has no correlation with $\mathbf{Q}$. Specifically, one has $\langle F_i^{\mu}(t)F_j^{\nu}(\tau)\rangle = mk_B\zeta(t-\tau)\delta_{ij}\delta^{\mu\nu}$, $\langle F_i^{\mu}(t)\rangle = 0$, and $\langle F_i^{\mu}(t)q_j^{\nu}(t)\rangle = 0$. If $\zeta(t) = 2\zeta\delta(t)$, Eq. (1b) reduces to Eq. (1a)."

Q: (1) If $i = j$ and $\mu = \nu$, $\langle F_i^\mu(t) F_i^\mu(\tau)\rangle = mk_B\zeta(t-\tau)$, that is, considering anyone bead and the one of its coordinates, the damping constant depends on time and the damping constant measures the time correlation of the Gaussian noise operating on it.

(2) How does Eq. (1b) reduce to Eq. (1a) if $\zeta(t) = 2\zeta\delta(t)$?

M: (1) That's sort of standard, you know, it is a random force, of the medium on there, and frequently taken to be, seen to be, Gaussian, sometimes not. Well, usually it is, but people have investigated equations like this and all sorts of noise.

(2) I don't know why is that factor of 2 in there. I mean, if $\zeta(t) = \zeta\delta(t)$, I don't know why he put a factor of 2 in there…oh, one problem is that is coming at the end…if you write a $\delta(t)$ that is at an end point there is then a subtlety, because you are just integrating over half of the range that would be covered by the delta function and not the whole range, so that probably give you a 1/2, then you multiply by 2. See, normally when you use a delta function, it is in the interior of the integration range, but here he has got a delta function at one end of the integration range, so it is like a half delta function so he multiplies by 2.

3. Following: "We first consider Eq. (1a) where ζ is a constant friction coefficient and treat Eq. (1b) later. Defining $\Gamma = 4\omega^2/\zeta$ one can show that the pairwise correlation of displacement vectors satisfies

$$\frac{d}{dt}\langle \mathbf{q}_0(t)\cdot\mathbf{q}_i(0)\rangle + \frac{\omega^2}{\varsigma}[\langle \mathbf{q}_0(t)\cdot\mathbf{q}_j(0)\rangle - \langle \mathbf{q}_1(t)\cdot\mathbf{q}_j(0)\rangle] = 0\ldots\text{''} \tag{2}$$

p. 2 2nd column middle: "If ζ is time dependent and has a memory, one needs to use Eq. (1b) with a retarded memory kernel $\zeta(\tau)$.

By replacing Γ in Eq. (8)

$$\overline{C}_Q(s)/\overline{C}_Q(0) = 2/\sqrt{s}(\sqrt{s+\Gamma} + \sqrt{s}) \tag{8}$$

with $\overline{\Gamma}(s)$ or $4\omega^2/\overline{\varsigma}(s)\ldots$"

Q: I suspect that in the above there has been a confusion between the Greek letters ς (the form of σ to be used in ancient Greek at the end of words ending by sigma) and the letter ζ (the Greek z). I believe the above ς's are to be substituted by ζ's.

M: It should be ζ. That's a common symbol used in the treatment of the Langevin equation. I can see it can be confusing once one is knowledgeable, as you are, about the Greek…

4. p. 2 1st column middle: "Because $\ell(t) = |\mathbf{q}_{n_1}(t) + \mathbf{q}_{n_A,eq} - \mathbf{q}_{n_D}(t) - \mathbf{q}_{n_Deq}|$ is the distance of the D-A at beam index n_D and $n_A\ldots$"

Q: I believe it should be $l(t) = |\mathbf{q}_{n_A}(t) + \mathbf{q}_{n_A,eq} - \mathbf{q}_{n_D}(t) - \mathbf{q}_{n_Deq}|$.

M: OK.

5. p. 2 1st column bottom: "To analyze the fluctuations in single molecule ET rate one is concerned with about small distance changes from that of the *equilibrium conformation*. We assume that the ratio of the D-A distance fluctuations with respect to the equilibrium D-A distance is small and is primarily due to stretching or contraction of the springs."

Q: Legend of Fig. 1: "Schematic diagram of a Rouse model with a chain of beads coupled by a spring to their nearest neighbors, containing a donor (D) and an acceptor (A).

M: Yes, it is the first figure in the paper on chain dynamics, those balls are points. I think this chain model is probably not original with Jau, he probably took it from somewhere. The chain model is probably of that type, you just think of these things as being idealized points with some interaction between them, and so you don't get into all the subtleties of what these idealized points do in terms of rotations, I mean you just focus on the other motions and not on the internal motions.

6. p. 4 2nd column bottom: "If $\delta\ell(t)$, the *D*-*A* distance fluctuation from its equilibrium value, is not small, higher order terms in Taylor's expansion need to be included. Because of the *slow* ℓ^{-6} *distance dependence*, for FRET efficiency, one can explore a *D*-*A* pair at much larger separation (25–80 Å).

Q: Why is the ℓ^{-6} dependence considered as slow?

M: *Slow meaning that it is slow relative to if it were an exponential decay. Any time it's a power's law, it is long range compared with an exponential.*

M352 Determination of Energetics and Kinetics from Single-particle Intermittency and Ensemble-averaged Fluorescence Intensity Decay of Quantum Dots

Jau Tang and R. A. Marcus

1. p. 1 1st column bottom: "Ensemble studies, in which fluorescence intensity $\langle I(t)\rangle$ can be measured at much longer times (10^4 s or much longer), can be used to obtain complementary information not readily accessible from single-QD studies."

Q: Why one cannot study longtime behavior of *single* QDs?

M: The signal is too small at long times, the signal at long times becomes a tiny fraction of what it started to be. Noise is huge. It is too faint for single molecule.

2. p. 2 1st column top: "We consider $|F_2\rangle$ in Fig. 1(a) or $|1\rangle$ in Fig. 1(b) as a neutral dark exciton state, which has a total angular momentum projection J of $\pm 2\ldots$"

Q: Does a hole have angular momentum?

M: *Yes, sure, if you had a fluorine . . . if you had a neon atom, and if you took an electron away, you would have a hole there, you would have angular momentum . . . I mean, a hole is just the absence of an electron, and if you look at the wave function with the absent electron you would have angular momentum properties.*

Q: But intuitively a hole is something that doesn't exist in a sense.

M: *That's right, but when you think of it in terms of the entire electronic representation then you will find that entire electronic representation, that you also say has a hole, is really a group of electrons in different states, that electronic representation has angular momentum.*

3. Following: "These band-edge exciton states $|A_0\rangle$, $|A_1\rangle$, $|F_1\rangle$ and $|F_2\rangle$ in Fig. 1(a) are denoted in Refs. 8 and 9 as $|0^U\rangle$, $|\pm 1^U\rangle$, $|\pm 1^L\rangle$, and $|\pm 2\rangle$ respectively."

Q: Which is the meaning of the superscripts U and L?

M: I assume it is upper and lower, but I don't know. I assume but I don't know, It could be something else.

4. Following: "A rapid population recycling mechanism is induced by continuous photoexcitation...and the fast thermal process... between $|3\rangle$ and $|1\rangle$ (Fig. 1)"

Q: Are the symbols $\longleftrightarrow$ and $\rightleftharpoons$ appearing in the Figure equivalent?

M: I don't know why he has that double-pointed arrow going like that, maybe he thinks that it could be thermally populated, from that trap state, I guess that is a trap state F, but I don't think that there should be a double pointed arrow, I don't think it because it looks as if it is going too much uphill for it to be any kind of a probable process, it looks though that it is downhill, so why is that a double arrow I don't know. I suspect that it should be a single arrow. Let's see...if you look at that small perpendicular arrow below, at the end of the double arrow, does he have it going up and down there? A double arrow there, vertical?

Q: Yes, it goes up and down.

M: It shouldn't be going up and down...you know, that is thoroughly populating those states, that is too high up to be populated. I think, depends on what they are, *if states are very close then they can be populated one way thermally*, and so on, but if they are not very close then he shouldn't have an up sign there. But you see, it depends on what those states are, there are some phonon states which should be certainly thermally populated, and it may be that he is taking this F be *just a little bit below, some millivolts below, in which case you can go up*, if those upper states there are phonon states but if there the state is a $1P_e$ state instead of the $1S_e$ state, and that's about 0.3 eVs, you wouldn't populate those thermally.

Q: Does double pointed arrow mean thermal population?

M: Well, I would think so, yes. I think he's thinking of thermal population, of there going up and down, thermally.

Q: Are then the double pointed arrow and the two opposite arrows two equivalent symbols? Why did he use the double pointed arrow?

M: Here it would have been cluttered if he used two lines.

Q: OK, just a matter of graphics.

M: Yes, I think so. *The picture we have now is different from this, here the trap state is off the conduction band, now for various reasons we have it off the valence band.*

5. Following: "Although $|1\rangle$ or $|F_2\rangle$ is a *dark* exciton state in nature, the fast population recycling among $|1\rangle$, $|0\rangle$, and other optically active states ($|A_1\rangle$, $|A_0\rangle$ and $|F_1\rangle$) makes this state *appear light* in the relevant time scale of usual intermittency measurements (greater than milliseconds). For this reason, $|1\rangle$ is referred here as a *light* state."

M: Well, I guess what he is probably saying, I don't agree with that, I don't think, but anyways... I think what he is probably saying is that dark exciton state isn't living for a long, it is rapidly recycling, changing to light states, and so on, so if there is a rapid interchange between a light state and a dark state, that state would appear as light, *but I don't buy that actually, I don't buy that model*, but I understand it... roughly.

Q: I read on Chemical and Engineering News of some experiments done in Los Alamos on QDs using both electrochemical and spectroscopic methods.

M: I think Guyot Sionnest was the first one to do it, and then I think... at Los Alamos sort of carried it on, yes, that's right, I'm trying to remember whether they've got information on energetics having an extra electron number there, two extra electrons there, I forgot, there was interesting information that I forgot.

6. Following: "For the *charged* dark state $|2\rangle$ due to fast Auger relaxation from the higher excited dark states (Fig. 1)..."

Q: Why is the $|2\rangle$ state charged?

M: Maybe he is putting an electron there and there are several possibilities, and I don't know which one he means, but one possibility is that you have an electron in that trap, then you have a hole, in the rest of the dot, so the dot as a whole isn't charged in this case, but there is a charge separation with a charge in the edge of the dot, a surface state. *The picture we now have is rather different from this.*

Q: Do you remember the wonderful preface of the book of Thermodynamics by G. N. Lewis? There Lewis writes that "The labor of generations of architects and artisans has been forgotten, the scaffolding erected for their toil has long since been removed, their mistakes have been erased, or have become hidden by the dust of centuries."

M: I don't remember the preface but I remember the book, yes.

Q: We are discussing now things while they are being done.

M: Oh, yes, *ideas evolve and they are evolving all the time now.* For example, well, as a just most recent thing, there seems to be an anomaly that Bawendi found, that the nonradiative rate of this dark state seems to be very very fast compared with what you would expect by an Auger process. Looking at the light state, there seems to be a real anomaly and in a draft that one of my students is writing up, the work she has done, I asked her to really put it down for me what the anomaly really is. She did and then it occurred to me that when you have a trap state, with a *local* charge, and then *a delocalized one*, when you compare with a state of a biexciton where you have *two delocalized* charges, the big difference in those two electronic states is that since one charge is highly localized, the distances around it are going to be different from in the other, you have a large Huang Rhys factor, the potential energy curves are horizontally spaced from each other a lot, and that makes for a very rapid nonradiative decay, compared with the other case. So, you know, it is interesting how ideas evolve. This paper hasn't been published but I think it does explain what Bawendi thought an anomaly in recent work that he has done.

M352 Determination of Energetics and Kinetics from Single-particle Intermittency and Ensemble-averaged Fluorescence Intensity Decay of Quantum Dots

(Continuation)

Jau Tang and R. A. Marcus

1. p. 1, 2nd column: "In our previous treatment, for simplicity we neglected the *Stokes shift effect.* These effects are not that small for some QDs."

M: *In the excited state the atomic positions are only slightly different from that in the ground state, so the Stokes shift is very small. When you have some molecules instead, say in water or other polar solvents, when you have some dyes that in the excited state have a large charge separation, then you get a huge Stokes shift, but here, you know, the atoms are pretty rigid and it is very small.*

2. Following: "With the improved model, we correct the values for the free energy gap between the light and dark states, the activation energy, and *reorganization energy*..."

Q: Have you been the first to introduce the concept of reorganization energy λ in Physical Chemistry or to interpret the energy λ as reorganization energy? If not, who was the first? Is there a history of this concept and of its symbol?

M: I think so, at least people used λ, I think so, but, you know, surely I was the first to call it that, I think so.

Q: So concept, symbol and idea really.

M: Well, there is a Griffiths' paper that occurred around 1956, I think but I am not sure, I refer to it in some of the articles. Griffiths considered vibrations and a special model where you can get an energy barrier and you can extract a λ from that. The model would have the ingredients of reorganization energy, but I doubt if he called that, he certainly didn't use a symbol λ.

Q: Is there a reason why you decided to use a symbol λ?

M: Oh, *that's a good question*, no, no, I just used it, I don't know why, I've no idea, I just used it. Do you suggest I should have used a Roman letter instead of a Greek letter?

Q: No, no that's OK...

M: No idea, I had to call it something, and probably using a Roman letter would sort of have not making it special, I think there is no more than that. I never thought of it.

Q: I am your historian, you know...

M: Yes, you surely are... I should have been my own historian but...

3. p. 2 1st column bottom: "Although $|1\rangle$ (or $|F_2\rangle$) is a *dark exciton state in nature*, the *fast population recycling* among $|1\rangle$, $|0\rangle$, and other optically active states ($|A_1\rangle$, $|A_1\rangle$ and $|F_1\rangle$) makes this state *appear light* in the relevant *time scale of usual intermittency measurements (greater than milliseconds)*. For these reasons, $|1\rangle$ is referred here as a *light state*."

M: If you look at those four states roughly in time to the order of nanoseconds, and the time for going over to a dark state is longer than milliseconds, there is a lot of going on before there is any significant trapping, so you can regard those four states as being in a sort of a dynamic equilibrium and summarize them all, in some way, as a state $|1\rangle$. There are many cycles going on in the millisecond region, so there are lot of processes going in the short time scale, and that's probably what's I was talking about, you know, there is a lot of going on at much shorter times, a number of processes. You more or less establish a kind of steady state for those fast processes, even though everything is for slower processes changing with time.

4. Following: "In this model diffusion is light driven...For the charged dark state $|2\rangle$, *due to fast Auger relaxation* from the higher excited dark states (Fig. 1), *its diffusion is slower* and could have a *light-independent component.*"

"... the reaction between $|1\rangle$ and $|2\rangle$ is diffusion controlled."

M: *The dark state is dark because the fluorescence from that state can compete successfully with the Auger relaxation, that's why's dark. Now, whether the diffusion is light driven or not, is uncertain. At low temperatures the spectral diffusion was light driven, you can rescale things according to the intensity and all of points will follow on one line and that was in the first paper that Tang and I did, but that was at low temperature, we have no idea of whether it's light driven at room temperature or not. Bawendi might know, but I don't know.*

Q: What is the light-independent component?

M: *This would be the light independent component.* Yes, but, you see, we don't know if that is light dependent now, there are studies that are made at variable intensity but I'm not sure if they have been analyzed with that in mind, and that careful, and have analyzed how their effects occur.

5. p. 3 1st column middle: "Chung and Bawendi ... In their ensemble measurements of $\langle I(t)\rangle$ QDs are initially in darkness, presumably in the ground state. When light illumination was stopped after each cycle of $\langle I(t)\rangle$ measurements, the intensity recovered its *full intensity* after a long period (8000 s) until starting the next measurement cycle."

M: *Well, I think they have an ensemble, they shine light on it, they stop illuminating and watch the recovery. It takes 8000 seconds to recover, it takes about 8000 seconds until you reach the steady state.* I think so, the time to reach the steady state is about that time. So the ensemble experiments of Chung and Bawendi are very important.

6. Following: "Thus, we assume that QDs are in $|0\rangle$ initially in thermal equilibrium prior to each cycle of light illumination. Due to fast population recycling described above and *the presence of a Stokes shift between* $|3\rangle$ *and* $|1\rangle$, the excited population distribution is centered at $Q_{0,3}$ rather than at $Q_{0,1}$, the lowest point of $U_1(Q)$. *Here*, an initial distribution of QDs in $|1\rangle$ is assumed to have a Stokes shift as shown in Fig. (2b), after the initial transient period,

$$\rho_1(Q, 0) = (1/\sqrt{2\pi k_B T/\kappa}) \times \exp(-\kappa(Q - Q_{0,0})^2/2k_B T),$$

where $Q_{0,0}$ is the center position for the distribution."

Q: (1) What do you mean by Stokes shift between $|3\rangle$ and $|1\rangle$? Normally by Stokes shift one means the difference between positions of the band maxima of the absorption and emission spectra. Here then apparently because of Stokes shift in emission and absorption between $|3\rangle$ and $|1\rangle$ there is a shift in population between the two states. But is the energy transfer between $|3\rangle$ and $|1\rangle$ a radiative one? Please look at Fig. 1 and explain...

(2) Because of the Stokes shift "the excited population distribution is centered at $Q_{0,3}$ rather than at $Q_{0,1}$, the lowest point of $U_1(Q)$."

M: (1) Well, whenever anything relaxes, you know, something is excited *at the ground state geometry*, and then relaxes *to the excited state geometry*, that's a Stokes shift. In other words, *he is talking of the relaxation that you would have if you had a Stokes shift, then if that radiates, fluoresces, it is going to fluoresce at slightly different geometry, and that difference would be the Stokes shift.* Yes, the problem, one problem here is that *you're exciting from many different* *levels in the ground state, to many different levels in the excited state, so the absorption spectrum is very broad, because you're not just exciting from the highest level in the ground state, you have a whole group of other levels in the valence band, they're excited too...it makes the absorption spectrum very broad, but when it comes to the fluorescence spectrum, everything relaxes to one of the lowest states and now it can only radiate to the unoccupied states which are pretty much restricted to being right near the bottom of the band gap,...* anyplace in the valence band to any place in the conduction band, with some selection rules.

Q: The figure doesn't tell the whole story.

M: Oh, the figure doesn't tell you that.

(2) Yes, in other words after the excitation there may well be a small relaxation in geometry to a slightly different geometry that you have in the ground state.

7. p. 3 1st column bottom: "...where $\xi \equiv \frac{(Q_{0,1}-Q_{0,0})}{(Q_c-Q_{0,1})}$ and the Stokes shift $\Lambda = \frac{\xi^2 E_{A,1}}{2}$ ...with activation energies $E_{A,1} = \frac{(\lambda+\Delta G^0)^2}{4\lambda}$ ..."

Q: Please explain the above formula for the Stokes shift in terms of parabolic potentials, the Stokes shift parameter ξ and activation energy, that is, how and why the energy difference of the peaks of absorption and luminescence are mathematically expressed by that formula.

M: *Yes, now, as far as parabolic potentials...if you take two parabolas that are displaced from each other, and you have one above the other but displaced, OK?, now take a vertical distance from the bottom of the lower one to the other one and take a vertical distance from the bottom of the upper one to the lower one, and that difference of energies is the Stokes shift.*

Q: And the relation to the λ?

M: Yes, λ *is related to the Stokes shift.* I fact, if you think in terms of λ's, *I think 2 λ is the Stokes shift.* I think so, you can easily figure it out by drawing the things and by looking at the definition of the Stokes shift and express everything in terms of λ.

8. Legend of Fig. 3, p. 4 2nd column: "The lower potential W(Q) near the crossing has a round top and the electron transfer reactions become a Kramers-type reaction."

M: Well, a rounded top could be a Kramers' type potential. *Kramers' is where you take into account a slow diffusion in the vicinity, especially in the vicinity of the transition state. You really get a slow diffusion whether you're having intersecting parabolas or whether you're having a rounded top. And in fact I think if you look at Kramers' paper you'll see that he did both cases, if I remember right. He would call them nonadiabatic and adiabatic.*

9. Legend to Table 1, Note a: "Two possible solutions exist for λ from a quadratic equation, corresponding to the 'normal' and 'inverted' regimes."

Q: If the existence of the inverted regime is just simply related to the existence of the second solution for a quadratic equation why was it so difficult to accept for some people?

M: I don't know, you should have to ask those people... probably some of them didn't accept the whole formalism or something, I don't know.

Q: Because until now you never put the story in such a simple way.

M: Yes, basically that's what it is. They may not have accepted that a quadratic equation is involved.

Q: By the way... when I was at Illinois you happened to talk to Uhlenbeck. Which one of the other great theoreticians did you know?

M: I never knew Landau, Kramers, Schroedinger, Born, but Aron Kuppermann knew Born, I never knew Heisenberg, I knew Fritz London, and in fact I almost applied to become a postdoctoral, I knew Mulliken, of course I knew Pauling.

Q: In that meeting at Caltech in honor of Pauling you weren't in the picture in the book edited by Zewail...

M: You know, that upset Zewail, I was supposed to go for a picture and I completely forgot it and he was very upset, I'm sure he will never forgive me for that.

Q: There was also no contribution of yours.

M: But I wasn't a Nobel laureate, was I? Maybe I wasn't asked to contribute because Ahmed was upset.

10. p. 5 1st column middle: "The value for V_{ex} cannot be determined from our analysis of the data of Chung and Bawendi. Its determination and the question about the distance between the electron and hole in the trap state have to rely on *other means* that remain to be explored."

Q: Have you an idea about the possible "other means"?

M: Yes... the answer is yes. A little bit obscure though. *There's a transition in the slope, in the Jau paper, and that transition in the slope is related to the rate constant for transition from one state to the other, is related to Landau–Zener rate constant. If one knew the diffusion coefficient, and maybe one knows it roughly from spectral diffusion studies, although... that's too fast a thing to come through, but supposing for a moment one knew, then if one could locate that transition in the slope, then one would get that rate constant, that Landau–Zener rate constant.*

Q: And from that the coupling.

M: *Yes, that would give you the coupling. Well... if you knew the slopes in the potential energy.*

11. p. 5 2nd column top: "According to Eq. (7)

$$\Gamma_k \tau_k = E_{A,k}/2k_B T, \quad 2\Gamma_1\tau_1 - 2\Gamma_2\tau_2 = \Delta G^0/k_B T \tag{7}$$

the *bending rate* Γ_k is predicted to increase linearly with light intensity W."

Q: Why is Γ_k a bending rate? Bending of what?

M: Bending of the curve, the curve has $e^{-t/\Gamma}$.

12. p. 6 1st column top-middle: "According to Table I a *bending in* $P_{\text{off}}(t)$ for CdSe of 25 Å is expected to occur by 100 s at a light intensity of 2 kW/cm^2."

Q: How does one use the data in the table to determine the bending?

M: That Γ is the bending, yes. That comes from seeing where the curve bends. You know, $e^{-3/2}$, $e^{-3/\Gamma}$ ...

13. p. 6 1st column middle: "According to our previous analysis of spectral diffusion data, the second moment follows $\sigma^2(t)/\sigma^2(0) = 1 - \exp(-t/\tau_1)$. Because $1/\tau_k$ is linearly proportional to the light intensity W, a similar dependence of $\sigma^2(t)$ is expected on a *universal variable of energy density*, i.e. *Wt*, in accordance with earlier experiments"

Following: "For QDs with negligible Stokes shift, treated previously, a *universal dependence* is expected for the exponent and time constant on several kinetic and energetic parameters in the stretched exponential decay fit to ensemble-averaged fluorescence intensity. Such a simple universal dependence is expected to break down when Stokes shift effects are included."

Q: What do you mean by "universal variable of energy density" and by the above universal dependence?

M: By the way, this occurs with low temperature data, I don't think they measured spectral diffusion at room temperature. Well, yes, probably *it means that it depends on time according to the quantity Wt, depends on light intensity and on time, not in a random way but in the form of the product of those two variables, of course it depends on the parameters but in terms of the variables light intensity and time regardless of the dependence on W and regardless of the dependence on t, but universally depends on one quantity, Wt*. That was in a particular paper of Jau Tang's and I, and in one of the things we plotted, it depended on...I think it may have been the width or something, I'm not sure, but something that we had...in the width of something, that you could plot vs. time or vs. intensity, you could get a whole set of curves but you can get the curves to follow on the same line if you plotted vs. the product of the intensity W times the time. Universal means that it doesn't matter whether you vary W or you're varying time or you're varying both, if you plotted vs. Wt then you would get all the points to follow on the same plot.

14. p. 7 1st column middle: "The diffusion equation involving such an effective potential $W(Q)$ can be written as

$$\frac{\partial}{\partial t}\rho(Q,t) = \frac{\Delta^2}{\tau_L}\frac{\partial}{\partial Q}\left[\frac{\partial}{\partial Q} + \frac{1}{k_B T}\frac{\partial W(Q)}{\partial Q}\right]\rho(Q,t). \qquad \text{(A3)}$$

By defining a dimensionless variable q for the reaction coordinate with $q = Q/\Delta$ and a dimensionless potential $w(q) = W(Q)/k_B T$, one can express Eq. (A3) as

$$\frac{\partial}{\partial t}\rho(q,t) = \frac{1}{\tau_L}\frac{\partial^2}{\partial q^2}\rho(q,t) + \frac{\rho(q,t)}{\tau_L}\frac{\partial^2}{\partial q^2}w(q) + \frac{1}{\tau_L}\frac{\partial w(q)}{\partial q}\frac{\partial}{\partial q}\rho(q,t). \tag{A4}$$

The diffusion-controlled reaction in the lower W-shaped potential $W(Q)$ is essentially equivalent to the Kramers reaction."

M: Well, maybe to the Kramers' equation that Kramers used to the overdamped limit to treat the reaction rates.

Q: Why do you change from Eqs. (A3) to (A4)?

M: *Well, whenever you can you introduce clear variables, dimensionless variables.* I think that's what we tried to do, I think they're dimensionless, I'd have to look how $W(Q)$ is defined. *The main point is to have as few constants in there as possible left over, like if you have a* Δ^2 *and a* τ *and now you just have a* τ.

15. Following: "If the mixing due to electronic coupling is strong, the states near the round top actually represent a quantum mixture between the light and the dark states."

Q: Is this phenomenon analogous to the fact that when you have strong coupling between the *R* and *P* states in ET you don't really have an abrupt ET but rather a gradual charge density change in going from one state to the other?

M: *Yes, I think that's true*, of course if you have some interaction between two things you can't say it is just one or just the other, you have something between. I am not sure what the process is here but in one case it may be electronic state where the system is in a bright state and the other electronic state is a dark state, namely there has been an electron transfer to a trap.

16. p. 7 2nd column top: “The blinking statistics $P(t)$ is equivalent to the probability flux at the round top of the Kramers potential.”

M: I suppose what that is, that is really sort of *rate of change of the survival probability, there should be a better way to describe it*. . . because you see, *the probability flux . . . the flux is sort of a rate of change of probability too*, so it may be that they are actually equal, I haven’t thought about that, but whether that is a rate of change survival probability, that is it.

CHAPTER 12

Summarizing Lecture: Factors Influencing Enzymatic H-transfers, Analysis of Nuclear Tunneling Isotope Effects and Thermodynamic versus Specific Effects; Enzymatic Catalysis and Transfers in Solution. I. Theory and Computations: A Unified View; H and Other Transfers in Enzymes and in Solution: Theory and Computations: A Unified View. 2. Applications to Experiment and Computations

Interviews on M353, M354, M357

M353 Summarizing Lecture: Factors Influencing Enzymatic H-transfers, Analysis of Nuclear Tunneling Isotope Effects and Thermodynamic vs. Specific Effects

R. A. Marcus

NOTES

1. p. 1446 2nd column top: "In the case of an $\alpha - type$ path the deuterium/tritium KIE is attributed primarily to the difference of zero-point energies in the reactants and the TS"

Q: Why deuterium/tritium and not just H/D?

M: Because people in enzymes, some people in enzymes, are very conscious of the importance of doing deuterium and tritium experiments, that plays an extra constraint, and there is something called the Swain equation, by Gardner Swain, where he explained the relation of the KIE for H to D to T. So, some people are very conscious of that, of using all three, H, D, and T.

2. Following: "an A-H stretching vibration has disappeared in forming the TS AHB†, and bending vibrations have been modified."

Q: (1) Is the stretching vibration disappeared just because A-H is disappeared and AHB has been formed?

(2) The bending vibrations refer evidently to bending vibrations of AHB. What do you mean by "have been modified"?

M: (1) Well, there's no longer a simple vibration of H with respect of A, in other words you have here a concerted vibration involving an unstable system, involving H going from A to B, it is an asymmetric stretch now, before it is just a stretch of A-H, that has become an asymmetric stretch, that of A-H-B.

(2) It means changed. The way nature is changed, you know, the H now is bending in a A-H-B someway.

Q: But when you have only A-H there's no bending.

M: A stands for a complex system, not an atom. Maybe, say, the oxygen and everything that is attached to it, or a carbon and everything that is attached to it.

3. Following: "In this case, the maximum KIE without tunnelling is perhaps a factor of about 7... at room temperature. The KIE is maximum at $\Delta G^0 = 0$ and so the maximum KIE for an $\alpha - type$ path may be less than 7, depending on the value of ΔG^0."

M: Supposing that the ΔG^0 was so negative that the transition state looks like the reactants, then if the transition state is looking like the reactants there is essentially no isotope effect. Similarly, if you go the other way around thinking from the products point of view.

Q: There are a lot of subtle things in kinetics...

M: Subtleties in everything.

4. Following: "In a $\beta - type$ path, unless β is close to α, the light particle (H) typically tunnels from the reactants' well to the products' well, and the KIE can considerably exceed 7."

M: Yes, it can even be a 100 or 200, anything.

Q: Why so?

M: Because the barrier depends exponentially on the mass, tunneling.

5. p. 1446 2nd column bottom: "For any given total energy E and H-vibration quantum state, the corresponding classical trajectory for a bound quantum state fills the *distorted rectangle* in figure 2..."

Q: Are the distorted rectangles just convenient approximations for the potential wells? How does one choose one among many possible distorted rectangles?

M: No, it's actually the region that describes the trajectory, that the system in that potential well will occupy. It is the semiclassical mechanical treatment that will go in that potential well.

Q: How does one choose one among many possible distorted rectangles?

M: You start out with the one at the zero-point energy and start anyplace where the energy is the zero-point energy, and then you can start drawing, but of course you are going to get different curves depending upon what the quantum number is for the two vibrations. There are ways of doing it, *to choose trajectories that correspond to the quantum numbers*, and all of that, a lot of that, is in the semiclassical papers that I wrote with Don Noid and others, back around 1970s.

Q: One should write a third book about that.

M: The third book will be a book in heaven ...

6. p. 1447 1st column top: "The amplitude of the side of the 'rectangle' nearer to the origin is larger than that of the opposite side, because of the decreasing H-vibration frequency at fixed $\mathcal{F}_\theta$ as the system moves closer to the TS region (decrease in R)."

Q: Is it so because the closer H moves to the TS, the larger is the distance between the points of motion reversal and so the oscillation frequency decreases?

M: If you are talking about the transverse motion, yes, if you are talking about the other reaction coordinate, no.

7. p. 1447 1st column middle: "For the given *H-quantum state*..."

Q: (1) Is there a quantum state *of H alone* or a vibrational quantum state of the H-A molecule to which H belongs? Or maybe when one considers the motion of H inside the complex $A - H \cdots B \to A \cdots H - B$ along a chord or an arc between, say, X and Y on Figure 2, one considers A and B fixed and just H moving?

(2) The above motion is a translational motion of H moving from A to B or a particular vibrational motion of the AHB system?

M: (1) No, not the quantum state of H alone that is a convenient way of describing it.

Q: Or maybe when one considers the motion of H inside the complex.

M: Well, it depends, normally one thinks of the transverse motion, transverse to the reaction coordinate.

(2) Well, since it's so distorted, you know, it's like going along a double well in effect, you will call that *a vibrational translation*, you just call it a motion. *That because it's sort of vibration initially and then when that thing at the top gets really flat is like a translation, and then you reverse to an anharmonic vibration on the other side.*

8. Following: "and the given energy E, the wave function decays exponentially outside these distorted rectangles. In wave mechanical terms, the boundary of the distorted rectangle forms the 'caustic.'"

M: Yes, that is where, beyond the caustic, the wave function decays exponentially to zero, and all of that is part of semiclassical theory of bound states. It is understood in terms of Kramer's theory. In fact we did a lot of that sort of theory in 1970s with Noid in the semiclassical work.

Q: But why "caustic," a concept that comes from optics?

M: That's what they are called, caustics, it actually comes from light waves, and it is where some light wave sort of suddenly becomes dark, sort of there is light on one side and it is almost dark on the other, it is called the caustic.

9. Following: "The system tunnels in figure 2 from any point X or X' on the side of the distorted rectangle nearest to the products' well to a *corresponding point* on Y or Y' on the nearest side of the distorted rectangle in the products' well... The tunnelling analysis is made coordinate independent by *choosing the path from X' to Y' variationally*, instead of *along an arc of θ or along a chord*."

Q: Are the paths along an arc or a chord just the most simple to consider or are they also the paths of highest tunneling probability?

M: They're the most simple ones to consider, that's right.

Q: Not the ones of highest probability?

M: They're close to each other, they're probably the ones of highest probability, starting at the end points, but you have a whole distribution of paths.

Q: Is there some relation between semiclassical mechanics with all these paths and the Feynman formulation of Quantum Mechanics?

M: Probably, except not all possible paths but paths that minimize something. *Feynman considers all possible paths*, here one considers going from one of those rectangles to the one on the other side, by a whole bunch of possible paths.

Q: So, in semiclassical mechanics you consider only a subset, somehow.

M: Yes, and you sort of would calculate some distribution about the subset... It depends, even in semiclassical mechanics if you are using what is called the integral representation, where the reaction probability, or really S-matrix, is written as an integral, then you are considering a whole group of paths, what you do in any wave function.

10. p. 1447 2nd column top-middle: "If $P(E)$ is the probability of an H-transfer for any given initial *H-quantum state* and total energy E, then the *unimolecular rate constant* k_{rate} for this bound AH-B system..."

Q: The bimolecular H-transfer reaction is the AH + B $\rightarrow$ A + HB. Are you here considering instead the unimolecular reaction $A - H \cdots B \rightarrow A \cdots H - B$ which happens in the TS?

M: *Yes, but they're essentially fixed in position, partly moving, it is not as if you have two objects coming from afar colliding, is sort of* *fixed, almost semifixed in position, so it is like a pseudo unimolecular process.*

11. Following: ... is obtained by integration over all E ...

$$k_{\text{rate}} = \int P(E)e^{-E/kT}dE/hQ, \tag{2.1}$$

where Q is the product of the partition functions for the R-vibration (Q_R) and for the H-motion (Q_H) in the reactants' well. If we write Q_R approximately as a classical harmonic value, $\frac{kT}{h\nu_R}$, and Q_{H} as $\exp(\frac{-E_0}{kT})$ when the H is in its lowest quantum state in the reactants, E_0 being the zero-point energy of the H-motion ..."

Q: When you write of the zero-point energy of the H-motion you implicitly consider it to be a vibrational motion. Why the R-vibration and the H-motion have different forms of the partition functions?

M: *The R motion itself involves principally a motion of A with respect to B,* right? And yes, *that is sort of a vibrational motion, but very heavy, with very little zero-point motion for that, and then the H follows a symmetric sort of stretch, with the H going from the A to the B, so that is a real H vibrational motion, but for the A-B part, when A is moving toward B, H is playing a very minor role there*, as far as the masses are concerned, it is mainly the reduced mass of the AB that plays a role.

Q: Why do you have those two different forms of the partition function?

M: It depends on where you are, if you are in the initial reactants stage, *the initial reactant stage when the A and B are fixed and don't move, you just have the H motion.* But now if you have a little A-B motion, if it's sort of vibrational initially, and then it goes over to a kind of translation, then there is going to be a little zero-point motion associated with it too, but, you know, *the R motion, which is the motion of A with respect to B, is very different from the H motion, of H with respect to A.*

Q: And why does one have the form $kT/h\nu_R$ and the other one is exponential $\exp(-E_0/kT)$?

M: The exponential is right from the bottom of the well to the TS. Are you talking about the E_0 being the zero-point motion? Well, it depends on details. If for example, you're talking about the system not tunneling but going over the top of the barrier, then there is no zero-point energy there, except the transverse motion, but there is no zero-point energy associated with what is formally the A-H vibration becoming translation. $kT/h\nu_R$ that's the classical harmonic partition function.

Q: And the $\exp(-E_0/kT)$?

M: I don't know why I wrote that, that term E_0 is the energy barrier to go from the zero-point energy of the H motion to the top of the barrier, in the TS. That is the contribution of the zero-point motion, I don't know why I had that...

Q: Q is the product of Q_R and of Q_H

M: The Q_R *is treating the particle as moving not only with its regular vibration but in a kind of a well.* So that is what that Q_R is. And now the Q_H, *that's the vibrational partition function of the H-motion in its lowest quantum state where you start off at zero-point energy in it.* When the reactants system is in its lowest state, its partition function is $\exp(-E_0/kT)$. The partition function $kT/h\nu_R$ there for the R motion is like that for a little vibration.

12. p. 1448 1st column top-middle: "For any given $\mathcal{F}_\theta$, the higher the E, the more easily the system can reach smaller Rs, an effect that *contributes to the activation energy*."

Q: In the sense of decreasing it?

M: It would actually go in deeper, it needs more energy, but the effect of the tunneling factor is higher, so there's a compromise.

Q: You mean you go higher in energy but at the same time the tunneling becomes easier.

M: Yes, that's right.

13. Following: "In addition to the integrand in equation (2.3),

$$P(E) \cong \int_{R_X}^{\infty} 2e^{-K(R)} \nu_{\mathrm{H}} dR/\dot{R}(R) \tag{2.3}$$

There is also, for a triatomic collinear system, an *off-resonant factor*, given in Klippenstein *et al* (1986). This factor introduces into the rate a factor of approximately 0.6 in the systems studied."

Q: Please describe this off-resonant factor...

M: *Let's look at that β path, to the zero-point energy, let's go from zero point to zero point. The zero-point energy in the final well may be different from the zero-point energy in the initial well, unless it is symmetrical, that means you're going to go off-resonant, because your vibrational energy is not the same before and after, it has to go someplace else, go into the R motion of the new well, that is called off-resonance.*

14. Following: "These off-resonance energies Δ are not very large, since protonic states in each well at any *R* are relatively closely spaced compared with the several electron volt spacing of electronic states."

M: In the electronic problem you can get big differences in ΔG^0, big difference of energy in the electronic states, in fact if you think about going to an excited state instead of the ground state of the products, that's a couple of electron volts higher. In the case of the protonic states though, if you start off at one protonic state and you go to a different protonic state of the products, they are just separated by, you know, 5 or 6 kcal, instead of 100 or 50, so the product states, the protonic states, are much closer to each other than the electronic states, so there's much less chance of an inverted effect. *So, the energy* *differences are not that important* and there is less chance of getting the inverted effect. *You get the inverted effect because the electronic states are widely separated*, so if you have something that is very very exothermic, you can't easily access the ground state, so you go to excited state.

15. Following: "They are less significant when there are many other coordinates, as in a protein, since the *coordinates can 'reorganize'* so as to make Δ very small."

M: Basically, what happens in the electron transfer case, when *you get* intersection, same thing here for the protonic states. You have different potential energies and different protonic states, if you don't go into one you go into another.

16. p. 1448 1st column bottom: "Where... $w^r(R)$ contains the '*gating term*' for the reactants plus any change in the (e.g. conformational change) in the protein-substrate complex that cannot be overcome by a favourable ΔG^0"

M: Well, gating term is a term which is sometimes used in the past twenty or so years, when something has to rearrange in some way or another, like maybe reorient in order for the reaction to occur.

17. Following: "In equation (2.5)

$$P(E) = \int 2\nu_{\mathrm{H}}(R)e^{-K(R)}e^{-\Delta G^*/kT}dR/\dot{R}(R) \tag{2.5}$$

we have neglected the off-resonant factor, whose effect is reduced by the reorganization."

M: Yes, that's right, *if you think of the electron transfer problem and the free energy curves, if you transferred at any point other than the intersection, it would be an off-resonant transfer.*

M353 Summarizing Lecture: Factors Influencing Enzymatic H-transfers, Analysis of Nuclear Tunneling Isotope Effects and Thermodynamic vs. Specific Effect

(Continuation)

R. A. Marcus

01: p. 1445 2nd column top: "We are also reminded in several articles that for a multi-step reaction the H-transfer step need not be the rate limiting one, and so may cause an H/D kinetic isotope effect (KIE) *on the rate* to be closer to unity than a single-step mechanism."

Q: I believe that by "rate" you mean the overall rate. Is it so?

M: The answer is yes.

02. p. 1445 2nd column middle: "...environmental effects, in the form of the 'reorganization energy'...influence the rate, as do factors controlling the distance of closest approach of the heavy atoms participating immediately in the H-transfer (sometimes called *gating*)..."

M: Gating is not a concept that I introduced, it's a concept that others introduced and I think the idea is this: that in addition to the reorganization, in addition to the tunneling, there is also, you may call it *preparation*, something like having to rotate in order that the reaction can occur, and that sort of rotation, that sort or reorientation, is independent of the free energy of reaction ΔG^0, so it's what I called the work term, the w^r term,...it occurs beforehand, it has to occur before the reorganization, so I think that's what they mean by "gating." This answer is just guessing. It's an answer about something that I know nothing about.

03. Following: "The reorganization serves to bring the H-energy levels immediately before and after the H-transfer rate into resonance (same total energy)."

Q: I have some general questions:

(1) Consider the gas phase H-transfer reaction: $H_\alpha - H_\beta + H_\gamma \rightarrow H_\alpha + H_\beta - H_\gamma$. There the H-transfer happens when reactant and product systems have the same energy at the TS. Your discovery of the reorganization energy in solution extends then the resonance requirements to reactions happening in solution through the participation of the solvent to establishing the resonance condition prior to the transfer. Is it so?

(2) Is there a fluctuation frequency of the solvent so that one can say that for a given ET transfer in solution there is a certain number of possibilities for it to happen in a second, that is, a certain number of times per second when the resonance condition is fulfilled by the suitable solvent fluctuations? If so, how are such frequencies calculated?

(3) Please think of the three Pauling's resonant structures for benzene, two Kekulè forms of equal energy ("perfect resonance") and a "Dewar" structure of different energy. Pauling says that the "resonance" (what Herschbach calls "Pauling's resonance"...) is maximal for the Kekulè's forms and lower between a Kekulè's and a Dewar's form. Can we say that in the case of the perfectly resonating Kekule's forms there is a higher probability of *electronic structures transfer*?

(4) Can the Pauling's electronic structures be considered diabatic states?

M: (1) Let me answer your first question there, the products and the reactants only have the same energy at the TS in a nonadiabatic reaction, in the case of an adiabatic reaction you don't distinguish the two, it is a new species, do you see what I am saying?

Q: No, please explain it better.

M: OK: *when it is a nonadiabatic reaction you use diabatic curves and then you can speak of reactant state and product state and the* *transition state, they are just simply different things. But if you have an adiabatic reaction you no longer have a reactant state and a product electronic state, you have some mixture. So you can't say in the adiabatic case that the reactants and the products have the same energy.*

Q: Because you are on one single surface

M: *That's right, you no longer have two intersecting surfaces.*

(2) Instead of having solvent frequency that is in high frequency regime, you speak of a solvent relaxation. Yes, when you bring in solvent, normally you don't speak of a solvent frequency, or maybe you do in a very crude way. Solvent sort of decays, it fluctuates, but a frequency implies an oscillatory motion, which it doesn't have, except in the very high frequency regime, so you speak of fluctuations but you speak of *relaxation of the fluctuations*, rather than of a fluctuation frequency.

Q: But aren't there a certain number of fluctuations, on the average, every second?

M: Well, it is highly irregular.

Q: Does it make any sense to think of an average frequency in which you reach the resonance?

M: No, I guess not, it is really like in Kramers' theory, if you think of going over the barrier, and if you think of solvent relaxation, if you happen to be in that regime of solvent fluctuations, the dynamic part is slow, then the so many fluctuations per second just don't enter in the case of the solvent, that concept isn't there. If you look at the equations involved, you speak then of the system diffusing to that region, and there is a certain relaxation time associated with diffusion, but if something is diffusing you don't normally speak of fluctuations. I mean it is true that one model of diffusion is to go from site to site with a certain average frequency, but that is a sort of model, you know.

(3) That is a thought experiment because you don't have the experiment in which you would start with one type of Kekulè form and measure the frequency of the system going to the other forms, it is a purely thought experiment, so if you think of it purely as a thought experiment, then the system going to the other Kekulè form would be a nonresonant sort of transition, the system going to that other dual form would be an off-resonant transition.

(4) Yes, *each of those would be a diabatic state,* that's correct, of course that is not a great description, is it? But it is a simple description, each of those are diabatic states.

04. p. 1445 2nd column bottom: "We . . . first introduce in §2, a coordinate-free semiclassical view of nuclear tunnelling in multidimensional systems."

Q: If you look at your Figures 1, 2, and 3 one sees graphs with coordinates. In which sense then is the description coordinate-free?

M: *Let me focus first on the quantization, at each of those regions. If you calculate the* $\int pdq$ *going from one side of those rectangles to its opposite side, and making a cycle, then it doesn't matter in what coordinates you calculate, the result is going to be the same. You can go diagonally or follow other curves, the results are the same. Similarly, if you calculate a tunneling integral, from one point on one of those curves to another point on one of those curves, it wouldn't matter what line you took, you get the same answer. So, I mean that, of course, you have to use coordinates there and the whole thing is in coordinates, but certain results are invariants.*

Q: Semiclassical invariance?

M: *Yes, like for example if you took the $\int pdq$ going on a complete cycle, say transverse, in either direction of those rectangles, take one of the rectangles, that will give you a quantum number for that state.*

05. p. 1447 2nd column bottom: "For small tunnelling probabilities, the tunnelling can occur in both the *forward and backward direction*, and so a factor of 2 is then introduced into equation (2.3) below.

$$P(E) \cong \int_{R_X}^{\infty} 2e^{-K(R)} \nu_{\mathrm{H}} dR/\dot{R}(R) \tag{2.3}$$

$$K(R) = 2\int_{\theta(X')}^{\theta(Y')} |p_\theta| d\theta/h \tag{2.4}$$

Q: Forward and backward direction of what?

M: I saw that and I think I got confused there, because I think there I was thinking of something that Hsu did, where she treated the problem for sort of resonance situation where she had a rate coming in both channels, and coming out, in other words she used a symmetric and asymmetric combination of adiabatic descriptions, for the two channels, and there you have forward and backward, so I think somehow I must have been looking at that and got confused, attractor of two would come in, into something like that Hsu has treated, for her thing, but that is where she is using both channels for the entrance and getting symmetric and asymmetric combinations, where Hsu is going one channel and coming out the other, and there is no factor of 2. *You know, I hope that when you write this book that you'll have something, appendix or whatever, that smooths the errors*, and this should be one of them, and you can also mention that confusion.

Q: After all I congratulate with myself that with this extremely careful reading of your papers I'm finding something interesting after all . . . .

M: Yes, definitely.

Q: Modern scholars reading Einstein's papers have found mathematical errors there . . .

M: I had a colleague of mine who was also a professor at Brooklyn Poli, he died long time ago, and he found an error in one of Einstein's papers.

Q: When Galileo was speaking with the Pope Urban VIII, the Pope was an extremely intelligent man and he was very curious about Galileo's ideas, and there was one topic though in those discussions where Galileo's ideas about the tides was completely wrong and the Pope was right.

1. p. 1448 2nd column middle: "One reaction coordinate for the entire system is the energy difference ΔE of the energies of two electronic valence bond states, introduced initially for electron transfer (Marcus... Warshel...) and is used as a reaction coordinate for H-transfers here... In Warshel analysis, the *non-separability* is treated using a one-dimensional QM treatment for H and a classical treatment for the remaining coordinates."

Q: How about this "nonseparability"?

M: I don't know why I used the term nonseparability... but really *for the H-transfer you don't have intersecting curves*, those are the diabatic curves, and the fact that you can't do that is the reason why, I guess, I called that a nonseparability but I don't know why I called it that.

Q: Why can't you have diabatic curves for the H-transfer?

M: *You can have them but that's not what the motion is on, the motion is really on an adiabatic curve, and so is not really on a nonadiabatic.*

2. p. 1448 2nd column bottom: "The reaction coordinate more closely related to the MEP..."

Q: What does "MEP" mean?

M: Minimum energy path, I think. Coltrin and I discussed it.

3. Following: "Use of the reaction coordinate ΔE will result in some recrossing of the dividing surface (surface for which $\Delta E = 0)\ldots$"

Q: Are there choices of reaction coordinates for which there are no recrossings?

M: *That would be the best choice of a reaction coordinate, and there is no guarantee that it exists, the one with the fewest recrossings would be the best choice.*

4. p. 1449 1st column top: "Wigner . . . pointed out that a key assumption in classical TST is that there is a hypersurface in the full *space phase across which there are no recrossings of classical mechanical trajectories.*"

M: Well, *if classical transition state theory be valid, you could find such a hypersurface, it may not exist though. I mean, if it exists it's extremely complicated.*

Q: But it would be the best.

M: *And be extremely complicated.*

Q: Because there would be no recrossings

M: *Yes, it has to be a surface with no recrossings. Now, usually recrossings are few, and so TS theory works pretty well*, but . . . , *for example, if you go to a regime* where the reaction rate *doesn't just depend on the energy barrier but depends on a kind of diffusion constantly crossing the barrier. When you have diffusion you go back and forth many times so you are crossing the TS many many times there and the rate is less than than what it would be for pure TS theory.*

5. Following: "In some simple double well problems, there could be many recrossings of the H-motion though *they are much less likely in the tunnelling regime.*"

Q: Why so?

M: If you go cross once in the tunneling regime, there is a small probability of crossing, so usually there is less chance to going back, but if you are going over the barrier then it is easier that you can go back too, it depends on the grade of loss to other degrees of freedom.

Q: But why in the tunneling regime is difficult to go back?

M: *Because to go back would have to tunnel again, another small probability.*

6. Following: "No separation of variables is needed and none was assumed by Wigner in the classical mechanical form of TST. However, because of the extension of the relevant wave function of the TS outside the hypersurface (uncertainty principle) in the QM or semiclassical description, one needs to treat the dynamics of several coupled non-separable coordinates, as, for example, in equations (2.3) and (2.4). Fortunately . . . in these equations . . . one can modify the TST so as to include this non-separable aspect."

M: *Wigner didn't say anything about separation variables, he didn't have to, so, if you don't have recrossings TS theory is always valid, without assuming any separation of variables you show that it is valid.* There's a simple argument about motion in phase space really. . . . *Now, that is classical.* And *in quantum* you can't use that description, so you have to use a different description for crossing that region, so Wigner's statement doesn't really imply the quantum description, but *you could use sort of major local quantum treatment of the various degrees of freedom there*. Miller has done this, and *has calculated actually the dynamics of crossing the surface, and so you can treat it that way and make a kind of correction to TS theory that way*, but for doing it right you are doing a full blown *calculation in that region for the flux that crosses that region, and for that you don't need to assume that the reaction coordinate is separable from the other coordinates.*

7. p. 1450 1st column middle: "In addition to the $k_{\mathrm{H}}/k_{\mathrm{D}}$ ratio, there is the ratio of the corresponding Arrhenius pre-exponential factors $A_{\mathrm{H}}/A_{\mathrm{D}}$. At sufficiently low temperatures, the reaction occurs by tunnelling from the zero-point state of both H and D, and so $A_{\mathrm{H}}/A_{\mathrm{D}} \gg 1$ and the activation energy difference, $\Delta E_{\mathrm{act}}^{\mathrm{HD}} = -k \mathrm{d} \ln(k_{\mathrm{D}}/k_{\mathrm{H}})/\mathrm{d}(1/T)$ is close to zero."

Q: (1) Can you please give me a short tutorial on the preexponential factor? From your words it clearly includes tunneling. How is it included in A and what else is included in it?

(2) Moreover, which one is the relation between tunneling, activation energies difference and zero-point energies?

M: (1) Let's speak of lowest temperatures, down to zero Kelvin. The reactants are going travel from the zero-point energy level and they are not get to do any climbing over the barrier, they are just going to tunnel over to the products. *So, since they're not going to climb over the barrier, they're just tunneling, there's zero activation energy for that.* Then you are going to get a certain tunneling factor, and since the activation energy is zero there, you can change the temperature and so going from the lowest state, if you are going up temperature, the rate constant is actually A_H, the preexponential factor. There is no exponential factor.

Q: Just A_H.

M: Yes, and similarly for A_D, they are not going to change with the temperature, at very very low temperatures, and so the activation energy, for the difference of H and D is zero. It is really so, if you plot the rate constant vs. temperature, at very low temperatures the rate constant is almost constant. Then you start getting an activation energy, and there the rate constant starts increasing with the temperature increase, but at low temperatures, it is constant.

Q: But in the preexponential factor is there only the tunneling factor?

M: There are only some kind of entropic contributions, the rearrangements, entropic rearrangements.

Q: It is after all a rather complex quantity.

M: Yes, it can be complex because it involves some changes in entropy that comes in.

Q: Which is the general relation between tunneling, activation energy and zero-point energies?

M: Well, when you're only tunneling you have no activation energy, but there is no general relation between those things.

Q: By the way: You speak here of uncertainty principle. I have a curiosity for you. Did you know Asher Peres? The guy was a graduate student of Nathan Rosen.

M: I took a course of Nathan Rosen at Carolina.

Q: He proposed for a Nobel Prize in Physics a general of Israel, Ariel Sharon. He hates the concept of uncertainty principle.

M: Einstein didn't like it either.

8. Following: "At somewhat higher temperatures, *the reaction for D will be over-the-barrier, but H still tunnels* and so $A_{\mathrm{H}}/A_{\mathrm{D}} \ll 1$ and there is a large $\Delta E_{\mathrm{act}}^{\mathrm{HD}}$ for $k_{\mathrm{D}}/k_{\mathrm{H}}$."

Q: Isn't the reaction for H also over the barrier as for D? And why $A_{\mathrm{H}}/A_{\mathrm{D}} \ll 1$?

M: H can tunnel more easily than D, so it doesn't have to spend the energy to go over the barrier as much.

Q: So, tunneling is prevailing for H.

M: Well, it depends on the conditions, you can imagine some condition where that is the case, and other conditions where the H is going over the barrier if you have high enough temperatures.

Q: But why $A_H/A_D \ll 1$?

M: Because if the D is not tunneling A_D is sort of normal, maybe 10^{12}, 10^{13} per second for unimolecular processes.

M353 Summarizing Lecture: Factors Influencing Enzymatic H-transfers, Analysis of Nuclear Tunneling Isotope Effects and Thermodynamic vs. Specific Effects

(Continuation 2)

R. A. Marcus

1. p. 1450 2nd column bottom: "*Examining the classical trajectory or trajectories* corresponding to an initial quantum state, this three atoms or so cluster with environmental configurations that permit the energy of the products' state to equal that of the reactants' just before and after the H-transfer, can provide *additional insight into tunnelling in several dimensional QM calculations.* If one uses three coordinates in the quantum mechanics such as the three coordinates of the H-atom . . . one uses three in the corresponding semi-classical analysis . . ."

M: Well, that may refer to some work of Sharon Hammes-Schiffer, she does things quantum mechanically but by having the H atom treated quantum mechanically, not the other atoms. There are three coordinates for that, I don't know if that is what I meant or not.

Q: But what about this additional insight into tunneling?

M: If you draw a picture like I've drawn there, it shows you where the likely tunneling object is coming from, using semiclassical ideas. Well, first of all it is going to have to roughly conserve energy, so you can go from one rectangle on one set of contours to another rectangle that is of the same energy, you get different rectangles for different energies, and so that will give you a picture of where the tunneling is dominantly occurring from one corner to the other corner. Not exclusively, but a lot of it is occurring there, so that gives you a picture but you have to know the background...

2. p. 1451 2nd column middle: "As many authors have recognized in discussing electron transfers in proteins, *the protein serves to reduce the 'reorganization energy.'* The same effect occurs for H-transfers in proteins."

M: *That should refer to a reaction in a protein compared with reaction in water, the reaction in water has a lot of reorganization energy, and Warshel for one has pointed out that if you don't have much water around, as you don't inside a protein, then you don't have as big a* λ.

Q: Why so?

M: *Think of it in terms of a strongly polar medium, whenever the medium is strongly polar you have a big reorganization energy, water is a strongly polar medium, a protein is much less polar.*

3. p. 1452 1st column top: "To explain the trends and why the two KISs change over the same pressure range, we first recall that all masses in AHB contribute to the effective mass for the tunnelling or for going over-the-barrier, not just H. A change in the mass of either A or B changes the contours of the mass-weighted diagram in Figure 2, the distorted rectangles there, and the acute angle. *Phrased slightly differently, it is not just the H that tunnels or goes over the barrier. Since the nearby atoms contribute to the 'effective mass' they do also...*"

Q: Is it so that what does go above or below the barrier is really the point in configuration space representing the whole system?

M: The two KISs change... *Yes, that's right*, it's dominantly the motion of the H, but there are contributions from the other masses.

4. Following: "... but their contribution is considerably smaller, a KIE of a few percent instead of a factor of the order of 10 for deuterium! So, it is perhaps not surprising in hindsight that the ^{13}C and deuterium KIEs change in the same pressure range."

Q: How do you measure the contributions of the different atoms? By their contribution to the effective mass?

M: You look at the rate of ^{13}C compared with ^{12}C, you just measure the reaction rate with ^{13}C, you substitute one of the carbons with ^{13}C, measure the rate, that gives you the isotope effect.

5. Following: "An interpretation of the decrease in the KIEs with increasing pressure is that an increase in pressure favours a smaller R and so increases the ratio of α to β paths, thereby reducing the tunnelling and so decreasing k_H/k_D, and since all masses contribute to the effective mass, decreasing the k_{12}/k_{13}."

Q: What is the k_{12}/k_{13}?

M: ^{12}C compared with ^{13}C.

6. Following: "In passing, we note that pressure can also affect each H-transfer k by affecting ΔG^0 for the H-transfer step (*the internal* ΔG^0), as in Equations (2.5) and (2.6)."

$$P(E) = \int 2\nu_H(R)e^{-K(R)}e^{-\Delta G^*/kT}dR/\dot{R}(R) \qquad (2.5)$$

$$\Delta G^* = w^r(R) + (\lambda + \Delta G_R^0)^2/4\lambda \qquad (2.6)$$

Q: What do you mean by "internal ΔG^0"?

M: *There's a certain overall* ΔG^0 *but part of it may be associated with bringing the reactants together, and another part of it associated with separating the products, so the internal is the difference of that. Yes, corrected and uncorrected for work terms.*

One corrected for work terms is the internal and the total ΔG^0 *is the one that has the work terms in it.*

7. Following: "This ΔG^0 can affect the KIEs when there is *a differential effect for different isotopes*."

M: *If you know* ΔG^0 *for one isotope and you know it for another, if they are the same there won't be any* ΔG^0 *effect, if they're different that will affect the rate.*

8. Following: "A determination of *the* ΔV^0 for the H-transfer step (*the internal* ΔV^0) *for each isotope* would be of interest in understanding the pressure effect."

Q: What exactly is meant by ΔV^0?

M: ΔV^0 is a change in standard molar volume in a reaction. That'd be corrected for the work terms dependence on V. In other words, what counts is the whole reaction is to be corrected by the work terms, if you are looking at some volume change you take the volume change and you also correct for work terms. Like the ΔG^0 that comes in is always corrected for work terms, if you take the three terms $\Delta G^0 + w^p + w^r$ together, then the last two terms are correcting the previous one for work terms. So, the same thing happens for volume, the standard volume change should be corrected for work terms. You know, every time you have a free energy change, and work is a free energy change, there can be always volume changes. If you take the derivative of the free energy with respect to pressure you get a volume. So, if you take the derivative of the free energy the volume changes.

9. p. 1452 2nd column top: "In order to resolve *the question of a statistical versus a dynamical effect*, a definition is needed and can be made in terms of correlations, e.g. a correlation between a vibrational excitation at one point and reaction at another. If the correlation is substantially larger than what one would obtain from a statistical theory, *subject to whatever constrains one might impose on the statistics*, then the correlation can be termed dynamical."

Q: How about these constraints?

M: Let's see . . . I wonder what I had in mind there . . . well, maybe a constraint is that the transfer be at a crossing point.

10. p. 1452 2nd column middle: "We note that TST is a theory of rare events. For example, based on a typical vibration frequency of 10^{13} s^{-1}, the probability of reaction for a first-order reaction that requires a millisecond is $10^3/10^{13}$, i.e. 1 in 10^{10}. So the rarity of an event does not in itself require a dynamical origin."

Q: Do you mean that a theory can be statistic in character—like the TST—even if is a theory of rare events?

M: Absolutely, *most of the time there are rare events, to get up that potential energy wall.*

11. p. 1453 1st column bottom: "In contrast to electron transfers, an H-transfer is typically a short-range effect, and so a question arises as to whether one can still vary ΔG^0 by varying a substituent without causing other changes such as in H-bonded networks. *At low* ΔG^0*'s will the slope of a plot of* $-kT \ln k_{\mathrm{rate}}$ *versus* ΔG^0 *have the value of* $1/2$ *found in electron transfers and in Equation* (2.6)*?*"

M: When the TS is like the reactants, the Brønsted plot you describe when you consider $-k \log k$ vs. ΔG^0, would have a slope of 0, when it is like the products would have a slope of unity, when is sort of in between in some sense, like for example when $\Delta G^0 = 0$, is in between, then the behaviour would be in between, with a slope of 1/2, both in the ET theory and the bond energy bond order model that I described in 1968, the same thing comes up, and so you expect a slope of 1/2. Now, in the case of these proton transfers your changes are not just breaking a bond, you are breaking a whole ray of hydrogen bonds, you form a new ray of hydrogen bonds, but roughly the same thing carries over, that when $\Delta G^0 = 0 \ldots.$

12. Following: "Ultimately, for H-transfers the limiting behaviour at extremely large $|\Delta G^0|$ would differ from that for electron transfers, namely a large 'inverted effect' at sufficiently high driving force, $-\Delta G^0 \gg 0$, should not occur in the adiabatic (*i.e. no H-tunnelling*) case."

Q: Why no inverted effect for H-transfer?

M: *There is a very sensitive interaction, you no longer have the cusps at the crossing of two parabolas. And so, the topology is very different.* Nuclear tunneling occurs really between nonadiabatic curves.

13. Following: "The barrier now occurs in the reactants channel, as in Figure 3. The question of an *inverted* effect in the H-tunnelling case is still, I believe, *an open question*."

M: I don't understand that, I don't know why I said that. I don't know what I had in mind, I don't know why I said that, I probably had something in mind that I don't remember what it was...

14. Following: "An aspect where ΔS^0 played a major role in electron transfers was in understanding the negative activation energy of an electron transfer... However, the measurement of the internal ΔS^0 in enzymes seems to be rare or non-existent—perhaps due to the challenges in obtaining *the 'internal' equilibrium constants and their temperature derivative*?"

Q: What are these "internal equilibrium constants"?

M: *That's when you correct the equilibrium. See, normally an enzyme reaction involves something going into the enzyme and something* *coming out of the enzyme, and you have to correct for those things, you want to know the free energy of reactions in the enzyme*, when the reactant is inside the enzyme, and when a product is inside the enzyme, *so that's the internal thing, normally the reactant is outside and goes in, and then the product comes out, so there is an overall* ΔG^0 but you don't know the ΔG^0's for the *enzyme bound for reactant and enzyme bound for product, that's the internal.*

Q: Some kind of precursor and successor complex?

M: Yes, that's right.

15. Following: "For many electron transfers, these experimental data were often readily accessible. If one knows the internal ΔS^0, one can determine to what extent the high-negative entropy of activation of an enzyme (*one which operates at low temperatures in the millisecond range*) is due in part to a very negative ΔS^0."

Q: Why the low temperatures and the millisecond range?

M: *Most of the enzymes operate in the millisecond range*, and the experiment I was referring to was so operating, and I don't think that ΔS^0 was actually measured. The enzyme was one that normally operates at high temperatures and then when you make it go at low temperatures then the behavior is very different, so there you have an enzyme that is operating outside its normal region... I am not quite sure what I was referring to, either I was referring to some reaction that was known having a very negative ΔS^0 or was presumed to have it because, when you look at the charges of the products, they were more charged than the reactants were inside the enzyme, and that was the cause why it would have a very negative ΔS^0, and I don't remember if I was thinking about that or what... *Certainly in ET in solution you can have some very negative* ΔS^0 *and that's when the products are more heavily charged than the reactants, and of course they crush out the water, or what have you, that makes the* *entropy very low so the* ΔS^0 *is very negative*. Now, *to the extent that you can have that in an enzyme, it would be for an enzyme where you get a charge separation, but it might not be as big because the enzyme is not as polar as the water* is, so maybe that was what I was thinking about. And now about the lower temperature behavior. I think that whenever temperature, lower temperature, may have been... the operations are frequently in the millisecond range, and if you want to have something in millisecond range and have in that range because of a very negative ΔS^0, then probably the activation energy has to be relatively small, in order that a lot of the effect be due to an entropy effect. At low temperatures there is a huge negative entropy of activation. *If there had been a large energy of activation the enzyme wouldn't be operating in the millisecond region at low temperatures.* In other words, I guess I was trying to set the stage for thinking of *some enzyme reaction that would have a very negative* ΔS^0 *and one way of doing that is to say: OK, supposing it is in millisecond range, and is at low temperatures, the only way it can be in that way is if it has a very negative entropy of activation.*

Q: You see how many concepts are behind your words....

M: Yes, that's right...

Q: A Talmud is necessary to read your Bible.

M: Yes....

16. p. 1453 2nd column middle: "While the range of ΔG^0's for natural enzymes appears to be small for the overall reaction, for them *to operate in the millisecond regime, the variation in the internal* ΔG^0*'s may be large*."

Q: Which one is the relation between the millisecond regime and the variation in internal ΔG^0?

M: *The overall* ΔG^0 *that includes the binding, the unbinding and so on, is maybe fairly small. The biological system I think operates* *near equilibrium or something like that.* But indeed if you look at the internal equilibrium there is going to be a big difference in ΔG^0's, there has to be a big difference in the free energy of binding of the reactants, free energy of binding of the products, *there will be large internal ΔG^0 when the overall ΔG^0 is small.* Sometimes the overall ΔG^0 is corrected for some work terms.

17. Following: "Broad issues such as the effect of thermodynamics (internal ΔG^0 and ΔS^0) versus *specific factors influencing the KIEs...*"

Q: (1) Which are the effects of ΔS^0 on KIEs?

(2) Which are the specific factors influencing the KIEs?

M: (1) *If ΔG^0 is very negative the reactions can be very fast, it isn't then possible to have much of isotope effect. Whereas for the reverse reaction, you would have a large isotope effect*, you know, the reaction is more up the barrier, but *once something is very fast then nothing is going to change much, the isotope effect doesn't change much*, when something is very small for the reverse reaction then of course you get maximum effect, or you might have it. *When the reaction is very fast, you can't get a big isotope effect, so if the ΔG^0 is very negative you are not going to get a big isotope effect*. The ΔS^0 part... I don't know why I put in ΔS^0, it should have just said ΔG^0, it is better than one talks of ΔG^0, and I don't know why I said ΔS^0.

(2) Well, *if there is steric hindrance, then it might be difficult for the reactant to come close, and so you may get a huge isotope effect*, that happens in some work, I think, I'm not sure, whether it was done at Rice University, or where, but I think there are some... and the fellow at Rice University may have been Lewis, I don't remember, has *a huge isotope effect, due to hindrance, the reaction probably had occurred at long distance, the reactants because of the steric hindrance can't get close to each other, so for a reaction to occur*

the H had to tunnel through long distance, so there is a huge isotope effect.

18. Following: "In this contest, there is an interesting study on *the effect of a change in reaction asymmetry* ($|\Delta G^0|/\lambda$) *on the KIEs in an enzymatic system*."

M: I think that I had something in a paper in 1968, I am not sure if it was there, where I may have discussed isotope effects and maybe with Audrey Cohen as a coauthor, and there *the isotope effect was a maximum with the* $\Delta G^0 = 0$, and it is like a parabola and I think this is *a common observation, and the idea is that normally isotope effects for the overall reaction are usually small, you break a bond, you form a bond, it is unlike at the transition state where you have some partially broken bond, where you may have to tunnel and so on, but for the overall reaction the isotope effect is normally relatively small. So that means that other conditions being equal, when you do some plot vs.* ΔG^0, *it has a maximum at* $\Delta G^0 = 0$. I think that Ronald (Ronnie) Bell has a graph on it.[1]

19. p. 1453 2nd column bottom: "In an asymmetrical system, *R* becomes a more dominant contributor to the reaction coordinate and so the ^{13}C KIE increases while the deuterium KIE decreases, in agreement with the data."

M: Well, *in a symmetrical system the TS is at that usual saddle point, near the corner of the skewed diagram and at the corner there is a lot of the tunneling, whereas in exothermic or endothermic reactions the TS is in one of the channels, the reaction coordinate there is more of an R motion, there is not much of an isotope effect.*

[1]Fig. 4.3 p. 103 in R.P. Bell "The Tunnel Effect in Chemistry," Chapman and Hall (1980).

20. p. 1454 1st column middle: "We note . . . that the TS of AH+B → A + HB is not simply the saddle-point, but instead is an $N - 1$ dimensional space in a space of N coordinates, *which may or may not contain the saddle-point*."

M: *What determines the TS is sort of an N − 1 dimensional set of surface states and is the surface where the total number of states is a minimum, and that may or may not occur at the saddle point. If the vibrations were largely harmonic, the TS would occur at saddle point, but if there are locations where the reaction is partially hindered, there occur displacements at some point and it may not be so. Because it is the free energy that counts, it is the number of states that counts, at a particular energy and not the potential energy alone. So it all depends where you put Wigner's dividing surface, and you take it at the surface that corresponds to where the total number of states for the transition thing is a minimum.*

Q: I have a naïve kind of sacred respect for the saddle point.

M: Yes, I know, and also I.

21. Following: "More generally, as Wigner (1938) pointed out, it is really a $2N - 1$ subspace in a $2N$-dimensional phase space (coordinate-momentum space). *In some cases this distinction between coordinate space and phase space is important . . .*"

Q: When and why is important?

M: I read a paper of Wigner around 1936, I think was then, as I troubled understanding, at the time when I was a postdoc, and tried to read it and seeing what was all about, . . . on the combinations of atoms and the presence in the process of a third body, because of course for the atoms to recombine you have to remove some of the energy, that means to have to get rid of some of the momenta, so the recombination depends on the energy removed. So, in the TS and in the states you are talking about, he had both coordinates and momenta and I had difficulty in understanding it, and that's an example where you have to remove energy as such, and some of it is kinetic energy.

22. Following: "...but typically is not expected to be so for the present enzymatic system, except when 'viscous' effects occur."

M: Well, *when viscous effects occur, when of course the usual TS theory has been replaced by something like Kramers' theory of the overdamped limit or something related to that, not just the properties of the TS are to be considered, other effects come in too then.*

M354 Enzymatic Catalysis and Transfers in Solution. I. Theory and Computations, a Unified View

R. A. Marcus

1. p. 1 1st column middle: "The discussion in this article is intended to apply both to reactions in enzymes and to reactions in solution when, in the latter case, one converts the usual second-order rate constant to a first-order one for the 'collision complex.'"

M: Well, *if you have a second-order reaction, you might say*, as Sutin has said and some just said, that in the reaction *you first form what you might call collision complex, the collision complex may just live for a half of vibration, and then from there on now it's a first order rate constant, so the second-order rate constant is really an equilibrium constant times the rate of reaction of the complex, and the idea was to reduce everything, you combine bimolecular and unimolecular to just simply unimolecular, by presuming somebody has already taken the rate constant and converted to a first order one, by dividing it by the equilibrium constant for the collision complex.*

2. p. 1 2nd column top-middle: "...fluctuations in the structure of the enzyme and in solution (in electron transfer called 'reorganization') so as to reach the transition state (TS) for the reaction."

Q: (1) Does the reaching of the TS, from the distance at which the reactants are more likely to react, *always* imply a fluctuation?

(2) What about fluctuations and reorganization in the gas phase? Consider for instance the classical

$$H_\alpha + H_\beta - H_\gamma \rightarrow H_\alpha - H_\beta + H_\gamma$$

Do fluctuations play here any role?

M: (1) In some enzymes the structure is such that the system really doesn't need a stretch... if H or whatever has been transferred, it is all fixed in place and it just needs that the fluctuation of the environment around is such that the proton can then jump without change in energy, like an electron transfer, at the crossing point. So, I think some enzymes adopt that way and the reason why I say that is that that can explain why a number of enzymes, a fair number, in their natural condition, have *a KIE that is independent of temperature*. If, for example, the enzyme wasn't already fixed in position, as far as the atoms immediately involved in the transfer, and in one bond, one H had to stretch quite a bit, *that would change the zero-point energy* and *that would give rise to a temperature effect*, for the H/D reaction, the KIE for H and D.

(2) Well, there is a fluctuation of the $H_\beta - H_\gamma$ distance, and without the fluctuation you wouldn't get a reaction.

Q: And so you consider fluctuation even a normal vibration.

M: Yes, sure.

3. p. 2 1st column middle: "...the vertical energy difference ΔE of the two lowest diabatic empirical valence bond (EVB) structures (e.g. DH, A and D, HA) is used as a reaction coordinate... The free energy of an ensemble of configurations for each of the EVB structures is calculated as a function of ΔE. When $\Delta E = 0$ the system is taken to be in the transition state."

Q: Why with this method does one get free energy *parabolas*?

M: Well, *there are several ways of answering that. One, the simplest one, is that the free energy is assumed to be a quadratic function of some variable and that means it is a parabola. Another way of answering is to say that in the canonical ensemble integral one uses a cumulant expansion, defined in a certain way, and stops to second order, that immediately makes it a parabola. Another way is if one calculates the free energy in terms of some parameter, then makes an expansion and that stops at the second-order term, that's a parabola. There are all sorts of equivalent ways of doing it. When I did it, I did with the last method above and only later did I learn that what I was doing was a cumulant expansion stopped at second order.*

Q: Well, in more than one respect you had in your life a behavior similar to that of Heisenberg...

M: Well...I wouldn't put it that way...he is at a higher plane.

Q: Gerald Holton says that very often the very original scientists rather liked, at the beginning at least, to formulate their theories in terms of a more simple or an older theory.

M: Yes sure, definitely.

Q: A typical example is Newton. When he wrote the Principia he already knew calculus and he could have put his theory in a more modern way, in the form of infinitesimal calculus somehow, but he rather liked to put it in the old way of the Euclidean geometry, and so somehow you started with electrostatics, so you have more than one point of contact with old masters.

M: Something worked out well.

Q: Another thing that Holton was telling is that he distinguishes between a science that he calls S_1 and a science that he calls S_2, that is a science in the making and a science of the textbooks.

M: Sure science in the making can be different from how it did evolve.

4. p. 2 1st column bottom: "If V denotes the matrix element coupling the two lowest EVB diabatic electronic states, the energy of the TS is reduced in magnitude by the 'resonance energy' V of the two structures, as in Fig. 1. When this EVB approach was used to treat experiments in enzymes, the value of V was adjusted so that the free energy barrier ΔG^* for the corresponding reaction rate in solution agreed with the experimental value. This result for V was then used to calculate the $\Delta G^\dagger$ for the rate of the enzyme catalyzed reaction."

Q: Please check my understanding: from the experimental reaction rate in solution one gets V, one makes then the hypothesis that the value of V is the same for the enzyme catalyzed reaction and so one uses this value to calculate the free energy for the enzyme catalyzed reaction.

M: That could be. I guess I looked at Warshel's work there, and was summarizing Warshel's work. I forgot now what Warshel's work really was, but *the main point is that V is a parameter and if you calculate everything right in solution you can fit the value of V to the activation energy remembering that the activation energy is not the same thing as the free energy of activation, it is related to a temperature derivative. So, anyway, if one does things right, one can get from the activation energy in solution the value of V, and then make the assumption that it's the same value in the enzyme. One has to be very careful because normally in TS theory you calculate a free energy of activation, that's not the same as the energy of activation, so either you convert free energies of activation or you compare energies of activation, I mean you just have to be careful*, not to make substitute.

Q: Just a curiosity: is the energy of activation an enthalpy?

M: *It's an enthalpy but what it really is, is defined, both for the theory and the experiment as* $\frac{d\log k}{d(1/T)R}$, *that is the energy of activation, so is* *defined operationally as the slope of the log of the rate vs. 1/T, divided by R.*

Q: And thermodynamically?

M: Thermodynamically it turns out that's a way of calculating heat content change when you have a free energy change, they do turn out to be related. Now, *there is something a bit different because in the TS theory there is a* $\frac{kT}{h}$, *so that affects the relationship a little bit, but one can handle that all right, one has just to be careful*, so.

Q: So one cannot say that it is an enthalpy, directly so easily.

M: You mean an enthalpy change, an internal energy change, but sometimes that can involve the solvent, when there is a large entropy change in the reaction due to ions changing their charge, say from + to 0. *Whenever there is a change in charge, there is a huge entropy change*, and correspondingly there is also quite an energy change in that too, and all of that has to be taken into account. So, that all comes in when you calculate the free energy of activation and you can obtain from that in the usual thermodynamic way the heat of activation, and following you can relate that to the energy of activation. The energy of activation and the heat of activation differ by kT because of that $\frac{kT}{h}$ factor. Because, so to speak, the heat of activation is associated with the free energy of activation, is related to the $e^{-\frac{\Delta G^\dagger}{kT}}$, and there is also in the rate a $\frac{kT}{h}$, so that means that the overall heat of activation has to include what the effective $\frac{kT}{h}$ is on that definition. It depends not just on the ΔG, on the $\frac{\Delta G}{kT}$, the $\frac{d \log k}{d(1/T)R}$, the $\frac{kT}{h}$ comes in too. It's all pretty straightforward actually, one just has to be careful.

5. Following: "In the EVB approach a reorganization of the system occurs. There are fluctuations in the positions of all the atoms in the protein so as to reach the TS, a state in which the two lowest EVB states have the same energy. On the other hand, *in the bond length* *difference approach*, there are again fluctuations in the enzyme structure to reach the TS, but *a functional form for any such reorganization is not explicitly singled out*."

Q: Please expand on the two approaches, in particular the last point above about the functional form for the reorganization.

M: When we talk about functional form there are two functional forms, one is the quadratic and the other is a log cosh one, the Bond Energy Bond Order (BEBO) model. There are two quantitative equations, one is the quadratic equation, the other was based... was much later based on the BEBO model, and that has ΔG^* is a function of ΔG^0, doesn't have the quadratic term, it does have the log cosh term instead. Now those are the two separate equations. What I think I used mostly in these reactions is the usual equation, the quadratic, but of course that's only an approximation, for reactions that are electron transfers. For proton transfer reactions, as in some enzymes, it is only an approximate replace, it wasn't derived for that, on the other hand there are papers that show how well it works for some reactions that aren't ETs, and so I don't remember now whether in this enzyme paper I only used the quadratic or whether I also mentioned the BEBO model too. And one key point is that where the quadratic model has been applied to a bond breaking forming system it worked pretty well, surprisingly, it worked pretty well experimentally as in the work of Lewis at Rice University for transfer of methyl cations between aryl sulfonates and also as in the work, computational work, of several people who tested it for Bond Energy Bond Breaking reactions and the quadratic equation worked very well for it too. So, that is a surprise, so it was on that basis that even though these enzyme reactions are not electron transfers it may be that the quadratic equation is OK as a first approximation. I mean, there are key differences, there is no inverted region in BEBO, the topology is totally different. So, one may only use it provided that ΔG^0 is not comparable with λ, it's smaller. Once you get near the inverted region the quadratic model can't be used for reactions other than ETs.

6. p. 2 2nd column top-middle: "...EVB yielded very useful pioneering results. A merit of the EVB approach is that in the numerical calculations it showed a simple *quadratic behaviour* for the free energy barrier $\Delta G^{\dagger}$ *due to the fluctuations of the protein leading to the TS (Gaussian fluctuations)*."

Q: Please explain the relation between quadratic behavior and Gaussian fluctuations.

M: They are the same. Because the Gaussian fluctuations mean that if you have a Gaussian ensemble, and that Gaussian ensemble depends on a parameter, then you have sort of a quadratic free energy change. And I always wondered about why the quadratic change looked so well, and I know it's connected in some way, somebody may have discussed this, to the central limit theorem. So, when you have a lot of objects but those are normally a lot of identical and noninteracting objects, it turns out that there is also a central limit theorem when, you know, your number of interactions is sort of relatively small, each neighbor just interacts with each neighbor, in terms of a volume element, and not with everything else, apparently the central limit theorem works well there. So, whenever the central limit theorem works, then you have a free energy which is quadratic in some parameter. I haven't thought through in detail with all of its connections, but sure they are connected. That would mean, for example, the probability of having some polarization that depends quadratically, exponentially quadratically, on the difference between that polarization and some equilibrium polarization. That would be an example of quadratic fluctuation.

Q: So Gaussian fluctuations, quadratic fluctuations, central limit theorem, and cumulant expansions are all bound together.

M: Yes, I think so.

7. p. 31st column top: "The equation for a 'unimolecular' rate constant k_{et} of an electron transfer reaction occurring in a reaction complex between two molecular species...

$$k_{\mathrm{et}} = Ae^{-\Delta G^*/kT} \tag{1}$$

Where

$$\Delta G^* = w^r + (\lambda + \Delta G^0(R))^2/4\lambda \tag{2}$$

and

$$\Delta G^0(R) = \Delta G^0 + w^p(R) - w^r(R) \tag{3"}$$

Q: (1) Is w^r the work necessary to bring the two moieties of the complex from infinity to their equilibrium state in the complex?

(2) Is $w^r(R)$ the work to go from that equilibrium state to the TS? I mean, is the separation distance R the distance between the equilibrium geometry of the complex and the TS geometry?

M: (1) w^r does depend on R, R should be present in both Eqns. (2) and (3), I don't know why I didn't put it in Eq. (2), it should be there.

(2) *It is really intended to be the value in the TS, but of course you are not in the TS because all of the fluctuations around haven't occurred.*

Q: *And the meaning of R what would exactly be*?

M: *You have a TS and you look at the distance between the two reacting centers.*

Q: *At the TS*?

M: There things are a little bit loose in terms of precise wording,... you could ask yourself if, once reactants *start* strongly interacting, the getting of things closer together is really also part of the reorganization energy. The work for *bringing the reactants to the distance R but not having a reorganization of energy, that would give you $w^r(R)$, and now there would be fluctuations.*

Q: So, fluctuation is following.

M: Yes, I presume that, *what is intended by* $w^r(R)$ *is that part of the free energy barrier which is not affected by* ΔG^0, *you haven't started breaking a bond or forming a bond, so that part is not affected by* ΔG^0, it says sort of: *bring close together to maybe a steric effect, to maybe part of it, or maybe a Coulombic effect, to some repulsion, and so on, to interactions that are hardly immediately related to the stretching of a bond.*

8. Following: "The preexponential factor A depends on the electronic coupling V of the two reactants. When the reactants are weakly coupled electronically in the transition state the reaction is nonadiabatic and A is proportional to $|V|^2$. When the electronic coupling V is instead strong in the TS region (adiabatic reaction) *A is of the order of a vibration frequency.*"

Q: Please explain this last point. Why in the nonadiabatic and in the adiabatic case does A have different physical units?

M: *A depends on the square of the energy, is not equal to that, is* $\frac{2\pi V^2}{\hbar}$.

Q: Why is A of the order of a vibrational frequency?

M: *That's true actually only when there is no major entropy change going on*, I should have called by that. *In normally adiabatic transition state theory* one way of thinking of it is the following: that *the reactant has one vibration that the TS does not, and that means in the* $\frac{Q^\dagger}{Q}$ *if you remove that factor from the Q reactant, you have a* $\frac{kT}{h\nu}$ *in the denominator, now the* $\frac{kT}{h}$ *cancels part of that* $\frac{kT}{h\nu}$ *and the* ν *comes upstairs*, comes in A...Now, of course *if the reaction itself involves a major* ΔS^0, *then the true preexponential factor will involve also definitely a* $\frac{\Delta S^0}{2}$ (when the slope is 1/2), the way Sutin and I discussed in the 80s, I think, and anyway I discussed in earlier papers ... so I should have been more careful then in my phrasing. What I should have said is that the rate constant equals $\frac{kT}{h}$.

Q: In the papers that you write yourself you are in general extremely careful.

M: Yes, I usually am, but not always.

9. Following: "The $w^r(R)$ and $w^p(R)$ denote *that part of the free energy barrier that is not overcome by a favourable* $\Delta G^0(R)$."

M: What I had in mind there was that *if you think two reactants coming together and they have the Coulomb repulsion, that Coulomb repulsion is the same no matter what* ΔG^0 *is, so but yet that repulsion is part of the barrier, so that is a part of the barrier which is independent of the* ΔG^0.

10. Following: "In a weak-overlap reaction usually one electron per successive electron transfer step is preferred energetically over a single-step multiple electron transfer, i.e., usually $\Delta e = e$."

M: *If there is a large change in the free energy barrier roughly comes in as* $(\Delta e)^2$, *so if* Δe *is 2e then you'd have four times a barrier*. The idea is that *for weak overlap reactions if we transfer two electrons then you might avoid some unstable intermediate, but on the other hand you need a far bigger reorganization.*

11. p. 3 2nd column top: "Quantum expressions for the high frequency normal mode vibrations of the reactants have been introduced ... The latter can be particularly important in the 'inverted region' ..."

M: *If you think of the inverted region and you think of the two free energy curves, it's like a half moon, right? And that's a very thin barrier. If, on the other hand, in the normal region you write the two parabolas, you get a thick barrier going from one to the other. So, if* *the curvatures are of the same sign, your barrier is very thin, then it's more apt to be tunneled through.*

Q: Why are the high frequency modes more important?

M: *Well, the high frequency modes have the lighter masses, they tunnel more.*

12. p. 3 2nd column middle: "The electronic polarization of the environment is again a 'fast variable' leading, in a continuum description, the presence of D_{op}... instead of unity that would occur in the absence of a *correlative shielding effect* of the fast (electronic) polarization."

Q: What do you mean by "correlative shielding effect"?

M: *At any point in the medium there are lots of change required going from reactants to products. The electronic polarization reduces the change of the field it's very fast, so the polarization doesn't have to change as much to satisfy both masters.*

Q: And why do you call correlating shielding effect?

M: I don't know why I called it correlated, I could have just said shielding effect.

13. Following: "...for the H transfer systems, this $2V$ in Fig. 1 is frequently large, perhaps ~1 eV instead of ~0.02 eV, because chemical bonds are breaking and forming. So we can no longer invoke an energy equality to define the TS in Fig. 1."

Q: How is then the TS to be defined in this more general case? Is it may be defined as the point of the lower adiabatic curve at which the free energy reaches a maximum and at which the vertical energy difference ΔE of the two lowest adiabatic states reaches a minimum?

M: *As the maximum*...the first criterion, and in a way that's called variational transition state theory.

14. Following: "The DHA reacting pair is defined as consisting of the atoms directly or closely involved in the covalent bond breaking-bond forming reaction, frequently some 50 or so atoms."

Q: Why so many?

M: Well, what they do is that they treat quantum mechanically those 50 or so atoms . . . 3 atoms that are usually involved, but they're connected to other atoms as part of structures, so quantum mechanically you have to treat that whole immediate region. For example supposing one treats the interaction of . . . let's say . . . an alcohol molecule, and another alcohol, or something like that, with some interaction between them, then you wouldn't just break up the alcohol into one H and the other part with no H. I mean, what they do is that they make some cutoffs at bonds but they try to keep as many atoms as they can, as long as the calculation is practical.

15. p. 4 1st column top: "At each value of ΔE there is a *thermal ensemble of configurations* in the many-dimensional coordinate space and as in electron transfer it yields the free energy."

Q: Why the ensemble of configurations with a fixed value of ΔE is called "*thermal* ensemble of configurations"?

M: Because if you fix a ΔE, you have an $n - 1$ dimensional phase space, an ensemble of configurations in there, and you can define a formal ensemble, where $e^{-\frac{E}{kT}}$ is the weighting factor.

16. Following: "weak-overlap H transfer."

Q: When does one have a weak overlap H-transfer? Can you give an example?

M: Yes. *If the H has to jump a long distance because of steric reasons. And there are some reactions where I think that the isotope effect may be as large as a factor of 100. In that case one has* *to tunnel a long distance, and that condition corresponds to weak overlaps.*

Q: But because of this long trip of the H isn't the interaction bigger?

M: Well, if the two reactants are close together, then you may not have to do any tunneling, but there may be steric restrictions which prevent that. Because of such possible restrictions, which may be present, the reactants may be far apart and there may be a long tunneling distance.

17. p. 4 2nd column bottom: "When H approaches midway between the heavy atoms ($n \rightarrow 1/2$), the analog of the two very different charge distributions present in 'weak overlap' electron transfer disappears. Accordingly there can be some deviation from the harmonic behaviour, as in the dashed line in Fig. 1. In Fig. 1 a 'rounding off' arose from the coupling term V. However, now in addition the rounding off is *augmented by the merging of the two charge distributions, a merging that modifies the protein reorganization term* when n is in the vicinity of ½."

M: Well, remember the reorganization in the nonadiabatic case is such that *you could have two configurations in the electron distribution that have the same energy, at the crossing point.* Well, *you no longer have that now, you don't have two distinct charge distributions, you have one charge distribution that peaks in a certain region with a certain probability, but you have one charge distribution, so that whole nonadiabatic treatment disappeared and somehow became modified.*

Q: But is it a one charge distribution or a continuity of charge distributions?

M: *In the nonadiabatic case* there is a one charge distribution with the probability of being near unity when the charge is near one reactant and near unity when it is in the other, so it's kind of *bimodal* *charge distribution*, and in this case you can always go from one to the other there by jump, but in this case everything sort of merges, and you have a gradual change in charge distribution. The distribution doesn't just suddenly change by getting through a crossing point.

18. Following: "For simplicity we...use the quadratic form to approximate $G_{\text{prot}}(m)$...and use as its reaction coordinate the m in ET theory...We than have:

$$\Delta G_{\text{prot}}(m) = m^2\lambda_o \quad (m < m_c) \tag{12}$$

and for the reverse reaction, starting from the products p,

$$\Delta G^p_{\text{prot}}(m) = (1-m)^2\lambda_o \quad (m > m_c) \tag{13}$$

where λ_o is the reorganization energy of the protein and m_c is the m which satisfies the intersection condition

$$\Delta G_{\text{prot}}(m) = \Delta G^p_{\text{prot}}(m) + \Delta G'_{\text{prot}} \tag{14}$$

Thereby, m_c satisfies

$$(1-2m)\lambda_o = -\Delta G'_{\text{prot}} \tag{15}$$

For symmetry, we replaced m...by $-m$, so that m in the normal region now lies in the interval (0, 1) instead of $(-1, 0)$. From Eqns. (13) and (14) we have, *for the lowest adiabatic state* when $m > m_c$, *instead of Eq.* (13),

$$\Delta G_{\text{prot}}(m) = (1-m)^2\lambda_o + \Delta G'_{\text{prot}} \quad (m > m_c). \qquad (16)\text{''}$$

M: There are a couple of enzyme papers where I applied the quadratic model. In one of them I think I applied it separately to the proton transfer apart from the other part, in another paper I combined everything all together, and I think is this second form, presentation that I prefer over the one in the first paper. In the first paper, I treated a couple of the things quite separately, and later I thought it be better that I combine them, which is what I did in the second paper on the enzymes. The first paper should be replaced by the second paper. So, it'd be better to go with the second paper rather than the first paper. I think is best to go directly to the unified view which combines the two things in one. I wrote several papers, forget about the very first one because I think in the next one I replaced it with what I thought was the better way of doing it.

M354 Enzymatic Catalysis and Transfers in Solution. I. Theory and Computations, a Unified View

(Continuation)

R. A. Marcus

0. p. 1 1st column bottom: "Among the factors contributing to the reaction rate in a bound substrate-cofactor enzyme complex are (1) the work required for the reactants in the bound complex to approach each other closely enough to optimize the H-transfer rate, (2) an additional energy barrier accompanying the bond rupture and bond formation..., (3) fluctuations in the structure of the enzyme and in solution... 'reorganization'... so as to reach the TS..."

Q: Suppose one has a macroscopic sample of the system. How do I calculate how many times per second (on the average) does a fluctuation (3) allow one bound substrate–enzyme complex in the sample to reach the TS? In other words, how does one compute the average fluctuation frequency or the probability per unit time of reaching the TS?

M: *What you're calculating is a rate constant, in a standard way. You refer to the fluctuations, the fluctuations produce a rate constant.*

01. p. 4 2nd column top: "We have, *relative to the value of* $E_{\text{DHA}}(n, R)$ *at* $n = 0$,

$$E_{\text{DHA}}(n, R) = -\frac{\lambda_i}{4 \ln 2}[n \ln n + (1 - n)\ln(1 - n)] + n\Delta E^0_{\text{DHA}}(R) \tag{8}$$

where n is the bond order of the newly forming H-A chemical bond and $\Delta E^0_{\text{DHA}}(R)$ is the electronic energy of reaction of the bound substrate-cofactor at the given R. One sees in Eq. (8) that *at* $n = 1$, $E_{\text{DA}} = \Delta E^0_{\text{DHA}}(R)$."

Q: If one inserts $n = 1$ in Eq. (8) one has $\ln(1 - n) = \ln 0$. On the other hand, for $n = 0$ one has the same problem for $\ln n$. Please explain... There is a typo, I believe, the last equation should be $E_{\text{DHA}} = \Delta E^0_{\text{DHA}}(R)$.

M: Yes, but if you do some sort of expansion, the (1 − n) dominates the ln(1 − n). In other words, if you look at the limit... I'm not quite sure, I'd have to think about how you do that, but basically the (1 − n) dominates the ln(1 − n).

1. p. 5 1st column bottom: "... the approximation leading to Eq. (18),

$$\Delta G^* = [\lambda_o + \Delta G^0_{\text{prot}}]^2/4\lambda_o \tag{18}$$

the assumption of no rounding off, leads to this being no (n, m) cross term. The free energy barrier then the sum of an n−dependent term, m−dependent term, and w^r."

Q: Can you give a physical explanation of why there are no cross (n, m) terms when there is no rounding off?

M: I suppose that *if you think of some reaction parameter that describes progress of a reaction, then along that parameter you have fluctuations of the solvent. In the enzyme you have fluctuations of the bond, those are just fluctuations that occur separately.*

Q: A very general question: imagine you have a normal harmonic oscillator, now when the oscillator oscillates going away from the equilibrium position, can these oscillations be called fluctuations?

M: Ja, you can call them fluctuations, I mean, they're usually not oscillatory fluctuations, you know... I mean, *anytime you get a system in a state different from equilibrium you might call it a fluctuation, but in the case of a simple harmonic oscillator would just have a sheer mechanical oscillation, I mean it is a fluctuation but normally you think of a fluctuation as having a more random path to it.*

Typo: p. 5 2nd column top-middle: $\Delta G^*_R = w^r + \lambda/4$ should be $\Delta G^*(R) = w^r + \lambda/4$.

2. p. 5 2nd column middle: "When $\Delta E^0_{\mathrm{DHA}}(R) \rightarrow -\infty$ then the $\Delta E^*_{\mathrm{DHA}}(R)$ term vanishes but the $\Delta G^*_{\mathrm{prot}}$ would show an inverted effect if $|\Delta G^0_{\mathrm{prot}}/\lambda_o| > 1$. However, this effect neglects the rounding off mentioned earlier. Further, the *spacing of the protonic quantum states of the products* is small compared with that of the electronic states of the products and there can be expected to be *a smaller inverted effect when reaction into the excited protonic states of the product HA is included.*"

Q: (1) Which is the effect of the rounding off on the inverted effect?

(2) What do you mean by "protonic quantum states." Aren't there just different electronic quantum states of DH or of HA? Or do you maybe refer to excited vibrational states?

(3) Which one is the effect of the small spacing of protonic quantum state of the products on the inverted effect?

(4) How does the smallness of the inverted effect manifest itself?

M: (1) First of all there shouldn't be any inverted effect for proton transfer reactions. If one uses the quadratic equation for the proton transfer one should not use it in the inverted region. So, forget about the inverted effect, when the system is in what would be the inverted region don't use that equation, not for the proton transfers, there is no inverted effect.

(2) No, no, because *they're just excited states of the DH or HA vibration.*

Q: Excited vibrational states

M: Yes.

(3) *You get the inverted effect if there is no other electronic state nearby, and of course if there's one nearby then the system goes over to that electronic state instead of trying to do the inverted effect, but now when those states are very close, as the protonic states are, compared with the electronic states, then there's no chance for that to happen.*

(4) Well, if there is no inverted effect there is no maximum.

3. p. 6 1st column middle-bottom: "The KIE affects ΔE_{DHA} *and perhaps* w^r, *depending on the origin of the latter.*"

M: *If* w^r *didn't involve any bond stretching, then it wouldn't contribute to the KIE*, but if there's *some part of the reaction barrier which doesn't depend upon* ΔG^0, *that's the definition of the* w^r *part*, and involves bond stretching, then you'd have an isotope effect but typically you don't expect that to be the case. It's only when the bond stretching starts, and you expect that when the bond stretching starts, that could be part of the... sort of reorganizational term, but *you don't expect a* w^r *to have any bond stretching associated with it*. Now, if there is some kind of steric effect, because *steric effects are included in* w^r, and in an enzyme you don't expect that because everything is pretty well ordered, to start with, but *if you had some* *sort of steric effect, then that could be different for H, D. It depends on detailed geometry, and then the w^r should be affected, but that's probably unlikely.*

Q: In Wikipedia there is an article about the Nobel Prize for the Higgs boson. Six people predicted it theoretically in 1964.

M: My goodness, why did they name after him? I noticed that one of them, I don't know if he is still alive, was R. Brout and if I remember rightly I knew him when he was a chemist at the faculty of Columbia.

Q: And then Philip Anderson claims he discovered it in 1962. Higgs was a graduate student of Charles Coulson.

M: Coulson was in the mathematical faculty.

4. p. 6 1st column bottom: "In principle, a charge transfer (CT) spectrum is possible, as seen from Fig. 1. However, *if the D-A distance has to become significantly smaller in order to facilitate a charge transfer, then the intensity of this CT absorption spectrum will be small...*"

M: During the optical act, the proton isn't going to move, there would be a gross violation of the Franck–Condon principle to get a charge transfer spectrum out of that. If you think of the charge transfer spectrum involving initially the H on one reactant and the H on the other, that involves a movement of the H, so that's kind of a violation of the Franck–Condon principle for the H atom, but maybe it could happen a bit, I don't know if there are any charge transfer spectra associated with H-transfer, are there?

Q: You write that as yet they have not been observed.

M: Because you have a big problem overcoming the Franck–Condon principle there. The charge transfer spectrum involves as initial state the H on one reactant and the H on the other is the final state, and it all that happening by light because of a charge transfer spectrum, that's highly unlikely.

Q: And what about this "the D-A distance has to become significantly smaller…"

M: If the distance is extremely small then the difference of the position of the H in the initial and final charge transfer states may be just light, and so, there is a small violation of the Franck–Condon.

Q: Yes, because the H is in between there.

5. Following: "A *vertical transition* to the upper curve in Fig. 1 with its sharply repulsive nature, *may spread out the intensity*, while a *non-vertical transition may have small intensity because of a poor overlap of the H wave function*."

Q: (1) Until now I have known only vertical transitions…please explain when, how and why one can have nonvertical ones (maybe because of the "heavy" H compared to the light electron?).

(2) Why the above spreading out of the intensity?

(3) Which one is the H wave function showing small overlap?

M: (1) In general there're also what they call non-Franck–Condon factors,…that's a nonvertical, so there is a little contribution from that, that happens in some cases, I mean not for these systems, for other spectroscopic systems.

Q: Maybe nonvertical because we have H instead of e?

M: Well, that's one reason, *yes, because you probably want to have the H in another position, that means that's nonvertical.*

(2) Because probably *the total extent of a band is constant, that's called the oscillator strength, the band cannot be wider or narrow, the*

total oscillator strength is sort of the integrated band, so if something is extremely weak, has to be extremely wide.

(3) Well, in the initial state the H is attached to one molecule, in the final state the H is attached to the other, and the overlap of those two H wave functions may be small.

6. Following: "The isotopic results on pressure effects . . . suggest that pressure tends to convert β to α paths . . . *it would increase the intensity of any CT spectrum.*"

Q: Why?

M: *It brings things close to each other, so there is less distance to travel.*

7. Following: "The protein/solvent contribution to a CT spectrum (*vertical for these heavier mass coordinates*) . . ."

Q: Which is the relation between verticality and heavy masses?

M: *If heavy masses are involved, there is less need for the Franck–Condon principle, you don't have that disparity of masses.*

Q: *So the Franck–Condon principle applies to small masses . . .*

M: It applies for electrons, when there's change of electronic state, *if you try to change the position of the nucleus then it's a nonvertical transition, optically, and that has a lower intensity.*

8. Following: ". . . if one energy 'surface' of the reactants has the form $\frac{1}{2}(k_a q^2 + k_b Q^2)$ and products' surface is displaced from it but with the same force constants, $\frac{1}{2}(k_a(q-a)^2 + k_b(Q-b)^2 + \Delta)$, then the energy difference ΔE^*_{tot} is $k_a q_a + k_b Q_b + \text{const.}$"

Q: I believe there are **two typos**: the energy difference should be ΔE_{tot} without asterisk and one should have $k_a aq + k_b bQ + const$ for it.

M: Yes, you're right.

9. Following: "Similar remarks apply when $\frac{k_a q_a^2}{2}$ and $\frac{k_a(q-a)^2}{2}$ describe free energy curves..."

Q: **Typo** The first formula should be $\frac{k_a q^2}{2}$.

M: You right, yes.

10. p. 7 1st column top: "Thus, from Eq. (8),

$$E_{\text{DHA}}(n, R) = -\frac{\lambda_i}{4\ln 2}[n \ln n + (1-n)\ln(1-n)] + n\Delta E^0_{\text{DHA}}(R) \tag{8}$$

the E_{DHA} of the upper adiabatic state contains a term

$$+(\lambda_i/4\ln 2)[2\ln n + (1-n)\ln(1-n)."$$

Q: Another **typo,** I believe: it should be $+(\lambda_i/4\ln 2)[n\ln n + (1-n)\ln(1-n)$.

M: Yes, I think you right.

11. Following: "A consequence is that the vertical energy difference of the DHA pair at any n, $\Delta E_{\text{DHA}}(n, R)$ is

$$\Delta E_{\text{DHA}}(n, R) = \left(\frac{\lambda_i}{2\ln 2}\right)[n\ln n + (1-n)\ln(1-n) + n\Delta E_{\text{DHA}}(R) - (1-n)\Delta E^p_{\text{DHA}}(R) \tag{28}$$

where $\Delta E_{\text{DHA}}(R)$ and $\Delta E^p_{\text{DHA}}(R)$ denote the vertical excitation energies of the reactants ($n = 0$) and of the products ($n = 1$) at R, respectively."

Q: **Typo.** I believe in the first term to the right there should be a $-\frac{\lambda_i}{4 \ln 2}$ instead of a $\frac{\lambda_i}{2 \ln 2}$, please cf. Eq. (8) above.

M: Yes, there should be a (–) sine.

M354 Enzymatic Catalysis and Transfers in Solution. I. Theory and Computations, a Unified View

(Continuation 2)

R. A. Marcus

1. p. 7 1st column middle-bottom: "For a series of related reactants tunnelling is maximum when $\Delta G^0 = 0 \ldots$ it is a maximum in the symmetric case..."

Q: Why so?

M: *At either ends the reaction is essentially diffusion controlled, or collision controlled, it's going at every collision, or a certain fraction of them, and so there is no room for any isotope effect. In other words, as the stretching of the bond isn't the factor there, that's relatively fast compared with, say, the approach of the reactants to each other. When ΔG^0 overcomes the activation energy, then there's no barrier, and so the barrier is that in the approaching or, if it's in a complex, the vibration itself.*

Q: But here you speak of tunneling.

M: *The tunneling is a maximum then when you're not at either of those limits, the limit for the forward reaction and the limit for the back reaction, when the forward reaction is fast then the system doesn't have to tunnel, and similarly thinking of the back reaction when it's so fast and so it doesn't have to tunnel, so at either extreme there is no tunneling....*

M357 H and Other Transfers in Enzymes and in Solution: Theory and Computations: A Unified View. 2. Applications to Experiment and Computations

R. A. Marcus

1. p. 6644 2nd column top-middle: "In addition to the H/D tunneling considered later *any R-dependent term appearing in Eqs.* (1)–(7),

$$\Delta G^* = w^r(R) + \Delta G^*(R) \tag{1}$$

$$\Delta G^*(R) = \Delta E^*_{\mathrm{DHA}}(R) + \Delta G^*(\mathrm{prot}) \tag{2}$$

$$\Delta E^*_{\mathrm{DHA}}(R) = \frac{\lambda_i}{4} + \frac{\Delta E^0_{\mathrm{DHA}}(R)}{2} + \frac{\lambda_i}{4\ln 2}\ln\cosh y \tag{3}$$

$$y = (2\Delta E^0_{\mathrm{DHA}}(R)\ln 2)/\lambda_i \tag{4}$$

$$\Delta G^0(R) = \Delta E^0_{\mathrm{DHA}}(R) + \Delta G^0(\mathrm{prot}) \tag{5}$$

$$\Delta G^*(\mathrm{prot}) = [\lambda_o + \Delta \mathrm{G}^0(\mathrm{prot})]^2/4\lambda_o \tag{6}$$

$$\Delta G^0(R) = \Delta G^0 + w^p(R) - w^r(R) \tag{7}$$

such as $w^r(R)$, may be isotopically *sensitive, since R may be isotope-dependent.*"

Q: Please explain the above R's isotopic dependence ... how and why does the zero-point energy vary along the reaction coordinate?

M: The point is that the R is some compromise between getting as close together as one can and not to have to do too much tunneling, therefore R can be expected to differ. Since R is determined by the steric repulsion vs. getting close enough to tunnel, the tunneling properties are different for H and D, which means that the R that balances those two situations has to differ.

Q: Now, why does the zero-point energy vary along the reaction coordinate?

M: Typically with interactions the vibration frequency of participating vibrations changes because all vibrating changes when moving along the reaction coordinate except for those vibrations that are far from the center of the reaction.

Q: So, the nature really of the bonding changing...

M: *Yes, the whole bonding.*

Q: And with it the zero-point energy...

M: *Well, yes because every single frequency is changing along the reaction coordinate, therefore every single* $h\nu/2$ *is changing along the reaction coordinate...in any potential energy surface when you are bringing two reactants together, all sorts of vibration frequencies that are newer, the bond is forming and bond is stretching and changing.*

2. p. 6645 1st column middle-bottom: "In model B we first employ the general symbol $E_{\mathrm{DHA}}(n, R)$, for the DHA term instead of the specific form used to derive Eq. (3), and then we use a single coordinate n so as to satisfy a single equation for n in the TS. One approach, *prompted in part by the second half of eq 10*

$$-(1 - 2m)\lambda_o = \Delta G^0(\text{prot}) \tag{10}$$

is to choose n in the TS to satisfy

$$\frac{\partial}{\partial n} E_{\mathrm{DHA}}(n, R) + \lambda_o(1 - 2n) + \Delta G^0(R) = 0 \quad (n\,in\,\text{TS}) \tag{11}$$

Q: I see a similarity between $\lambda_o(1-2m)+\Delta G^0(\text{prot}) = 0$ in Eq. (10) and the second part of Eq. (11). As a matter of fact, if one remembers Eq. (5) above, that is, $\Delta G^0(R) = \Delta E^0_{\mathrm{DHA}}(R) + \Delta G^0(\text{prot})$ one would be tempted to rather write the second part of Eq. (11) as $\lambda_o(1-2n) + \Delta E^0_{\mathrm{DHA}}(R)$...But please explain how you get Eq. (11)...

M: First point, I use the same sort of reaction coordinate for the reorganization as for the change of bonding, that was done out of simplicity, I don't remember if that's the way I ended up or if I used an n_1 for one and an n_2 for the other, you know, separate properties, I don't remember but I did tie the two together and I don't remember if that was a good idea or a bad idea.

Q: The point is this: let me ask you question 2 in detail: if you look at Eq. (11), OK? I see a similarity here with Eq. (11), the second part of Eq. (11). As a matter of fact, if one remembers Eq. (5) one would be tempted to rather write the second part of Eq. (11) this way, that is adding a $\Delta E^0_{\text{DHA}}(R)$, but please explain how did you get to electron 11, if it is just a clever guess.

M: I guess what is I saw is that I could get the previous results for the bonding using Eq. (11) and that equation looked to me like Eq. (10), so I thought I would use the bonding part in that form because of the analogy to Eq. (10), so I could put it more or less on the same basis. *In one case you have parabolas, in the other case you have a smooth curve, so strictly speaking we are not on the same basis but, because of the way the formulas look, I thought I tried to put them on the same basis, by using something mathematically equivalent to something else*. What I don't remember now is whether is successful or not in its use or whether I ended up using something else in the paper.

3. Following: "This equation for n in the TS is equivalent to finding the maximum of an expression

$$\Delta G(n, R) = E_{\text{DHA}}(n, R) + \lambda_o n(1 - n) + n\Delta G^0(R) \quad (n \, in \, \text{TS}) \tag{12}$$

It satisfies the initial condition $\Delta G(0, R) = 0$, taking $E_{\text{DHA}}(0, R) = 0$. This equation and eq 11 are *an interpolation* in that the maximum of $\Delta G(n, R)$ gives the correct values of n in the TS in the two limits, one where $E_{\text{DHA}}(n, R) = 0$, and the other where $\lambda_o = 0$. As noted in the equation, it is intended only to determine n in the TS *and not to serve as a profile of the free energy change along the reaction coordinate n*."

Q: (1) Please explain the "interpolation" point.

(2) Why it cannot serve as a profile of the free energy change along the reaction coordinate n?

M: (1) We know something about the two ends, that's in the first sentence there. And we know something about the maximum, and so using those three points we give a curve, that curve is essentially bound on either ends and in the middle is an interpolation.

(2) It may be, I don't know, I'd have to look at that, but it may be that it was just designed to fit three points rather than everywhere, it wasn't chosen to fit everywhere, it was chosen to fit three points for some convenient reason or other that I have forgotten now. I mean one can plot to see whether it looks like a profile, but often something which gives the correct maximum doesn't mean it gives every point elsewhere correct. But it may be a convenient way of defining a maximum, that's all the purpose is, so it may be locally OK around the maximum, but not everywhere.

4. Following: "We first write $E_{\mathrm{DHA}}(n, R)$ in the form of eq 12. $\partial \Delta G(n, R)/\partial n = 0$; then yield the value of n in the TS,

$$-\left(\frac{\lambda_i}{4 \ln 2}\right) \ln[n/(1-n)] + \lambda_o(1-2n) + \Delta G^0(R) = 0 \quad (n \text{ in TS}) \tag{13"}$$

Q: The last two terms in Eq. (13) are the derivatives, with respect to n, of the last two terms in Eq. (12). The first term in Eq. (13) should be the derivative of the first term, $E_{\mathrm{DHA}}(n, R)$, in Eq. (12) with respect to n. From Eq. (8) in **M354** we have:

$$E_{\mathrm{DHA}}(n, R) = -\frac{\lambda_i}{4 \ln 2}[n \ln n + (1-n)\ln(1-n)] + n\Delta E^0_{\mathrm{DHA}} \qquad \text{(8) in } \mathbf{M354}$$

If one derives $\partial E(n, R)/\partial n$ one gets $-\left(\frac{\lambda_i}{4 \ln 2}\right) \ln[n/(1-n)] + \Delta E^0_{\mathrm{DHA}}$. You see that the term $\Delta E^0_{\mathrm{DHA}}$ is missing in Eq. (13). Why?

M: Yes, the $\Delta G^0(R)$ is really the sum of two terms, one that ΔE^0_{DHA} term, and the other a corresponding term for the other part. $\Delta G^0(R)$ has two terms in it, one is ΔE^0_{DHA} and the other is the standard free energy change associated with the solvation.

Q: But the point is this: if you take Eq. (12) and you derive it, you do the partial derivatives of $\Delta G(n, R)$ with respect to n, the last two terms in Eq. (13) are OK, then you get the first term. Now this one you would get when you do the derivative of Eq. (8) of the previous paper. If you derive now, partial derivative of $E_{\text{DHA}}(n, R)$ with respect to n, you get something very similar to the first term in Eq. (13) except that there is this ΔE^0_{DHA}.

M: The point is this: one should add to that $E_{\text{DHA}}(n, R)$ a solvation free energy barrier, and then one would have then an n and sort of an ΔG^0 contribution from the solvation, and the sum of the ΔG^0 for the solvation plus the ΔE^0_{DHA} is equal to ΔG^0, so maybe there is a missing term in there but that's OK but if you look at the solvation part, you will have a contribution to $\Delta G^0(R)$, the bonding part will also have a contribution and this contribution is ΔE^0_{DHA}, so $\Delta G^0$is the sum of two contributions, one is ΔE^0_{DHA} and the other I'll call ΔG^0 of solvation. There is a barrier for the DHA part, right? Now write down a barrier for the solvation part, and on the right hand side write an $n\Delta G^0(solvation)$, add those two contributions together and you'll have the derivative and take the derivative, you'll have Eq. (13) then add it to the Eq. (8) there and you will get Eq. (13).

5. Following: "This eq 13 reduces to previous expressions for the two individual results for DHA and prot, when λ_o or λ_i vanishes, respectively. For example, when $\lambda_o = 0$, eq 13 reduces to eq 8."

Q: Equation (8) is $(\lambda_i/4\ln 2)\ln[n/(1-n)] = \Delta E^0_{\text{DHA}}(R)$, whereas in Eq. (13) there is $\Delta G^0(R)$. Do you tacitly assume $\Delta G^0(R) = \Delta E^0_{\text{DHA}}(R)$ because in general we have $\Delta G^0(R) = \Delta E^0_{\text{DHA}}(R) + \Delta G^0(\text{prot})$ and here, where we have supposed $\lambda_o = 0$, we also tacitly suppose $\Delta G^0(\text{prot}) = 0$?

M: There is a problem I guess, I'd have to look at it, and try to use the same n for both things, and so things are getting a little bit mixed up, since in a way one wants the two things separately, *so if you use the same n for both things as I did in Eq. (13) you run into that kind of problem if you try to break it up into parts.*

Q: One should not do that.

M: Probably not. If you're tying together one thing to the other is like a marriage in the old days, is bound forever.

Q: A very humane explanation.

6. Following: "For the present case eq 12 yields for n in the TS

$$\Delta G(n, R) = -\left(\frac{\lambda_i}{4\ln 2}\right)[n\ln n + (1-n)\ln(1-n)] + \lambda_o n(1-n) + n\Delta G^0(R) \quad (n \text{ in TS}) \qquad (14)\text{"}$$

Q: If we compare Eq. (14) with Eq. (12) and with Eq. (8) in **M354**, we see that the term $n\Delta E^0_{\text{DHA}}$ is missing. Why?

M: $n\Delta E^0_{\text{DHA}}$ is really a contribution to the $n\Delta G^0$, so it may be that some of the other things that we put in together should have been kept separately, that's possible. In other words, if you look at Eq. (8) there things are focused on the bonding, probably I should have written a similar thing focused on the solvation part and I would have had an $n\Delta G^0$(solvation), and then I could have combined the two to make it look like Eq. (13)....

7. Following: "The free-energy barrier for *this particular form for* $E_{\text{DHA}}(n, R)$ *in model B* obtained from eqs. 13–14 gives

$$\Delta G^*(R) = -\frac{\lambda_i}{4\ln 2}\ln(1-n) + n^2\lambda_o \quad (n \text{ in TS}) \qquad (15)$$

M: The first term looks like that bonding paper that I had in 1968, and the second term looks like the free energy barrier that I had in 1956, so this now combines both, so what's particular about that form, is that it managed to combine both things. Now whether it's best to do that way or not I don't remember now, but it's possible to do it that way. May be that it's not best to do it that way, I've forgotten now. In other words, I've gotten in a form where in either limit I get a result that I had obtained previously.

Q: How do you get Eq. (15) from Eqs. (13) and (14)?

M: Probably by maximizing.

8. Following: "While $\Delta E^0_{\text{DHA}}(R)$ no longer occurs explicitly in eqs 13 and 15, and so no longer *solely* controls the n for DHA in the TS for model B, it does contribute to the driving force $\Delta G^0(R)$ *in eq 2*."

Q: (1) What do you mean by "solely"?

(2) A **typo**: not in Eq. (2) but rather in Eq. (5).

M: The term has been subsumed into $\Delta G^0(R)$ term that contributes there, and the only way of making into affecting things individually is not by having a single n for solvation and the same n for bond. In other words, once you try to make the same n for both, as I did in this model, then the control, the complete control that the $\Delta E^0_{\text{DHA}}(R)$ had for the bonding part is gone, it doesn't have that complete control anymore.

9. Following: "When $\Delta G^0(R)$ is small, eqs 13 and 15 can be expanded about $n = \frac{1}{2}$ to yield for the TS

$$n = {}^1\!/_2 + \Delta G^0(R)/2\Lambda_2 + \cdots \qquad (16)\text{"}$$

Q: Do you get n considering that in Eq. (13) it is given as an implicit function of $\Delta G^0(R)$? Look at Eq. (13) and you want to expand the function n of $\Delta G^0(R)$.

M: Yes, I just look at Eq. (13)... is that the same as Eq. (13)"?

Q: The "just means that the statement finishes there.

M: When $\Delta G^0(R) = 0$ in Eq. (13), you could see that the solution of that problem is $n = \frac{1}{2}$, $0 = 0$ then. You see that?

Q: Yes.

M: OK, so when $\Delta G^0(R) = 0$ then $n = \frac{1}{2}$. And now you write $n = \frac{1}{2} + \varepsilon$ and make the expansion, and that will give you an expansion for values of $\Delta G^0(R) \neq 0$. Eq. (13) is an equation for n, and if you expand in that equation n about the value 1/2 you'll end up with Eq. (16). For small $\Delta G^0(R)$s if you want to get the value of n write it as $n = \frac{1}{2} + \varepsilon$ and solve for ε, and the you get Eq. (16).

10. p. 6646 1st column top: "The vertical energy difference ΔE of the two diabatic states has been a useful reaction coordinate. Whether the vertical energy difference of two *adiabatic states* is also useful remains to be explored. Instead, for a treatment using adiabatic states it may be necessary to first transform them into diabatic states to define a suitable ΔE coordinate."

Q: Which one is in general the relation of diabatic and adiabatic states in electron transfer theory?

M: The diabatic states are the two parabolas, the adiabatic states would be the correction of those two parabolas by taking into account the electronic interaction, and that will result into a small splitting of the two parabolas at their intersection, and those would be the adiabatic curves. I did give some thought to what to do for a general reaction coordinate in the adiabatic case but I've never written...

Q: So, it is something that is in elaboration somehow.

M: Yes, I thought of it maybe a couple of years ago now, I got around it but in fact I forgot about it.

11. p. 6646 middle: "An equation for the corresponding entropic contribution to the reaction barrier ΔS^* and for the standard entropy of reaction at R, $\Delta S^0(R)$, is obtained, in principle, from

$$\Delta S^* = -\partial \Delta G^*/\partial T, \quad \Delta S^0 = -\partial \Delta G^0(R)/\partial T \qquad (20)$$

Q: Why "in principle"? Is it so just because there may be not enough data of $\Delta G^*(T)$ or $\Delta G^0_R(T)$?

M: That's what one should do, but I imagine what I was thinking of, although I don't remember now exactly, was that there really wasn't a lot of data around to do that, but in principle that's what one should do, but often they don't give all that data. . . I mean you can do in some cases. Sutin and I did something like that in one paper we wrote but often there isn't a lot of discussion of the ΔS^0, for example, people may speak of ΔG^0, but often they don't speak of the ΔS^0.

12. Following: "It can be shown that ΔS^* differs from the entropy of activation $\Delta S^\ddagger$ by a small amount, k"

Q: How or where do you show that?

M: *It all depends on how you define ΔS^* and $\Delta S^\ddagger$. You define $\Delta S^\ddagger$ by $k = \frac{kT}{h} e^{-\Delta G^{\neq}}$ then $\Delta S^{\neq} = -\frac{\Delta G^{\neq}}{R}$. If you equate that to the rate constant expression that I wrote, it is not in terms of the $\Delta S^\ddagger$ but in terms of the ΔS^*, then you'll see that they differ by small value of k or something like that, because they're defined a little bit differently, one has in part of its definition a $\frac{kT}{h}$, the other doesn't.*

13. Following: "Nuclear tunnelling can affect the numerical values of both A and E_a."

Q: You already discussed the influence of tunneling on A at 0 K. How about the influence on E_a?

M: *If you have a reaction that's all tunneling, doesn't go over the top of the barrier, then that means its activation energy is zero, so the* *nuclear tunneling has affected the activation energy. The activation energy is defined as* $\frac{d \log k}{d(1/T)}$, *that is its definition, . . . so its operational definition is taken from plotting* $\log k$ . . . *I mean the activation energy is defined* $k = e^{-\frac{E_a}{kT}}$, *that's the definition of it. Now, if it's nuclear tunneling isn't going to have any temperature dependence, it's only going from the lowest state, if it's completely clear tunneling, that means its activation energy is zero.*

Q: In this sense.

M: *Yes, in the sense that it's defined by an Arrhenius equation. You write an Arrhenius equation for low temperature and you write a different Arrhenius equation for room temperature. In other words, the activation energy strictly speaking is defined as* $\frac{d \log k}{d(1/T)}$ *but* $\log k$ *vs. 1/T is not necessarily a straight line, eventually it flattens out at low temperature, and the slope becomes zero, so the activation energy as defined by Arrhenius becomes zero. In the sense that the activation energy is defined not by the energy barrier but by* $\frac{d \log k}{d(1/T)}$.

14. Following: "When there are two or more reaction steps rather than one, and when the reaction step from the reactants to a reaction intermediate is rate determining, the overall $\Delta G^0(R)$ is not the relevant quantity to correlate with the observed rate but rather the $\Delta G^0(R)$ for the rate-determining step. For example, *if variation of the overall* $\Delta G^0(R)$ *has little effect on the first step*, but the first step is the slow one..."

Q: But if the overall $\Delta G^0(R)$ contains the contribution also of the $\Delta G^0(R)$ of the rate determining step why should the variation of the overall $\Delta G^0(R)$ have little effect on the rate determining step?

M: The overall ΔG^0 will have effect... *when the first step is rate controlling, in other words it's not the reorganization part, not the rate bonding part, when the first step of approach is rate controlling, then if you vary the overall standard free energy of reaction, maybe you vary the substituents, maybe you vary the energy barrier for the* *intermediate thing by varying ΔG^0, but if it's the approach that's rate determining then it has no effect on the rate. The overall ΔG^0, the usual way to vary it is by varying a substituent, that is you're varying the part not of the approaching the reactants to each other but the whole reorganization business.*

Q: But doesn't it contain also the partial ΔG^0?

M: *That's probably relatively little affected when you vary substituents, it's the bonding part, the solvation, free solvation part that's affected, but not so much the approach, like, typically, if you make the products bonds stronger than the reactants, that affects what happens after the approach, when the bonds start changing their lengths, the approach part doesn't involve any change of lengths of bonds, doesn't involve any of the kind of reorganization of solvation, that's involved in electron transfer, so doesn't involve the thing that is usually specifically changed by systematically changing ΔG^0.*

Q: That is the ΔG^0 for the determining step remains the same, basically.

M: Yes, *suppose you change a substituent, usually all change work require to bring the two reactants together, usually you're changing maybe the solvation or you're changing the strength of a bond that's being formed, or you're changing the strength of a bond that's breaking, but the approach part is before any of those bond changes, any solvation changes occur. In other words, when I wrote w^r by w^r I meant the part of the free energy barrier which is independent of these other processes, that make ΔG^0.*

15: Following: "... the rate plotted versus $\Delta G^0(R)$ will reach a limiting low value..."

M: *The limiting value being that due to the first step, in other words when the first step becomes rate determining, doesn't matter what ΔG^0 is, in ΔG^0 there's all that change in structure that's going on* *after the first step, that's tied up with the change, what normally changes ΔG^0, but that has no effect on the rate if the first step in this particular case is the slow step. If you have two successive steps, and the first step is the slow step, it doesn't matter how fast the second step is.*

16. p. 6647 1st column middle: "The P's in eqs 23–24

$$k_{\mathrm{rate}}(R) \cong \nu \int P(E)\exp(-E/kT)\mathrm{d}(E/kT) \qquad (23)$$

$$P_{00}(E) = A'\exp(-2K(R_0)\exp(-\Delta G^*/kT) \qquad (24)$$

tacitly contain a term w^r via the ΔG^* in eq 3 that by definition contains orientation and distance R effects. This factor $\exp(-w^r/kT)$ could be expressed in more formal statistical mechanical terms, but we have simplified the notation."

Q: (1) A **typo**: there is no ΔG^* in Eq. (3).

(2) Where is it possible to find the more formal statistical mechanical expression for $\exp(-w^r/kT)$?

M: (2) In terms of certain partition functions. In Eyring's book when he gives a rate, when he's looking at some reactions and he gets the rate... *maybe you have a rotation going over to vibration, that's the equivalent of a steric factor. In other words, there is less freedom in the transition state.* Nowadays of course one can get some of these partition functions more accurately from the potential energy surface so that's what's done when one applies transition state theory, one captures that.

17. p. 6648 2nd column top: "The dependence of ΔG^* on ΔG^0 in a series of ET reactions is studied with a series of related reactants in which one of the reactants is systematically varied at more or less fixed λ. *In some cases using self-exchange data it has been possible to correct for changes in λ, usually via testing the 'cross-relation'...*"

M: Well, for example, Lewis who did a lot of proton transfer work at Rice University, might have been hydrate transfer, measured the self-exchange reactions, he measured the cross reaction, so if the λ did the same for all self-exchange reactions, they would have had the same rate. In fact they had slightly different rates, so by using the actual rates he did make no assumption that λ was the same for all.

18. Following: "Prior to and following the H transfer there is a binding of the reactants (substrate and cofactor) or products to the enzyme. *The 'commitments' (involving the two equilibrium constants) need to be measured or calculated in order to obtain the* ΔG^* *and* ΔG^0 *for the H-transfer step, but to determine* ΔG^0 *it is still necessary to know the commitment of the product.*"

M: These are things that are defined in the isotopic literature, and they're related to the pre-equilibrium post-equilibrium constants, involving the formation of the complex and the destruction of the complex. The *commitments* are really certain equilibrium constants. It's a terminology that the isotope people use. Isotope reactions where you have complexes formed, you're not just looking at one reaction but you form a complex, you study it, the complex dissociates, those equilibrium constants involve the formation of a complex, and the dissociation, they are called commitments in the literature.

M357 H and Other Transfers in Enzymes and in Solution: Theory and Computations: A Unified View. 2. Applications to Experiment and Computations

(Continuation)

R. A. Marcus

1. Following: "When the binding constants can be determined, or otherwise overcome, and when the binding steps are not rate-limiting, both the ΔG^* and the ΔG^0 of the H-transfer step can be measured."

M: Those binding constants are probably the ones involved in the equation for the rate constant, and if the binding step is rate limiting then doesn't matter what the rate constant is for the catalytic step, if the binding constant is rate limiting, then everything after that doesn't contribute to the rate because it is already the binding process that affects the rate. So, in the case that the binding is rate limiting then by definition nothing else affects the rate.

2. p. 6649 2nd column middle: "Knowledge of ΔS^0 for this reaction step could assist in understanding ΔS^*."

M: Well, there are several contributions to ΔS^*, and one of them, certainly ΔS^0, plays a role. As you go along the reaction coordinate the reaction developed that ΔS^0. If the TS is close in structure to the products, then the ΔS^* will have a large component of ΔS^0, if it's close to the reactants ΔS^0 plays very little role, the more usual situation is roughly half way in between. ΔS^0 is one of the contributions, in the form of some fraction of ΔS^0, to the entropy of activation. Undoubtedly, Sutin and I wrote about that in the 1985 BBA article.

3. Following: "In a proton transfer in the dehydration of HCO_3^- catalysis (in the form of $\log k_{\rm rate}$ vs ΔpK_a) by carbonic anhydrase, Silverman found that a plot of ΔG^* versus ΔG^0 reached a limiting value at a relatively low value of the rate constant. *If one assumed only a two-states description, the result would imply a very large* w^r, *the residual barrier at a quite negative* $\Delta G^0(R)$."

M: *If you reach a limiting value, in other words if you lower the* ΔG^0 *as much as you can do, you can't decrease the intrinsic barrier, the barrier* ΔG^*, *anymore. Then the only barrier that's left is the work, is the part of the barrier that doesn't depend upon* ΔG^0, *and by definition that's* w^r, but you see, that's a two states approximation.

4. p. 6650 1st column middle: "A puzzling experimental result for $\Delta S^\ddagger$ is the very large negative value (-56 cal mol^{-1}deg^{-1}) seen in the $H^{\bullet}$ transfer in soybean lipoxygenase. The tunneling factor of 80 only accounts for -9 cal mol^{-1}deg^{-1}. It appears to be due, at least in part, to an electronically nonadiabatic—vibrational nonadiabatic H^+ transfer."

Q: Please explain the relation between very negative $\Delta S^{\ddagger}$, tunneling and nonadiabaticity...

M: *Various factors affect the preexponential factor. The $\Delta S^{\ddagger}$ is an Arrhenius term, is related to the preexponential factor, and not to the heath of activation, and a ΔS^0 that's quite different from zero affects that, as we just discussed, and also tunneling affects that, a small tunneling probability that translates into a negative ΔS^* or $\Delta S^{\ddagger}$.*

Q: And what about the last point, nonadiabaticity?

M: *That's an interesting question... if you think of the H-transfer as a transition*, then, *if you start discussing with transition in terms of vibrational overlaps, the overlap of the H-motion of the reactants with the H-motion of the products, then that's discussing the reaction not in the way in which a vibration smoothly evolves from one into the other, but rather in terms of a jump, a vibrationally nonadiabatic process.* So probably I meant something like that, that *when you had a vibrational transition, from one well to the other, for example, supposing it is largely at the zero-point energy, the zero-point state of the reactants well, and it goes into a high vibrational state of the products well, which would happen if the reaction is quite exothermic, then the vibrational wave function of the product is rapidly oscillating and it just has a small amplitude outside, so the oscillations shouldn't come in the tunneling region.* It may be that one would have to think semi-classical. *If you're in a high vibrational state, the wave function may extend only just very slightly beyond the turning point*, that's possible, I'm not sure, I would have to think that through. If that would the case, there'd be very little vibrational overlap.... *one can see it actually just by looking at some of these vibrational overlap factors,* *in nonadiabatic treatments, for example, for electron transfers. In a high vibrational state of the products the vibration potential energy function is rising very rapidly, you know, it's quadratic so it rises very rapidly, that means that if you're in a high state, the wave function decreases very rapidly in the classically forbidden region, and so . . . once you get into the vibrationally forbidden region the overlap of the vibrational wave functions, the H vibrational wave functions, is small,* so I think probably that's what I meant, or should have meant.

5. Following: "If a protein is more 'rigid' its λ_o will be larger. (The analogue to the λ for a chemical bond in ET reactions is $ka^2/2$, so that when λ is larger the equivalent to the bond force constant k is larger, for a given a. That is, the bond is stiffer)."

Q: Why don't you then use here the λ_i instead of the λ_o?

M: I think that the immediate environment of the transferring H in a protein has contributed to the λ_i, and everything outside of that is contributing to λ_o. That is, the rest of the protein is serving as a "solvent."

6. p. 6650 2nd column middle-bottom: "Some comparison of enzyme-catalyzed reactions and the analogous reactions in solution are given in Liang and Klinman and Olsen et al. In one example a large difference in KIE for the two media was found. The dramatic effect of mutations does not have its counterpart for reactions in solution."

Q: Which mutations are you referring to and how do they influence the KIE?

M: There's a series of experiments by a number of people, you know, especially I think Stephen Benkovic, at Penn State. He finds that *when you have those mutations you change all sorts of things and, whenever there is, a KIE. A change in hydrogen bonding structure can make the bond have to stretch more, because the two reactants are* *not in quite the same arrangement they were. The natural conditions are normally the best ones, and as you disturb them, if you make some mutation that cause the two reactants sort of non being in close contact with each other, then the H-bond will have to stretch more than it ordinarily would, and that brings in a larger isotope effect.*

7. p. 6651 1st column top: "The experimental slope of the KIE in hyperthermophilic dimeric DHFR is small or zero in the physiological temperature domain but fairly large in the nonphysiological domain, namely, at lower T values."

M: Well, if you look at some of the thermophilic enzymes, then you see that there is a breakpoint, where the slope of the log KIE with $1/T$ changes with that temperature, and has one value above that point, above that temperature, and another value below that point, correspondingly the preexponential factor changes at this breakpoint, this temperature is a transition point for the thermophilic enzymes. At temperatures below the breakpoint they no longer are operating the way they would at temperatures above the breakpoint. DHFR is the most commonly studied thermophilic enzyme. The origin of this transition, I saw a poster of one of Warshel's student maybe three or four years ago, in which he may have explained that big change in slope has been due to ionic effects.

8. Following: "A bond-length-based computation for a thermophilic DHFR has been made and gave a modest T-dependence but did not show *a break in the slope* of log KIE versus $1/T$ observed in the experiments."

M: Yes, a break of the slope is when you plot a log KIE vs. $1/T$, the slope has one value at a temperature above, at a given temperature, and a different value below.

9. p. 6651 1st column bottom: "Another study is that of Hay et al. on a hydride transfer from NADH by a morphinone reductase. The deuterium KIE was about 4 at 25°C. Both k_H and k_D depended on temperature such that the KIE, $\frac{k_H}{k_D}$, was also temperature dependent. The simplest interpretation is that it is a $H_\alpha D_\alpha$ case, and thus the temperature effect on $\frac{k_H}{k_D}$ is primarily due to *the differences in zero-point energies between the TS and the reactants*."

M: *If the system existed before the jump with a zero-point energy and then there wouldn't be changes in zero-point energy throughout the reaction, no stretching of a bond... then the KIE would be independent of chemistry. If there wouldn't be differences in zero-point energies between the reactants and the TS, if they'd be fixed in their zeroth order vibrational states, there would be no chemistry dependence. But if in the course of forming a TS you have to stretch that bond, then you're going to change the zero-point energies, and so the zero-point energy on the TS would now be different from that of the reactants.*

10. Following: "There was also an accelerating effect on k_H and k_D individually when the pressure was increased, but the $\frac{k_H}{k_D}$ decreased somewhat."

Q: Please check my understanding: An increase in pressure reduces the widths of the barriers to be penetrated by tunneling and the decrease in width favors both H and D but, *relatively* more D.

M: If you decrease the width of the barrier then there's going to be more tunneling, the barrier is thinner, there is a shorter distance to go, normally you can tunnel more through a thin barrier than you can through a thick barrier, so there is actually more tunneling. Now, with a wider barrier there is a bigger discrepancy between H and D because H tunnels much more than D. But there are subtleties. If you make the barrier much wider, then the system will tend to go over the barrier. It depends on if one has the choice between tunneling and not tunneling what the answer is to *D being more or less favored.* Other things being equal, the thicker the barrier the less is the tunneling.

The smaller the distance to jump, the less the difference between the rates with D compared with H since there is less tunneling needed.

If the tunneling goes to zero the $\frac{k_H}{k_D}$ is closer to unity, than when tunneling occurs.

11. Following: "The pressure effect is consistent with the $H_\alpha D_\alpha$ behaviour, since pressure would affect this ratio if the reaction were of the $H_\beta D_\beta$ or $H_\beta D_\alpha$ type..."

Q: Probably "in a different way" is missing...

M: You right, yes, correct.

12. Following: "Proton-coupled electron transfer in lipoxygenase and in nonenzymatic systems can have very large isotope effects because of the long distance for the H-tunneling. In the latter case the deuterium KIE was 400. *The role of electronically adiabatic versus nonadiabatic behavior is discussed by several authors*..."

Q: Please explain the adiabatic vs. nonadiabatic role in tunneling ...

M: If there's very weak coupling, then you really are talking about going from one potential energy surface to another. If there's strong coupling, you always stay on the lower surface. So no tunneling then.

13. p. 6651 2nd column middle: "The pressure effect depends on how close the reaction is to $n \approx \frac{1}{2}$. For model B, $n \approx \frac{1}{2}$ requires $\Delta G^0(R) \cong 0$, whereas for model A it requires $\Delta E^0_{\text{DHA}} \cong 0$. The pressure effect would thus be different for the two models.

M: Since ΔE^0 is different from ΔG^0, because of the entropy, those conditions are different, $\Delta G^0(R) \cong 0$ and $\Delta E^0_{\text{DHA}} \cong 0$ are two different magnitudes when there is a finite ΔS^0. $n \approx \frac{1}{2}$ depends on which condition is fulfilled, and so the pressure effect is related, among other things, to the two different conditions that are going to affect the rate. In other words, the pressure effect among other things affects ΔE^0 and ΔG^0 in different ways, it is so because the pressure effect depends on how close is n to the 1/2, and the conditions are different for B and A, the pressure effect is going to be different.

14. Following: "When the enzyme is of the *wild type* and has its natural substrate *and operates at its physiological temperature, the T-dependence is zero or weak*... Of the several cases considered in an earlier section, $H_\beta D_\beta$, $H_\beta D_\alpha$, and $H_\alpha D_\alpha$, *only the* $H_\beta D_\beta$ *can have a T-independent KIE, and then only when the work term* w^r *is either small or is about the same for the D and H reactions.*"

Q: (1) What does it mean "of the wild type"?

(2) Why the zero or weak temperature dependence at the physiological temperature?

(3) Why only the $H_\beta D_\beta$ can have a T-independent KIE and why only when the work term w^r is either small or is about the same for the D and H reactions?

(4) How and why can a work term be different for a D and an H reaction?

M: (1) That's the standard terminology for the natural enzyme, without making imitations.

(2) Enzymes are designed to operate in certain temperature ranges, and you see an example of one that was designed to operate at high temperatures. When you make to operate at low temperatures you get a quite different behavior.

(3) You always will get a temperature independent KIE as if you are not stretching bond, but you do the jumping, that's a β for both.

(4) *The work term is defined as the part of the barrier which occurs prior to the actual jump. Now, in principle if a D is little bit closer*

before it starts making its move, because it doesn't want to jump as much, in principle it can have a different work term.

15. Following: "Indeed, in examples of thermophilic enzymes, the KIE is T-independent in its natural temperature range, *but outside that range the KIE has an activation energy either reflecting the* $H_\beta D_\beta$ *becoming* $H_\beta D_\alpha$ *or* due to a difference of the work terms for the H and D systems."

Q: Does temperature independence automatically mean zero activation energy?

M: Temperature independence is equivalent to zero activation energy. They say the same thing.

16. p. 6651 2nd column bottom: "...they calculated a KIE of 2.81 without tunnelling and of 3.1 with tunneling... The calculated value of 2.81 is *far less than the semiclassical value of 6 to 7 that occurs when one vibration of the reactants disappears in forming the TS and so suggests some tightening of other vibration frequencies on approaching the TS.*"

M: *In one model that produces the 7 isotope effect, 6 or 7, you simply break the C-H bond, do nothing else, and the difference of those two bonding things is the zero-point energies, the difference of those two C-D, C-H gives rise to a factor of 6 or 7, so you have a difference of zero-point energies, but clearly that's an unrealistic picture of what's going on, because as you're breaking you're forming something the net isotope effect is different. In other words, the 6 or 7 is the highly idealized model, but probably doesn't imply anything.*

17. p. 6652 1st column middle: "In EVB calculations of enzyme reaction rates the H-transfer is treated as a transition from one electronic state (DH, A) to another, (D, HA). In the EVB formalism the vertical energy difference ΔE is usually used as a reaction coordinate in the many dimensional DHA/protein coordinate space. The computational results typically yield a pair of parabolic free-energy curves from which a λ can be extracted from the results. Since a DH stretching and an HA compression both contribute to this calculation, the λ obtained in the fit is the sum $\lambda_o + \lambda_i$. ΔG^0 is also obtained or fit to the experimental value. *Perturbation theory is then used to obtain a modified free-energy curve along the reaction coordinate ΔE, using a distance-dependent off-diagonal matrix element describing the interaction between the two valence bond states, and so yielding ΔG^* for the H-transfer step.*"

Q: Do you here describe how to go from diabatic states to adiabatic ones?

M: That's basically what Warshel does . . . he uses perturbation theory to go from the diabatic to the adiabatic states, that's right, and in the process he has an interaction term that appears in addition to diabatic states.

18. Following: "In practice, the work term w^r is not studied explicitly, but could be evaluated, as discussed earlier, *by changing a substrate or a cofactor in a way that ΔG^0 is made increasingly negative. Then only a w^r will remain.*"

M: The w^r is the part of the barrier that's independent of ΔG^0, so if you make the part of the barrier that depends on ΔG^0 small, you're left with the w^r.

CHAPTER 13

Evidence for a Diffusion-controlled Mechanism for Fluorescence Blinking of Colloidal Quantum Dots, Universal Emission Intermittency in Quantum Dots, Nanorods, and Nanowires, ET Past and Future, Beyond the Historical Perspective on Hydrogen and Electron Transfers, Interaction between Experiments, Analytical Theories, and Computation, Interaction of Theory and Experiment: Examples from Single Molecule Studies of Nanoparticles, Theory of a Single Dye Molecule Blinking with a Diffusion-based Power Law Distribution, Extension of the Diffusion Controlled Electron Transfer Theory for Intermittent Fluorescence of Quantum Dots: Inclusion of Biexcitons and the Difference of "On" and "Off" Time Distributions

Interviews on M358, M366, M368, M373, M374, M379, M389

M358 Evidence for a Diffusion-controlled Mechanism for Fluorescence Blinking of Colloidal Quantum Dots

Matthew Pelton, Glenna Smith, Norbert F. Scherer, and Rudolph A. Marcus

1. p. 14249 1st column middle: "Widespread interest in this blinking phenomenon was stimulated by the *surprising observation* that the durations of the bright and dark periods follow power-law statistics."

Q: Why was the observation surprising? What was to be expected?

M: *An exponential decay.*

Q: Why so?

M:*If* the diffusion process *were just emission, it'd be exponential decay. If it would be just ejection* of the electron*...from one place to another or jumping from a particular site back into the thing, it'd be two exponential decays.*

Q: Like in radioactivity.

M: Yes, exactly.

2. p. 14249 2nd column middle: "To calculate the blinking-time probability densities, the recorded photon arrival times are converted into a time series by grouping the photon detection events into a time series by grouping the photon detection events into time bins of fixed width, as illustrated in Fig. 1. The minimum bin size that can be used is dictated by the need to clearly distinguish bright and dark states. More specifically, the average number of detected photons per time bin must be *significantly greater than unity*, or it will not be possible to differentiate between the detection of photons emitted from the dot and *background counts.*"

Q: (1) What do you mean by "background counts"?

(2) How much is it "significantly greater than unity"?

M: (1) In any measurement there is always some noise, and so when the system is in what's called the dark state and emits very few quanta, then there's still some noise, and in any machine you're going to get some electrical response, whether it's coming from light pulses or not, there is always some noise, so somehow you have to make some assumption about discriminating noise from an actual signal, and they do in a somewhat primitive way. They say that everything below a certain magnitude is noise. In other words, if there would be no emission of light during those start periods, none, then of course you would be at the base line, but if any measurement involves noise, then there is always going to be some sort of a signal above the base line which has nothing to do with the light itself, with a fluorescence, just noise in the electrical circuit.

(2) I suppose that it depends on your sensitivity. I mean, I'm not sure, whether by counts one means what you've actually observed or what the actual number of events are. I don't know what is meant there. By the way, if you just have one count in there, it may not be differentiated much from the noise because the noise provides a signal too, your thing is some sort of signal, maybe some current or something like that... so you need to have enough of a signal that's bigger than the noise. I think there is some technique, maybe not used for quantum dots, I don't know, a special way of somehow getting out the noise, and some sort of correlation technique I think I've seen, but I don't know anything about it, but I think there are some special ways of handling things when there is noise, and I don't know that he [Pelton] used them, probably he didn't, but there may be something there, maybe I was at some seminar and somebody was talking about it, I just don't remember.

3. p. 14250, legend to Fig. 1: "Blinking of quantum dots and diffusion-controlled model. (a) Intensity of fluorescence as a function of time measured from a single CdSe/ZnS core-shell nanocrystal QD. The two tracings show the same data on different time scales, grouped into time bins of 100 msec (*Upper*) and 5 msec (*Lower*)"

Q: Please look at Fig. 1. I have two remarks: (1) The time coordinate is reported in the figure as "sec." while from the legend it should be "msec."

(2) In the lower intensity graph the time bins are clearly of 12.5 msec and not of 5 msec as written in the legend.

(3) How can you distinguish on and off periods from the messy system of spikes in Fig. 1? There you see intervals in which there are no spikes, like for instance in the upmost set of spikes to the left of number 200 or to the left of number 600. It is then easy to identify those periods as "off." But in general there are groups of spikes of different height. Look for instance at number 600 in the abscissa. There at the left one finds a typical "off" period, then at the right a typical "on" period, but then how would qualify the farther group of lower spikes to the right? Are we there in an on or in an off period?

M: (1) *Yes, absolutely, it is milliseconds, it's wrong to call it seconds.*

(2) I don't know how many bins there are . . . the bins are much smaller than the 25 ms . . . . I don't know how what the size of the bin is . . .

(3) *That's one of the worst figures I've ever seen . . . I mean, I've never seen such a terrible figure, if you look a Bawendi's figures, if you look at Nesbitt's, or if you look at Jau's, you don't see such terrible figures. If I knew what I know now, I would have never allowed them to publish that, it's terrible*, it is, compared with the figures of Bawendi, you see, it's very hard to distinguish there, I think he must have events occurring in each bin, otherwise you wouldn't see that sort of stuff.

Q: There you see intervals in which there are no spikes, like for instance in the upmost set of spikes to the left of number 200 or to the left of number 600.

M: Yes, that looks cleaner, it occurs that's a very gross time scales. I think what he got is multiple events occurring in each bin.

Q: So you wouldn't pay much attention to this figure.

M: I wouldn't now, I wasn't that aware of things before, but we looked a lot into things and especially in the work with Wei Chen and Joy Anne, and whether there are good things to work on, and something like that we wouldn't touch at all, there is one particular one that we would have allowed to attach, but it looked a bit like that and so we didn't.

Q: I'm sorry to be so nit-picking . . .

M: *Oh, no, I hadn't looked to this paper . . . that kind of detail, . . . on a long long time, by I'm shocked.* But you see, *the main thing that he did in that paper though, was not this, the main thing he did was the frequency analysis, and the idea behind that was that if you do the frequency analysis you don't do any binning. To the best of my knowledge, when he does the frequency analysis and plots some quantity on a frequency analysis basis, there's no binning in that, there is no threshold . . . you just do Fourier analysis of the signal at various times.* In what Pelton did, was that he did a Fourier analysis, so you don't assign bins and count the number of points in the bins, you just simply do a Fourier analysis, so it's a different way of handling data.

Q: By the way . . . what is this bin and binning?

M: That's when you observe the crystal for a certain length of time, very small length of time, you observe the crystal, the fluorescence for a certain time, and that time that you're collecting data on is a bin, then you go to another time and get data in a bin, or to a later time you get data in a bin, so you're collecting data in bins, short periods of time when you're getting a number of pulses of signal.

Q: So they are just short intervals of time.

M: Yes, that's right, and you are doing some observation during those intervals. This is a way of calculating the intensity of the fluorescent light.

Q: It is similar to the beta emissions of an atom . . .

M: Yes, only this is a multiatom that is giving off fluorescence.

4. Following: "**Autocorrelation Function**. The photon autocorrelation function, defined as

$G^{(2)}(\tau) = \langle i(t)i(t+\tau)\rangle$, where $i(t)$ is the rate at which photons are counted experimentally, can be calculated directly from the photon counts without *the need for additional binning*, as described in the SI. To allow comparison with theory, we calculated the normalized correlation function,

$$g^{(2)}(\tau) = G^{(2)}(\tau)/(\langle i_1(t)\rangle\langle i_2(t)\rangle).\text{''}$$

Q: Which is the meaning of $\langle i_1(t)\rangle$ vs. that of $\langle i_2(t)\rangle$ and of their product?

M: That's the main focus of the paper, to calculate that.

Q: The autocorrelation function.

M: Yes, that's right. We calculated the normalized correlation function, that's what one usually does. The $i(t)$ is sort of a signal that you're getting. Now, if you average it over time, that's the $\langle i(t)\rangle$. $\langle i_1(t)\rangle$ and $\langle i_2(t)\rangle$ are probably equal, the same thing, it depends on what $i_1(t)$ signifies. If it's sort of the rate of photons counting at any time, then the rate of that counting at time 1 would be the same as that counting at time 2.

Q: But the time is the same.

M: I don't know why he put the 1 and the 2 into there, why he didn't just write $i(t)$ and $i(t)$, because that's what you have. *As far as I can tell, you should drop the 1 and the 2.*

5. p. 14251 1st column top: "The power spectrum is defined as $P(f) = |I(f)|^2$, where $I(f)$ is the Fourier transform of the photon count rate, $i(t)$, for frequency f. As for the autocorrelation function, no additional *binning of the data* is necessary to calculate the power spectrum."

Q: What is meant by "binning of the data"?

M: The alternative to calculating the power spectrum and from calculating the correlation function is to use binning data as they are found and do a Fourier transform of the data without putting them into bins, as best as I understand it.

6. Following: "Eq. 2

$$P(f) \propto \begin{cases} f^{\nu-2}, & 0 < \nu < 1 \\ f^{-2}, & -1 < \nu < 0 \end{cases} \tag{2}$$

holds exactly only if the on and off blinking periods are described by the same power law probability density. In most cases, including the present data, they are described by two separate power laws, with different exponents ν_{on} and ν_{off}, as illustrated in Fig. 2. In this case, the power spectrum is dominated by the process that contributes the larger noise. For our data, this is the on-time probability density . . ."

M: *The way he's looking at his "off" state, of course that's when he did the binning, what you would see would be those signals for different lengths of time, and that would affect your spectrum. I mean that would affect when you look at the correlation function, that affects that correlation function, even if it is not emitting, is just that i(t) at that time is very, very small. . . so is affecting the correlation of i(t) and i(t'), and even if the QD is not emitting is affecting that correlation function, is giving a zero at that point.* But I think that his binning has been absolutely too coarse and normally, instead of having what he has there with a slope of -1.9, maybe if he uses a finer binning would have at the early times a slope of -1.5 and then later on an exponential tail, that's the more usual result. There's something messy about what he's done here with the bin stuff.

7. Following: "**Distributed-trapping models.**... The key difficulty with this model is that, if blinking could simply be described by electron transfer at fixed rates to and from a single surface trap, then the durations of bright and dark periods would follow *exponential probability densities.*"

Q: Why so?

M:*If you had a single rate constant, in other words if, say, you had only one trap and something maybe came out of that trap, or went into that trap, with a single rate constant, then its formation or disappearance would vary with a single exponential for a unimolecular process, but if you had many traps of different strengths, different rate constants for being filled and different rate constants or for being empty, then your overall rate is a compositive of many independent events, each of which has its own exponential decay, so that gives you a sum of exponentials.* In fact you can even imitate a square wave form with a bunch of exponentials, you can expand in a set of exponentials and so on, so that gives you a rate, an *overall* rate, which is not to have a single exponential, it is more complicated and it can even give rise to what's called a *power law*, under special circumstances. That's not the explanation we used for a power law, but that's the explanation that some people used.

Q: From what you say one trap means one rate constant, two traps two rate constants...

M: Yes, if the traps are different, yes, if they have different capture ability or different dissociation ability, and normally one assumes in solid state physics, going back to the 1950s anyway, that near the band edges one has traps, one has an exponential distribution of traps, each with its own rate constant.

Q: By the way, how are these traps physically made?

M: There are probably any number of traps, they may be missing ions, an interstitial intermediate, a site that has an ion of a different valence state from that of the other ions of similar type size surrounding, there are various sources of them, you may call discontinuities or trap-like behavior, whereby that atom is not the same as all the other atoms of the system, there is something special about it and in a book of solid state physics you'll see these various kinds of dislocations.

Q: And these dislocations act as traps...

M: Yes, anything that isn't a periodic can act as a trap.

8. Following: "A static distribution of trap states, with a corresponding distribution of electron-transfer times, could explain the power-law statistics of off times, but *would still produce an exponential probability density of on times.*"

Q: Why the two different probability densities for off and on times?

M: *"On" times is when it decays, so if it decays exponentially, then it would be single exponential, but it may be that it's thrown out at different distances, and then the return, coming from distributed places, with distributed tunneling lengths, gives distributed rates.*

Q: The two processes are not symmetrical.

M: That's correct, being thrown at different distances, you are recovering from different distances.

9. p. 14252 1st column, legend to Fig. 5: "Schematic of models for QD blinking. (a) Energy levels in the DCET model. Transitions from $|G\rangle$ to $|L^*\rangle$ or from $|D\rangle$ to $|D^*\rangle$ are driven by incident light. Transitions from $|L^*\rangle$ to $|G\rangle$ are *primarily radiative*, whereas transitions from $|D^*\rangle$ to $|D\rangle$ are *primarily nonradiative*."

Q: (1) Is this due to some selection rule not perfectly obeyed?

(2) Does the above mean that the off "dark" states are really just emitting a fluorescence but of lesser intensity, I mean they are just less bright light states?

M: (1) No, I don't think that selection rules are brought in there, *just what resonances are possible.*

(2) *Yes, by maybe a factor of thousand.*

Q: So, they are not completely dark.

M: *No, there'd be a certain probability of emission, but see, the point is that if you have an electron trapped, and an extra electron in the conduction band and then you shine a light on it, then there is a certain rate of the Auger process, which I think it's of the order of a picosecond, and the fluorescence is of the order of nanosecond, actually 15 ns... say nanoseconds, so fluorescence is much slower than the Auger process, than the Auger can occur, roughly a factor of a thousand, though it can still occur, it can still have a light, but very little.*

10. Following: "**Diffusion-Controlled Models.**... The duration of a blinking period is thus given by the time that the trap energy takes to diffuse away from and back to this resonance condition. *The first passage time is known to yield a probability density with a universal value of* $\nu = \frac{1}{2}$."

M: Yes, that's right, where you're coming to the intersection.

M: "*The first passage time is known to yield a probability density with a universal value of* $\nu = \frac{1}{2}$." Well, *if you have a particle that starts at a point on a straight line and it diffuses away, and diffuses back, then the time of its first crossing, the derivative of the survival time, gives you a* $-3/2$ *slope and in this notation is* $\nu = \frac{1}{2}$. *But starting from a point and return on a line is a standard diffusion problem, there is a certain distribution of times, a certain distribution of rate of times.*

11. Following: "The idea of *diffusion-controlled electron transfer* (DCET) was developed into a detailed model by Tang and Marcus... It involves four states ... with $|D\rangle$ *lying slightly lower in energy than* $|L^*\rangle$."

Q: Why so? Is there some reason for that choice?

M: I think that $|D\rangle$ is the dark state. That is what we thought at the time, now we think differently, or I think differently, what we thought at the time was that the trap is actually just off the conduction band, and in the band gap, so would be just lower than the conduction band state. Now we put it just off the valence band, now it's a hole trap instead of an electron trap, so there's that difference between what Tang and I envisioned there. Now we think in terms of a Silicon atom having lost an electron, that's easier than to think of a Cadmium ion gaining an electron, but I don't know which idea is correct. $|D\rangle$ is lower because it's in the band gap, all those states are in the band gap, only now we still have it in the band gap but it's off the valence band instead of the conduction band. We were just thinking in terms of electrons moving, we weren't thinking in terms of holes, that's simpler.

Pavel and I are still in touch with each other, of course, Tang and I are still in touch too, all that reminds me that he's coming in June so I have to write him a letter.

Q: By the way, with all your experience with autocorrelations of all kinds, did you ever think of applying this deep knowledge of nonequilibrium statistical mechanics, chaos, and so on. to financial problems and become an extremely rich man?

M: If I applied it to financial problems I'd probably become an extremely poor man . . .

Q: I was reading about this selling and buying 10^6 times a second, unbelievable.

M: I know, I know. You know, when you're doing theory if you make a mistake, if you break a pencil point, no big deal, . . . but with all that money if one is not careful . . . Besides, what I would do with all of that money when I would come to visit St. Peter or his Jewish equivalent, or something like that?

Q: But you are still a long time away from that....

M: Yes, I know, I know, but with all that money what could I tell, "I made a lot of money"? He might tell me to go elsewhere.

Q: You are the typical scientist not much interested in money. You know that when Einstein went to Princeton he was asked by the president there how much he was asking for a salary and he asked for an amount of money so small that the president told him: but professor Einstein, this is the amount of money we pay for a teaching assistant.

M: I heard a different story and I don't know what's is true. I think that when Millikan offered Einstein to stay was not very much, in any case Einstein didn't stay. I don't know, you know there are stories going around and God knows what the truth is.

12. Following: "The transition between the two states $|D\rangle$ and $|L^*\rangle$ occurs at the crossing point $|Q^*\rangle$ of these two energy curves. Motion on these parabolas and transitions between them provide *the classical counterpart of quantum phonon-assisted transitions.*"

Q: When was that symbol for the transition state made up of two +'s one on top of the other first introduced?

M: I think that if one looks way way back, it was around 1935 or so, it may be that that's when that kind of a symbol came from.

Q: Do you know who was the first to use it?

M: I wouldn't be surprised if it was Eyring, but probably either be Eyring or Evans or Polanyi, because they were the first to sort of doing the transition state in 1935, you might check Eyring's book Theory of Rate Processes and see what he uses.

Q: As one of the essential characters of your theory is the use of the two parabolas, can one say that your theory is—in part—a classical counterpart of the theory of quantum phonon-assisted transitions?

M: *Yes, sure.*

Q: And why is it the classical counterpart?

M: *Because it's a classical description and it has no phonons in it as such. Phonon is a quantum concept. For example, the simplest way of seeing that is to look at the treatment of Levich and Dogonadze in 1959, but that's in Czech, so you can look at one of their treatments shortly after that, and what they have is, say, a one dimensional quantum treatment where they have phonons, and now they may have had many dimensions all of the same frequency, I've forgotten.* There's a translation and I have it some place, that was made by the National Research Council of Canada, around 1960 and I got a hold of that, and so it's around some place, I don't know where it is, but I certainly have it some place. I'll look for it, in recent years I have written to the Research Council and ordered various things from it online. If I can find I shall be glad to send to you.

I am going this coming week to McGill, they named some Chemistry laboratory the Rudy Marcus Chemistry Laboratory, I was not that all good in the laboratory but I'm not going to tell them that.

13. Following: "A validation of the usefulness of this approach is seen in the fact that spectral diffusion, which is due to the diffusion of the vertical energy difference between the parabolas for the $|G\rangle$ and $|L^*\rangle$ states, exhibits a *nearly classical behaviour* at all but the lowest temperatures."

M: Yes, if you look at, I think, the first Tang's paper, the first or second paper that we wrote with Tang, we treated the spectral diffusion there, and we used a quantum model, and you see that at very low temperatures the classical model gives over to a quantum description.

14. Following: "*The model thus predicts a broadening of single-QD emission lines . . .*"

Q: Please explain the relation between "nearly classical behavior" and "broadening of single-QD emission lines."

M: That nearly classical behavior spectral diffusion and the broadening of the emission lines . . . there's a certain quantum version of that and there's also a classical version of that quantum version, *there is a certain quantum expression for the width with a* $h\nu/2kT$, *and there's a classical expression for the width that contains a kT.*

15. p. 14252 2nd column top: "If we approximate the rate constant for electron transfer at the crossing point by $k_r(Q)\delta(Q - Q^{\neq})\ldots$"

Q: Is the physical meaning of the above formula that only the Q's in the neighborhood of $Q^{\neq}$ participate in determining the rate constant?

M: *Yes, that's right, if you look of, say, the derivation of the Landau–Zener theory, because that's truly almost Landau–Zener-like, if you look at the derivation of that, the rate depends only on the slopes of the intersection and on the magnitude of the matrix element at the intersection, in other words it depends on certain properties of the intersection.*

M 358 Evidence for a Diffusion-controlled Mechanism for Fluorescence Blinking of Colloidal Quantum Dots

(Continuation)

Matthew Pelton, Glenna Smith, Norbert F. Scherer, and Rudolph A. Marcus

0. p. p. 14250 1st column middle: "**Autocorrelation Function.** The photon autocorrelation function, defined as $G^2(\tau) = i(t)i(t + \tau)$, where $i(t)$ is *the rate at which photons are counted experimentally . . .*"

Q: If the autocorrelation function is high, does it mean that the state of the QD remains the same after time τ is elapsed? I mean: *is the value of the autocorrelation function a measure of the permanence of a system in the same state or in similar states?*

M: *Yes, that's right*, but he really normally normalizes it by dividing in this case by the $\langle i^2 \rangle$, so you normalize it to unity, to start off with. The question then becomes: supposing that $i(t)$ remains near unity for a long time, that that'd be your real statement then, and yes, a measure for example of it is a velocity autocorrelation function, it's a measure of the velocity of a particle, say, a measure of the velocity not changing much with time.

Q: I am a guy who needs intuition in order to understand.

M: Ok, well, so I'm I.

01. Following: "Fig. 3 shows representative experimental correlation functions... Interpretation is complicated, however, by the nearly flat form of $g^{(2)}(\tau)$, which changes by only a factor of ≈ 2 over seven orders of magnitude of time delay."

Q: (1) Note that the ≈ 2 refers to a logarithmic scale. Is a 100 factor small? Why?

(2) Why is the interpretation complicated? Is it because for some reason a seven orders of magnitude time during which the state of the system remains the same is strange?

M: (1) There's still a big change going on there, but compared with an exponential decay there isn't, so he sure made that statement comparing with exponential decay or something like that.

Q: At first sight a change of 100 seems big.

M: Yes... I'm just wondering whether that "seven orders of magnitude" was almost a typo... because where's almost flat, is over, let's say, one to two orders of magnitude, about two orders of magnitude... so I think that should be two orders of magnitude time delay, it'd be my guess because you can see a change by seven orders of magnitude later on. That's not small, but the initial part is small.

(2) Yes, that's a subjective statement, one can then say that it's complicated because it is not a single exponential decay. It means complicated in that sense, that for a single exponential decay you wouldn't say that. Of course you wouldn't plot that way either, you would make a log of something vs. t instead of this... but that's kind of a subjective treatment, when you look at that you need a detailed theory to account for, and that's what's contained in the paper. I don't think it helps to say it's complicated.

1. p. 14252 2nd column middle: "When the two transitions are in resonance, there is a *continuous cycling* from $1S_e$ to $1P_e$ and from the VB to the trap, followed by a relaxation..."

Q: Does each cycle refer to a single photon?

M: Yes.

2. p. 14253 1st column top: "Differences between alternative diffusion-based models enter into the *meaning of* t_c and its dependence on physical parameters."

Q: How and why does the above meaning change?

M: *You have a certain behavior at very short times during which the concentration of the system near intersection is decreasing toward zero, its probability decreasing toward zero, and then you have a kind of steady state after that, so that's the reason for the change in slope. Now, another mechanism is that if you have a distribution of times, that's another way of getting a power law, that mechanism actually doesn't have a change in slope that I know of, this change in slope is sort of a unique to this diffusion mechanism.*

Q: And is it unique of the Tang-Marcus, not of the Frantsuzov?

M: Yes, that's right, Frantsuzov never considered that, to the best of my knowledge, so it's unique to the Tang-Marcus theory.

3. Following: "The merits of using an Auger-assisted transition are several-fold. The Auger coupling is relatively weak, so that small values of k_r are not unreasonable. The k_r value for the bright-to-dark transitions can now be different from the k_r for the dark-to-bright transitions, because of a different mechanism for the transitions. For example, as illustrated in Fig. 5d, a $1S_e$ state can be produced by the usual optical absorption to higher-energy states and downward relaxation, with a transition *en route* from $1P_e$ to $1S_e$ in resonance with a hole transition from trap state to VB state. In this case, *the lifetime of the dark state would be prolonged,* and the transition would be sensitive to excitation energy. *This mechanism might apply only to long-lived traps,* in the millisecond rather than the nanosecond regime."

Q: (1) Why would the lifetime of the dark state be prolonged?

(2) Why might the transition apply only to long-lived traps?

M: (1) *If you just had that recycling going on, then you're not changing the dark state back to a light state, you probably need to have a reverse kind of Auger transition . . . in order to go back to not having something in the trap, recycling just means that the state continues and doesn't change its nature.*

(2) Well, it *depends on why the short lived traps would be short lived. If the short lived traps were so short lived that they wouldn't be affected by that, didn't come in that recycling, then they're not affected, they're just there decaying spontaneously, maybe. I don't know what role short lived traps are playing, it may be that if you have short lived traps, you don't even see them. You know, if a trap is so short lived that it's not hanging around long enough to wait for the next photon to come, the trap never knew that it was momentarily dark.*

4. Following: "**Autocorrelation Functions.** The form of the blinking-time probability density predicted by the diffusion-based models removes the need to apply *arbitrary time bounds in the derivation of the autocorrelation function.*"

M: *When you do the binning, instead of the autocorrelation function, when you bin, then you have to decide whether what you got binned is a light or a dark state, so it means that you establish some lower bound for the state and if it is below that it's dark, if it is above of that is light, something like that. So, in the case of binning you have to establish some minimum that you regard as being bright enough to be regarded as a bright state. When you get instead the autocorrelation function, you just count the number of counts, you don't say whether it's light or dark or anything like that, you're just counting, that's all. Let me just add that the advantage of binning is that you can tell the difference between light and dark states, the disadvantage of the autocorrelation function is that you can't, you're lumping them all together. So, in a sense the autocorrelation function gives you less information, on the other hand it turns out that you can move, apparently you can go to earlier times with it and so get something meaningful.*

5. Following: "However, a difficulty remains to compare these predictions to the experimental results, *the measured correlation functions must be normalized by the mean photon count rates.*"

M: *I don't know why that's a disadvantage. I mean, that's sort of normalizing, you always normalize, you almost always normalize your correlation functions, so I don't know why that could be a disadvantage*...I don't see how it could be a disadvantage. *Normally some theories have trouble at very short time because they diverge, if we didn't have that* $t^{-1/2}$ *at short times and we only had* $t^{-3/2}$ *at large times, we would get a divergence,* I mean, in terms of some of the properties, some integrated property, for example, but that's kind of an aside. You almost always normalize correlation functions anyway, so I don't see why that's a disadvantage . . . we didn't know the mean photon count rates, we normalized by the square of them . . . unless there's something technical at there, he's doing photon count rates and . . . he's dividing, I guess, by the square of that. That's maybe an extra complication that may be more an experimental question . . .

6. Following: "Indeed, an accurate quantitative estimate of the correlation function cannot be obtained *unless the duration of the experiment exceeds the saturation time by at least two orders of magnitude.*"

Q: Two orders of magnitude means a time 100 times greater than the saturation time . . . why the need of such a long measurement?

M: *That is an interesting question Francesco . . . in the early days of course, with the exception of one paper, I did all of my work on my own, the theoretical work, and later on after I went to Illinois and theoretical students became available, then I started publishing with students a lot more, and of course it's certainly true that I went into things the most deeply when I did it on my own.*

7. p. 14253 2nd column middle: "The significance of the critical time can be seen by examining the physical interpretation of Eqs. 3–5

$$\rho(t) \approx \frac{1}{\sqrt{\pi t_c}} t^{-(1-\nu)}, \quad t \ll t_c \qquad [3]$$

$$\rho(t) \approx \frac{\sqrt{t_c}}{2\sqrt{\pi}} t^{-(1+\nu)}, \quad t \gg t_c \qquad [4]$$

$$t_c = \frac{4k_B T}{\kappa t_{diff} k_r^2}, \qquad [5]$$

Q: Please explain how the significance of the critical time can be seen from the physical interpretation of the above equations . . .

M: If you use the simple value for ν, which is 1/2 or whatever... the value from the simple theory, with normal diffusion, then the physical interpretation of those equations, or the equivalent of those equations, is given in the paper with Tang, and in one case you're sort of building up a steady state, you have the case where at longer times you establish a steady state, and I think we did something where we showed physically, I think we did in one of the papers with Tang, where you look in the time critical... the critical time, what sort of origin are you using up to sort of, almost, reduce something from unity to zero, and that's sort of the part of the physical interpretation. You're building up a system where you began to have zero concentration at the intersection point, you're building up a steady state.

8. Following: "... and we note that the diffusion constant, D, for the motion along Q is approximately equal to $(\Delta Q)^2/2t_{diff}$."

M: Yes, the Einstein relation.

9. Following: "From Eq. 5, it then follows that

$$k_c t_c = 2\sqrt{\frac{k_B T t_c}{\kappa t_{diff}}} \cong \sqrt{2Dt_c}, \tag{11}$$

We can approximate the delta function $\delta(Q - Q^{\ddagger})$ in the expression for the rate constant at the intersection crossing point at time t_c by a Gaussian of width $\sqrt{2Dt_c}$. *Eq. 11 then corresponds to a statement that t_c is the time for the population*, which has expanded *in this time* to occupy a width $\sqrt{2Dt_c}$, to largely disappear into the 'sink' at $Q^{\ddagger}$, forming in the process a population gradient there, with an effective population at the intersection of approximately zero."

Q: How does one read the above story in Eq. (11)? Apparently you read the above story from the form of Eq. (11).

M: Yes, that's right, $k_c t_c$ is sort of the amount of population disappeared, roughly, in time t_c and in that equation is approximately the

$\sqrt{2Dt_c}$, so *that's sort of tells you that there is the region in which it disappeared, during that time*. In other words, how long does it take to effectively form *a little region around the origin* in which the population is essentially disappeared? So effectively you have a small boundary condition at zero concentration there, *a new steady state.*

10. Following: "The power spectrum for high frequencies, corresponding to times less than t_c, thus has the $1/f^2$ form characteristic of Brownian motion, whereas the power spectrum at low frequencies has a power-law form, reflecting the *modified diffusion process* after a concentration gradient has developed."

M: If *you can speak of just mean squared deviation, just not depleting anything, just mean square deviation, that apparently gives rise to that $1/f^2$ form, and then you can speak of a process in which you've got some depletion in that system, so the mean square deviation doesn't talk about depletion, because it talks about mean square fluctuations.*

11. Following: "We can, however, attribute the significant *dot-to-dot variation in t_c* to variation in the physical parameters of Eq. 5..."

Q: What do you mean by "dot-to-dot"?

M: *Different QDs may have little differences in t_c.*

12. p. 14254 1st column middle: "The *direct quenching* of QD luminescence by charges in trap states would account for the recently observed collective blinking of several closely packed QDs: charging and neutralization of traps would cause the blinking of all nearby dots."

Q: What is this "direct quenching"? Is it related or not to your diffusion-controlled models?

M: I'm not sure what direct quenching means, but what he may be talking about there is this . . . that *if you have something in a trap you have the possibility of these Auger processes and that you can regard as a direct quenching, what is within the dot. Now, if you have another quantum dot that's coupled to it, and the coupling is strong enough, then I suppose that if that the other excited dot maybe transfers its energy . . . . if that second dot is electronically coupled to the dot that has a trap in it, then the dot which has a trap in it has an electron in the 1P state and that electron can go up while one electron in the nearby dot goes down, quench them all.*

Q: Some kind an Auger process in two dots.

M: I didn't even know that happened, does he cite a reference to that? It's an interesting point I wasn't aware of it. I have to look it up, it's interesting. I'm glad you mentioned it, I didn't even noticed that. You know, there's been a big change, Francesco, one has involved those papers with Pavel and with Tang, I really didn't know much about the experimentalists, didn't really know them, for the most part, other that Pelton, I think I've met him at that meeting in Notre Dame, that's around that time, and so the papers that we wrote then, with Tang and so on, I think carried some of that flavor. And because I noticed now, when Joyen Zhou is writing up the paper, that I'm far more demanding over to explain things, than when I was working with Tang, and in part I think maybe because I know a number of the experimentalists now like, at the same meeting, Bawendi, a couple of months ago, I know Nesbitt well now, and I know some of the others, so I'm now far more conscious of trying to make sure that everything is understood, in the mathematics and so on, than I was then.

Q: But do you have still a new theory coming up?

M: Yes, there's some additional theory, that's right, because one of the things we didn't understand, or I didn't understand, really, was the difference between the "on" and "off" periods, why the "off" period was a straight line for so long, and the "on" period wasn't, the "on" period had an exponential tail, and I think that now thanks to what Joyen has done and to what other people have found experimentally, and in their interpretation, I think we have a better inside on that now. Before, what always bothered me was that there's a certain asymmetry, and with Tang we couldn't really explain it, and we said that the exponential tail was due to getting away from the intersections, that it causes an exponential tail, but what bothered me there was that our theory was just not symmetrical between "on" and "off," because when you get away from the slope and the "on" state you still should then get an exponential tail, and what Tang suggested was: "Well, maybe the diffusion coefficient is very different in the dark and light state." Maybe it is, but it seemed to me arbitrary, so now with what Joyen is doing, we are not attributing it to the slopes at the intersection, not at all, we attribute it to different processes. I don't know that we have either the right answer now, but I feel better about it than with what we had before.

Q: Did you read the mail where I was asking if you knew that wonderful composer of 1500s Jacques Arcadelt?

M: You know, at one time I used to play the piano a little bit, my mother played so well . . . and one of the pieces that I remember played and I have because was a slow piece, was by Arcadelt.

Q: Maybe the Ave Maria then.

M: I don't know, I don't remember, I have in a big book, I still kept all used books even though I haven't looked to them for years . . ..

Q: Look on YouTube . . . also "Il bianco e dolce cigno."

M: That I don't know, can you repeat that one?

Q: Next time we shall discuss about the paper "Universal emission intermittency in QDs, nanorods . . ."

M: Oh, yes, that was largely written by Pavel, I made a few comments but it's really Pavel's.

Q: After that some papers of yours on enzymes.

M: There electron transfer plays a little role, you know...

Q: Or proton transfer.

M: Ja, proton transfer, hydride transfer pays a role.

Q: Thanks a lot Maestro, for your enlightening words. Every time I speak with you my mind expands.

M: Well, I hope you don't have growing pains... now, is the link to Arcadelt on your mail?

M366 Universal Emission Intermittency in Quantum Dots, Nanorods, and Nanowires

Pavel Frantsuzov, Masaru Kuno, Boldizsàr Jànko, and Rudolph A. Marcus

M: Pavel hasn't found a professorial position and he's back to Russia, at Novosibirsk, so for a while he was to the University there, and now he is active in some sort of politics, I guess, connected with trying to bring technology to Novosibirsk. You know, they have a lot of it there, it used to be absolute first rate basic research, technically, but now it really needs to get more into some technology, so apparently he still has some papers to write but he's so busy with the other, he hasn't enough time... anyway I see his name here, so if you want to call him ask some questions. You find with Skype an easier connection to United States than to Novosibirsk. Actually it wasn't bad, I think eventually we turned off the video.

1. p. 1 top: "Few problems of early quantum mechanics remain unsolved today. Fluorescence intermittency is one exceptions."

Q: Does he maybe mean that fluorescence intermittency cannot be explained *only* on the base of quantum mechanics (because statistical mechanics and kinetics also come in?)

M: Certainly no quantum mechanical problem . . . the paper was written for Nature, so it's written to excite the reader, ... Pavel did most of the writing of this . . . and I made some comments out of some points but Pavel is the one who really did most of writing.

Q: The statement is extremely impressive.

M: Yes, but I think it's just really to excite . . . In fact it is possibly a Jànko's statement, because he is most favorable to publicize things, in fact this article I think was accepted because Jànko knew somebody at Nature. In other words, there was some preliminary discussion that went on to submit it.

Let me just comment on this though. Fluorescence was an early study, fluorescence in simple systems is understood, but fluorescence in this QD system with its many possible surface states, and God knows what, is not well understood, but it's not fluorescence per se really that is not well understood. It is well understood, but fluorescence *of these QDs* certainly not . . . . In fact, I'm not sure that fluorescence intermittency itself was an early problem in quantum mechanics, I don't know if it was studied then, I don't know if there is reference to an early study of intermittency, but fluorescence itself is understood, but in these QDs with intermittency and with all sorts of surface states, that's certainly not well understood.

2. Following: "Such jumps, where the fluorophore literally stops emitting light under continuous excitation, are very different from those predicted by Bohr."

M: Yes, well, Bohr didn't say anything about traps and anything like that. Normally if you have a fluorophore and if you're shining light on it then it will emit, then it will continue emitting, so that'd be the standard behavior, it's only that now you got a more complicated situation, and Bohr was not involved in that complicated thing at all. Bohr wasn't treating fluorescence in the presence of surface states.

3. Following: "Even more intriguing is the statistics of intermittency. Whereas Bohr would have postulated exponential distributions of 'on'-times and 'off'-times, universal power law probability densities are actually observed. For colloidal QDs, this power law extends over an extraordinarily wide range that spans nine orders of magnitude in probability density and five to six orders of magnitude in time. This phenomenon is remarkable for many reasons. First, the experimentally observed distribution *refuses to yield a time scale.*"

M: For simple systems you get the exponential decays, that's pretty standard, it's just for some of these QDs that show intermittency that you get power law. *In exponential decay the time scale is the half time, you have no such thing here, if you would calculate an average it wouldn't converge.*

4. Following: "Second, *universality on a lesser time scale has revolutionized our understanding about phase transitions.*"

M: Yes... there I assume he's probably talking about critical phenomena, and... where you have probably time scales which are universal... I don't know enough about that, but I assume he's probably talking about that, I don't know. I think he just wants to talk about universality, and make it more popular. I mean, don't forget you're in Nature here, so it has to be somewhat popular, touching on all sorts of things.

5. Following: "While the underlying mechanism for answering such questions remains *a mystery*, we argue in this perspective that many, but not all, key experiments have already been conducted and that substantial theoretical progress has been made."

Q: Such was the situation in 2008 . . . what does remain still mysterious today?

M: The underlying mechanism.

Q: So, the mystery is still there.

M: The mystery is still there, yes, I mean, there are models . . .

6. Following: "*When a threshold is used* to distinguish on- from off-states, second-to-minute off-times are *apparent.*"

Q: Why is the threshold necessary or useful to distinguish on- and off-states?

M: *Well, you have to have some intensity above which we'll call "on" and below which we'll call "off."* You need to have something like that to have a definition of what's "on" and what's "off."

Q: Because you don't really have completely light and completely dark states.

M: Yes, that's right . . . you have noise anyway, and then also you may not have completely "on" and completely "off" states. If you look at some of the intermittency, you'll see that there are not just high or low that there is a lot of that but there is also some that's in between, for various reasons . . . it's unclear.

Q: How then can you decide where to put a threshold?

M: I think is Haw Yang who discussed that problem and to find what is the best way of doing that, I think one should contact Haw Yang, Haw Yang was now at Princeton. I haven't used this method, but I think my former student Zhu knows more about than I do.

Q: Is there some book about QDs that you could advice?

M: Yes, I'm not sure, there may be books on QDs but they are edited volumes, for example, Viktor I. Klimov has edited some books on

QDs. Now edited books are not necessarily the best way to learn because they are written by different authors and often specialized in various chapters, but certainly if one wants to get some sort of view of quantum dots one might look at a book that he has edited, I know is recently updated, after about 5 or 10 years, and I'm sure there are other books around that . . . actually a review chapter might be useful too, is usually written by one person or a pair of coauthors, and so it's more uniform and one might look for reviews of QDs, the trouble is that they are normally for specialists rather than for some general learner, they're invaluable sources of references, but it may be that if Klimov's book is edited I would recommend it, because he's one of the leaders in the field. Did you ever meet Pavel?

Q: No, I didn't.

M: He was here for a couple of years, and he has now got I guess a faculty position in Moscow, and he recently moved there from Novosibirsk and he's been involved in several reviews, I have been a coauthor in one of them, he's the one who really did the writing, and so you might look at a review by him, but written for people in the field.

7. p. 2 bottom: "Truncation times ('cutoffs') were discovered in on-time distributions. Such cutoffs occur on the second timescale and may represent *a competing physical process* which interrupts power law blinking. A corresponding off-time cutoff in QDs has *not* been reported, *although it is speculated to occur on an hour timescale.*"

Q: (1) Which is the nature of the "competing physical process"?

(2) Why is the situation different for on-time and off-time distributions?

(3) Why an hour timescale for off-time?

M: (1) The cutoff, when it occurs and how it occurs, seems to depend on the light intensity, that's a fairly recent work of Nesbitt, and it may be that you get biexcitons, and they can affect the cutoff, certainly biexcitons is an additional thing, some of the cutoff now is by exciton formation at high intensity. What I don't know is whether it's associated with the whole cutoff. There's a recent work of Nesbitt where he showed how the tail end, this exponential, depends on the square of the light intensity, and it could be that too is by exciton formation, implying that if you went to zero light intensity you wouldn't get a cutoff. So, the answer may be that it's the biexciton.

(2) I can tell you what we think, we're writing a paper that's related to that, and maybe other people think the same thing, I don't know, but we're saying that when it's off you can get all sorts of Auger processes possible, that's why it's dark, it's off, because of the competing Auger that competes with fluorescence. That means that objects like biexcitons have less chance of forming, which means that you get less chance to get the product, this is something in a paper that we haven't submitted yet.

Q: But why is the situation different for on-time and off-time distributions?

M: On "on" time you don't have extra electrons that are all sticking ... remaining around, so there's less chance of an Auge process.

(3) They don't at long times, and they haven't got a cutoff for that, but yet we know that does have to exist, in fact I and Zhaoyan Zhu use some sort of statistical mechanical theorem to show that since, when you shine light for about, say, an hour, you get a steady state between "off" and dark, then the only way you can explain that is if you have a cutoff because if you didn't have a cutoff you would never get a steady state, and she writes I think with Wiener–Khinchin theorem something to show that. I think that was in her thesis.

9. p. 3 top: "Intermittency is light-induced, as indicated by experiments revealing *statistical 'aging' of emission trajectories.*"

M: I'm not sure about that, at low temperatures certainly is true, it was found that the nature of the curve and the tail at the onset depended on light intensity, and in fact *if you use as a coordinate the intensity times the time instead of just the time, the different intensity curves at low temperatures fall on each other*, so that may be that the diffusion process was light induced. I don't think there's that kind of evidence at room temperature, because what was involved in that other thing was some sort of spectral diffusion and I don't think they've been able to observe spectral diffusion. *At room temperature we don't know whether the spectral diffusion depends on intensity, at low temperature it does.* When you plot something you use for the x coordinate anything you want, some coordinate makes more sense than the other, and frequently we plot the intensity vs. time, but instead you could plot intensity vs. the product of intensity times time, in other words something proportional to the total amount of quanta that goes in. It shows that curves of different intensity, in that paper of Jao, fall in the same line when instead of plotting each vs. time we plotted each vs. the intensity times the time, in other words there are some more universal variables which may be products of other variables and you bring that out plotting vs. universal variables for the given product.

Q: What about this "aging," statistical aging of initial trajectories?

M: I forget now what that is, there is a paper written on it, at least one paper, and I'd have to look that at . . . I never really studied that paper, but I know that there is a paper on statistical aging, it's a well-known process, and what I'm not sure of is if it's just a power law or what, but I never did look it up, I'm not sure.

10. Following: "The ensemble emission intensity decays under continuous excitation and *recovers in the dark.*"

Q: Do you mean that under continuous excitation the emission intensity becomes fainter and fainter but if one switches off the excitation the emission intensity becomes strong again after a while?

M: If you shine the light for a long time gradually over a period of time you get a steady state of the dark states that takes quite a while.

Q: And if you switch it off?

M: It will take an hour to recover. *You turn off the light and if you wait an hour now it shows its original behavior.* Some of these QDs are pretty stable, organic dyes are less stable, so those dyes don't seem to exist as long but in inorganic QDs they seem to be pretty stable, so the emission intensity recovers...

11. Following: "In existing experiments, the on-time distribution cutoff is inversely proportional to the excitation intensity."

M: *The exponential moves to shorter times*, the decaying parable moves to shorter times and I think ... I'm not sure if it is inversely proportional to the square now, it's a work of Nesbitt, it may depend on how you define things.

Q: So if you excite strongly...

M: The power law will last for less time and you'll get the cutoff more.

Q: An extremely interesting system.

M: Yes, but it's a messy system, you know, with all sorts of surface states...I think that you're going to have some simple theoretical ideas and expressions like a power law that has an exponential tail, and that expression is there in the paper with Joyanne Zhu, but then there can be other phenomena occurring that complicate the whole process, that you haven't data to account of, and they may generate roughly the same kind of curve qualitatively, so you always have to be alert to that possibility.

12. p. 3 middle: "Large shifts in the spectrum coincide with equally rare jumps of the intensity. This suggests a direct correlation between spectral diffusion and emission intermittency *through the redistribution of charges on or nearby the QD surface*."

Q: (1) Are the shifts frequency shifts?

(2) Why is the correlation suggested between spectral diffusion and emission intermittency?

(3) Why the correlation through the above redistribution of charges?

M: (1) Yes.

(2) Yes, that's really Bawendi's work, related to the diffusion, and I'm trying to remember what the key point was there... Well, first of all he showed that *if you look at the spectrum as a function of time... that it shifted around and occasionally there is a large shift, presumably reflecting an occasional real change of charges around there*... but *he showed that these QDs underwent spectral diffusion, and if you think at what may be causing this transition from light to dark, off to on, on to off, if you bring two electronic states into resonance you'd get a transition, one of those is the on state and the other one is the off state... so spectral diffusion provides a mechanism for bringing two electronic states into resonance and them having a transition from one to the other.* That's Bawendi's work.

(3) Normally a spectral diffusion occurs around some little region, but then you can get occasionally a pretty big jump and so I think that I've seen it where you almost have two peaks, part of the time to one, part of the time to the other, but when you get a big shift in that fluorescence emission, in the spectrum, then that suggested that your charge distribution there in the QD is a bit different, say there are

different sized QDs which emit at a different fluorescence frequency. I mean these are all small shifts, by the way, so it's not dramatic, but it's there.

13. p. 4 top: "The first QD blinking model was developed by Efros and Rosen . . . Electrons are ejected from the dot to the surrounding acceptor-like states. This leaves behind a positively charged QD. Subsequently, electron–hole pairs experience rapid Auger-like non-radiative relaxation to the ground state, quenching any emission and thereby rendering the particle dark. This process continues until the QD is neutralized."

M: (1) "*The first QD blinking model was developed by Efros and Rosen*" That's right, early 1990s.

Q: (1) Please check my understanding: the excited QD emits electrons (by photoelectric effect?) and becomes positively charged. Subsequently these electrons come back (why?) and recombine with the holes previously left behind. The recombination happens by a non-radiative relaxation and so the QD appears to be dark during this process.

(2) The recombination process is Auger-like, you write. But in an Auger process there is a compensation between two processes, for example, an electron is excited and a hole lowers its energy. Which is the other part of the process in this case?

M: (1) We no longer believe that, we think now they're not fully ejected from the dot but are put in traps on the surface.

"*the excited QD emits electrons* (*by photoelectric effect*?)" Well, *it's questionable that it actually ionizes.* That was a postulate that Efros and Rosen . . . it's something that we are using now but something that's questionable. *The main point is that the electron has been put in some trap, whether that trap is outside the dot or on the surface of the dot may be the question there.*

"*Subsequently these electrons come back (why?)*" They're in the neighborhood and there is an extra positive charge there, so they come back. *It is called geminate recombination.* Geminate recombination happens really when you have two objects in essentially the same pocket and they recombine, or they reform at the same time, maybe drift apart a little bit and recombine rather than having a redistribution and a recombining the things at random, that's what geminate recombination is. You remember from your Latin that it comes from gemini, the twins.

(2) I don't think that they thought that the recombination process is Auger-like.

Q: But that is what is written there.

M: That's not what Efros and Rosen thought.

Q: Yes, because by the way, which is the other part of the process in this case?

M: Well, the point is that for the mechanism of how the dark state becomes light again there are several possibilities. We have one view, Pavel has a different view, and so on. So the mechanism of how that comes out... that's one point, OK? That's a separate issue. *The mechanism of why is dark when the electron is either ionized or in a surface state. We believe that the hole state is the main thing, is because you've got an extra particle kicking around, either it's electron or a hole, and once you excite the molecule under those conditions Auger processes become possible, and the system is dark instead of fluorescent.*

Q: Because in an Auger process you have two processes basically.

M: Yes, so *the Auger process originally is to explain why when you've got an ejection of some particle, whether it is an electron or a hole, the system is dark, that's why Auger is invoked. But subsequently Pavel in a paper he and I wrote together, said that: well, one way of* *getting something going into a trap is by an Auger process, but that's a different Auger process than that which explains why an ionized QD is dark.*

14. Following: "Due to *a static distribution* of trapping and detrapping rates, varying with distance, and/or trap depth, power law*off-time distributions are naturally obtained.*"

Q: (1) Can there be also a *dynamic* distribution of trapping and detrapping rates?

(2) Why from this static distribution is the power law naturally obtained?

(3) Why only for the off-time and not also for the on-time?

M: (1) I guess that a dynamic distribution means that they're changing with time, sure that's possible, in other words you look at the trapping and detrapping rates that are possible at one time and they could be different at some other time, but normally, if you have series of traps, you think of a trap as more or less fixed.

(2) Well... that part is sort of clear for the reverse process, for the electrons, say, coming back, because if you have a distribution of sites when the electron jumps back into the dot, it depends on the tunneling distance, so that gives you *a distribution of rate constants for coming back, and any time you have a distribution of rate constants that is broad enough you get a power law.*

Q: Is this is to be found in chemical kinetics books?

M: I haven't seen it in chemical kinetics books, no, maybe I've seen it in noise, when trying to explain $\frac{1}{f}$ power noise instead of $\frac{1}{f^2}$ power noise. A Lorentzian distribution is something like a constant divided by a constant square plus omega square, where ω is a frequency. When ω becomes very large you essentially have that intensity of

$\frac{1}{f^2}$, one over frequency squared, so that's $\frac{1}{f^2}$ noise, but under other circumstances where the distribution is not Lorentzian, is more complicated, you can get a $\frac{1}{f}$ noise. I think there are various ways in the literature that you can get $\frac{1}{f}$ noise, you commonly hear in the literature about so many processes having $\frac{1}{f}$ noise, in other words the intensity of the noise varies inversely as the frequency, so with a very small frequency you get a lot of noise.

(3) When the particle is jumping back, that's when you get the distribution of trapping and detrapping rates. For the other way, is sort of jumping the other direction, and you haven't a distribution of times there. If you have an "off" state, that means you have a particle outside that comes back, so it's tunneling back, so there you have a clear process, that's for an "off" time, but for an "on" time the particle has to jump out... Do you have a jumping to a series of traps or what do you have? That's really not spelled out.

15. Following: "**Spectral diffusion model**... Tang and Marcus... A key prediction is a change in the slopes of both on-time and off-time power laws from 3/2 at long times to 1/2 at short times."

Q: Can you summarize the predictions of the Tang and Marcus and of the Pavel and Marcus models?

M: *With Pavel we never looked at very short times*, we looked at something that affected longer times, so we never looked at the very short time process, we didn't go into that kind of detail. With Tang we used a somewhat different model, because the Pavel model is really quite different, he doesn't actually have traps in there, he says that when states are in resonance with a band of states, then you have a recycling going on all the time, and don't get fluorescence start. So, I don't know if he believes in traps, no, he doesn't believe in traps, then he has just a recycling, permits recycling back and forth, and so it couldn't be dark. So, that was his view. *So, he never looked* *though into short enough times with enough detail, that would bring out this* 1/2. What Jau did was that he solved the diffusion equation for all of times essentially, short and large.

16. Following: "**Spatial diffusion model**: Margolin and Barkai suggested that any ejected electron performs a 3D diffusion in space about the QD prior to its return. While the model naturally predicts a $t^{-3/2}$ distribution of off-times, deviations from the -1.5 exponent require the introduction of *anomalous diffusion processes*."

Q: Which ones are the merits and the explanation powers of the anomalous diffusion processes?

M: The usual things that you see in the literature on diffusion are depending on the dimensionality, there is $\sqrt{t}$ or t or $t^{\frac{-3}{2}}$. That's very simple but there is something called fractal dimension where particles are moving in a much more complicated kind of space, is not just simply 1 or 2 or 3... God knows what, *they may speak of fractal dimension and then the theory of diffusion under those conditions gives you a fraction different from the usual 1/2, 1, or 3/2 depending on the dimensionality.*

17. Following: "**Fluctuating barrier model**: ... a model where emission intermittency involves fluctuations in the height or width of a tunneling barrier between an electron within the QD and an external trap state ..."

Q: What happens in the QD in time that can justify such a jiggling of barriers widths and heights?

M: I've no idea, I've seen that sort of model used in objects other than QDs... you know, some particles are far apart and ... well, *the in-between material does fluctuate, and in principle that could affect the height of the barrier in a very small way, so if the intermediate material is fluctuating certainly the height of the barrier could vary,* *but its variation is probably very very small.* I've seen occasionally in the literature where people talked about fluctuations outside of the QDs literature.

18. Following: "...a key problem shared by all models is the general difficulty in explaining *a continuous distribution of relaxation times.* Alternative models which don't invoke long-lived (> 1 s) electron traps, have, in turn, been suggested to account for this distribution of relaxation times."

"Frantsuzov and Marcus...This mechanism *naturally explains* a continuous distribution of relaxation times."

Q: Why is the explanation natural?

M: I don't know if it is natural, I mean I think the diffusion mechanism is natural too...in the sense of giving you the power law distribution...the model that Pavel had, is that you have this resonance between states in which one set of states was a band, a band of the surface states, and if the other state was within that band, then you have recycling and, you know, the QD would be dark...Now, in that band *if the resonance state happen to be near one of the peaks of the band you get one relaxation time, if it was in-between peaks you get another, and so on, so you can get a distribution of relaxation times that way.* And actually I think that there's a figure in the paper with Pavel which goes into that.

M367 Beyond the Historical Perspective on Hydrogen and Electron Transfers

R. A. Marcus

1. p. vi top: "These reactions form the simplest class of reactions in all of Chemistry, no chemical bonds being broken or formed in some cases and there being zero chemical 'driving force'—standard free energy of reaction."

Q: We know that $\Delta G^0 = -RT \ln K$, so the equilibrium is driven in one direction or another by ΔG^0. But is then the meaning of ΔG^0 "equivalent" to that of K or is it more general than that of K?

M: I think it's more general, the thermodynamic quantities you would calculate directly from statistical mechanics, or you can calculate any equilibrium property, but quantities like free energy difference, as that, is a fundamental quantity, the K just happens to be one of its effects on the ratio of constants, under such and such conditions.

2. Following: "In 1960 this work was extended using statistical mechanics instead of the dielectric continuum theory, and now included changes in nuclear configurations of the reactants . . . to this end a global reaction coordinate was *needed* . . ."

Q: Why was it "needed"?

M: *Anytime you do statistical mechanical calculations of reaction rates, unless you use very special techniques invented recently, computationally very expensive, you need a reaction coordinate.*

3. p. vii top: "In the case of the transfer of H^+, H^-, or $H^\bullet$, we again have a transfer of a light particle. The Franck–Condon principle applies, though more weakly, when the mechanism is that of 'jumping' of the H from one reactant to the other, a so-called nonadiabatic H transfer."

Q: Were you the first to introduce this concept?

M: I doubt it, because the Russians Levich and Dogonadze tacitly introduced it when they treated proton transfer reactions in the same way that electron transfer reactions were treated.

Q: So they were the first ones.

M: They could well be the first ones, yes, they could well be. By the way, it's just nonadiabatic proton transfers which would apply, but probably many of them are nonadiabatic. But, you know, if you have two curves and they are intersecting with the right separation, then of course the whole nonadiabaticity breaks down a bit, so it depends on how widely separated the curves are, so it becomes increasingly approximate, but for something it is... may be OK. I can try to remember... a nearly example was a study of H + O_3 in the gas phase, to produce OH, and the OH was very highly vibrationally excited, and the way you can understand that is if you apply the Franck–Condon principle to the movement of the H, but that occurred long before the electron transfer theory, if I remember rightly, and occurred before Levich and Dogonadze, before the electron transfer theory, and I don't remember how explicitly that was described but I remember the phenomenon, it's very striking, very highly vibrationally excited OH, and you can understand that in terms of the Franck–Condon principle for H transfer, but I don't remember how well that was discussed, or anything, I just remember it happened.

Q: Did it inspire you for your ET theory?

M: I doubt it, I wasn't thinking in terms of H transfers... well, I was, but not in this context. And it may even have happened after the ET theory, I don't know, I just don't remember the details, but I do remember the phenomenon. It's a chap at Princeton, I think, who did that work, maybe David Garwin working with somebody, I forgot, I remember the work, it was really a very striking thing.

4. Following: "When the H transfer is not as sudden, for example in an 'adiabatic H transfer' ..."

M: If the two potential energy surfaces are highly distorted, the intersection is far removed, then of course having a system where you have two different electronic configurations, one is completely inaccessible, the higher one, so some parts of the formalism have to differ.

5. p. vii middle-bottom: "The KIE itself is usually smaller than a factor of 100, and *part of that* is due is *often* due to zero-point differences for the H and D systems."

M: It's usually a factor of between 3 and 15, far smaller, I should have said "far smaller." Occasionally there is an example of a factor of 100. Generally the deuterium starts at a lower point, its zero-point energy is lower than the other, but the frequency is higher than that in the TS, kinetic effect of that is that the deuterium then has a higher barrier to go through than the H does, because the zero-point difference of the reactants in simple cases anyway, is frequently larger for the reactants than for the TS, but maybe there are exceptions to that, so I don't know, so that's one point. But the dominant effect probably is that for D is more difficult to tunnel.

Q: And why do you say "part of that"?

M: *Because part of the KIE is due to zero-point differences and part to tunneling differences.*

Q: Which are bound to each other though.

M: *They're all part of the same potential energy surface.*

6. p. viii middle-bottom: "In addition in the field of electron transfers one has the relation of the kinetic properties to charge-transfer spectra..."

Q: Were you the first to discover the relation?

M: I don't think so, I think that Hush may have been the first, I think it's very possible that Hush was the first, I'd have to look back at the literature... my guess is that Hush was the first, but I'd have to look back... By the way, surely Hush was the first to work out the details of it, I don't know whether somebody else also... made some statement about ... but for the best of my memory Hush should be the first.

7. p. viii bottom: "For the electron transfers for two approaching ions in solution w^{r} consisted of the electrostatic repulsion of the ions (and of *any other free energy barrier that* could not be reduced by favourable chemical alteration of the ΔG^0 for the actual transfer step)."

Q: Which "any other free energy barrier..."?

M: A steric effect for example...

8. p. ix top: "w_{r} can be entropic or energetic in nature."

Q: (1) Is it so because w_{r} is a free energy and $\Delta G = \Delta H - T\Delta S$?
(2) When is its nature energetic and when entropic?

M: (1) Yes, and *so can have different contributions. For example, there could be the Coulombic contribution, which was the original one, the* $\frac{e^2}{R}$ *term, that happening before the actual ET, then it has that* $\frac{e^2}{DR}$ *term which has both entropic and energetic contributions, part of that free energy change is energetic and part of it is entropic because D has a temperature coefficient and anytime that happens you have an entropic contribution. Part of the* w_r *may be also in some cases a steric effect, and that could be largely entropic.*

(2) "*When is its nature energetic and when entropic? Can one distinguish it clearly?*" *Well, no, because for example in the Coulomb repulsion case you had both, you can't separate effects, I mean you can say what they are, by getting the temperature coefficient, but they both can occur. Probably in most cases even steric effects have both energetic and entropic contributions, but it's true that in most phenomena there is predominantly one or the other.*

9. Following: "The distance *sampling* in w_{r} is sometimes called *gating ...*"

Q: What do you mean by "distance sampling" and why is it called "gating"?

M: Well, I don't know if that's a correct statement, actually. *Gating is some process that occurs before the electron transfer, for example, if something has a rough take getting into place and then electron transfer, that'd be an example of gating, maybe there are many other types of examples of gating, I don't know, but some process that occurs before the electron transfer, it has to occur before the ET can happen.*

Q: And the distance sampling?

M: No, I don't think it's distance, that's simply approach of the reactants, *gating is more something you turn on and off, and I think that's more usually a rotation* but I don't know.

Q: So this distance sampling?

M: I would hope I didn't say that, maybe I did but I wouldn't. I don't know why I called that distance sampling, why I did that.

10. Following: "The protein reorganization is *approximately* a harmonic function of the relevant coordinates, a parabola. The role of bond breaking-bond forming in AHB has been taken into account in several alternate ways. In one approach (empirical valence bond) it is treated *via* a pair of free-energy profiles that are *approximately* a pair of parabolas for the free energy of reactants and products..."

Q: Why approximately? Shouldn't the free energy profiles be just parabolas?

M: Because there's high distortion at the intersection in the case of strongly interacting...

11. p. ix middle-bottom: "The equation for the rate constant of the reverse reaction $k_{\mathrm{H}}^{\mathrm{rev}}$ is obtained similarly, and the theoretical expression can be tested to see if *the resulting equilibrium constant* $\frac{k_{\mathrm{H}}}{k_{\mathrm{H}}^{\mathrm{rev}}}$ *is independent of* w_{r} *and* w_{p} *as it should be.*"

Q: Is it so because the ΔG^0 in $\Delta G^0 = -RT \ln K$ refers to reactants (and products) at infinite distance from each other?

M: Yes, *at arbitrary distance from each other, it's independent of the path and so on. You know, w_r involves a description of the path.*

12. p. x top-middle: "At temperatures above a breakpoint the KIE is temperature independent and at temperatures below the breakpoint the KIE is temperature dependent."

M: Yes, *explaining that was the essence of one of the papers I wrote, and the idea is this: that if everything is beautifully arranged in the enzyme, so that no bond has to stretch, the H doesn't have to stretch in order to transfer, it can rather just jump, once the environment is favorable, then there's no change in zero-point energy, and that means that at all temperatures you are going tunneling from the same ground state. I mean, that's typically the case, you go from the ground state there, in these tunneling processes, and the barrier is mainly due to reorganization, so at all temperatures, so that if you are tunneling from the same state at different temperatures then the energy of that state is the same at all temperatures, and so you get no temperature effect.*

M 367 Beyond the Historical Perspective on Hydrogen and Electron Transfers

(Continuation)

R. A. Marcus

1. p. x middle: "At lower temperatures, the commitments presumably became more important and, being isotopically insensitive, reduced the KIE towards unity."

Q: Is it so because being the commitments are equilibrium constants and so they are isotopically insensitive?

M: Yes, that's right, the commitment is really an equilibrium constant. In enzyme kinetics they speak of something called commitment, which was new to me, and it's related to the binding of the enzyme with either the substrate or the product, and they speak of that binding as being a commitment, I don't know that literature well but I remember seeing that...

2. Following: "The small A is attributed to poor overlap of the relevant vibrational and electronic wave functions."

Q: How does the above overlap enter A? Is it through Franck–Condon factors?

M: Yes, exactly.

3. p. x bottom: "The high A for $T < 30\,°\mathrm{C}$ cannot be explained by sampling a small subset of reactive conformations. The sampling would create an A *much less* than $10^{13}\,\mathrm{s}^{-1}$ rather than much greater."

Q: Why much less?

M: I think that the small sampling meant that some of the conformations were reactive, some inactive, so they wouldn't contribute, so that would make for small probability, I think that's what it was.

4. p. xi top: "There is another factor that influences A, and can be expected to arise in some cases... If the standard entropy of reaction ΔS^0 for that step is different from zero, then when $|\Delta G^0|$ is small $\Delta S^\dagger$ contains a term $\frac{\sim\Delta S^0}{2}$ arising from this contribution, using the two-parabolas formalism for the protein reorganization."

Q: How does the $\frac{\sim\Delta S^0}{2}$ arises?

M: Yes, the simplest thing is to look in the paper with Sutin and you see expressions for the entropy there, the entropy of activation, and that's how it comes in, if you think of the free energy of activation, and then you expand a $\lambda + \Delta G^0$, you'll get a $\frac{\Delta G^0}{2}$ as a term, you can differentiate to get an entropy and so you get $\frac{\Delta S^0}{2}$.

5. p. xi middle: "In a comparison of two intrinsic KIEs, $\frac{k_H}{k_D}$, and the carbon isotope effect, $\frac{^{12}k}{^{13}k}$, do they show the same trend as a function of pressure, at least at high pressures.... (in particular they have a similar value of the *intrinsic volume of activation*) perhaps reflecting that the C atom is a component of the reaction coordinate in the TS. (*Only in a special case is the H-tunneling a purely H motion*).

M: Well, *the key point is this,* irrespective of the terminology: *the carbon atom motion participates in the overall motion of the hydrogen, it is not just the hydrogen jumping by itself, all the atoms are interconnected, so it's not just the hydrogen that is jumping, it's the whole structure that's jumping,....* other atoms playing very low role, so... *it could be said that carbon is contributing to that, due to the fact that carbon is coupled to hydrogen, there may be pressure effects*, in terms of them would be that part which is really associated with where atoms have got into position and ready for tunneling.

6. p. xi bottom: "...if the protein at temperatures below the breakpoint is sufficiently *rigid* that the '*internal viscosity' of the motion* along the reaction coordinate becomes rate controlling, this motion would be *diffusive in nature*, leading to many (*diffusive*) *recrossings* of the transition state region..."

Q: The more the protein is rigid, the more its motion is analogous to that of a viscous liquid, OK. If the motion is hampered by viscosity there would be slow diffusion. The diffusion would then become the slow rate controlling step. But can you please explain what are the "diffusive recrossings"?

M: *Diffusing means that it is not only diffusing as it's approaching the standard TS, diffusing is involved in motions going back and forth* *and here and there and all that, it's not a ballistic thing, automatically diffusive motion would involve recrossing in the TS.*

Q: But normally when one thinks of diffusion, for instance in a liquid, in a liquid you have two regions with different concentrations, the diffusion has a very precise direction.

M: *That's right, that's a macroscopic view, and I was giving you a molecular insight into what it does.*

7. p. xii, top: "The classical results in Kramers' article are for a simple one-coordinate model, but they can be extended approximately to a multidimensional system by introducing a free-energy curve instead of the original one-dimensional potential-energy curve."

Q: Like you did in the 1960 *Discussions* paper?

M: No, I don't think so because... *it'd be more like the Sumi paper, because in the 1960 paper I wasn't using slow diffusion as part of the theory,* and that in 1960 was a multidimensional reduction to one coordinate, yes, but not in the sense we did later.

8. Following: "His TS expression for the rate constant is then $k = \nu \exp\left(-\frac{\Delta G^*}{kT}\right)$... At the other limit, the limit of high internal friction coefficient ζ along the reaction coordinate, we have the Kramers' equation for the '*overdamped' limit,* $k = \left(\frac{2\pi\nu\nu'}{\zeta}\right)\exp\left(-\frac{\Delta G^*}{kT'}\right)$, *when* $\frac{\zeta}{2} \gg 2\pi\nu'$. *Here,* ν' *is the frequency of the inverted parabola at the top of the TS barrier.* With ν and $\nu' \sim 10^{13}\, s^{-1}$ and $\frac{\zeta}{2} \gg 2\pi\nu'$ in the overdamped regime, this $\frac{2\pi\nu\nu'}{\zeta}$ factor in the rate constant k is less than $10^{13}\, \mathrm{s}^{-1}$, in agreement with an interpretation in terms of recrossings of the TS in this regime."

Q: (1) What does "overdamped limit" mean?

(2) What is this "inverted parabola"?

(3) Why in the equation for k in the overdamped limit a T' appears . . .

M: (1) Well, *that is a standard in problems where you'd have frictional motion.* In other words, *if you have an equation where there is acceleration plus a frictional term, which is proportional to velocity,* plus other terms and so on, then when you're in the overdamped limit *the damping is reducing the importance of acceleration, that describes the overdamped limit. Underdamped is quite the opposite, but there is a certain rule for how the frictional coefficient compares with something or another depending on whether you're either in the overdamped limit or in the underdamped limit . . . .*

(2) *Inverted parabola means is a parabola upside down,* instead of the usual way it's done. *If you look at the free energy curve, it is sort of rising at the TS and in the immediate vicinity at the top of the hill it's an inverted parabola.*

Q: Especially if there is some splitting there.

M: Yes that's right, I'm really talking more of an adiabatic process.

(3) **Typo**: that T' should be T.

9: Following: "A value of ζ can be estimated from a diffusion constant D and the Einstein equation, $\zeta = \frac{kT}{mD}$, or from the viscosity η and the Stokes equation, $\zeta = \frac{6\pi\eta r}{m}$, where m and r denote the molecular mass and radius in the liquid (or $\zeta = \frac{4\pi\eta r}{m}$ for *a 'stick' boundary condition*)."

Q: What's a "stick boundary condition" with that equation so similar to that of Stokes, just a 4 instead of a 6?

M: Well, *if one looks that in hydrodynamics physics books, in the derivation of the Stokes equation, you find the stick formula and the slip formula, and you impose boundary conditions on the fluid at the surface of the particle. In the case of the stick boundary condition, the velocity at the surface is zero there, sticking,* and then, of course, the local velocity of the fluid for the slip boundary condition is different, . . . in books that derive the dynamics of the shear motion into fluid there are two different boundary conditions, sometimes is called stick and slip.

10. Following: "For a nonviscous liquid like acetonitrile ζ is about $10^{12}\,\mathrm{s}^{-1}$, and so the reacting system is not 'overdamped.' For a viscous liquid like glycerol at 30° C ζ is about $10^{15}\,\mathrm{s}^{-1}$, so $\frac{2\pi\nu\nu'}{\zeta}$ *is significantly less than* ν and the system is overdamped."

Q: If $\frac{2\pi\nu\nu'}{\zeta}$ is significantly less than ν, then $\frac{2\pi\nu'}{\zeta} < 1$, that is, $\zeta > 2\pi\nu'$. But please look at your words on p. xii. There you write the condition $\frac{\zeta}{2} \gg 2\pi\nu'$. Why do you divide there ζ by 2?

M: ν, that would be the case where there is no friction, but then it's said that the $\frac{2\pi\nu\nu'}{\zeta}$, that's the overdamped limit, so the ν has more friction, the other thing has friction significantly less than a factor of 2. . .

Q: So both equations are OK.

M: Yes, it comes from looking at the general expression when you ask when it is reduced to a simple expression and so that's maybe the condition, in other words on one hand when you just compare ν and the $\frac{2\pi\nu\nu'}{\zeta}$, on the other one looks at the rigorous full expression, so maybe it comes about that, probably that's the point.

11. p. xiii top: "Current computations do not treat as yet the dynamics of the slow timescales of milliseconds for the overall H-transfer."

M: In the enzymes system that's slowest, not the H-transfer itself but certainly the conditions for it, and that can be of milliseconds.

13. Following: "Nevertheless, they may eventually be able to treat the dynamics of the short individual *diffusive steps*. . ."

Q: What is exactly meant by "diffusive step"? Do you mean that the diffusion path is made up of a series of zigzagging steps?

M: Yes, that's right, steps means that.

M368 Interaction between Experiments, Analytical Theories, and Computation

R. A. Marcus

1. p. 14601 1st column top: "The break in the slope at T_b in Fig. 6 can be viewed in a sense as a phase transformation, a transition somewhat similar to a glass transition, the glass becoming more fluid like above the glass transition temperature T_g. For example, in the latter the *activation energy for viscosity* is smaller at temperatures above T_g than below T_g."

Q: What do you mean by "activation energy for viscosity"? Is it an activation energy to overcome viscosity?

M: If you look at *the viscosity of the solution, and you can usually write it in an Arrhenius form, an Arrhenius expression, it is certainly related to the experiment.*

2. p. 14601 1st column bottom: "In the present instance, where there may be two instead of one slow coordinate, it would be useful to learn from *suitable spectroscopic measurements* how much X_1 and X_2 are each slowed down in a system where $T < T_b$."

Q: To which kind of suitable spectroscopic measurement do you refer?

M: *There are some spectroscopic measurement where you look at some relaxation of some excited state and, depending on the environment, that relaxation can change enormously.*

3. Following: "If the distance X_1 before the transition state has to be a smaller for the D than for the H system in order for the D to tunnel effectively in a system where the change of X_1 is important, the *extra* X_1 *motion* causes the D rate to be more adversely affected than the H rate by the extra sluggishness of the system below T_b."

Q: Do you mean that, because in order for D to tunnel effectively, the distance is to be small enough, and to reach this distance is for the system more difficult in viscous systems?

M: Answer is yes.

4. p. 14603 2nd column top-middle: "In the next time regime, the effect of a finite slope of the curves at the intersection on the diffusion becomes apparent: it causes a 'forced diffusion' that enhances the rate of loss from the intersection region and the survival probability decreases exponentially. In this regime the rate varies as $t^{-3/2}\exp(-\Gamma t)$. (This functional form was later confirmed in experiments)."

Q: Should one consider this result as one of your theoretical prediction later confirmed in experiments?

M: Yes, but I think that now the answer for the exponential is quite different from the one we formerly had, by now I think it's due to biexcitons... The answer for the exponential that Jau and I gave, was that it was related to the slope of the two electronic states, to the two slopes, to a difference of slopes, of the slopes themselves at the intersection of those two curves, so we had a kind of a forced diffusion which then uses an exponential times the power law. But all what always bothered me about that, is that experimentally you only get the exponential for the bright states and not for the dark states, and that wasn't explained by the ideas that Jau Tang and I had, but then in the paper with Joyanne Zhu there was an explanation because at high intensities you can form biexcitons and if the system were in a dark state you couldn't really form a biexciton because the first exciton would rapidly decay by a nonradiative process, so you wouldn't form biexcitons, only in the bright state you can form biexcitons and they can react with each other to produce a state with a very high energy, in other words twice the band gap energy, and it's an energy so high that they can be ejected. A Fermi golden rule gives rise to an exponential, so that's a better description.

Q: How would you define a biexciton?

M: A biexciton is when you have two excitons present, in other words you excite one that gives you an electron and a hole, and *before that the system has a chance to decay* you excite with another, because you are shining light steadily, and at high intensity there is a reasonable chance of doing that, so you have two electrons and two holes, that's called a biexciton.

M373 Interaction of Theory and Experiment: Examples from Single Molecule Studies of Nanoparticles

R. A. Marcus

1. p. 1112 middle: "Inorganic semiconductor nanoparticles, such as quantum dots (QDs), have been widely used in biological imaging. *Their high spectral absorption and narrow fluorescence spectrum are attractive features.*"

M: *The high spectral absorption is part of their properties, the QD can absorb a lot of light because it's large, all parts of it can absorb light. There is a narrow fluorescence spectrum because if you look at the Stokes shifts . . . going from the upper state to the lower state, there are very small shifts, the parabolas are standing over each other and there is a very small rate.*

2. Following: "Each semiconductor nanoparticle is typically encapsulated in a thin layer of another semiconductor that has a larger bandgap *and so prevents a photoexcited electron or hole in the core semiconductor from escaping too far.*"

M: Yes, *the wave function of an electron in a semiconductor can peel over to outside the semiconductor and, random, the electron could escape, but if you have a larger band gap the barrier to go into the outside is greater,* so its chance of escaping is smaller.

3. p. 1114: "... the distribution of times $P(t)$ when the nanoparticle goes from 'on' to 'off' in time...does not follow an exponential decay with time and so is not a simple decay process."

Q: Could one define a single decay process as one with a constant half-life?

M: *A half-life is exclusively one point on the decay curve, so you can't say it is constant, it is just one point, but what you are asking is whether the decay is exponential or not, whether you have a single exponential, because that has one time constant.* It is more complicated than that, is a different story. Typically if you have traps on a surface off a valence band for the conduction band there is usually an exponential distribution of traps and it gives rise in the absorption spectrum to what's called Urbach tail, and that reflects the role of these trap states in affecting the absorption, it is supposed that they could affect the fluorescence, but I don't think that's what I am talking about, maybe one's talking about multiple exponentials, like if you have a distribution of exponentials that's like a multiple exponential, and you can tail over any function you want that way, including a power law.

4. p. 1115 middle: "If a fluorescing organic molecule undergoes an electron transfer and becomes an ion, because of an electron ejection to another molecule or to another *plot* of the system..."

Q: What do you mean by "plot"? Is it just another part of the system?

M: Plot is a **typo**.

5. Following: "This new radiationless process that is competitive with the fluorescence, new to QDs *but old to atomic physics*..."

Q: Were the radiationless transition first observed in atomic spectroscopy?

M: I'm just wondering how that can happen, in atomic spectroscopy. In principle you can have a fluorescence state, you could also instead have some sort of the arrangements of the electrons such that eventually a state will photofluoresce...I don't know if they have seen that or not, they may have, I don't know.

6. p. 1116 bottom: "Before considering the trapping and detrapping processes we first recall the explanation (Efron & Rosen 1997) as to why a trapped state in the QD causes it to be 'off.' We do so using the specific model just presented, and depicted in figure 5. It is seen there how for an 'off' state the photoexcitation of *a second electron* from the valence band to the conduction band results in a situation where there is now a radiationless alternative to the fluorescence. There are many electronic states of the QD where the *second excited electron* can go while the other electron *in the conduction band* goes into the newly created hole in a valence band in this radiationless transition."

Q: Please look at my Fig. 1* where I'm trying to represent your words. I begin with one electron in the VB and one electron in the CB. I excite then the electron in the VB bringing it in the CB and leaving behind a hole in the VB. The last panel in the figure is the same as that in the second panel of your Fig. 5 except that I believe the VB and CB symbols there should be inverted. Moreover, in your first panel there appears an excitation of a hole.

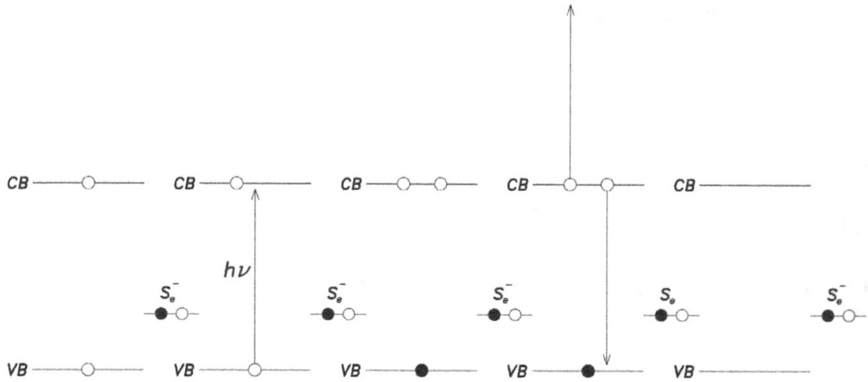

Fig. 1*.

M: I have the figures. The electron in the Selenium is filling a hole that was caused through the electron having been excited to C state. That hole is created in the VB by the excitation.

Q: In the second panel of Fig. 5 there is an evident **typo.**

M: You see, one of the electrons goes down and for energy conservation one of them goes up.

Q: But there the indications VB and CB are inverted with respect to the first panel

M: **Typo**, All right, that's a typo.

Q: Look at my Fig. 1* where I interpret your Fig. 5 somehow.

M: All right.

Q: My Fig. 1* compared with your Fig. 5

M: Yes.

Q: This is somehow more detailed, is my figure OK?

M: Yes, *your Fig. 1* is perfect.*

Q: The first panel of your Fig. 5 gives the impression that you excite a hole.

M: Yes.

Q: I think this is a better description.

M: Yes, *I think you right.*

Q: I have also a further panel that comes after your last panel.

M: *What you have is correct.*

7. p. 1117 middle: "A key question is how does the trapping transition occur?.... When the energy difference of the two lowest energy levels in the 'conduction' band, S_e and P_e, in *figure 6* matches the energy change there for the transfer of an electron from dangling Se^{2-} ion to the *newly created hole in the state* S_h in the 'valence band' in figure 6 then a transition may occur and then the QD becomes 'off.' This 'resonance' is between these two states of the system, the state specified by (1e in S_e, 1 hole in S_h, Se^{2-}) and another state *(1e in* S_e*, 0 hole in* S_h*,* Se^-*)*. The transition can be written as

$$P_e(1) + S_h(1) + Se^{2-} \rightarrow S_e(1) + S_h(0) + Se^-, \qquad (4.1)$$

where (1) and (0) denote the number of electrons or holes in the specified state."

Q: (1) Please look at my Fig. 2*. I think the first two panels are more clear than the first panel of Figure 6;

(2) In the text above in the "another state *(1e in* S_e*, 0 hole in* S_h*,* Se^-*)*, the state should really be (1e in P_e, 1e in S_h, Se^-);

(3) The reaction in Eq. (4.1) should be substituted by:

$$S_e(1) + S_h(0) + Se^{2-} \rightarrow P_e(1) + S_h(1) + Se^-$$

(4) In the Auger process there is an energy balance between two processes, see for instance the two processes in the second panel of Fig. 6. But how do they know of each other?

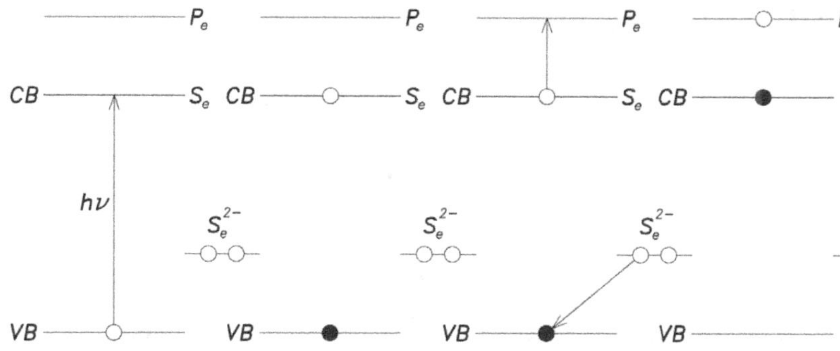

Fig. 2*.

M: (1) *What you got is better, yes.*

Q: Please look at Fig. 2* fourth panel.

M: *Yes, that's correct what you have there*. The one in P isn't the last story, that P_e then one electron eventually goes to the... *What I did is very much shortened from this, so your way is definitely better.*

(2) Yes, *that's fine.*

(3) I agree, *Fig. 2* is clearer than Fig. 6.*

(4) Yes, *that Eq. (4.1) doesn't look right. That's fine, yes, good. You're clarifying to Rudy... Well, certainly the way you spelled out is certainly a better way, a clearer way of describing what's going on.*

Q: But, you know, you are in the tradition of the best theoreticians. Frank Wilczek writes that Maxwell was rather messy in his exposition, I'll send you a copy of Wilczek's paper in Scientific American. The way we learn it was the work of Lorentz, of Heaviside and Hertz.

M: I think Heaviside...he did write in some sort of engineering fashion, you know, Laplace transform he made more rigorous, but I think he did something intuitively, I never looked at Heaviside's...

8. p. 1119 top: "If $\rho_1(Q)$ denotes a probability density...

$$\frac{\partial}{\partial t}\rho_1(Q,t) = L_1\rho_1(Q,t) - \frac{2\pi|V_{\mathrm{ex}}|^2}{\hbar}\delta[U_{12}(Q)][\rho_1(Q,t)] \quad (4.2)$$

where the diffusion operator is given by

$$L_1 \equiv D_1\frac{\partial}{\partial Q}\left[\frac{\partial}{\partial Q} + \frac{1}{k_{\mathrm{B}}T}\frac{\partial}{\partial Q}U_1(Q)\right] \quad (4.3)$$

Here, D_1 *is a diffusion coefficient in this energy space on curve1* ... the last term in equation (4.2) is a statement that this transition from curve 1 to 2 is *a weak transition* and is given by a Fermi's Golden Rule expression."

Q: (1) How does one measures/computes D_1?

(2) How do I see from the last term in Eq. (4.2) that the transition is a weak transition?

M: (1) Yes,...I'm trying to remember, I don't know that one has measured it, I mean *there is something called spectral diffusion, and you have a line and you can watch it gradually fall up because of slow small changes in the crystal, sorry in the nanoparticle, the line is moving around and, maybe because charges are jumping around in the dot, or whatever have you, anyways the positions of the line is sort of diffusing, it goes back and forth and so on. So, at low temperatures you can measure the diffusion* but I don't know if you can extract the diffusion constant by that way, it would be one of the processes contributing, I just don't remember, but at high temperatures I think it is too fast...so that one can measure, but it has to be *from spectral diffusion measurements*.

(2) *The delta function means is occurring at a point, and if you think of Landau–Zener, Landau–Zener is occurring at a point. And if you think of Fermi's golden rule where you have a whole group of states and you have one state going to a continuum or a group of states that is essentially occurring at a point.*

9. p. 1120 top: "In somewhat oversimplified terms the survival probability distribution near the sink behaves as *a well known* $(Dt)^{-\frac{1}{2}}\ldots$"

Q: Why "well known"?

M: Because in diffusion it is random walk, that's a common expression you see. *It's a law of random walk steps.*

Q: Yes, it is so also in the Feynman's Lectures on Physics.

10. Following: "In the next regime for this model, given by equation (4.7),

$$P(t) \sim \frac{t_c}{4\pi} t^{-3/2} \exp(-\Gamma t), \quad t \gg t_c \text{ (region 3)} \qquad (4.7)$$

the effect of a finite slope of the curves at the intersection on the diffusion *becomes apparent: it causes a 'forced diffusion'* that enhances the rate of loss from the intersection...."

M: *If you have a diffusion equation, then if you have a forcing term, a potential energy gradient, which sometimes you have, that is part of the motion of the ions under a field. Then, when you have that, you have what is called a forced diffusion, in other words the diffusion is affected by the force, you have a potential energy curve that has anything other than a zero slope, it's a forcing term.*

Q: The potential itself forces.

M: Yes, over free energy.

11. Following: "In the final regime, given by equation (4.8),

$$P(t) = A \exp(-\gamma t), \quad t \to \infty \text{ (region 4)} \qquad (4.8)$$

the calculated survival probability distribution is a pure exponential and is due to *escape from the bottom of the free energy curve* in which it resides."

Q: Shouldn't everything happen on the free energy curves?

M: Yes, supposing you have a particle in a well, and it can go over a barrier, it gets there not suddenly but somehow diffusing in energy space, *you have fluctuations*, so it gets some energy, loses some energy, and so on, eventually it gets enough energy to go over the barrier, so there is a kind of *diffusion in energy space*. A particle that has overcome a barrier has to have enough energy to overcome the barrier, it does not get that all in one shot, it collides, gets a little bit, loses a little bit, and it is almost like diffusing in energy space, Kramers had a theory on that in 1940, a classic paper, he had many many things at that, a part of it involves a diffusion in energy space.

12. Following: "This reaction-diffusion model serves to explain ...However, how does one explain the asymmetry—why is there not a cut-off for the 'off' times, or at least not an easily discernible cut-off in the time scale of intermittency?"

M: Yes, that's right, *that's what we think we now understand in terms of the biexciton,* which is relatively new, biexcitons have come into the literature more recently, not by us but by others who have made use of that.

Q: OK Maestro, without you it would have been impossible to write this book, it would have been a booklet somehow if I had to write it on my own. So, I think that your authorship would be natural.

M: OK, well *I'm sure it will be an interesting book that you write, a good combination of history and science, it's a tremendous task though, you know, it's because there is so much that you study, so much to learn about, it's a challenge, I'm sure you're writing an interesting and really useful book, because electron transfer is just so wide spread now and so many people want to learn about.*

Q: I don't really know how to express my gratitude for this long time, for so much time you have devoted to this work, I hope the book will satisfy you.

M: Well, I'm sure it will and I'm a teacher anyway you now...

M379 Theory of a Single Dye Molecule Blinking with a Diffusion-based Power Law Distribution

Wei-Chen Chen and R. A. Marcus

1. p. 15782 1st column middle: "Recently there have been single molecule studies of the fluorescence intermittency of a single photoexcited dye molecule on a semiconductor nanoparticle system or on a crystal surface."

Q: (1) Until now you have studied fluorescence intermittency of QDs. Apparently now there is also the intermittency of the dye. How does the fluorescence of a dye compares with that of a QD?

(2) Is the semiconductor nanoparticle a QD or not necessarily so?

M: (1) They both fluoresce, you shine light on and they both fluoresce. A dye fluoresces, at least if it is in a fluorescent state it fluoresces, and a quantum dot fluoresces, if it is in a fluorescent state.

(2) Usually they are quantum dots but there are quantum rods, you have different dimensions....

2. Following: "The switch between light and dark periods appears to be of two types: (1) a system that has "on" and "off" periods ... and (2) a system that is also "partially on" and "partially off."

Q: What does it mean to be "partially on" or "partially off"?

M: *That's an interesting question.* If one looks at light intensity of the fluorescence, one sees periods where it's way up, and periods where it's way down, and occasionally, depending on the system, one would see periods where the intensity is in between, that's a partially on, partially off dot, and the reason why it's in between . . . that's another story.

3. Following: ". . . (1) distribution of tunneling barriers for the ejected electron to return to the dye cation from a distribution of trap depths."

Q: (1) barriers from where to where?

(2) How's the nature of these traps?

M: If one looks at solid state physics going back to the 1950s, one would see that off the conduction band or off the valence band there are series of traps, exponentially distributed, same thing here.

Traps can be different things, they can be a defect, you know, some missing ion, they can be just a discontinuity, such that somehow you can put an electron there, you can put a hole there, they can be of various kinds.

If you have a trap, then the electron may go from the trap to the dye. If it's going from a trap to another trap, because you can have surface conductivity between traps on the semiconductors, then that's where the electron is going from where to where, it depends on the process and the nature of these traps, they can be anything, it's hard to say.

4. p. 15782 2nd column top: "Here the electron injected from the dye into the semiconductor can diffuse in an assembly of nanoparticles, perhaps via trap-to-trap tunneling."

Q: (1) How many traps in each nanoparticle?

(2) Which is the nature of such traps? Are they some kind of particular potential wells?

M: Every nanoparticle has many defects, so there will probably be many traps. Some of the traps have been studied by spin resonance methods I think by Efrat Lifshitz of Technion in Israel, but I haven't looked into the detailed description of the traps.

5. Following: "(1) If an excited electron has an energy high enough to be injected into the conduction band, the injection is followed by a trap-to-trap diffusion back to the dye cation."

Q: Is this process the inverse of the preceding one?

M: No, the process isn't the inverse of the preceding one, because in the case where the electron can be injected into a conduction band, it's not injected into a trap, it's injected into the conduction band, then often you can relax from that conduction band into a trap just below the conduction band and the system can now go from the trap diffusing all over the place, including back to where it was in the dye.

6. Following: "A power law for the survival probability density in the dark state is expected, but *not for the injection itself, i.e., not for the light state.*"

M: In other words, if you just inject into the conduction band, there is no diffusion, or motion of any kind, so you don't expect a power law. It is possible only when you have a diffusion, either of the particle itself in the case of a dye on the semiconductor surface, or diffusion of structures, so to speak, structural diffusion, spectral diffusion as it's called in the case of a QD.

7. Following: "... an acceptor trap *can diffuse* and ultimately, if close enough, accept the electron from the excited dye..."

Q: But aren't the traps by their nature rather fixed in the structure of the crystal? How do they diffuse?

M: Yes, the trap is fixed but the electron or hole can diffuse from one trap to another. There are many traps and what precisely the traps are

I don't know, there are maybe some studies of them, over the years they have been going on for more than 70 years, about 70 years, so there have been a lot of studies in one form or another. The system can diffuse from trap to trap, the traps are not diffusing.

8. Following: "This electron injection into the band gap may proceed via surface states that typically exist in the band gap.

Q: (1) What is, physically, a surface state?

(2) Apparently a surface state is located on the surface of the crystal but its energy is located in the band gap. Is it so?

M: (1) There are various kinds of surface states, any kind of a defect on surface is a surface state, there is a lot of literature on surface states in the semiconductor literature, for example, if you have a dangling Selenium ion, a doubly charged Se^{-2} ion, it isn't highly stabilized, then you can easily loose an electron from there and that's a hole trap now, I mean there is a hole there.

(2) Usually surface states would have any stability if they were in the band gap, but occasionally you have surface states that are in the conduction or valence band gap, that are called evanescent surface states because a system can't stay there, it's coupled in a sense to the conduction band and eventually goes into the conduction band or the valence band, depending on where it is. So, if you want to have some kind of stability of the state, it has to be in the band gap and not connected with a continuum.

9. p. 15783 2nd column top-middle: "We denote by $\mathscr{P}(t)$, the probability that the electron is outside the dye molecule..."

Following: "The survival probability $\mathscr{P}$ at time t is

$$1 = \mathscr{P}(t \to 0^+) = \int_{r_0}^{\infty} dr\, 2\pi r \rho(r, t \to 0^+) \tag{4}$$

Q: I believe the above formula specifies the fact that the electron may be at all the possible distances.

M: Yes, it can be at all possible distances, that's right. Now, initially it probably goes initially into something heavy localized and then, after that, it diffuses all over the map, but after a while it can be at all possible distances.

10. p. 15784 1st column middle:

$$\tilde{p}(s) \approx \frac{r_0 k}{D} \frac{K_0(u_0)}{u_0 K_1(u_0)} \tag{11}$$

$$= \frac{r_0 k}{D} \left(I_0(u_0) K_0(u_0) + \frac{I_1(u_0) K_0^2(u_0)}{K_1(u_0)} \right) \tag{12}$$

$$\approx \frac{r_0 k}{D} (I_0(u_0) K_0(u_0) + I_1(u_0) K_1(u_0)) \tag{13}$$

The property that $K_0(u_0)/K_1(u_0)$ approaches unity when $u_0 \to \infty$ was used in deriving the above approximations.

Q: I believe there is a **typo**: the last formula should be

$$\frac{r_0 k}{D} (I_0(u_0) K_0(u_0) + I_1(u_0) K_0(u_0)).$$

M: They seem to be the same thing... since one is talking about this limit, it doesn't matter whether $K_1(u_0)$ or $K_0(u_0)$, either one will do, in the limit they are equal.

11. p. 15784 2nd column bottom: "The half-life of the absorption recovery of dye molecules in the Wang et al. ensemble experiment is around 50 ps."

Q: What is it this half-life absorption recovery?

M: Supposing you're shining light on the dye, and then you monitor the fluorescence of the dye. Well, when the electron goes in the nanoparticle that dye doesn't fluoresce anymore, it's a cation, and typically the cations don't fluoresce. Eventually the electron comes back and so the fluorescence then would change as a function of time, it would increase somehow as a function of time, if it comes back... when the cation becomes a neutral molecule and you have light shining on it, it will start fluorescing, so you look to where the fluorescence becomes half as much as it was at the start. See, that would be half-life of the absorption recovery, because the dye molecules can't fluoresce unless they absorb.

12. Following: "The radius of the dye molecule is around 1 nm. Assuming *the k value is around the dye size divided by the half-life* (50 ps), then k is about $2 \times 10^{10}\,\text{nm} \cdot \text{s}^{-1}$."

Q: How do you get the k dividing the dye size by the half-life?

M: The inverse of 50 ps would be 2 times $10^{10}\ \text{s}^{-1}$. Inverting the 50 ps you get a rate, an inverse of time instead of a time.

13. p. 15785 1st column top-middle: "A numerical simulation of reaction-diffusion in a Coulombic interaction confirms that the recovery ratio for the encapsulated molecule..."

Q: What do you mean by "encapsulated molecule"?

M: The fellow did two times an experiment, one where there was the dye itself on the surface, the other where he had some sort of a shield, another molecule that was encapsulating the dye.

14. p. 15785 1st column bottom: "Wu et al. studied the blinking of single rhodamine-type dye molecules spin-coated on a TiO_2 nanoparticle film."

Q: Please explain, what does it mean spin-coated?

M: That's a technique he used, I don't know exactly what it is, there are lots of ways to spread particles, some films, by spin-coating them,

it's a special technique that is widely used, that's all I know about it. Nothing to do with nuclear or electron spin, not to my knowledge, it's just a materials technique for deposit, you are probably spinning something on.

15. p. 15785 2nd column top: "... dye-sensitized colloidal TiO_2 nanoparticles..."

Q: How are they made?

M: I don't know how they are made... Often these covers are made chemically by mixing things in solution, something like that.

16. Following: "Because the attenuation factor β due to the tunneling effect in electron transfer at dye-sensitized nanocrystalline TiO_2 films is ~ 1 Å^{-1}, and the half-height of the capsulate molecule is ~ 4.6 Å, the estimated back-electron-transfer rate for encapsulate molecule for the encapsulated dye is reduced to 2×10^8 nm $\cdot$ s^{-1}."

Q: What is the capsulate molecule, its half-height, the encapsulate molecule and the encapsulated dye?

M: Yes that's right, he did experiments with both. The height would then be 9.2 Å, I don't know why the half-height is referred there, I just don't know.

17. Following: "The experimental electron diffusion coefficients in non-sintered TiO_2 nanoparticles are around 4×10^{-7} to 4×10^{-6} cm$^2 \cdot$ s^{-1}, which is lower by ~ 2 orders of magnitude...."

Q: I believe in the nonsintered case the nanoparticles are separated from each other whereas in the case of the sintered they are attached to each other.

M: If it's sintered you would end up with the particles sort of sticking to each other.

18. p. 15786 1st column top: "the experiment of Wu et al . . . a single dye molecule on a TiO_2 nanoparticle surface . . . "

Q: Are these nanoparticles QDs?

M: Usually the nanoparticles are QDs, but they can be fine rods or have other shapes too.

19. p. 15786 1st column middle-bottom: "Accordingly, the neutral dye molecule may require many photoexcitations *and fluorescence* before the electron-trap on the surface has diffused to the dye molecule . . ."

M: Supposing you have electron traps on the surface, and they carry an electron. So, there is an electron in the trap and maybe the electron is going from trap to trap, so there is the electron on the surface of the QD moving from trap to trap, and it may take a while to exit to the dye molecule. I would remove the words "and fluorescence" in the second line, the idea is that there can be many photoexcitations, I don't know if that dye cation absorbs, so I don't know if it is photoexcited or not, surely there can be photons that are bombarding the dye cation, but *it is not fluorescing because it's a cation*, so I think just that there can just be many photoexcitations. Of course, there may be always photoexcitations if the dye cation absorbs when the light is impinging on it, but it's certainly not fluorescing, that dye is not fluorescing, until the electron goes back to it, so I would omit the "and fluorescence."

20. p. 15786 1st column bottom: "The energetic scheme of the injection into the conduction band for the dye-TiO_2 or into the band gap for dye-ZrO_2 is shown in Figure 5 . . . All of these dyes have a photoexcited state between 3.2 and 4.5 eV. One would thus anticipate, on the basis of current theory, that in a single molecule experiment the injection into TiO_2 would have a power law only for the return to the light state, whereas it should have a power law for both the formation and the disappearance of the dark state in the case of ZrO_2 under

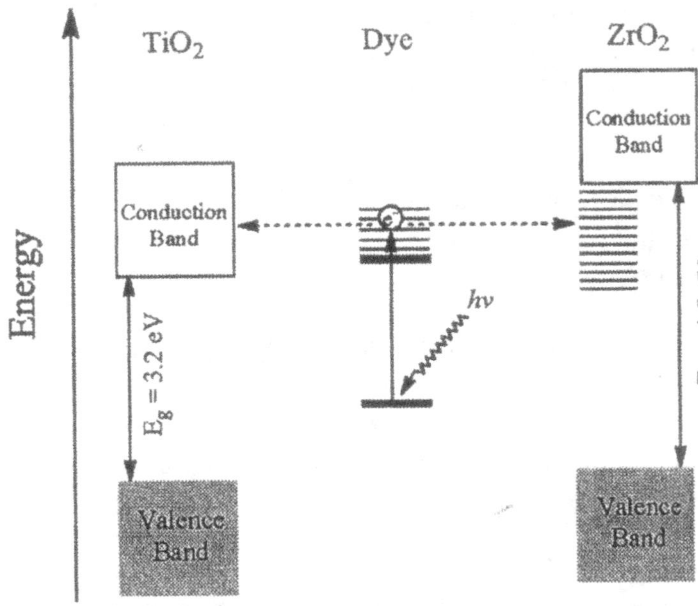

Fig. 5. Energetic scheme of the dye–TiO_2 and dye–ZrO_2 system. The dashed arrows show the injection of a photoexcited electron into the conduction band of TiO_2 nanoparticles or into the states in the band gap of ZrO_2 nanoparticles if the excited-state energy of the dye molecule is between 3.2 and 4.5 eV. Examples of dye molecules include alizarin,[39] quinizarin,[40] and 7-hydroxycoumarin 4-acetic acid.[41]

the condition where in the former the injection is into the conduction band and in the latter into the band gap."

M: If you look at the figure, going on the left the electron is going into a conduction band. But going on the right we're dealing with a system with a higher band gap and it's going directly into the band gap. All right, so now the only way it can diffuse back, diffuse some place, is going from trap to trap, but it's moving in traps now, it's in the band gap. Now how it can get up back into the conduction band, it may gradually go down, eventually reach the valence band and it should diffuse around in the space on the surface and eventually comes back to near the dye, and then it can reform the ground state of the dye molecule.

21. p. 15787 1st column bottom: "If the photoexcited electron from the dye molecule is injected into a localized surface state near the surface/molecule interface, and then diffuses among the surface states . . . "

Q: Does the electron diffuse in the coordinate space among surface states with the same energy or does it diffuse also among states of different energy?

M: My guess is it's moving from trap to trap so its energy is probably fluctuating, any the diffusing particle is taking up and losing energy, not much but a little bit, so there is no reason for a constant energy, it is just moving a lot, taking and losing, as systems are there in contact with each other.

M389 Extension of the Diffusion-controlled Electron Transfer Theory for Intermittent Fluorescence of Quantum Dots: Inclusion of Biexcitons and the Difference of "On" and "Off" Time Distributions

Z. Zhu and R. A. Marcus

1. p. 1, 1st column middle: ". . . absorption of light and fluorescence recycling is followed by sustained periods of darkness . . . "

Q: What do you mean by "fluorescence recycling" and by "sustained periods"?

M: Fluorescence recycling just means that you keep on shining light, it fluoresces, then you're still shining light, it fluoresces, so you're recycling the state of the QD going from ground state to upper state, back to ground state and so on, that's what recycling means. And sustained periods means that it's bright for a long time and suddenly it's dark, it doesn't fluoresce, and it's dark for a long time, a sustained period, that's what that means.

2. Following: "Memory in subsequent fluorescence or dark episodes . . . "

M: There is some evidence, although we didn't treat it directly, that the system has some memory, namely after a bright fluorescence and then dark, it will tend to have a bright one again that I don't remember if of the same length, but there is some kind of memory occurring, some evidence has been presented.

4. p. 1 1st column bottom: ". . . two different mechanisms for the intermittency, resonant and Fermi's golden rule, depending in the energy . . . "

Following: "A physical reason, based on amount of excess energy, is given as to why there are two forms of change of state, one that is a resonant transition (~ 0.3 eV) and the other is a Fermi's Golden rule transition (~ 2.2 eV)."

Q: Please explain the two mechanisms . . . Oh, I skipped Question number 3 . . .

M: Oh, you don't agree with what the Germans are saying, alle gute Ding sind drei . . . You're probably thinking of what happened in the first world war, when it was bad to have three on a match because the one was seeing the match, would shoot at the first one, take in on the second one and the third one would be shot.

Well now, if you're thinking about transitions quantum mechanically you're going to have a transition at the intersection of two potential energy surfaces, that's one, and that's resonance transition, yes, the two points are on resonance.

Q: Same energy.

M: Yes, that's right, same energy. So that's a *resonance transition* so, for example, *the famous Landau and Zener are talking about resonant transition*. But then there is also *a transition where you can go from a discrete state not to another discrete state but to a whole* *continuum of states, and for that a Fermi Golden rule is used*, so that's a different kind of process. So, when you have a lot of extra energy around you might have a high density of states there available because maybe you can get outside the QD, we have a large number of states.

Q: By the way, when you say "outside the QD"...

M: I mean outside the core of the QD, it maybe into the shell, maybe inside the organic layer that's attached to it, something or other, the surface of the QD... because you have layers to protect the QD against....

5. p. 2, 1st column top: "... the bright and dark cycling ... "

Q: What do you mean by "cycling"?

M: Just repeated going shining: illumination, illumination, illumination, that's cycling. And bright and dark means that during the course of the cycling one has some bright periods, some dark periods, some bright periods... so cycling just means repetition.

6. p. 2 top-middle: "The diffusion considered here differs from that discussed in ref 49, which considers diffusion of electrons rather than "spectral diffusion."

Q: Please describe the two types of diffusion...

M: There is a *structural diffusion*, where there are *no particles diffusing*, instead there is *structural changing in a diffusional-like fashion*, and that's called structural diffusion, and if you look at the difference of the two levels you see that it fluctuates with time, there is no particle diffusion, that's kind of a structural diffusion.

7. p. 2 2nd column top: Please look at the list of assumed reaction mechanisms. You write: "The k_{eg} and k_{be} are both a sum of the radiative and nonradiative rate constants"

Q: (1) I believe you mean that the process $|e\rangle \rightarrow |g\rangle$ may be both radiative and nonradiative. Is the radiative process the one responsible for fluorescence? Same question for the process $|b\rangle \rightarrow |e\rangle$. Are the two radiative processes the ones responsible for fluoresce? If so are the two processes the only ones responsible for fluorescence in the list of reaction steps for a bright QD?

(2) Looking at the list of reaction steps for a dark QD, I see that the process for going from dark to bright is represented by $|d^*\rangle \rightarrow |e\rangle$ but there is not the $|d\rangle \rightarrow |e\rangle$. Why so?

M: (1) The $|e\rangle \rightarrow |g\rangle$ is radiative, that's the fluorescence... I forget what $|b\rangle$ was ... $|b\rangle \rightarrow |e\rangle$, that can be radiative too, *radiative and fluorescent are the same thing.*

(2) There is no $|d\rangle \rightarrow |e\rangle$ because the dark state by definition... The $|d^*\rangle$ is the excited dark state and the $|d\rangle$ state is maybe way low down, so it can't just go to an excited state, in other words, if you take a dark state, it's excited, and then if somehow it remains a dark state but loses its extra energy then remains a dark state, then it won't fluoresce. You show two curves in Fig. 1, there is a third curve... that $|d\rangle$ dark state... I think that $|d^*\rangle$ is the excited dark state and if you go back to where it came from, namely an excited fluorescent state... they

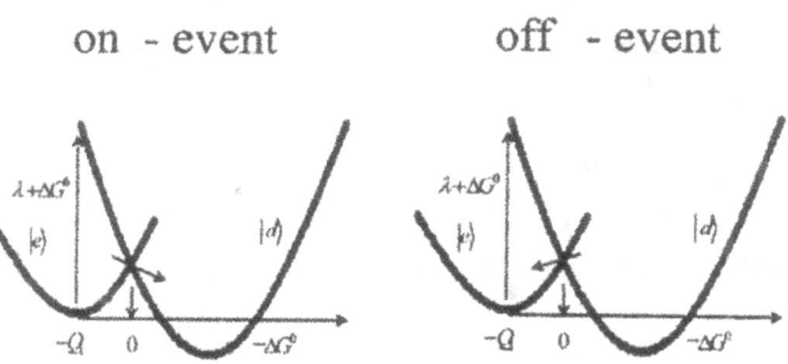

Fig. 1. Diffusion on the parabolic potential surfaces |1⟩ and |d⟩ across a sink at the energy-level crossing governs the intermittency phenomenon (corresponding to Fig. 2(b) from ref. 38).

are about the same energy, $|d\rangle$ itself is a low energy state, has lost its energy, it differs from $|d^*\rangle$ by maybe 2.5 eVs, that $|d\rangle$ in the figure should be $|d^*\rangle$ instead of $|d\rangle$, you are going from an excited state, an exciton state, and you are going to a high energy $|d\rangle$ state, there probably should be $|d^*\rangle$ there. That's my guess, I mean check it out but I think that $|d^*\rangle$ is the excited dark state, it's got 2.5 eV energy more than the dark state. The dark state cannot easily go back to $|e\rangle$ which is about 2.5 eV higher in energy. The figure is misleading, it should have a $|d^*\rangle$ there.

Q: I'm reading carefully your papers...

M: Something I should have done.

8. p. 2 2nd column middle: "$\frac{\partial \rho_e(Q,t)}{\partial t} = I_{ge}\rho_g(Q,t) + L_e\rho_e(Q,t) + k_{be}\rho_b(Q,t) - \left(k_{eg} + I_{eb}\right)\rho_e(Q,t) - k_{ed}\delta(Q)\rho_e(Q,t)$ (2)

Following, bottom: $L_e \equiv D_e\frac{\partial}{\partial Q}\left[\frac{\partial}{\partial Q} + \frac{1}{k_BT}\frac{\partial}{\partial Q}U_e(Q)\right]$"

Q: (1) The equation is clear to me when considering the terms depending on the various rate constants but less so when considering the diffusion term. The velocity with which ρ_e increases at Q is proportional to ρ_g, OK. I believe the term $L_e\rho_e(Q,t)$ means that that velocity also increases at Q with the diffusion toward Q in time and this happens when $\rho_e(Q,t)$ is greater in an immediate neighborhood of Q than in Q. Is it so?

(2) What does it mean the term $\delta(Q)$ which looks like a δ function at Q? Does it restrict $k_{ed}\delta(Q)\rho_e(Qt)$ to considering only the Q at the crossing of the free energy curves?

M: (1) *It is a spectral diffusion, not an ordinary diffusion*, we are talking about here... with Wei-Chen we were talking of ordinary diffusion, but here we are talking about spectral diffusion. That L_e, that's spectral diffusion. In other words, *the energy levels are diffusing, not the particles but the energy levels*, due to structural changes, the fluctuations that are going on and that's very common, standard in diffusion problems, you have a pure diffusional term and a forced diffusional term, whenever there is a driving force sometimes, so those are the two terms there.

(2) That's identifying that you're talking about a transition at the intersection of two curves.

9. Following: "... the dark state doesn't diffuse in the absence of light (otherwise it would reach the intersection of potential energy curves)"

Q: (1) In Fig. 1 you use the free energy surfaces as in the usual ET reactions. There the fluctuations of free energy are fluctuations of polarization energy. Which kind of energy is the one that fluctuates and why does it fluctuate? Does it fluctuates around every Q because of photons absorption or emission or because of random nonradiative processes happening at Q?

(2) What does it mean that a state diffuses? In the classical ET the intersection is reached thanks to an appropriate polarization fluctuation but there is no diffusion from the previous state before fluctuation to the new one after fluctuation. Please explain... is maybe the diffusion in relation with photons diffusing inside the system where they were absorbed?

M: (1) Spectral diffusion occurs only when you have light shining on it, because only that causes little structural changes to occur, so if you have a dark state and you're shining light on it, it will only get spectral diffusion, but if you are not shining light on it there is no spectral diffusion, so the spectral diffusion depends on the light that has been shone, that is moving charges around. The energy levels are diffusing and if there is no light they are not diffusing.

(2) Yes, these are different... well, they are maybe related, the fluctuations of maybe charges there, around there, fluctuations of

the structure a little bit, and, yes, that could be in part related to fluctuations in polarization, but is not restricted to it...

10. p. 3 1st column top: "Unlike the resonant transition terms in eqn (2) given by the delta function (Fig. 1), we assume that because of the large energy present in the biexciton state an electron or perhaps even a more localized hole can be ejected to any site outside the core QD"

Q: Ejected where? On the QD surface?

M: It may go further than that, it may go into the surface, it may go into the shell, and may go beyond the shell, there is some evidence that the nature of the organic material outside the shell has some effect on the process, somebody got some evidence of that.

11. p. 3, 1st column middle: "... the total 'on' population $\rho_g+\rho_e+\rho_b$ denoted by ρ_L ..."

Q: What does it mean? Maybe that the state $e\rangle$ may fluoresce with the process $e\rangle \to g\rangle$ while the other two states $g\rangle$ and $b\rangle$ may produce the fluorescing $e\rangle$ state through the $g\rangle + h\nu \to e\rangle$ and $b\rangle \to e\rangle$ and, moreover, only from $e\rangle$ it is possible to go to the dark state through the step $e\rangle \to d\rangle$?

M: The main point is that the "on" population is something which when excited it fluoresces and not be dark. Now, some of the population is in the ground state, there is a little of the population that's in the excited state, and there is a little of the population that's in the biexcitons state, and all of them are in a state such that they can fluoresce.

Q: Even the one on the ground state?

M: Well, it's the "on" state, if it absorbs light it can fluoresce, all either can absorb light and fluoresce or can fluoresce on their own. In other words, none of them are in the dark state.

12. p. 3, 2nd column middle: "For pulsed laser excitation, there are two time frames, one arising at the start of the experiment, t, and the other, the time between two consecutive pulses counting from the start of every pulse, t_r, where $0 \leq t_r < T$, T being the interval between two consecutive pulses."

Q: From your definition I believe that T is the time interval between the beginnings of two consecutive pulses. Is it so? Or is it the time interval between the end of a preceding pulse and the beginning of a successive one? Can you better define t_r, and so why $0 \leq t_r < T$?

M: If you consider some time integral between two consecutive pulses that means that one pulse had stated at time 0, right? Because if it is between two consecutive pulses it had started at time 0 so the second pulse would occur at some time. . . oh, wait a minute, here. . . I think that's not well phrased, I think.

Q: No, no.

M: Definitely not, I blame on Joy Anne and my not catching it. That's definitely not well phrased. I mean one can think there is a time between consecutive pulses, and of course that's variable.

Q: Exactly.

M: Yes, and so one can define two time scales related to that, but this way of phrasing is not right. The first pulse occurs at 0, the second pulse occurs at t_r, and a third pulse occurs at T.

Q: So, three pulses?

M: Well, if you are talking of the time between two consecutive pulses. . . this is not clear at all. I don't know what it is. It's certainly unclear that way. Why don't you contact Joy Anne? I can give you her e-mail. That would be the simplest way.

13. Following: "p_{ionb} is the ionization efficiency of a biexciton."

Q: How do you define it?

M: I think that what she means by that is that if you have a biexciton it can undergo certain processes, one it can ionize, but then it can do an all set of several other processes, so the rate of the ionization process compared with the total rate would be the efficiency.

14. p. 4 1st column top-middle: "Comparing eqn (11) and (17)

$$\frac{I_{\mathrm{ge}}}{k_{\mathrm{eg}}}\frac{\partial \rho_L(Q,t)}{\partial t} = L_{\mathrm{e}}\rho_L(Q,t) - k_{\mathrm{ed}}\delta(Q)\rho_L(Q,t) - \rho_L(Q,t)I_{\mathrm{ge}}\frac{k_{\mathrm{bd}'}}{k_{\mathrm{be}}} \quad (11)$$

$$\frac{\langle M_X\rangle}{1+\langle M_X\rangle+\langle M_X\rangle^2/2}\frac{\partial \rho_L(Q,t)}{\partial t} = L_{\mathrm{e}}\rho_L(Q,t))\frac{\tau_{\mathrm{X}}}{T} - k_{\mathrm{ed}}\delta(Q)\rho_L(Q,t)\frac{t_{\mathrm{X}}}{T} - \rho_{\mathrm{L}}(Q,t)P_{\mathrm{ionb}}\frac{\langle M_X\rangle}{2} \quad (17)$$

with eq (4a) we then see that the solution of eqn (11) and (17) equals that of eq (4a) multiplied by $e^{-k_{\mathrm{c}}t}$, where $k_{\mathrm{c}} = I_{\mathrm{ge}}^2 k_{\mathrm{bd}'}/[k_{\mathrm{eg}}k_{\mathrm{be}}]\ldots$ "

Q: What is the meaning of the "c" subscript?

M: It probably means composite rate constants, it's defined that way, composition of rate constants.

Q: Is it not the inverse of the critical time t_c?

M: No, I don't think it has anything to do with that. It is just a symbol that Joy Anne hasn't used before and so she uses.

15. Following: "This $P_L(t)$ for a QD, defined in eq (19)

$$P_L(t) = -\frac{d}{dt}\int_{-\infty}^{\infty} dQ\rho_L(Q,t)$$

as the derivative of the survival probability, is the probability of a QD that has been in the 'on' cycling for t and transfers to an 'off' cycling during dt per unit dt."

Q: Why do you add "per unit dt"?

M: Because you want to have... when you have something divided by dt that's per unit dt.

Q: You consider it a unit.

M: Yes, it has units, it has a magnitude dt, seconds, or microseconds ... There are maybe better ways to phrasing, I'm sure of that.

16. p. 4 bottom: "The solution for $\rho_L(Q, t)$ and hence for $P_L(t)$ is given by

$$P_L(t) = \frac{1 + 2k_c t}{\sqrt{\pi t_c t}} \left[1 - \sqrt{\frac{\pi t}{t_c}} e^{1/t_c} \mathrm{erfc}\left(\sqrt{\frac{t}{t_c}}\right)\right] e^{-k_c t} \qquad (20)$$

Q: Normally one considers $t = \frac{1}{k}$, is then here $t_c = \frac{1}{k_c}$?

M: I don't think that $t_c = \frac{1}{k_c}$. Yes, because that t_c is associated with something else occurring. The k_c is associated with the exponential decay. The t_c is associated with doing something achieving a steady state at the intersection point.

17. Following: "where t_c is the critical time in which the population has largely been depleted near the sink (time to set up a steady state) due to disappearance into the sink at the crossing and

$$t_c = 4D_e I_{ge}/[k_{eg} k_{ed}^2] \qquad (23)\text{"}$$

Q: Please explain the whole story... what exactly is the "sink," why the setting up of the steady state, and please give a physical interpretation of the equation...

M: And that is what t_c was t_c and k_c are unrelated quantities.

The sink is the point where reaction occurs, like at the intersection of two energy levels, that intersection is the sink.

Initially, when you start shining light and at first you have no excited states, but then you form some excited state you reach a certain concentration of excited states, and then the pulses continue . . . their number decrease because they radiate and eventually you reach a steady state of excited states.

18. p. 4, 2nd column bottom: "there are now two extra charges of the same kind in the conduction band or in the valence band that can participate in the Auger process."

Q: Are the charges in the conduction band those of the electrons and the ones in the valence band those of the holes?

M: Yes.

19. Fig. 3

Q: Please explain . . . by the way, from the figure it is impossible to see dashed lines.

M: I can see the dashed lines here . . . you may be confusing with some of the dots, if you look, say, at the lowest curve, then you'll see at the end there are a couple of dashes that go toward higher values of the time.

Q: You right, that small thing there . . .

M: That's right, that's a dashed line. You need have a sharpshooter's eye . . .

Q: So the dashed line is the straight line . . .

M: That's right, that's the power law.

20. p. 5 2nd column top-middle: "the trajectories would look 'ragged' for a light state, instead of an ideal 'picket fence' type trajectory."

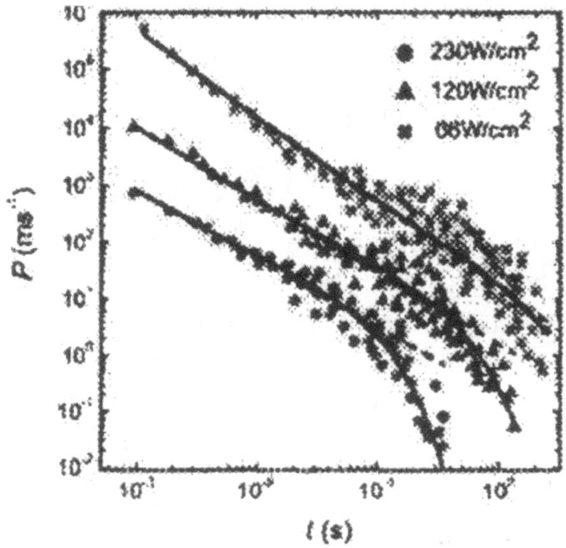

Fig. 3. On-time probability distributions measured under pulsed laser conditions at $\lambda_{\text{exc}} = 434$ nm and laser intensities of 230 (red ●), 120 (blue ▲) and 66 W cm^{-2} (green ■). The solid lines are fits to the data of power law with exponential cutoffs k_c ($k_c = 11.3, 2.2, 0.8\,\text{s}^{-1}$, respectively) and bumps according to eqn (22) and the dashed lines are fits to the data only of power law. Here the slopes are not fixed at 3/2. The curves[41] had been vertically shifted by unspecified amounts. Data reprinted with permission from ref. 41 Copyright (2009) American Chemical Society.

M: In the simplest situations you'll see this on off periods where the on period corresponds to a line that goes up to certain height and it's the same height of all of these lines that you see, that's a picket fence like thing. But in some other systems you see it going up all to heights all over the map, and so it's not a simple on off process, you have a more complicated map, and whenever looking at some photosynthetic systems I was hoping they would be of this kind, but they're not, it shows that the mechanism is just more complicated than the sole on off for those cases. So, we tried to apply the theory only to the cases where it seems to be a relatively simple on off behavior, and not where you have a whole group of states in between.

Q: I saw a paper of some guy in Sweden and there he was considering also Si QDs and there you had a more regular behavior...

M: Ja, I think that was Linros.

Q: Whereas in CdSe is more complicated...

M: Well, in some lovely work of Bawendi, he has same nice on off behavior but he shows what happens in the intermediate case, where you have occasionally something that is just half way, and then he looks at the behavior for that instead of a single exponential on the nanotime scale for that line is biexponential, in other words the mechanism is more complicated when you have a line which is biexponential.

21. Following: "In general, it would be helpful to decrease the bin size when analyzing high excitation intensity experimental results and see if the power exponent converges to a constant value, signal intensity permitting."

M: In the interpretation of most of the experiments they use a certain bin size and calculate all points that occur in that time bin, and so the answer may depend upon the size of the bin and so you want to work in a regime where your power exponent is independent of that, otherwise one has got an artifact in it.

www.ingramcontent.com/pod-product-compliance
Lightning Source LLC
LaVergne TN
LVHW021055220926
848484LV00021B/920

* 9 7 8 9 8 1 1 2 1 7 5 6 2 *